D0076103

HAWKEYE COMMUNITY COLLEGE
3 7944 1008 3538 4

QUANTUM GENERATIONS

QUANTUM GENERATIONS

A HISTORY OF PHYSICS IN THE TWENTIETH CENTURY

HELGE KRAGH

PRINCETON UNIVERSITY PRESS

PRINCETON, NEW JERSEY

Published by Princeton University Press, 41 William Street,
Princeton, New Jersey 08540
In the United Kingdom: Princeton University Press, Chichester, West Sussex

Library of Congress Cataloging-in-Publication Data

Kragh, Helge, 1944–
Quantum generations : a history of physics in the twentieth
century / Helge Kragh.
p. cm.
Includes bibliographical references and index.
ISBN 0-691-01206-7 (cloth : alk. paper)
1. Physics—History—20th century. I. Title.
QC7.K7 1999
530′.09′04—dc21 99-17903 CIP

This book has been composed in Times Roman

The paper used in this publication meets the minimum requirements
of ANSI/NISO Z39.48-1992 (R 1997) (*Permanence of Paper*)

http://pup.princeton.edu

Printed in the United States of America

10 9 8 7 6 5 4 3 2

CONTENTS

PREFACE

THIS WORK WAS written between 1996 and 1998, at the suggestion of Princeton University Press. Originally, when I accepted the invitation to write a book about the development of physics during the twentieth century, I thought it would be a relatively easy matter. I soon became wiser. I should have known that it is simply not possible to write a balanced and reasonably comprehensive one-volume account of twentieth-century physics. What follows is a substitute, a fairly brief and much condensed and selective account of what I believe have been the most significant developments in a century of physical thought and experiment that can well be called *the* most important century of physics.

The book is structured in three largely chronological parts. The first part covers developments from the 1890s to about 1918, the end of World War I. The second part concentrates on developments between the two world wars, roughly 1918–1945, and the third part takes up developments in the remainder of the century. The chosen periodization should be uncontroversial, and so should the decision to start in the mid-1890s rather than in 1900. It is generally accepted that "modern physics" started with the great discoveries of the 1890s and not with Planck's introduction of the quantum discontinuity in 1900.

I have endeavored to write an account that goes all the way up to the present and so includes parts of very recent developments that normally would be considered to be "not yet historical." There are problems with writing historically about recent developments, but these are practical problems and not rooted in contemporary science being beyond historical analysis. The book is of a type and size that preclude any ambitions of comprehensiveness, not to mention completeness. At any rate, a "complete" history of twentieth-century physics would probably be as pointless as it would be impossible to write from a practical point of view. Like most historical works, this one is selective and limited in scope and content. The selections can undoubtedly be criticized. The material I have included has been chosen for a variety of reasons, one of them being the availability of historical writings and analyses. The book's goal is to give an account of the development of physics during a one-hundred-year period that is digestible, informative, and fairly representative. There are, unavoidably, many interesting topics and subdisciplines that I do not include, in part because of lack of space and in part because of lack of secondary sources. Among the topics that I originally contemplated to include, but in the end had to leave out, are optics, materials science, chemical physics, geophysics, medical physics, physics in third-world countries, and the post-1950 discussion concerning the interpretation of quantum mechanics. Yet I believe that what is included does not, in spite of the more or less arbitrary selection criteria, misrepresent to any serious extent the general trends in the development of modern physics.

The problem of a balanced account is a difficult one, not only with regard to subdisciplines and dimensions, but also with regard to nations. Physics is and has always been international, but of course, some nations have contributed more to scientific progress than others. My account is essentially a history of physics in Europe and North America, with some mention also of contributions from Japan. This is simply a reflection of how the important contributions to physics have been distributed among nations and geographical regions. Whether one likes it or not, most of the world's nations have played almost no role at all in the development of modern physics. One of the significant trends of the postwar period has been the dominance of American physicists in a science that was originally European. Because of this dominance, and because of the strong position that American scholars have in the history of science, the historical knowledge of modern American physics is much richer than it is in the case of Europe and other regions, including the former Soviet Union. It is quite possible that the relative predominance of writings about American physics has caused my account to focus too much on the American scene, but under the circumstances, there was little I could do about it.

Taken together, the twenty-nine chapters cover a broad spectrum of physics, not only with respect to topics and disciplines, but also with respect to the dimensions of physics. We should always keep in mind that physics (or the physical sciences) is a rich and multifaceted area that has implications far beyond the purely scientific aspects related to fundamental physics. I have wanted to write a broad book, although not so broad that it loses its focus on what is distinctly the world of physics. The present work is not devoted solely to the scientific or intellectual aspects of physics, but neither does it concentrate on social and institutional history. It tries to integrate the various approaches or, at least, to include them in a reasonably balanced way. I have also paid more attention to applied or engineering physics than is usually done. To ignore the physics-technology interface, and concentrate on so-called fundamental physics alone, would surely give a distorted picture of how physics has developed in this century. Not only are many of the world's physicists occupied with applied aspects of their science, and have been so during most of the century, but it is also largely through the technological applications that physics has become a major force of societal change.

The intended audience of the book is not primarily physicists or specialists in the history of science. It is my hope that it will appeal to a much broader readership and that it may serve as a textbook in courses of an interdisciplinary nature or in introductory courses in physics and history. With a few exceptions I have avoided equations, and although the book presupposes some knowledge of physics, it is written mainly on an elementary level. My decision to avoid the note apparatus that is often seen as a hallmark of so-called academic books is an attempt to make the book more easily accessible to readers not acquainted with the (sometimes rather artificial) note system of more scholarly works. In almost all cases of citations, I have included references in the text to sources where they can be readily found. Rather than referring to the original source, I have in most cases referred to a later, secondary source, quite often the place where I happened to pick up the quotation. In a book of this type, there is no point in numerous references to old papers in *Annalen der Physik* or *Philosophical Magazine*; the reader who might want to dig up the original source can do so via the source I have quoted. The entire book is, to a large extent, based on secondary sources, especially the many fine books and articles written by historians of the physical sciences. I have also drawn freely and extensively on some of my earlier works dealing with the history of modern physics, chemistry, technology, and cosmology.

The source problem is quite different with regard to physics in the last third or quarter of the century. Whereas there is an abundance of secondary sources dealing with older developments in physics, written by either historians or participants, there are only few historical analyses of post-1960 physics (high-energy physics is an exception). Within this part of the chronology, I have had to base my account on useful material that happens to exist, on physicists' more or less historically informed recollections, and on a not-very-systematic survey of what I could find in scientific articles and reviews. *Physics Today* has been a helpful source; references to this journal in Part Three are abbreviated *PT*. The bibliography and the appendix on "further reading" list a good deal of the literature that the reader may wish to consult in order to go more deeply into the subjects covered by this book.

The working title of the book was originally *Revolution through Tradition*. With this title I wanted to refer to the dialectics between existing theories and revolutionary changes that has been characteristic of physics during the twentieth century. There have indeed been revolutions in the theoretical frameworks of physics, but these have not been wholesale rejections of the classical traditions; on the contrary, they have been solidly connected with essential parts of the physics of Newton, Maxwell, and Helmholtz. Relativity theory and quantum mechanics, undoubtedly the two major revolutions in twentieth-century physical thought, were carefully constructed to correspond with existing theories in the classical limits.

The respect for traditions has likewise been a characteristic theme in all the major theoretical changes that occurred after the completion of quantum mechanics. To the extent that these may be termed revolutionary, they have been conservative revolutions. Changes have been much less important on the methodological level than on the cognitive level. There have been some changes, but not of a fundamental nature. Basically, the accepted methods of science of the 1990s are the same methods that were accepted in the 1890s. If we are looking for really radical changes during the last three-quarters of the century, we should not look toward the methods, the conceptual structure, or the cognitive content of physics, but rather toward the basic fabric of the world, the ontology of physics; or we should look toward the social, economic, and political dimensions. In terms of manpower, organization, money, instruments, and political (and military) value, physics experienced a marked shift in the years following 1945. The sociopolitical changes made physics in 1960 a very different science than it had been a century earlier, but they did not cause a corresponding shift in the methodological and cognitive standards. In any event, this is not the place to discuss these broader issues at any length. In the book that follows, I have described, rather than analyzed, important parts of the development of physics between 1895 and 1995. The reader who is interested to draw the big picture—say, to evaluate the revolutionary changes and make comparisons over the course of a century—should be better equipped with the material and information that are presented here.

I would like to express my thanks to my colleague Ole Knudsen, who read the manuscript and suggested various improvements.

Helge Kragh
Aarhus, Denmark

PART ONE

FROM CONSOLIDATION TO REVOLUTION

Chapter 1

FIN-DE-SIÈCLE PHYSICS: A WORLD PICTURE IN FLUX

THE PHILOSOPHER and mathematician Alfred North Whitehead once referred to the last quarter of the nineteenth century as "an age of successful scientific orthodoxy, undisturbed by much thought beyond the conventions. . . . one of the dullest stages of thought since the time of the First Crusade" (Whitehead 1925, 148). It is still commonly believed that physics at the end of the century was a somewhat dull affair, building firmly and complacently on the deterministic and mechanical world view of Newton and his followers. Physicists, so we are told, were totally unprepared for the upheavals that took place in two stages: first, the unexpected discoveries of x-rays, the electron, and radioactivity; and then the real revolution, consisting of Planck's discovery of the quantum of action in 1900 and Einstein's relativity theory of 1905. According to this received view, not only did Newtonian mechanics reign supreme until it was shattered by the new theories, but the Victorian generation of physicists also naively believed that all things worth knowing were already known or would soon become known by following the route of existing physics. Albert Michelson, the great American experimentalist, said in 1894 that "it seems probable that most of the grand underlying principles have been firmly established and that further advances are to be sought chiefly in the rigorous application of these principles to all the phenomena which come under our notice" (Badash 1972, 52). How ironic, then, that Professor Röntgen's new rays—the first of several discoveries that defied explanation in terms of the known "grand underlying principles"—were announced only a year later. And how much more important the new physics of the early twentieth century appears if compared with views such as Michelson's.

The received view is in part a myth, but like most myths, it has a foundation in fact. For example, Michelson was not the only physicist of the decade who expressed the feeling that physics was essentially complete and that what remained was either applied physics, more precise measurements, or relatively minor discoveries. When Max Planck entered the University of Munich in 1875, he was warned by a professor of physics that his chosen subject was more or less finished and that nothing new could be expected to be discovered. Yet, although such feelings certainly existed among physicists, it is questionable how widespread they were. Very few theoretical physicists of the 1890s seem to have accepted Michelson's view, and after the amazing discoveries of Röntgen, Henri Becquerel, J. J. Thomson, and the

Curies, even the most conservative experimentalist was forced to realize its fallacy.

What about the claim that physics a hundred years ago rested on orthodoxy and a complacent acceptance of Newtonian mechanics? Was there a mechanical worldview, or any commonly accepted worldview at all? Whereas the question of completeness can be discussed, it is squarely a myth that physicists stuck obstinately to the mechanical worldview until they were taught a lesson by Einstein in 1905 (or by Planck in 1900). The most important nonmechanical trend was based on electromagnetic theory, but this was only one of the indications of a widespread willingness to challenge the mechanical worldview and seek new foundations, either opposed to it or radical modifications of it. According to the classical mechanical world picture—the Laplacian version of Newtonianism (not to be confused with Newton's own ideas)—the world consisted of atoms, which were the sites of, and on which acted, various forces of long and short ranges. The gravitational force was the paradigmatic example of such forces acting at a distance over empty space. With the advent of field theory, the mechanism of force propagation changed, but Maxwell and most other field physicists continued to seek a mechanical foundation for their models. The most important conceptual shift was perhaps the rise to prominence—indeed, necessity—of a universal *ether* as the quasihypothetical, continuous, and all-pervading medium through which forces propagated with a finite speed.

In 1902, in the final part of a textbook on optics, Michelson declared his belief that "the day seems not far distant when the converging lines from many apparently remote regions of thought will meet on . . . common ground." He went on, "Then the nature of the atoms, and the forces called into play in their chemical union; the interactions between these atoms . . . as manifested in the phenomena of light and electricity; the structures of the molecules and molecular systems of which the atoms are the units; the explanation of cohesion, elasticity, and gravitation—all these will be marshaled into a single compact and consistent body of scientific knowledge" (Michelson 1902, 163). And this was the same Michelson who, eight years earlier, had suggested that physics was near its end. Was it the discoveries of the electron and radioactivity that caused the changed attitude? Or perhaps Planck's discovery of the radiation law based on the notion of energy quantization? Not at all; these recent discoveries were not mentioned in the book. Michelson's enthusiasm was rooted in "one of the grandest generalizations of modern science . . . that all the phenomena of the physical universe are only different manifestations of the various modes of motion of one all-pervading substance—the ether."

Maxwell considered the possibility of explaining gravitation in terms of his electromagnetic theory, but abandoned the attempt after realizing that he would then have to ascribe an enormous intrinsic energy to the ether. Other

Victorian physicists were less easily discouraged and there were, in the last quarter of the century, many attempts to either explain or revise Newton's divine law of gravitation. Some of these attempts were based on electrodynamics, others on hydrodynamic models. For example, in the 1870s, the Norwegian physicist Carl A. Bjerknes studied the motion of bodies in an infinite and incompressible fluid and was led to the conclusion that two pulsating spheres would give rise to forces between them varying inversely to the separation of their centers. He considered this a kind of hydrodynamic explanation of gravitation, or at least an interesting analogy. Bjerknes's work was taken up by some British theorists and, in 1898, was revived by the German Arthur Korn at the University of Munich, who developed a hydrodynamic theory of gravitation. At that time, however, electrodynamics was in focus, and complicated hydrodynamic models in the style of Bjerknes and Korn failed to arouse much interest.

Related to hydrodynamical thinking, but of more importance and grandeur (if, in the end, no more successful), were the attempts to construct the world solely out of structures in the ether. The most important of the nonelectromagnetic theories was the vortex atomic theory, originally suggested in 1867 by William Thomson (later, Lord Kelvin) and subsequently developed by a whole school of British mathematical physicists. According to this theory, the atoms were vortical modes of motion of a primitive, perfect fluid, usually identified with the ether. In his Adams Prize essay of 1882, young J. J. Thomson gave an elaborate account of the vortex theory and extended it to cover chemical problems, including affinity and dissociation. The theory was also applied to electromagnetism, gravitation, and optics and was an ambitious attempt to establish a unitary and continuous "theory of everything" based solely on the dynamics of the ether. As late as 1895, William Hicks gave an optimistic report on the state of art of the vortex atom at the annual meeting of the British Association for the Advancement of Science (BAAS). Hicks's view of the goal of theoretical physics is worth quoting at some length:

> While, on the one hand, the end of scientific investigation is the discovery of laws, on the other, science will have reached its highest goal when it shall have reduced ultimate laws to one or two, the necessity of which lies outside the sphere of our recognition. These ultimate laws—in the domain of physical science at least—will be the dynamical laws of the relations of matter to number, space, and time. The ultimate data will be number, matter, space, and time themselves. When these relations shall be known, all physical phenomena will be a branch of pure mathematics. (BAAS Report 1895, 595)

As we shall see, very similar views continued to play an important role throughout the twentieth century. Although many of Hicks's contemporaries would have subscribed to his philosophy, by 1895, the vortex theory of atoms had been abandoned by most physicists. Decades of theoretical work had led to no real progress and the grand vortex program was degenerating into sterile mathematics.

Much the same can be said of another hydrodynamic atomic theory, the "ether squirt" theory worked out by the mathematician Karl Pearson in the 1880s and 1890s. According to this theory, the ultimate atom was a point in the ether from which new ether continuously flowed in all directions of space. Like the vortex theorists, Pearson applied his theory to a variety of problems and believed it would be able to explain—in principle—gravitation, electromagnetism, and chemical phenomena. Although Pearson's theory did not attract the same kind of interest as the vortex theory, it is worth mentioning because it included not only sources but also sinks of ether—that is, a kind of *negative matter*. Gravitationally "negative" matter, which repels ordinary matter but attracts other negative matter, had already been discussed in the 1880s by Hicks within the framework of the vortex atomic theory, and the strange concept reappeared in Pearson's theory, as well as in other discussions of fin-de-siècle physics. For example, the British physicist Arthur Schuster speculated lightheartedly that there might exist entire stellar systems of antimatter, indistinguishable from our own except that the two stellar systems would be repelled rather than attracted. Not only did he introduce the names "antimatter" and "antiatoms" in 1898, but he also suggested that colliding matter and antimatter would annihilate each other, thus anticipating an important concept of later quantum physics.

In Pearson's version of antimatter, ether poured in at a squirt and disappeared from our world at a sink. Where, in the first place, did the ether come from? According to Pearson, writing in 1892, it would not simply appear ex nihilo, but would possibly come from a fourth dimension, to which it would again return. Here we have another concept, usually seen as an invention of twentieth-century relativity theory, turning up unexpectedly in the physics of the old worldview. In fact, ideas of hyperspaces and their possible significance in physics were not new in the 1890s. In 1870, the British mathematician William Kingdon Clifford used Riemann's notion of curved, non-Euclidean geometry to suggest that the motion of matter and ether was in reality the manifestation of a variation of the curvature of space. This general idea of a "geometrization of physics" was well known at the end of the nineteenth century and inspired several physicists, astronomers, and mathematicians, not to mention science fiction authors, such as H. G. Wells. For example, in 1888, the eminent American astronomer Simon Newcomb suggested a model of the ether based on hyperdimensional space, and in 1900, the German Karl Schwarzschild made much use of non-Euclidean geometry in his astronomical work. Although these and other works were often speculative and always hypothetical, at the end of the century there existed a small group of researchers who considered hyperspace models of the ether or otherwise attempted to link four-dimensional space to problems of physical interest. Speculative or not, such attempts were considered legitimate within the spirit of physics characteristic of the 1890s.

The hydrodynamic ether models differed from the Laplacian program in physics, but they nonetheless rested on mechanical ground and were not attempts to overthrow the Newtonian worldview. Hydrodynamics, after all, is the mechanical science of fluid bodies. Thermodynamics, the science of heat and other manifestations of energy, constituted a much more difficult problem for the classical worldview. This branch of physics was sometimes argued not only to be different from mechanics in principle, but also to have priority over mechanics as a more satisfactory foundation on which all of physics could be built. In the 1890s, together with electrodynamics, thermodynamics entered as a competitor to mechanics as far as foundational problems were concerned. In this decade, there was a continual discussion of the *unity* of physics, and it was not at all clear what discipline could best serve as the foundation of the unity that almost all physicists believed their science must have.

Whereas the law of energy conservation was successfully explained in mechanical terms, the second law of thermodynamics did not succumb so easily to mechanical principles. For one thing, the laws of mechanics are reversible, or symmetric in time, whereas the second law of thermodynamics expresses an irreversible change in entropy. In his famous statistical-mechanical theory of entropy, developed first in 1872 and more fully in 1877, Ludwig Boltzmann believed he had reduced the second law to molecular-mechanical principles, but his interpretation was challenged and became the subject of much controversy. One of his critics, the German physicist Ernst Zermelo, argued in 1896 from Poincaré's so-called recurrence theorem that the second law could not be derived from mechanics and thus was incompatible with a unitary mechanical world picture. Boltzmann denied the validity of Zermelo's argument and remained convinced that there was no deep disagreement between mechanics and thermodynamics.

According to the physicist Georg Helm and the chemist Ludwig Ostwald, both Germans, energy was the most important of the unifying concepts of the physical sciences. A generalized thermodynamics was therefore held to replace mechanics as the foundation of physics. Helm and Ostwald came to this conclusion about 1890 and called their new program *energetics*. The new science of energetics was, in many ways, contrary to the mechanical world picture and was thought of as a revolt against what was called "scientific materialism." This revolt included the position that mechanics was to be subsumed under the more general laws of energetics in the sense that the mechanical laws were held to be reducible to energy principles. Another aspect of energetics was its denial of atomism as other than a useful mental representation. Ostwald and some other physical chemists, including Pierre

Duhem in France, argued that the belief in atoms and molecules was metaphysical and that all empirical phenomena could be explained without the atomic hypothesis.

The energetics program was considered enough of a challenge to the traditional molecular-mechanical view that it was taken up as a discussion theme at the annual meeting of the German Association of Natural Scientists and Physicians in Lübeck in 1895. The meeting featured a famous discussion between Boltzmann, who attacked the energeticists, and Helm and Ostwald, who argued against the mechanical world picture. It is interesting to note that neither Boltzmann nor others present at the meeting simply defended the classical-mechanical worldview or fully subscribed to the views that Helm and Ostwald criticized. Boltzmann declared that the mechanical worldview was a dead issue and that the "view that no other explanation can exist except that of the motion of material points, the laws of which are determined by central forces, had generally been abandoned long before Mr. Ostwald's remarks" (Jungnickel and McCormmach 1986, 222). All the same, Boltzmann saw no merit in the energetics program and preferred to work on a mechanical foundation, feeling that it alone was sufficiently developed to secure scientific progress.

The energetics alternative received only modest support among physicists and chemists, but criticism of the atomic theory and emphasis on the fundamentality of the energy concept were repeated also by many scientists not directly associated with the energetics program. The leading French physicist, Pierre Curie—perhaps better known as the husband of Marie Curie—may be an example. In accordance with his positivistic view of science, Curie refrained from materialistic and atomistic hypotheses and favored a phenomenalism inspired by the laws of thermodynamics. He, and several other French physicists, held thermodynamics to be the ideal of physical theory. They argued that energy, not matter, was the essence of a reality that could be understood only as processes or actions. From the early 1880s onward, the Austrian physicist-philosopher Ernst Mach argued for a phenomenological understanding of physics, according to which physical theories and concepts were economical ways of organizing sense data. Mach admitted the usefulness of molecular mechanics, but considered it neither a fundamental theory nor one expressing physical reality. From a foundational point of view, he preferred the energy principles to the laws of mechanics. Again in agreement with Ostwald and his allies, Mach held that atoms were nothing but convenient fictions. Moreover, Mach criticized the very heart of mechanics, the idea of force as expressed by Newton's second law. A somewhat similar foundational criticism of mechanics from a positivistic point of view was undertaken by Heinrich Hertz in his 1894 reformulation of mechanics, building only on the fundamental conceptions of space, time and mass. However, this kind of critical analysis of mechanics did not necessarily involve a wish to abandon the mechanical world picture. In the case of Mach it did, but to Hertz, the new versions of mechanics merely affirmed this picture of the world. In fact, a major aim of Hertz's force-free mechanics was to establish a mechanical theory of the electromagnetic ether.

The mechanical worldview was no longer considered progressive in the 1890s, and even traditionalists had to admit that it was not universally successful. Apart from the troubled relationship between mechanics and the entropy law, there was an older problem related to the kinetic theory of gases. As early as 1860, Maxwell had observed that the measured ratios between the specific heats of diatomic gases at constant pressure (c_p) and at constant volume (c_v) did not agree with the equipartition theorem based on the mechanical theory. According to this theory, $\gamma = c_p/c_v = 1 + 2/n$, where n is the number of degrees of freedom of the molecule. The problem was that the predicted result for diatomic gases matched experiments (which gave $\gamma = 1.4$) only if it was assumed that the molecule was rigid and had no internal parts; this assumption seemed inconsistent with the results of spectroscopy, which strongly indicated internal vibrations exchanging energy with the ether. The problem was recognized as an anomaly, but of course, it needed more than a single anomaly to shatter the mechanical world view. Yet the apparent failure of the equipartition theorem was considered serious enough to figure as one of the two clouds in a famous lecture, "Nineteenth Century Clouds Over the Dynamical Theory of Heat and Light," that Lord Kelvin delivered before the Royal Institution in April 1900. The other threat was the failure to explain the motion of the earth through the ether, as exhibited by the ether drift experiment of Michelson and Edward Morley. (For this, see chapter 7.)

The new physics that arose in the early years of the twentieth century was not a revolt against a petrified Newtonian worldview, something analogous to the revolt of Galileo against Aristotelianism. By 1905, the mechanical worldview had been under attack for more than a decade, and for this reason alone, there never was much of a clash between Einstein and Newton. Even more important than the opposition inspired by thermodynamics and energetics was the new and vigorous trend in electromagnetic theory that characterized the 1890s. We shall deal in more detail with this so-called electromagnetic worldview in chapter 8, and here we only emphasize its importance and key elements. The basic problem of physics in the late nineteenth century was perhaps the relationship between ether and matter: Was the ether the fundamental substratum out of which matter was constructed? Or, on the contrary, was matter a more fundamental ontological category of which the ether was just a special instance? The first view, where primacy was given to structures in the ether, became increasingly more common at the turn of the century, when mechanical ether models were replaced by electrodynamic models.

If electromagnetism was more fundamental than mechanics, it made sense to try to derive the mechanical laws from those of electromagnetism, and this was precisely what many theoretical physicists aimed at. Electromagnetism was considered a unifying principle of all science, not unlike the role assigned to energy in Ostwald's energetics. The way in which Ostwald and his energeticist allies spoke of "matter as subordinate to energy" and "all events [being] ultimately nothing but a change in energy" was strikingly similar to the rhetoric of the electrodynamicists, only with "ether" or "electromagnetic field" substituted for "energy." In both cases, materialism was discarded and matter declared an epiphenomenon. Joseph Larmor, the eminent British theorist, had no difficulty imagining a world based on a nonmaterial, transcendent ether. He admitted, as he wrote in 1900, that this might seem "as leaving reality behind us," but defended his ether worldview by arguing that it described an inner reality not directly accessible to the senses (Larmor 1900, vi). This was an argument that Mach, Ostwald, and other phenomenalists would not have accepted. At the beginning of the new century, the monistic, electromagnetic worldview was accepted by a growing proportion of avant-garde physicists in Germany, England, France, and the United States. Physics consisted of the physics of matter and the physics of the electromagnetic ether, and the trend to avoid the unwanted dualism was to identify matter with ether, rather than the other way around. This is not to say that there were no dissenting voices or that the electromagnetic doctrine permeated the entire field of physics. There were physicists who—years before Einstein—rejected the ether as a metaphysical concept, and there were those who continued to search for mechanical models of the ether or even considered the ether as a special state of ordinary matter. Textbooks of this period usually rested on a mechanical basis and did not reflect the change in worldview discussed in frontier theoretical physics. This is how textbooks usually are: They are by nature conservative and cautious in their attitude toward modern ideas.

The trend of theoretical physics around 1900 was more than a shift in fundamental ideas from mechanics to thermo- and electrodynamics, and it was more than the result of a number of spectacular discoveries. It was part of a change in worldview that had ramifications outside physics and was nourished by the particular Zeitgeist of the period, a spirit of time sometimes characterized as neoromantic. The historian Russell McCormmach has aptly summarized the situation as follows: "The whole cultural configuration at the turn of the century was implicated in the change from mechanical to electromagnetic thinking. The immaterial electromagnetic concepts were attractive in the same measure that the inert, material imagery of mechanics was displeasing" (McCormmach 1970, 495). One important element of this cultural configuration was a widespread antimaterialism. Taking different shapes in the different scientific nations, the antimateralistic doctrine amounted to the belief that "matter is dead." If matter was not the ultimate reality, but merely some manifestation of an immaterial ether, it would not seem unreasonable to challenge other established doctrines derived from the physics of matter, including the permanence of chemical elements and the laws of conservation of matter and energy. Indeed, in some quarters, the very qualities of permanence and conservation were considered suspicious within a world view emphasizing transformation, evolution, and becoming.

As an example, consider the French psychologist and amateur physicist Gustave LeBon, who, in 1896, reported the discovery of what he called "black light"—a new kind of invisible radiation that he believed was distinct from, but possibly related to, x-rays and cathode rays. Although LeBon's discovery claim did not fare well, his general ideas of matter, radiation, and ether were received favorably by the public and were, to some extent, representative of the period's Zeitgeist in scientific circles. In his book *The Evolution of Matter*, a hit that went through twelve editions and sold 44,000 copies, LeBon concluded that all matter is unstable, constantly emitting radiation or "effluvia." Material qualities were held to be epiphenomena exhibited by matter in the process of transforming into the imponderable ether from which it had once originated. According to LeBon, there was no dualism between energy and matter, which just represented different stages in an evolutionary process, the end result of which was the pure ethereal state. Among his many arguments for the continual degradation of matter into ether, LeBon considered radioactivity to be particularly convincing. He shared the view of many physicists that radioactivity is a property exhibited by all matter. Then, if all chemical elements emitted ether-like radiations, would they not slowly melt away and would this not prove the nonmateriality of matter? Or, as LeBon expressed it rather more dramatically, would the ether not represent "the final nirvana to which all things return after a more or less ephemeral existence" (LeBon 1905, 315)? LeBon and many of his contemporaries believed that this was indeed the case.

LeBon's quasiscientific speculations had considerable appeal among the many scientists who were dissatisfied with positivistic ideals and longed for an undogmatic, more youthful science that would better satisfy what they associated with the human spirit. His ideas struck a chord in a period that has been described as a "revolt against positivism" and even—but less justifiably—a "revolt from reason" (MacLeod 1982, 3). Among those who listened with sympathy to LeBon's arguments was the great Henri Poincaré. But LeBon was not a theoretical physicist and his views, although fashionable, were not particularly modern. If he was not taken too seriously by physicists outside France, it may have been because he failed to incorporate the electromagnetic ether into his speculations. Although it is reasonable to speak of a general spirit of physics about the turn of the century, the degree of consensus should not be exaggerated. There were considerable differences among the views of leading individual physicists and there were also important national differences. In Germany and France, for instance, the reaction against the mechanical worldview was more often associated with positivistic virtues and thermodynamic ideals than was the case in England. British physicists typically had no sympathy for the positivistic, fact-oriented view of science championed by Pierre Curie, Duhem, Ostwald, Mach, and others. In 1896, in a critical comment on Ostwald's energetics, the Irish physicist George FitzGerald distinguished between the metaphysically receptive British style and the inductive, unphilosophical German style. "A Briton wants emotion," he wrote, "something to raise enthusiasm, something with a human interest" (Wynne 1979, 171). The battle cry for emotions, actions, and evolutionary processes "with a human interest" was repeated by French neoromanticists, who contrasted their style with that of the supposedly unimaginative Germans. At the same time, they objected to the British style of physics, which they found too mechanistic and lacking in *esprit*.

Chapter 2

THE WORLD OF PHYSICS

PERSONNEL AND RESOURCES

WHO WERE THE physicists around 1900? How many were there and how were they distributed among nations and institutions? What kind of physics did they work with? How were they funded? Let us first look at the number of physicists—that is, people contributing to the advance of physics either directly, as researchers, or indirectly, as teachers. Not much is known about the social background of physicists at the turn of the century, but a study of German physics showed that the typical young German physicist came from the higher strata of society, the middle and upper classes. He—and it was always a he—was socially indistinguishable from the young humanist scholar. Chemists, on the other hand—and organic chemists in particular—more often came from the business community (table 2.1). The difference probably reflected the close connection between chemistry and German industry, a connection that had not yet grown strong with respect to physics.

The meaning of the term "physicist" has, of course, changed over time, but by 1900, the meaning of the word did not differ significantly from its modern meaning; physics had by then reached a professional level that bore more resemblance to what it would be in 1990 than to what it was in 1810. The large majority of physicists—i.e., contributors to the physics research literature—were professionals in the sense that they earned their living as faculty members of physical institutes at either universities or polytechnical colleges (such as the Technische Hochschulen in Germany). Gifted amateurs, secondary school teachers, and wealthy individuals still played a role, but it was small and rapidly diminishing. A large number of engineers and technical experts were engaged in applied physics (industrial and medical, for example) and could reasonably be classified as physicists as well. However, we will limit ourselves to those occupying teaching positions explicitly devoted to physics, so-called academic physicists. As in science generally, physics was almost exclusively a European-American phenomenon, notwithstanding the rise of Japan and the work done in the English, German, Dutch, and other colonies (mostly by whites).

From table 2.2 it appears, first of all, that physics in 1900 was a small world. The total number of academic physicists in the world was probably somewhere between 1,200 and 1,500. (In comparison, by 1900 membership

TABLE 2.1
Social Background of German Doctoral Scientists

Field	*Year*	*No. of doctorates*	*Middle and upper professional classes (%)*	*Business community (%)*	*Agriculture (%)*
Physics	1899	56	52	36	9
	1913	68	53	40	7
Math and	1899	19	53	39	16
astronomy	1913	8	50	50	0
Organic	1899	99	25	66	8
chemistry	1913	99	33	57	10
Nonorganic	1899	77	29	61	10
chemistry	1913	42	40	45	14
Humanities	1899	64	50	44	6
	1913	441	50	38	4

Note: Data for the year 1899 are means of 1896–1902. Data based on Pyenson 1979.

in the national chemical societies of England, Germany, and France had reached more than 3,500.) Moreover, this small world was dominated by a few countries, of which Britain, France, Germany, and the United States were by far the most important; their total of some 600 physicists made up almost half the world population of physicists. Second in the hierarchy were

TABLE 2.2
Academic Physics around 1900

Faculty plus assistants	*No. of physicists/*	*per million*	*Expenditures (1000 marks)/*	*per physicist*	*Productivity total (annual)/*	*per physicist*
Austria-Hungary	64	1.5	560	8.8		
Belgium	15	2.3	150	10		
United Kingdom	114	2.9	1650	14.5	290	2.2
France	105	2.8	1105	10.5	260	2.5
Germany	145	2.9	1490	10.3	460	3.2
Italy	63	1.8	520	8.3	90	1.4
Japan	8	0.2				
Netherlands	21	4.1	205	9.8	55	2.6
Russia	35	0.3	300	8.5		
Scandinavia	29	2.3	245	8.5		
Switzerland	27	8.1	220	8.2		
United States	215	2.8	2990	14.0	240	1.1

Note: The equivalent of 1,000 marks in 1900 was about $240. Data abstracted from Forman, Heilbron, and Weart 1975.

countries such as Italy, Russia, and Austria-Hungary; and then, as a third group, smaller nations followed, including Belgium, the Netherlands, Switzerland, and the Scandinavian countries. Notice that by 1900, the United States already had more physicists than any other nation, and that the density of physicists among the "big four" was the same (about 2.9 per million of population) and considerably less than in Switzerland and the Netherlands. Although the United States was quantitatively a leader in physics, it was far behind the three European powers when it came to productivity and original research. This was in part the result of a climate at the major American universities that was still foreign—indeed, often hostile—to the German ideals of research and scholarship as essential elements of the career of university teachers. In 1889, the president of the Massachusetts Institute of Technology (MIT) stated that "Our aim should be: *the mind of the student*, not scientific discovery, not professional accomplishment"—a view unlikely to lead to much scientific research (Kevles 1987, 34).

Ten years later, on the threshold of the twentieth century, Henry Rowland gave an address before the newly formed American Physical Society. A staunch advocate of pure science, Rowland shared the German ideal of physics as culture and free research. "We form a small and unique body of men, a new variety of the human race," Rowland told his audience. The new body was "an aristocracy, not of wealth, not of pedigree, but of intellect and of ideals, holding him in the highest respect who adds the most to our knowledge or who strives after it as the highest good." Rowland knew very well that his view was not shared by many of his compatriots. He deplored that "much of the intellect of the country is still wasted in the pursuit of so-called practical science which ministers to our physical needs [whereas] but little thought and money is given to the grander portion of the subject which appeals to our intellect alone" (Rowland 1902, 668). At the time of Rowland's address, things were rapidly changing and American physics on its way to become a major factor in world physics. In 1893 the first issue of *Physical Review* appeared, in 1899 the American Physical Society was founded, and two years later Congress authorized $250,000 to build the National Bureau of Standards. In that same year, Andrew Carnegie endowed the enormous sum of $10 million to found an institution to encourage basic research. The result, the Carnegie Institution, had a stimulating effect on research at American universities.

The membership of the American Physical Society indicates the growth of the physics profession in the United States. From fewer than one hundred members in 1899, ten years later membership had increased to 495. By 1914 it had passed 700, and during the following decade the growth rate was even higher (see figure 2.1). By 1910, American physics was moving out of the shadow of Europe, even though the country was still weak in theoretical physics and most of the contributions appearing in *Physical Review* did not

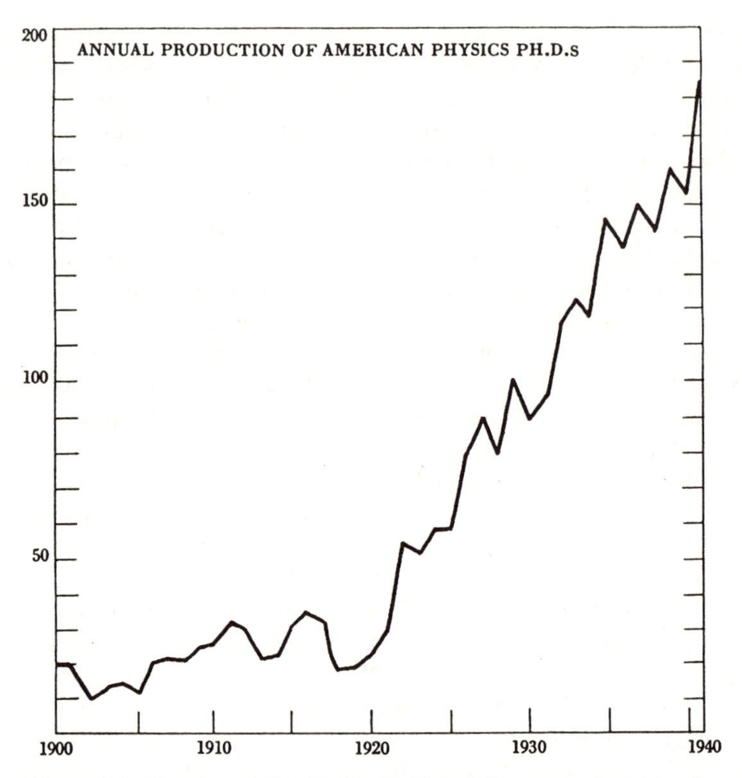

Figure 2.1. The rise of physics in the United States, as shown by the number of American Ph.D.s granted in physics between 1900 and 1940. *Source:* Weart 1979a, 296; N. Reingold, ed. *The Sciences in the American Context: New Perspectives.* Smithsonian Institution Press, copyright 1979. Used by permission of the publisher.

match the quality of papers in the *Annalen der Physik* or *Philosophical Magazine*. Notwithstanding eminent physicists such as Robert Wood, Albert Michelson, Robert Millikan, and Gilbert Lewis, American physics was still relatively provincial and dependent on what happened in Europe. Ludwig Boltzmann visited the University of California, Berkeley, in 1905. He found the nature beautiful and the ladies stout, but was less impressed by the local standards in theoretical physics. As he wrote in the popular essay "A German Professor's Trip to El Dorado," first published in 1905: "Yes, America will achieve great things. I believe in these people, even after seeing them at work in a setting where they're not at their best: integrating and differentiating at a theoretical physics seminar . . ." (Boltzmann 1992, 51).

Physics has always been international and contacts between physicists of different nationalities, including meetings and collaborations, were frequent in the nineteenth century. Organized conferences of a larger scope were rare, however, and it was not until the summer of 1900 that the first International Congress of Physics convened in Paris. At the same time, and in the same city, two hundred mathematicians participated in the International Congress of Mathematics, the second of its kind. The physics conference was organized by the French Physical Society in connection with the World Exhibition in Paris that year. The three-volume proceedings of the congress included seventy scientific review papers written by physicists from fifteen countries. Not surprisingly, the majority of the papers (36) were French. They were followed by 10 German contributions and 6 from the British, whereas the fourth of the major scientific nations, the United States, contributed only 2 papers. Also represented were physicists from the Netherlands (4 papers), Russia (4), Italy (3), Austria-Hungary (3), Denmark (2), Norway (2), Sweden (2), Belgium (2), and Switzerland (2); nonwhites were represented by papers delivered by one Japanese and one Indian. Of course the strong French representation at the Paris congress should not be taken as a correspondingly strong position in world physics—in fact, French physics, and especially theoretical physics, was declining at the time and was no longer competitive with physics in Germany and Britain. The subjects dealt with at the Paris congress covered a wide range, if not the entire spectrum of contemporary physics. Table 2.3 shows the subjects of the invited papers. There were lectures on metrology, elasticity, critical phenomena, mechanical and thermal properties of solids and liquids, spectroscopy, light pressure, Hertzian waves, magneto-optics, atmospheric electricity, solar physics, biophysics, and gravitation. Nor were the most recent discoveries forgotten: Henri Becquerel and the Curies lectured on radioactivity, Wilhelm Wien and Otto Lummer on blackbody radiation (theoretical and experimental, respectively), Paul Villard on cathode rays, and J. J. Thomson on the electron theory and atomic constitution. Villard's discussion of cathode rays could hardly have satisfied Thomson, nor most of the other attending physicists. Three years after Thomson's celebrated identification of cathode rays with

TABLE 2.3
Distribution of Subjects of Invited Papers at the 1900 Paris Conference

Subjects	*Number*	*Percent*
Electricity and magnetism	22	24
Mechanical and molecular physics	19	20
General questions; measure and units	15	16
Optics and thermodynamics	14	15
Cosmic physics	9	10
Magneto-optics, cathode rays, uranium rays	8	9
Biological physics	5	6

free electrons, the French physicist argued that cathode rays did not consist of electrons, but of hydrogen ions.

Compared with later standards, physics was a ridiculously inexpensive activity, but from the perspective of contemporary standards for science funding, things looked different. Adding to the expenditures for salaries and equipment were the many new laboratories that were created in the 1890s and the early years of the twentieth century (see table 2.4). Between 1890 and 1914, 22 new physics laboratories were built in Germany, 19 in the British Empire, 13 in the United States, and 12 in France. Although Germany erected the most institutions, British and American investments were substantially greater. While Germany invested a total of 10,300 marks per physicist per year, American investments were the equivalent of 14,000 marks, and British investments 15,500 marks. Of course these are average figures, and there were great variations from physicist to physicist and from institute to institute, with experimentalists being much more expensive to support than theorists. For example, whereas the institute of physics at Berlin University received 26,174 marks in 1909, Planck's small institute for theoretical physics had to do with a budget of 700 marks.

The rise in physics facilities was closely related to the marked change toward research as the prime function of higher education that had its roots in Germany and became fully accepted in most other countries only around 1900. At that time, the research student became a central figure in any physics institution of significance, and with the influx of research students came the need for more laboratory space. Data from Leipzig University are revealing: When the institute of physics was built in 1835, 12 percent of its

TABLE 2.4
Physics Institutes and Faculty

	No. of institutes	*Faculty, 1900*	*Faculty, 1910*
Austria-Hungary	18	48	59
Belgium	4	9	10
United Kingdom	25	87	106
British Empire, other	7	10	13
France	19	54	58
Germany	30	103	139
Italy	16	43	51
Japan	2	6	17
Netherlands	4	10	13
Scandinavia	7	18	26
Switzerland	8	17	23
United States	21	100	169

Note: Based on data in Forman, Heilbron and Weart 1975.

area was planned for laboratory use; in 1873 the institute was expanded by a factor of four, and laboratory space increased to 46 percent; another expansion, in 1904 and by a factor of three, resulted in 60 percent for laboratories and rooms for graduate students. Thus, during a period of seventy years, space for research had increased roughly sixty times. Research work as part of graduate studies went back to Justus Liebig's chemical laboratory in Giessen in the 1830s and became institutionalized in physics at many German universities in the 1870s and 1880s. But it was not a self-evident policy, even at major German universities. The physics institute at the University of Tübingen was erected in the late 1880s, and at that time, "it was not expected that students in the institute would carry out independent scientific investigations," as Friedrich Paschen reported in 1906, adding that "in this respect, the situation has been essentially altered in the last ten years" (Forman, Heilbron and Weart 1975, 103).

Until about 1890, in the United States and many other countries it was not even expected that professors would carry out independent scientific investigations. The Jefferson Physical Laboratory at Harvard University was completed in 1884. Housing laboratories and a large lecture room, it was intended for both instruction and research. John Trowbridge, the laboratory's director from 1888 to 1910, was well aware that it was high time to follow the German tradition in physics. In 1884 he warned against "the lecture or recitation system unsupported by laboratory work" still so common, and initiated laboratory-based courses for which the new laboratory was well suited. Twelve years later, he could look back on a period of change both in instruction and research: "To-day a certain amount of original work is expected of [the physics professor at Harvard]. During the past ten years more original research has been done in Harvard University than in the previous two hundred years" (Aronovitch 1989, 95 and 99).

Physics Journals

What, then, came out of the investments in physics? Physicists, like other scientists, produce papers; the number of research papers is the most direct measure of scientific productivity. By 1900 physics journals existed in more or less the same format as today, but of course in fewer numbers and much less specialized. Although the largest and most substantial part of physical research was published in special physics journals, there were also many articles that appeared in journals or yearbooks covering other scientific disciplines as well. Examples are *Comptes Rendus*, published by the Academy of Sciences in Paris, the *Göttingen Nachrichten*, published by the Göttingen Society of Science, and the *Proceedings of the American Association for the Advancement of Science*. Many countries had local journals devoted to physics, usually written in the local language and often published by the local physical society, but research papers of international significance rarely appeared in journals of smaller countries. Such papers were typically included in the core physics journals of the large Western nations, which were *Philosophical Magazine* (England), *Journal de Physique* (France), *Annalen der Physik* (Germany), *Nuovo Cimento* (Italy), and *Physical Review* (the United States). Table 2.5 gives the number of physics papers in 1900 and shows that Germany was clearly ahead of the other nations, both in the production of papers and in productivity per physicist.

Germany's strong position in physics, quantitatively as well as qualitatively, was unquestionable. As to the quality, or what later historians of science have judged to be important contributions to physics, the authoritative *Dictionary of Scientific Bibliography* lists 197 physicists who were more than twenty years old in 1900. Of these, 52 were Germans and six Austrians. The nationalities of other top physicists included in the *Dictionary* were as follows: Britain with 35, France with 34, the United States with 27, Russia with 9, the Netherlands and Italy both with 7, and Sweden with 5. This is very much the same list as appeared in other evaluations. To a significant extent, high-quality physical research was concentrated in the four economic and political leaders of the world. These accounted for 75 percent of the historically important physicists of the period. If Austria-Hungary, the Netherlands, and Scandinavia are included in a Germanic bloc, the figures are 86 percent for the "big four" and 38 percent for the Germanic bloc. Data from the Nobel prize awards speak roughly the same language: Of the twenty prize winners between 1901 and 1914, sixteen were from the big four (but only one American), and so were 77 percent of the suggested candidates.

The leading physics journal in the first decade of the twentieth century was probably *Annalen der Physik*, although in the English-speaking world it was rivaled by *Philosophical Magazine*. *Annalen* went back to 1799, when its title was *Annalen der Physik und Chemie*, but when Paul Drude took over

TABLE 2.5
Physics Journals in 1900

Country	*Core physics journal*	*Papers, 1900*	*Percent*	*Productivity*
Britain	*Philosophical Magazine*	420	19	2.2
France	*Journal de Physique*	360	18	2.5
Germany	*Annalen der Physik*	580	29	3.2
Italy	*Nuovo Cimento*	120	6	1.4
USA	*Physical Review*	240	12	1.1
All other	—	280	14	–

Note: The last column gives the average number of papers per academic physicist.

the editorship in 1900 the "Chemie" was dropped—there were already enough journals in chemistry and physical chemistry, and *Annalen* had long ago begun to concentrate on pure physics. After Drude's death in 1906, Wilhelm Wien became editor, with Planck continuing as co-editor and being responsible for theoretical physics. As *Annalen* was a German journal, naturally the large majority of the contributors were Germans, but there were also many contributions from Austrian, Scandinavian, and Dutch physicists. *Annalen* was a prestigious journal for pure research and almost all of its articles were written by university physicists or physicists at major physical laboratories, such as the Physikalisch-Technische Reichsanstalt (Imperial Institute of Physics and Technology) in Berlin. Prestigious as *Annalen* was, however, it was not particularly difficult to get articles published in it. The journal included a large number of dull and insignificant articles, often experimental doctoral dissertations only slightly abbreviated. In sharp contrast with later publication practice, in 1906 the rejection rate, according to Planck's estimate, was only 5 to 10 percent of the submitted papers. The growth in physics was reflected in the gradual expansion of *Annalen*, which increased its annual number of issues from twelve in 1900 to sixteen in 1912. In addition to the journal, where completed research was published in papers that were often very long, an abstracting journal, the *Beiblätter zu den Annalen der Physik*, was published as a sequel. It is well known that the characteristic pattern of physics literature, and scientific literature in general, is exponential growth. However, there are reasons to believe that the much discussed "law" of exponential growth is valid only for post-1920 science. As far as physics is concerned, the number of papers reviewed in the leading abstract journals was roughly constant during the the period 1900–10. The *Beiblätter* of 1900 reviewed 2,468 articles on its 1,358 pages; ten years later, the journal included 2,984 articles in its 1,312 pages. The number of papers appearing in *Fortschritte der Physik* likewise fluctuated between 1,872 and 2,825 during this period. There was no sign of systematic increase, exponential or otherwise.

During the first decade of the century, the *Annalen* became more oriented toward the kind of theoretical physics that Planck wanted to promote. It carried many articles on electron theory, relativity, quantum theory, thermodynamics, and theories of gravitation. For two decades, the journal was the flagship of the new trends in theoretical physics. These trends were also reflected in the biweekly *Physikalische Zeitschrift*, first published in 1899 and specializing in smaller articles. However, this journal did not have quite the quality of the *Annalen* and was often used by authors whose manuscripts had been declined by Planck and Wien. Yet another new German physics periodical was the *Jahrbuch für Radioaktivität und Elektronik*, founded in 1904 and edited by Johannes Stark. As indicated by its title, the *Jahrbuch* specialized in the physics originating in the great discoveries of the 1890s,

TABLE 2.6
Publications in Spectroscopy, 1890–1910

	1890	*1895*	*1900*	*1905*	*1910*
Germany	81	105	167	144	171
England	52	60	62	73	69
France	32	41	42	74	53
United States	31	37	44	70	67
Italy	5	11	11	13	11
Others	17	33	39	42	41
Total	218	287	365	416	412

Source: Kayser 1938.

especially radioactivity, x-rays, and electron physics. (*Elektronik* or "Electronics" at this time meant the physics of the electron and did not yet have the later engineering meaning.) Stark's *Jahrbuch* was almost purely devoted to experimental subjects and did not include electromagnetic theories of the electron or similar topics in its definition of "electronics."

If there was a field and technique of physics that characterized the entire discipline, perhaps spectroscopy would be a good candidate. Spectroscopy, both experimental and theoretical, was in many ways the leading field of physics, and it continued to occupy a very prominent position for two decades. Papers on spectroscopy, listed under the entries "spectra" and "Zeeman effect," made up 3.2 percent of all papers abstracted in *Science Abstracts* in 1898. In 1908 the figure had risen to 6.2 percent. The German spectroscopist Heinrich Kayser boasted that he had a complete collection of papers in his area of research. In 1938 the 85-year-old physicist published a statistical study of the development of spectroscopy (table 2.6). Not only do the data present a picture of the development of spectroscopy, but they also give a crude indication of the distribution of physics among different nations in the period 1890 to 1910.

A Japanese Look at European Physics

Physics was practically nonexistent in Japan before 1870; even at the turn of the century, institutes for scientific research were few and badly equipped. Physics was taught at Tokyo University, established in 1877, but research was given a very low priority. "There was no provision in the university budget for research," recalled a physicist about the conditions in the early years of the new century. "In the faculty of science of the University of Tokyo in those days, the operating budget was only about 5,000 yen [about $2,500] for physics and 4,000 yen for chemistry. This was hardly enough to pay for the instruments, chemicals, books, journals, specimens, charcoal, gas and electricity needed to teach students, much less to carry on research" (Nakayama, Swain, and Yagi 1974, 166).

Hantaro Nagaoka—who will be discussed further in chapter 4—was one of the pioneers of Japanese physics. He graduated from the new Tokyo University in 1887 and from 1893 to 1896 he went to Europe on an extended study tour, continuing his studies at the universities of Munich, Berlin, and Vienna. In July 1910, by then a professor at the Tokyo Imperial University (as Tokyo University had been renamed) and Japan's foremost physicist, he again set off to Europe for an extended study tour. He attended international congresses, visited several of the most eminent physics institutes, paid homage to the heroes of physics (dead or alive), and established contacts with important physicists. What Nagaoka saw on his study tour, and reported in his letters, was a fairly representative sample of what went on among first-rate European physicists—and at that time physics was still, to a large extent, identical with European physics. An active researcher in both experimental and theoretical aspects of physics (and competent in French, German, and English), Nagaoka was well qualified to give a reliable report on the state of physics in Europe, and perhaps one that was more objective than reports given by the Europeans themselves.

In a letter of November 1910 to the Italian physicist Augusto Righi, an experimentalist who worked in electricity and optics, Nagaoka mentioned that he went from Bologna (where he had stayed with Righi) to Rome, and from there to Geneva, Zurich, Munich, and Berlin. These were some of his impressions:

> There are many interesting magnetic works going on in Prof. Weiss's laboratory in Zurich, where I have seen an electromagnet giving more than 80,000 Gauss. In Munich, Ebert is engaged in works connected with atmospheric electricity, and several interesting investigations are being undertaken in Röntgen's laboratory and in [the] institute of theoretical physics under Sommerfeld. The Technische Hochschule in Munich has also [an] institute of technical physics, where Joule-Kelvin effect on the outflow of gases through porous plugs is investigated, and specific heat of steam under high pressure and temperature has already been examined in connection with the signification [*sic*] on the use of steam turbines. The only point of difference from the ordinary physical laboratory seems to lie in the use of machines of high power, and in the cooperation of mechanical engineers with the physicists. (Carazza and Kragh 1991, 39)

In the same letter, Nagaoka regretted the lack of interest that European physicists had shown his atomic theory of 1904 (see chapter 4) and intimated that the neglect was the result of J. J. Thomson's dominant position as the authority in atomic theory. "The ponderous name of the eminent English physicist seemed to me sufficiently weighty to suppress the work of an Oriental physicist," he complained. Nagaoka was always acutely aware of being an Oriental in a field dominated by whites, and he fought hard to prove that Japanese science was not necessarily inferior to that of the West. In a letter of 1888 he wrote, "There is no reason why the whites shall be so supreme in everything, and . . . I hope we shall be able to beat those *yattya hottya* [pompous] people in the course of 10 or 20 years" (Koizumi 1975, 87). To the new generation of Japanese scientists, scientific rivalry could not be wholly separated from racial rivalry.

The busy Japanese traveler also visited Ernest Rutherford in Manchester, Heike Kammerlingh Onnes in Leiden, Pieter Zeeman in Amsterdam, and Édouard Sarasin in Geneva; and in Germany he added to Munich and Berlin the universities or polytechnics of Bonn, Strassburg, Aachen, Heidelberg, Würzburg, Karlsruhe, Leipzig, Göttingen, and Breslau. To Rutherford he reported details about the many institutions. He was impressed by Kammerlingh Onnes's low-temperature laboratory and speculated whether radioactive decay would remain unaffected at the few degrees above absolute zero obtainable there. In Amsterdam, Nagaoka "saw Zeeman investigating the effect bearing his name," and in Leyden he listened to Hendrik Lorentz "discussing [Felix] Ehrenhaft's curious result on the charge of electrons," namely the existence of sub-electrons; when visiting the physical institute in Berlin, "[Erich] Regener was repeating Ehrenhaft's experiment and announced that the result was entirely wrong." It is obvious from Nagaoka's report that he considered Germany the leading nation in physics. He described Philipp Lenard's radiological institute at Heidelberg University as "perhaps one of the most active in Germany" and added that "Professor Lenard and most of his pupils are working on the phosophorescence and photoelectric action." Among the activities in Göttingen, he singled out Emil Wiechert's seismological research and Woldemar Voigt's laboratory, which "is justly celebrated for the numerous works, which are connected with the physics of crystals and the magneto- and electro-optics. There were more than 20 research students." He described Lummer's institute in Breslau, which specialized in optics, as "splendid." But not all was praise: "The physical institute in Leipzig is perhaps the largest in Germany; but I find that the largest is not always the best. However poor the laboratory may be, it will flourish if it has earnest investigators and an able director. The size and the equipment of the laboratory seems to me to play a secondary part in the scientific investigations" (Badash 1967, 60).

Nagaoka's account was limited to classical physics. He did not mention quantum theory, and relativity theory only in passing—"In Munich . . . there is also an institute for theoretical physics under Sommerfeld, who is working on the principle of relativity." This was not because Nagaoka was a conservative who had not discovered the revolution that supposedly took place in physics during the first decade of the century. Classical physics was completely dominant at the time and in a healthy, progressive state; the new theories of quanta and relativity were cultivated only by a very small minority of the physics community and were scarcely visible in the general landscape of physics. If there was a revolution under way, few were aware of it. This impression is confirmed by an analysis of the distribution of physics specialities taken from leading abstract journals. Thus, the 2,984 articles included in the 1910 volume of the *Beiblätter zu den Annalen der Physik* (see table 2.7) show that almost half the publications were in optics (16 percent) and electricity and magnetism (31 percent). The category "cosmic physics" might lead one to think of astrophysics or physical cosmology in a modern sense, but that would be anachronistic. Astrophysics and cosmological subjects did appear in the category, but so did astronomical spectroscopy, geophysics, meteorology, atmospheric electricity, and geomagnetism. In the abstracting journal of the Berlin Physical Society, *Fortschritte der Physik*, "cosmic physics" was introduced in 1890, at the same time as the term "geophysics." ("Astrophysics" was an older term, first appearing in 1873.) Cosmic physics was an interdisciplinary attempt to bring these and other phenomena into the domain of physics. It was a popular branch of physics around the turn of the century when Svante Arrhenius, the Swedish physical chemist, wrote his influential *Lehrbuch der kosmischen Physik* (1903). Ten years later, however, cosmic physics had been relegated to the periphery of physics.

The general composition of physics subjects given by the *Beiblätter* is confirmed by the American *Science Abstracts*, which listed 1,810 publications in physics for 1910. The differences are rooted mostly in different ways of classifying the subjects. For example, whereas the bulk of papers within the *Beiblätter* group called "constitution and structure of matter" dealt

TABLE 2.7
Distribution of Subjects in the 1910 Volume of *Beiblätter*

Subject	*Number*	*Percentage*
Electricity and magnetism	913	31
Optics	488	16
Constitution and structure of matter	351	12
Cosmic physics	321	11
Heat	293	10
Mechanics	264	9
Radioactivity	182	6
Measure and weight	60	2
General	59	2
Acoustics	27	1
Historical and biographical	26	1

with physical chemistry, *Science Abstracts* defined the area more narrowly and added a separate list on "chemical physics" (which was not the same as modern chemical physics, a post-1930 subdiscipline). Each of the main areas was divided into a number of subentries. The composition of the category "constitution and structure of matter" clearly shows that the content was not atomic and molecular physics, but rather what we would call theoretical chemistry. The subentries were given as follows: general; mass, density; atomic weight; molecular weight; [chemical] elements; compounds; reactions; affinity, equilibria; solutions; absorption, adsorption; colloids; solid solutions, alloys; crystals; and crystalline fluids.

In none of the abstract journals did papers on atomic physics count much, and one would look in vain for entries on topics such as "atomic structure," "quantum theory," "relativity," or associated names. Articles dealing with these topics did appear, of course, but they were hidden in other entries and were few in number. For example, papers on relativity were included mostly under the label "general" and those dealing with quanta appeared in the sections "heat" and "optics." An examination of the titles in the two abstract journals leads to the estimate that in 1910, fewer than forty papers dealt with the theory of relativity and fewer than twenty with quantum theory. The "new physics" was still a marginal activity, but would not be so for long. The impression of the dominance of classical, experimental physics is further confirmed by a count of the papers in the Italian *Nuovo Cimento*. In 1900–1904 classical physics (defined as mechanics, optics, acoustics, thermodynamics, and electromagnetism) and applied physics made up 33 percent and 22 percent of the publications, respectively; ten years later, the percentages were 32 and 17. Most of thc publications were experimental, namely, 86 percent in 1900–1904 and 77 percent in 1910–14. Papers on relativity and quantum theory were very few, less than ten in the five-year period 1910–14.

Chapter 3

DISCHARGES IN GASES AND WHAT FOLLOWED

ONE OF THE charms of physics is that it deals with nature, and nature is secretive. There are natural phenomena, or phenomena producible in the laboratory (which are nonetheless natural), that physicists at any given time do not know about. The discovery of new phenomena sometimes has important consequences for our worldview. The physicists of the 1890s, who believed that the foundation of physics was fixed once and for all, implicitly assumed that all the important phenomena of nature were already known. They were in for a surprise, as was dramatically shown by the series of discoveries between 1895 and 1900.

From chapter 1, one may get the impression that most physicists in the 1890s were occupied with deep theoretical questions about the foundation of physics. However, this was far from the case. The large majority were experimentalists investigating, for example, methods of electrical measurements, thermal phenomena, spectroscopy, and electric discharges in gases. Although the first experiments with discharges in evacuated glass tubes were made as early as the late eighteenth century, it was only in the 1850s that the area attracted serious interest, primarily as a result of the pioneering work of the Swiss Auguste de la Rive and the German Julius Plücker. Johann Wilhelm Hittorf, a student of Plücker's, investigated systematically the light appearing in the tubes and identified it with rays emitted from the cathode and moving rectilinearly through field-free space; that is, with cathode rays, a name first introduced by the Berlin physicist Eugen Goldstein in 1876. The subject had by then become an important part of physics; it received additional attention through the experiments performed by William Crookes in England. In 1879 Crookes hypothesized that cathode rays were corpuscular—a stream of electrified molecules, or what he called a "fourth state of matter." Young Heinrich Hertz was one of many who felt the new field presented an attractive challenge, primarily because it seemed to offer a quick way to recognition, but also because the discharge phenomena were enigmatic and simply fascinating to look at. "In the middle of it all," he wrote about his early experiments in 1882, "stands the luminous belljar in which the gases perform the maddest antics under the influence of discharges and produce the strangest, most varied, and most colourful phenomena. My place now really looks very much like a witch's kitchen" (Buchwald 1994, 132). The key apparatus in the witches' kitchens of Hertz and other cathode ray researchers were a series of specially designed vacuum tubes provided with electrodes, a Ruhmkorff induction coil to generate the necessary high voltage, and, not least, the all-important vacuum pumps.

As seen in retrospect, the question of the nature of the cathode rays was the most important one in this area of research. However, it was not initially considered an issue of crucial importance. Many physicists were occupied with other aspects of discharge phenomena, such as the patches or tubes of colored light and their behavior under the action of magnetic fields. In the late 1890s, some physicists believed they had identified a new kind of cathode rays, sometimes referred to as "magnetic rays," but the discovery claim was not generally accepted. Hertz's failure in detecting a deflection of cathode rays when the rays crossed the electric field of a plate capacitor made him conclude that the rays were electrically neutral and not a primary effect of the discharge. According to this view, an electric current flowed inside the tube, but the observed cathode rays were different from the current that generated the rays as some kind of flow or wave in the ether. Hertz's conclusion was contrary to the corpuscular hypothesis of Crookes and other British researchers, but was supported by most German physicists. It received strong experimental support by experiments made in 1894 by Hertz's student Philipp Lenard, who placed a thin metal foil as a window in the tube and showed that the cathode rays were able to pass through the foil. If the rays were beams of atomic or molecular particles, this permeability could not be explained. The whole area was confusing, with experimental evidence disagreeing with either of the rival views and none that could clearly distinguish between them. For example, in the same year as Lenard provided the ether-pulse hypothesis with experimental support, J. J. Thomson in England argued that the velocity of the cathode rays was only about one-thousandth of that of light; if the rays were electromagnetic processes in the ether, the velocity would be expected to be very close to the velocity of light. Lenard's experiments, and the undecided situation in cathode-ray research in general, caused many German physicists to take up the subject. One of them was Wilhelm Conrad Röntgen, a relatively undistinguished professor of physics at the University of Würzburg.

A New Kind of Rays

Inspired by Lenard's findings, Röntgen started investigating the penetrability of cathode rays in the fall of 1895. Like other physicists experimenting with the rays, he used a screen with barium platino-cyanide to detect fluorescence caused by the cathode rays. The screen, which was supposed to play no role in the particular experiment, lay some distance away from the tube, which was covered with black cardboard and operated in a dark room. On November 8, 1895, Röntgen noticed to his surprise that the screen fluoresced, a phenomenon that could not possibly have been caused by the cathode rays. Having confirmed that the phenomenon was real, he started a systematic investigation of what he soon began thinking of as a new kind of rays, different from both light and cathode rays. In Röntgen's own words, from his original communication of January 1896, his observation was this: "If we pass the discharge from a large Ruhmkorff coil through a Hittorf [tube] or a sufficiently exhausted Lenard, Crookes, or similar apparatus and cover the tube with a somewhat closely-fitting mantle of thin black cardboard, we observe in a completely darkened room that a paper screen covered with barium-platino-cyanide lights up brilliantly and fluoresces equally well whether the treated side or the other be turned towards the discharge tube" (Segré 1980, 22). November 8 is the official birthday of x-rays, but of course it is much too simplistic to claim that Röntgen just happened to discover the rays on the evening of this date. He became aware of a surprising phenomenon, and transforming the original observation into a discovery required hard work and much thinking. It was only at the end of the year that Röntgen was certain that he had indeed discovered a new kind of rays and then announced his discovery. To call it the discovery of the year, or even the decade, would not be an exaggeration. Contrary to the practice of a later generation of physicists, Röntgen did not call a press conference. In fact, he worked alone and kept his discovery a secret until it appeared in print.

Röntgen's new rays caused an enormous stir, especially among physicists and medical doctors, but also in the public at large. Question number one in the physics community concerned the nature of the new rays, which seemed even harder to classify and understand than the cathode rays. In his preliminary investigations, Röntgen had found that the rays shared some of the properties of light in that they followed straight lines, affected photographic plates, and were uninfluenced by magnetic fields; on the other hand, they showed neither reflection nor refraction and thus seemed to be different from both light and the Hertzian electromagnetic waves. He suggested tentatively that the rays might be longitudinal vibrations of the ether, an idea that for a time received support from several physicists. However, neither this nor other ideas, including the suggestion that x-rays were an extreme kind of ultraviolet radiation, won general recognition. For more than a decade, physicists worked happily with x-rays without knowing what they were working with. This lack of understanding did not prevent a growing body of empirical knowledge about the new rays to be established, nor did it prevent the rays to be used as tools in experimental physics. For example, it was quickly realized that x-ray tubes were ideal for ionizing gases and in this capacity, the new tubes were used widely in experiments.

The discussion of the nature of x-rays continued until evidence accumulated to the effect that they were a species of electromagnetic (transverse) waves, possibly with an extremely short wavelength. This belief was strengthened when the British physicist Charles Barkla concluded from scattering experiments in 1905 that x-rays showed polarization. Yet the matter was far from settled, and other experiments seemed to speak against the wave hypothesis and in favor of a kind of corpuscular radiation, as argued by William Bragg in 1907. With the advantage of hindsight, we can say that the confusing experimental situation reflected the wave-particle duality only clarified with the advent of quantum mechanics. X-rays are both particles and waves, so it was no wonder that physicists in the early part of the century obtained ambiguous results. About 1910, it looked as if x-rays might be simply electromagnetic waves of very short wavelength. If so, diffraction of the rays would require a grid with a correspondingly short spacing, and in 1912 Max von Laue realized that such spacings were found in the interatomic distances between the ions of a crystal. The experiment was made by the Munich physicist Walter Friedrich and his student Paul Knipping in the spring of 1912. The collaboration between the theorist Laue and the experimentalists Friedrich and Knipping resulted in the first x-ray diffraction pattern ever. This provided, at the same time, decisive proof that x-rays are electromagnetic waves, with a wavelength of the order of magnitude 10^{-13} m. Out of the experiments in Munich grew a whole new branch of x-ray research, the use of x-ray diffraction in crystallography. Pioneered by William H. Bragg and his son William L. Bragg, among others, the branch soon became of great importance in chemistry, geology, metallurgy, and even biology. For example, x-ray crystallography was a key element in the celebrated 1953 determination of the double-helix structure of the DNA molecule.

Back in the late 1890s, Röntgen's sensational discovery caused a minor revolution in physics and inspired a great many physicists to start investigating the new phenomenon. Rarely has a discovery been received so enthusiastically by scientists and nonscientists alike. According to a bibliography, in 1896 alone there appeared 1,044 publications on x-rays, including 49 books. Among the many physicists who joined the x-ray trend was Henri Becquerel, professor of physics at the Museum of Natural History in Paris.

From Becquerel Rays to Radioactivity

In January 1896, Röntgen's discovery was discussed in the French Academy of Sciences, where Poincaré suggested that the cause of the rays might not be electrical, but rather related to the fluorescing part of the glass tube. If so, there was reason to assume that all strongly fluorescing bodies emitted x-rays in addition to their luminating rays. Poincaré's remarks were taken up by Becquerel, an expert in fluorescence, and on February 24, he reported to the academy that a fluorescent double salt of uranium, uranium potassium sulfate, emitted what he believed were x-rays. He had placed the salt on a photographic plate wrapped in black paper, exposed it to sunlight for several hours, and then observed a distinct blackening of the plate when it was developed. Had radioactivity now been discovered? Not quite, for Becquerel believed the penetrating rays were a result of fluorescence and that the exposure to the sun was therefore crucial. A week later, when he repeated the experiment, but the sun failed to shine, he realized that the uranium salt emitted the rays even in the absence of sunlight. He now understood that he had made a discovery, namely that the uranium salt emitted penetrating rays without electric action and without the action of sunlight. The new phenomenon was apparently distinct from both x-rays and fluorescence. In addition, he quickly discovered that the rays were also emitted from other uranium salts and even more strongly from the non-fluorescing metallic uranium. Hence the rays were originally often referred to as "uranium rays."

Becquerel's discovery of radioactivity was fortunate, but not accidental. Contrary to Röntgen, the French physicist was guided by a hypothesis he wanted to test, namely that intensely fluorescent bodies emit x-rays. But most fluorescent materials are not radioactive, so why did he concentrate on uranium salts? His choice has traditionally been ascribed to luck, but it was hardly luck at all. Earlier, together with his father Edmond Becquerel (who was also a professor at the Museum of Natural History in Paris), he had studied the fluorescence spectra of uranium compounds and noticed that the spectral bands obeyed a remarkable regularity. It is likely that this regularity, peculiar to fluorescent uranium salts, inspired Becquerel to the idea that the visible light from the sun was transfomed into the much shorter wavelengths supposedly characteristic of x-rays. Such transformation is forbidden according to Stokes' law, discovered by Gabriel Stokes in 1852. According to this law, fluorescent bodies can emit radiation only of a wavelength larger than that of the exciting radiation. Stokes' law is correct—it follows from the quantum theory of radiation—but in the last part of the nineteenth century there were many reports of "anomalous fluorescence," that is, exceptions to Stokes' law. The law was not considered absolute and, according to a theory suggested by the German physicist Eugen Lommel, anomalous fluorescence should take place in substances that exhibited the kind of regular spectra that Becquerel had observed in uranium salts. If this was indeed Becquerel's line of reasoning, it is not so strange that he chose uranium compounds for his research. His original observation of penetrating rays emanating from uranium salts exposed to sunlight may only have confirmed what he more or less anticipated. The subsequent events did not.

The uranium rays did not cause nearly the same sensation as x-rays and for a year or two, Becquerel was one of a handful of scientists who actively studied the new phenomenon. After all, the effects of the uranium rays were weak and many physicists considered them as just a special kind of x-rays, although with an origin that defied explanation. From the perspective of Becquerel, who believed that the uranium rays were related to the peculiar spectra of uranium compounds, there was no reason to assume that the rays were also emitted by other compounds. It was only when Marie and Pierre Curie discovered substances much more active than uranium that radioactivity made headlines and became a phenomenon of great importance to the physicists.

The radioactivity of thorium was announced in the spring of 1898, independently by Marie Curie and the German Gerhard Schmidt. Later the same year, Marie and Pierre Curie discovered in uranium ores two hitherto unknown elements that they proposed to name polonium and radium. The extraordinarily active radium made radioactivity known to the public at large and initiated a new and exciting phase in the study of Becquerel's rays. Incidentally, the terms "radioactivity" and "radioactive substances" were first introduced by Marie Curie in the same year, 1898. During the next few years, an increasing number of physicists in Europe and North America took up the study of radioactivity, which soon became one of the fastest growing areas of physics. "I have to keep going, as there are always people on my track," Rutherford wrote to his mother in 1902. "I have to publish my present work as rapidly as possible in order to keep in the race. The best sprinters in this road of investigation are Becquerel and the Curies in Paris, who have done a great deal of very important work on the subject of radioactive bodies during the last few years" (Pais 1986, 62).

The early work in radioactivity was primarily experimental and explorative. Which substances were radioactive? How did they fit into the periodic system of the chemical elements? What were the rays given off by the radioactive bodies? Was the activity affected by physical or chemical changes? These were some of the questions that physicists addressed around the turn of the century—and not only physicists, for radioactivity was as much a concern of the chemists. Whether physicists or chemists, their approach was phenomenological and explorative; that is, focusing on the collection and classification of data. It was a period of great confusion and many blind alleys. For example, during the first eight years or so of the century, it was generally believed that all elements were radioactive. After all, it was difficult to believe that the property was confined to a few heavy elements, and the crude methods of detection seemed to indicate that weak radioactivity was indeed found everywhere.

Among the many topics within the study of radioactivity, the nature of the rays was one of the most important. By 1901, it had been established that the rays were complex, consisting of three species of different penetrability. Beta rays, easily deflected in a magnetic field, were quickly identified as swift electrons, whereas the neutral gamma rays were eventually (about 1912) found to be electromagnetic waves similar to x-rays. The nature of alpha rays was something of a mystery. Early experiments indicated that they went undeflected through electric and magnetic fields and thus were neutral, a view that Rutherford, among others, held for a brief period. However, further experiments, mainly made by Rutherford at Montreal's McGill University, showed that the particles were positively charged and with a mass comparable to that of the hydrogen atom. About 1905, evidence accumulated that alpha particles were doubly charged helium atoms, He^{2+}. The hypothesis was brilliantly confirmed in an experiment of 1908 that Rutherford, now in Manchester, made together with his assistant Thomas Royds. Rutherford and Royds proved spectroscopically that helium was produced from the alpha particles emanating from radon. Together with data from the magnetic deflection of alpha rays, this identification settled the matter.

Even more important than the nature of the rays was the insight that radioactivity is not a permanent phenomenon, but decreases over time. A radioactive substance transforms into another substance in the sense that the atoms change—transmute—from one element to another. This was the basic content of the law of transformation suggested by Rutherford and the chemist Frederic Soddy in 1902. According to this law, not only do atoms transmute, but they also do so randomly, which is expressed by the transformation having a certain decay constant (λ) depending only on the nature of the radioactive element. If it originally consisted of N_0 atoms, after a time t the number will be reduced to $N(t) = N_0\exp(-\lambda t)$. As Rutherford made clear, this means that the probability of an atom decaying is independent of the atom's age. This was a most peculiar phenomenon, and it became even more peculiar when it was found in 1903 that the energy continually liberated from radium was enormous—about 1,000 calories per gram per hour. Where did the energy come from? Granted that radioactivity consisted of subatomic changes, what was the cause of the changes? Such theoretical questions were avoided by most scientists, but they were nonetheless considered legitimate and several physicists and chemists were willing to speculate about the origin of radioactivity. According to a widely accepted hypothesis, based on the atomic model of J. J. Thomson, radioactivity was caused by changes in the internal configuration of the atom. From 1903, this kind of qualitative dynamic model was proposed, in different versions, by Thomson, Oliver Lodge, Lord Kelvin, James Jeans, and others. Rutherford had advocated a similar mechanism as early as 1900, and in 1904, in his Bakerian lecture, he argued that "atoms of the radio-elements may be supposed to be made up of electrons (β particles) and groups of electrons (α particles) in rapid motion, and held in equilibrium by their mutual forces." The accelerating electrons would radiate energy and this "must disturb the equilibrium of the atom and result either in a rearrangement of its component parts or to its final disintegration" (Kragh 1997a, 18). Although Rutherford soon decided that the state of atomic theory did not allow any definite explanation of radioactivity, neither he nor other researchers doubted that radioactivity could be causally explained in terms of subatomic dynamics. In fact, such futile attempts continued up to the mid-1920s.

We know that radioactivity is a probabilistic phenomenon that defies causal explanation and that the probabilistic nature is expressed by the decay law. This was vaguely suggested by Rutherford and Soddy in 1902 and argued more fully by Egon von Schweidler in 1905. From this point of view it seems strange that physicists, including Rutherford and Thomson, nonetheless looked for causal explanations in terms of subatomic changes. At the time, however, there was no reason to suspect that radioactivity was causally inexplainable in principle. The statistical theory was not associated with acausality, but rather with other statistical phenomena such as Brownian motion, where the statistical nature can, in principle, be resolved in deterministic microprocesses.

The attempts to design atomic models that would explain radioactivity on a mechanistic basis were not successful. By 1910, most physicists either ignored the problem or adopted a pragmatic attitude according to which phenomenological laws took priority over mechanistic explanations. But radioactivity's statistical nature was not interpreted as an irreducible feature that necessitated a rejection of causal models in principle. Such an interpretation came only with quantum mechanics and, for this reason, it would be a mistake to see radioactivity as the first known example of an acausal phenomenon.

Spurious Rays, More or Less

Radioactivity, x-rays, and cathode rays were not the only kinds of rays that attracted attention in the years around 1900. In the wake of these famous discoveries followed several claims of the discovery of rays that failed to be confirmed satisfactorily and do not, in fact, exist. Yet these claims, some of which were received with great interest and accepted by many physicists for a period of time, are no less part of the history of physics than are those discoveries that we label "real" today. Consider, for example, the "black light" that was announced in early 1896 by the French amateur physicist Gustave LeBon, a psychologist and sociologist who was an important member of one of Paris's intellectual circles, which also included several prominent scientists. LeBon claimed to have detected a new, invisible radiation that emanated from a closed metal box irradiated by an oil lamp; the radiation was able to produce an image on a photographic plate inside the box. He furthermore found that the black light, as he called it, could not be deviated in a magnetic field and concluded that it could be neither x-rays nor cathode rays. In papers and books published over the next decade, LeBon defended his discovery, made new experiments with the black light, and interpreted it as a manifestation of the gradual dematerialization of matter that he and many others believed in (see also chapter 1). Although LeBon's claims were received coolly by most physicists and chemists, they were not ignored in the scientific literature. In 1896 the Paris Academy of Sciences listened to fourteen papers on black light, as compared to three on Becquerel's uranium rays. Among those sympathetic to LeBon's claim were Henri Poincaré, the biophysicist Arsène d'Arsonvalle, and the physicist and later Nobel laureate Gabriel Lippmann. However, although these and several other scientists found LeBon's speculations about matter, electricity, and energy interesting, they considered the experimental evidence for black light to be less than convincing. About 1902, the subject disappeared from the scientific journals.

The false discovery of N-rays in 1903 has some features in common with the black light case, but is even more remarkable because of the significant amount of claimed confirmations of these nonexistent rays. Contrary to LeBon, René Blondlot of Nancy University was a highly reputed experimental physicist when he reached the conclusion in the spring of 1903 that a new radiation is emitted from discharge tubes. The N-rays ("N" for Nancy) could traverse metals and wood and could be focused, reflected, refracted, and diffracted. Blondlot and other N-ray researchers, many of them at Nancy, soon found that the rays were also emitted by gas burners, the sun, incandescent electric lamps, metals in states of strain, and—most sensationally—the human nervous system. As a detector Blondlot used variations in the brightness of a spark gap and, later, also screens with luminous calcium sulfide. With these he examined the spectrum of N-rays, concluding that the phenomenon was complex and that there were different kinds of N-rays with different refractive indices. N-rays do not exist, but for a couple of years this was not evident at all. In fact, at least forty people "saw" the rays and thus confirmed Blondlot's claim, and a substantial part of the about 300 papers that were published on the subject between 1903 and 1906 accepted that the rays existed.

How could this be possible? There is no doubt that part of the answer must be found in psychological factors and the social structure characteristic of the French scientific community. One of the N-ray researchers was young Jean Becquerel (son of Henri Becquerel), who many years later explained his early mistakes as follows: "The purely subjective method employed for testing the effects [of N-rays] is anti-scientific. It is easy to understand the illusion that deceived observers: with such a method you always see the expected effect when you are persuaded that the rays exist, and you have, a priori, the idea that such-and-such an effect can be produced. If you ask another observer to control the observation, he sees it equally (provided he is convinced); and if an observer (who is not convinced) sees nothing, you conclude that he does not have a sensitive eye" (Nye 1980, 153). That N-rays were really psycho-physiological (and psycho-sociological) effects was also what critics such as Heinrich Rubens and Otto Lummer in Germany, Jean Perrin in France, and Robert Wood in the United States concluded. Blondlot's experiments were done "with a radiation so feeble that no observer outside of France has been able to detect it at all," as one skeptic objected. The combined criticism and the failure to produce new consistent results with the N-rays had the effect that by 1905, a consensus was reached that the rays do not exist. At Nancy, the rays survived for a few more years. The whole episode has later be seen as a case of "pathological science," but this is not a reasonable characteristic. The way that N-ray research developed did not differ fundamentally from the early development of radioactivity or cosmic ray physics. A relatively large number of physicists believed that N-rays existed and that there were reasons for such belief, but in the end, the reasons were not strong enough and were contradicted by other experiments. The episode illustrates the power of subjectivity and illusion in science, but it also illustrates the power of objectivity obtained through critical repetitions of experiments.

Black light and N-rays were just two of several spurious rays reported in the period 1896–1910. The radiation produced in cathode ray tubes placed in a strong magnetic field had been studied since the 1850s, and at the end of the century, some physicists produced evidence for a new kind of cathode rays that behaved like pulses of magnetic emanation. These "magneto-cathode rays" attracted considerable interest; physicists discussed whether they were really a new kind of electrically neutral radiation or could be explained by the motion of ordinary cathode ray electrons. The distinguished Italian physicist Augusto Righi believed in 1908 that he had found undeniable evidence for what he called "magnetic rays" and explained them as streams of neutral doublets made up of electrons bound weakly to positive ions. From 1904 to 1918, 65 papers were written about magnetic rays, two-thirds by Righi and other Italian physicists. If N-rays were French rays, magnetic rays were Italian rays. Whereas almost all Italian papers supported Righi's view of magnetic rays, none of those written outside Italy accepted the claim. The controversy ended in 1918, when new experiments showed that Righi's claim could not be correct. Whereas the phenomena on which the N-ray claim was based were spurious, Righi's experiments with magnetic rays were not contended. Only his interpretation was proved wrong.

Reports about cosmic rays were not initially better founded than those about N-rays. Measurements of free ions in the atmosphere started about 1900, when it was generally believed that the ions had their origin in radioactivity, either from the earth or from gaseous substances in the air. There were occasional speculations about a possible extraterrestrial origin, but for a time, measurements were all too uncertain to warrant such a conclusion. If the source of the atmospheric radiation were the earth, the intensity should decrease with height; if the source were cosmic, it should increase. Experiments made about 1910—some on balloon flights, some on the top of the

Eiffel Tower—only brought confusion: Some measurements did not show any appreciable variation with altitude, some showed a decrease, and some an increase. It was only in 1912 that the Austrian physicist Victor Hess obtained more reliable results on a balloon flight that took him to the height of 5,350 m. He found that the intensity of the radiation first decreased but then, at heights more than about 1,500 m, began to rise markedly with increasing height. Hess concluded "that a radiation of a very high penetrating power enters our atmosphere from above" (Xu and Brown 1987, 29). The conclusion was confirmed by the German Werner Kohlhörster, who took his balloon to a maximum altitude of 9,300 m in 1913–14. With these measurements, the existence of a penetrating cosmic radiation of unknown nature was established, but it took a decade until most physicists became convinced that the radiation was real. When Millikan made unmanned balloon-flight measurements in 1922–23, he failed to confirm the results of Hess and Kohlhörster and concluded that the penetrating radiation did not exist. Two years later, however, he found the highly penetrating radiation, coined the term "cosmic rays," and failed in his reports to credit the much earlier results obtained in Europe. In the United States, the rays were often referred to as "Millikan rays," and Millikan did not protest. Yet Hess was eventually recognized as the true discoverer of the radiation and in 1936, he received the Nobel prize for his discovery. See table 3.1 for a summary of major discovery claims for the period of 1895–1912.

TABLE 3.1
Discovery Claims, 1895–1912, and Their Status by 1915

Entity	*Year*	*Scientists*	*Status, 1915*
Argon	1895	Rayleigh and W. Ramsay	accepted
X-rays	1896	W. Röntgen	accepted
Radioactivity	1896	H. Becquerel	accepted
Electron	1897	J. J. Thomson	accepted
Black light	1896	G. LeBon	rejected
Canal rays	1898	W. Wien	accepted
Etherion	1898	C. Brush	rejected
N-rays	1903	R. Blondlot	rejected
Magnetic rays	1908	A. Righi	doubted
Moser rays	1904	J. Blaas and P. Czermak	reinterpreted
Positive electrons	1908	J. Becquerel	reinterpreted
Cosmic rays	1912	V. Hess	uncertain

Note: Canal rays, or positive rays, were identified in cathode ray tubes by Eugen Goldstein in 1886; they were found to consist of positive gas ions. Etherion was the name for a chemical element consisting purely of ether that Charles Brush claimed to have discovered. Moser or metal rays, first proposed by the physicist L. F. Moser in 1847, were sometimes believed to be rays emanating from illuminated metals and other substances. The phenomenon was explained as photo- and electrochemical in nature.

The Electron before Thomson

Although 1897 is given as the electron's official year of birth, it is much too simplistic to claim that J. J. Thomson just discovered the particle in that year. The electron has an interesting and complex prehistory that can reasonably be traced back to the early part of the nineteenth century. According to the view of electrical theory favored on the Continent, matter consisted of, or included, electrical corpuscles in instantaneous interaction. This approach to electricity and magnetism goes back to André-Marie Ampère and Ottaviano Mossotti in the 1830s, and can be found in a rudimentary form as early as 1759, in a work by the Russian scientist Franz Æpinus. Later in the nineteenth century it was greatly developed by several German physicists, including Rudolf Clausius, Wilhelm Weber, and Karl-Friedrich Zöllner. In these theories, hypothetical electrical particles were considered to be the fundamental constituents of both matter and ether. In this respect, the particles corresponded to the later electrons. In the 1850s and later, Weber and others attempted to build up models of matter and ether from positive and negative unit particles of charge $+e$ and $-e$ where e is an unknown unit charge. However, with the increasing popularity of Maxwellian field theory (where electrical particles have no place), the theories were given up by most physicists.

A quite different source of the modern concept of the electron can be found in the corpuscular interpretation of Michael Faraday's electrolytic laws, especially as enunciated by George Johnstone Stoney in Ireland and Hermann von Helmholtz in Germany. In 1874 Stoney proposed the "electrine" as a unit electrical charge, and in 1891 he introduced the "electron" as a measure of an atomic unit charge. Independent of Stoney, Helmholtz argued the cause of "atoms of electricity" in his Faraday lecture of 1881. The Stoney-Helmholtz electron could be both a positive and a negative charge and was, contrary to the later electron concept, conceived as a unit quantity of electricity rather than a particle residing in all forms of matter. In some of his writings, however, Stoney associated his electron not only with electrolysis but also with the emission of light. In 1891 he suggested that electrons rotating in molecules or atoms might be responsible for the spectral lines, an idea that was close to the one accepted by later electron theorists.

A third version emerged in the early 1890s in connection with the "electron theories" introduced primarily by Hendrik A. Lorentz in the Netherlands and Joseph Larmor in England. (The theories of Lorentz and Larmor differed in many respects, but for our purposes, we shall ignore the differences.) Contrary to the other versions, the Lorentz-Larmor electron was part of electromagnetic field theory and conceived as a structure in the continuous ether. These particulate structures were usually named "charged particles" or "ions," but in his theory of 1894, Larmor introduced Stoney's term "electron" as denoting a singularity in the electromagnetic ether. Whatever the name, the concept introduced by Lorentz and Larmor was at first a highly abstract entity that did not necessarily manifest itself as a real, observable particle. Their charged particles were of unspecified charge and mass and usually not thought to be subatomic. Larmor's electromagnetic electron was primarily a free structure in the ether, but from about 1894, he began to think of the electrons also as primordial units of matter, either positively or negatively charged. For example, in 1895 he suggested "a molecule [atom] to be made up of, or to involve, a steady configuration of revolving electrons" (Larmor 1927, 741). Electromagnetic electrons had entered atomic theory, although so far in a loose way only. Until 1896, none of the three mentioned routes to the electron operated with particles with a charge-to-mass ratio (e/m) much larger than the electrolytically determined ratios of ions.

In 1896 a major change in the conceptualization of the electron occurred with Pieter Zeeman's discovery of the magnetic influence on the frequency and polarization of light. Zeeman observed a broadening of the yellow spectral lines of sodium, a result that was not only unforeseen by theory but also seemed to contradict the theories of light emission associated with the electron theories of Lorentz and Larmor. When Zeeman reported his result to Lorentz, his former teacher and current colleague at Leiden University, the latter responded that "that looks really bad; it does not agree at all with what is to be expected" (Arabatzis 1992, 378). However, what at first looked like a serious anomaly quickly turned into a brilliant confirmation of the electron theory. According to Lorentz, the emission of light was caused by vibrating "ions." When confronted with Zeeman's discovery, he calculated in a simple case what the effect should be when a magnetic field acted on the source of light. The result was that the original sharp frequency would split into two or three distinct frequencies with a widening depending on the ion's e/m. What Zeeman had observed was a blurring, not a splitting, of lines, but guided by Lorentz's prediction, found the separate lines in subsequent experiments. Moreover, from the observed widening, a remarkably large value of e/m followed from Lorentz's theory, namely, the order of magnitude 10^7 emu/g, or about 1,000 times the electrolytic value for hydrogen. (1 emu, or electromagnetic unit, equals 10 coulombs.) Whereas Lorentz's first reaction had been to consider the value "really bad" for his theory, he now realized that theory and experiment could be reconciled if the ionic oscillators (the electrons) had this large e/m ratio. Another important result followed from Lorentz's analysis, namely, that the observed polarization of the components required the oscillators to be negatively charged.

The Zeeman effect, along with phenomena such as electrical conductivity and optical properties of metals, led Lorentz and others to a more restricted and definite notion of the electron: It was now increasingly seen as a subatomic, negatively charged particle with a charge-to-mass ratio some 1,000 times larger than the numerical value of the hydrogen ion. The large value could mean either a very large charge or a very small mass, or some combination of the two. Prior to Zeeman's discovery, Larmor had attributed to the electron a mass comparable to that of a hydrogen atom; after the discovery, and just before J. J. Thomson's experiments, he was willing to consider an electron much smaller than the hydrogen atom. Within a few months, the whole question appeared in a new light, illuminated by results obtained from the experimental study of cathode rays.

The First Elementary Particle

As we have seen, when J. J. Thomson made his celebrated experiments in 1897, the electron was a well-known, if hypothetical, entity. However, this does not mean that Thomson discovered experimentally what Lorentz and others had predicted theoretically. Thomson's particle was at first thought to be different from earlier versions of the electron, and it took several years until the different pictures merged into a single and unified conception of the electron.

The question of the nature of cathode rays—were they corpuscular or etherial processes?—was not much discussed in England until 1896, when Röntgen's discovery forced cathode rays into prominence. It was largely this discovery that made Thomson take up the matter; and so it may be argued that not only did radioactivity follow in the wake of x-rays, but so did the electron, if in a less direct way. Convinced that cathode rays were corpuscular, Thomson decided to measure their velocity and *e/m* value. In his first series of experiments, he measured the two quantities by combining magnetic deflection with a calorimetric (thermoelectric) method of measuring the rays' kinetic energy. Because the experiments indicated that the *e/m* value was independent of both the cathode material and the gas in the tube, he suggested that the cathode ray particles were universal subatomic constituents of matter. At that time he had not yet deconstructed Hertz's experiment, which was an argument against the corpuscular hypothesis. But this did not worry him, for he was convinced that Hertz had failed because the discharge tube had not been sufficiently evacuated and that the observed lack of deflection was a result of the conductivity in the residual gas.

It was only in a later experiment that Thomson proved his case by deflecting cathode rays electrostatically in a highly evacuated tube. By manipulating the rays in a crossed magnetic and electrostatic field, he had at his disposal another method to determine *e/m*; the result he obtained was in agreement with that of the original method. In his famous paper published in *Philosophical Magazine* in October 1897, Thomson reiterated and amplified his earlier conclusion with regard to the nature and significance of the cathode rays. First, he found that *e/m* was of the order of magnitude 10^7 emu/g, which he first interpreted as a combination of a small mass and a large charge. He soon changed his mind and argued that the charge was equal to the electrolytic unit charge, meaning that the mass of the cathode ray particles was about 1,000 times smaller than that of the hydrogen atom. Second, he boldly hypothesized from the slender evidence that the particle was the constituent of all matter, the long-sought protyle, "which has been favourably entertained by many chemists." Thomson suggested that a dissociation of the atom took place in the intense electric field near the cathode, namely that the atoms of the gas (not the cathode) were split up into their constituent "primordial atoms, which we shall for brevity call corpuscles." For Thomson, the elucidation of the nature of the cathode rays was important primarily in the context of atomic theory: "We have in the cathode rays matter in a new state, a state in which the subdivision of matter is carried very much further than in the ordinary gaseous state: a state in which all matter . . . is of one and the same kind; this matter being the substance from which all the chemical elements are built up" (Davis and Falconer 1997, 169).

Thomson's claim of the universality of corpuscles was a bold hypothesis with only a meager experimental basis. It was far from a conclusion derived inductively from experiment, but it was a conclusion for which Thomson was well prepared. For many years, he had entertained ideas similar to those of "many chemists" and there were striking similarities between his subatomic corpuscles of 1897 and the vortex atomic theory he had entertained many years earlier. Less than one year before his first *e/m* experiments he suggested, in a discussion of absorption of x-rays, that "this appears to favour Prout's idea that the different elements are compounds of some primordial element" (Thomson 1896, 304). As far as inspiration is concerned, his discovery claim and concept of the corpuscle seem to have owed very little to contemporary electron theory. William Prout, Norman Lockyer, and Crookes were more important than Larmor, Zeeman, and Lorentz. The Zeeman-Lorentz conclusion of subatomic electrons agreed with Thomson's view, but did not influence it much.

Thomson named his primordial particles "corpuscles." Since the name "electron" was already in use and electron theory was on its way to become a fashionable branch of theoretical physics, why didn't he call the particles by that name? Briefly put, Thomson did not conceive his particle to be identical with the Lorentz-Larmor particle and he stressed the difference by choosing another name. According to Thomson, the cathode ray corpuscles were not ethereal—charges without matter, as the electron theoreticians would have it—but charged material particles, proto-atoms of a chemical nature. In his October 1897 paper, Thomson briefly considered the possibility of subjecting the corpuscular matter to "direct chemical investigation," but rejected the idea because the amount of the corpuscular substance produced in a cathode ray tube was much too small. The identification of corpuscles with free electrons was first suggested by George FitzGerald immediately after Thomson had announced his discovery. Characteristically, FitzGerald considered the reinterpretation an advantage because it "does not assume the electron to be a constituent part of the atom, nor that we are dissociating atoms, nor consequently that we are on the track of the alchemists" (Falconer 1987, 273). Yet, this was exactly what Thomson assumed. According to him, atoms were not merely made up of corpuscles, but they could also be broken down into corpuscles.

Thomson is celebrated as the discoverer of the electron because he suggested corpuscles to be subatomic constituents of matter, elementary particles; because he provided this suggestion with some experimental evidence; and because his contemporaries and later physicists accepted and substantiated the claim. He did not discover the electron simply by measuring the e/m value of cathode rays. Such measurements, more accurate than Thomson's, were being made at the same time by Emil Wiechert and Walter Kaufmann in Germany. Wiechert's first result was $e/m = 2 \times 10^7$ emu/g and Kaufmann initially obtained about 10^7 emu/g which, later the same year, he improved to 1.77×10^7 emu/g. Thomson's mean value was 0.77×10^7 emu/g, to be compared with the modern value of 1.76×10^7 emu/g. Although Kaufmann, like Thomson, varied the material of the cathode and the gas in the tube, he did not suggest from his data that the cathode rays were corpuscular. Wiechert did, but he did not make the same sweeping generalization as his colleague in Cambridge and thus missed one of the most important discoveries in the history of physics.

Thomson's idea of corpuscles or electrons received quick confirmation from the study of a wide range of phenomena. Electrons with approximately the e/m suggested by Thomson were detected in photoelectricity, beta radioactivity, and thermionic phenomena, and they were inferred from magneto-optics, metallic conduction, and chemical reactions. In order to settle the question of the electron's mass, its charge had to be determined. This was done in the final years of the century by Thomson and his associates at the Cavendish Laboratory, especially Charles T. R. Wilson and John Townsend. By 1899, they had obtained a value close to that of hydrogen in electrolysis, corresponding to the mass of the electron being 700 times smaller than the hydrogen atom. During the same short period the concept of the electron stabilized and by the turn of the century, the corpuscle-electron identification was generally accepted and Thomson's initial resistance to the idea was forgotten. Thomson was almost alone in using the term "corpuscle" for what other physicists called the electron. At that time, he conceived the mass of the particles to be electromagnetic, which helped form a consensus view of the electron. The electron had become a mature particle, but other changes waited in the future.

One important outcome of the events of 1896–1900 was a general acceptance of a dissymmetry between positive and negative charges. Zeeman's electron was negative and so was Thomson's corpuscle. The positive electrons of earlier electron theories became hypothetical particles and were abandoned during the first years of the twentieth century. A few physicists claimed to have found evidence for positive electrons, but their claims were not taken very seriously. "If there is one thing which recent research in electricity has established," Norman Campbell wrote in 1907 in his *Modern Electrical Theory*, "it is the fundamental difference between positive and negative electricity" (Kragh 1989b, 213). To summarize, the electron of the early twentieth century was negatively charged and with a mass about one-thousandth of the hydrogen atom; the mass was believed to be partly or entirely of electromagnetic origin. The electron could exist freely or be bound in matter and was recognized to be a constituent of all atoms—perhaps even the sole constituent.

Chapter 4

ATOMIC ARCHITECTURE

THE THOMSON ATOM

SPECULATIVE IDEAS about the constitution of atoms can be found many decades before the discovery of the electron, but only with the new particle did atomic models acquire a more realistic status. The electron was generally accepted as a building block of matter, which led directly to the first elaborate model of the interior of the atom. Thomson's important "plum pudding" model, consisting of electrons held in positions of equilibria by a positive fluid, was first enunciated in his 1897 paper, but it was indebted to work done many years earlier.

One of the important sources was the vortex atomic theory, according to which atoms were conceived as vortices in a perfect, all-pervading fluid. Another source was an experiment that the American physicist Alfred M. Mayer made in 1878. Mayer subjected equally magnetized needles floating in water to the attractive force of a central electromagnet and noticed that the needles took up equilibrium positions on concentric circles. Lord Kelvin (then still William Thomson) immediately realized that the experiment furnished a good analogy to the vortex atomic theory. The analogy was taken up by the young J. J. Thomson in his Adams prize essay of 1883, where he dealt in great mathematical detail with Kelvin's vortex theory. In this work, Thomson (from now on, "Thomson" will refer to J. J.) examined theoretically the stability of a number of vortices arranged at equal intervals around the circumference of a circle. For more than seven vortices, where the calculations became highly complex, he referred to Mayer's magnet experiment as a guide. Thomson assumed his elementary vortices to be of equal strength, which not only simplified the calculations but also agreed with his inclination toward a monistic theory of matter. There is a clear analogy between his vortex arrangement of 1883 and his later arrangement of electrons. Although Thomson, like most other physicists, abandoned the vortex atomic theory about 1890, the idea continued to appeal to him. In 1890 he linked the periodic system of the elements with the vortex atomic model and pointed out the suggestive similarity between an arrangement of columnar vortices and the regularity found among the chemical elements: "If we imagine the molecules [atoms] of all elements to be made up of the same primordial atom [elementary particle], and interpret increasing atomic weight to indicate an increase in the number of such atoms, then, on this view, as the number of atoms is continually increased, certain peculiarities in the structure will recur" (Kragh 1997b, 330). Clearly, Thomson was inclined to conceive of the atom as a composite system of primordial elements years before the discovery of the electron.

In his paper of October 1897, Thomson suggested that the atom consists of a large number of corpuscles (electrons), possibly held together by a central force. Relying on Mayer's experiment, he indicated that the electron configuration was a ring structure and that such a configuration might explain the periodic system. In this first version of the Thomson model, the atom was pictured as just an aggregation of electrons and "holes," and so, assuming Coulomb forces between the electrons, there was no attractive force to keep the atom from exploding. Two years later, Thomson presented a more definite hypothesis, in which he explicitly formulated what soon came to be known as the Thomson model of the atom: "I regard the atom as containing a large number of smaller bodies which I will call corpuscles. . . . In the normal atom, this assemblage of corpuscles forms a system which is electrically neutral. Though the individual corpuscles behave like negative ions [charges], yet when they are assembled in a neutral atom the negative effect is balanced by something which causes the space through which the corpuscles are spread to act as if it had a charge of positive electricity equal in amount to the sum of the negative charges on the corpuscles" (Kragh 1997b, 330). Thomson began developing this idea into a quantitative model only in 1903, shortly after he had become acquainted with a somewhat similar model proposed by Lord Kelvin.

The essence of the classical Thomson atomic model, as presented in books and articles between 1904 and 1909, was this: For reasons of simplicity, Thomson largely restricted his analysis to rotating rings of electrons restricted in a plane and subject to the elastic force from a homogeneous sphere of positive electricity. By direct calculation, he examined the mechanical stability of the equilibrium configurations, ruling out those that were not stable. Thomson's involved stability calculations were quite similar to those he had used in his vortex atom work more than twenty years earlier. The complex calculations, which he supplemented with a more approximate method for a larger number of electrons, showed that the electrons would be arranged in a series of concentric rings in such a way that the number of particles in a ring would increase with the ring's radius. Examples of his 1904 equilibrium configurations were 1, 8, 12, 16 for $n = 37$, and 1, 8, 12, 16, 19 for $n = 56$. Accelerating electrons will emit electromagnetic energy, and Thomson therefore had to make sure that his atoms would not be radiatively unstable and collapse. Applying a formula derived by Larmor in 1897, he was able to show that the radiation decreased drastically with the number of electrons in the rings and could therefore, for most purposes, be ignored. The original Thomson atom was mechanically, as well as radiatively, stable.

Thomson realized that his planar model atom would have to be generalized to a spherical model, but saw no reason to engage in the Herculean calculations such an extension would require. After all, the number of electrons in real atoms was unknown, so a detailed comparison with the physicochemical properties of the elements was out of the question. The model was undoubtedly the most popular atomic model in the period 1904–10, when many physicists considered it a good approximation of the real constitution of the atoms. In a lecture in Göttingen in 1909, Max Born praised it as "being like a piano excerpt from the great symphonies of luminating atoms." Rutherford and Lorentz also considered the model to be attractive and made use of it in their own works. The attractive features of the model were particularly related to its monistic nature, which promised a reduction of all matter to electrons, in agreement with the electromagnetic worldview. Moreover, Thomson's calculations provided the model with a good deal of mathematical authority, although most physicists paid little attention to the details of the electron configurations. When it came to empirical credentials, however, the model was less impressive. It was able to explain, in a qualitative and vague manner, phenomena such as radioactivity, photoelectricity, dispersion, the emission of light, the normal Zeeman effect, and, not least, the periodic system. In addition, it promised to illuminate many chemical facts and was, for this reason, popular among many chemists. But in most cases, the explanations were suggestive analogies rather than deductions based on the details of the model.

It was evident from the very beginning that the Thomson model was problematic, both conceptually and empirically. One of the weak points was the positive electricity, supposed to be frictionless and massless and ultimately a manifestation of negative electrons. As Thomson wrote to Oliver Lodge in 1904, "I have always had hopes (not yet realised) of being able to do without positive electrification as a separate entity, and to replace it by some property of the corpuscles. . . . One feels, I think, that the positive electrification will ultimately prove superfluous and it will be possible to get the effects we now attribute to it, from some property of the corpuscles" (Dahl 1997, 324). Thomson never succeeded in explaining the positive electricity as an epiphenomenon. On the contrary, from different kinds of experimental evidence, he concluded in 1906 that the number of electrons was comparable with the atomic weight, a conclusion that was soon generally accepted. During the next few years, a growing body of evidence indicated that the number might be even smaller, possibly corresponding to the ordering number in the periodic system. This was an uncomfortable conclusion, for then the positive sphere must account for the major part of the atom's mass, which contradicted the electromagnetic conception of mass. (The electromagnetic inertia varies inversely with the radius of the charge and so would be negligible for a body of atomic dimensions.) It was no wonder that Lodge called

Thomson's estimate of the number of electrons "the most serious blow yet dealt at the electric theory of matter."

The small number of electrons was a problem not only for the electromagnetic view of matter, but also, and in particular, for the credibility of the Thomson model. In 1904 Thomson did not need an exact match between his model atoms and those really existing, for with several thousand electrons even in the lightest atoms, there was no way in which such a match could be established. Around 1910, there was good reason to believe that the hydrogen atom contained only one electron, the helium atom two, three, or four electrons, and so on; this meant that Thomson's model calculations could now be confronted with the chemical and physical properties of the real elements. In the case of the lightest elements, it could no longer be argued that the number of electrons was too large or that three-dimensionsal calculations were not technically possible. Although Thomson proceeded as if the problem did not exist, it was clear that the desired correspondence between the model and the reality just did not exist.

There were other problems of a more empirical nature. In particular, Thomson's model was unable to explain, without artificial assumptions, the regularities known from line spectra, such as Balmer's law. Indeed, most spectra seemed to pose problems, for according to Thomson, light was emitted by the vibrations of electrons and this required that the number of electrons must be of the same order as the number of observed spectral lines. How could the tens of thousands of lines found in many metal spectra be understood on the basis of the vibrations of one hundred or fewer electrons?

New experiments added to the problems of the Thomson atom. Thomson explained scattering of beta particles by assuming multiple scattering, that is, that the observed scattering was the collective result of many individual scatterings on atomic electrons. On this basis, he managed to account fairly satisfactorily for experimental data. But Thomson's scattering theory failed when confronted with the results of alpha scattering experiments. On the other hand, these were nicely explained by Rutherford's idea of a nuclear atom; for this reason, the alpha scattering experiments have traditionally been regarded as the crucial test between the two theories of atomic structure. However, it would be misleading to consider the demise of the Thomson atom as simply the result of the Manchester alpha scattering experiments. A theory is not refuted just because it fails to account for some experiments. The refutation of Thomson's atomic theory was a gradual process, during which anomalies accumulated and it became more and more clear that the model could not be developed into a satisfactory state. By 1910, before Rutherford proposed his nuclear atom, the Thomson model had run out of steam and was no longer considered attractive by most physicists. A few, including Erich Haas in 1910 and Ludwig Föppl in 1912, continued to investigate the model, but the investigations were primarily mathematical and had no relevance to the real atoms studied by the experimentalists. In a 1910 work, Haas, an Austrian physicist, suggested relating Planck's constant to the dimensions of a Thomson hydrogen atom. Making use of some rather arbitrary assumptions, he obtained formulas for Planck's constant and Rydberg's constant in terms of the size of the atom and the electron's mass and charge. His model was the first attempt to apply quantum theory to the structure of atoms, but Haas's approach was essentially classical: He wanted to explain the quantum of action in terms of atomic theory rather than explain the structure of the atom by means of quantum theory.

At the second Solvay conference in 1913, physicists listened to the swan song of the Thomson atomic model. Thomson applied a suitably modified model in order to explain both alpha scattering and the linear relationship between the energy of photoelectrons and the frequency of incident light that experiments indicated. By assuming that the charge density of the positive sphere decreases from the center, and by making several other ad hoc assumptions, he was able to obtain the photoelectric law with Planck's constant expressed in terms of the electron's charge and mass. However, the whole procedure was so ad hoc and artificial that it must have appeared to his listeners what it was: a last attempt to save a once useful atomic model and to avoid the nonclassical features of quantum theory. Few, if any, of the physicists gathered in Brussels were convinced that Thomson's atom deserved to survive.

Other Early Atomic Models

Although Thomson's model was by far the most important atomic model of the first decade of the century, it was not the only one. The recognition of the electron as a universal constituent of matter stimulated physicists to propose a variety of models, most of them short-lived and some mere speculations. It was mostly in Britain that these atomic models had their origin and were discussed. In Europe and in North America, interest in atomic structure was limited. Common to all models was the inclusion of electrons; what distinguished them were the proposals of how to arrange the necessary positive charge. Thomson's suggestion of placing the electrons in a sphere of positive fluid was made independently by Kelvin in 1901. Kelvin's model had evidently much in common with Thomson's, but it was more qualitative and did not make use of the modern concept of the electron. Kelvin preferred to call his negative particles "electrions," possibly in order to distinguish them from the electrons of Thomson, Lorentz, and Larmor. In papers written between 1902 and 1907, the aging Kelvin (who died in 1907 at age 83) applied his model to a variety of phenomena, including radioactivity, which he believed was triggered by some external agency, perhaps ethereal waves. By introducing non-Coulomb forces into his model, he was able to avoid, to his own satisfaction, the "utterly impossible" conclusion that radioactivity was energy stored in the atoms. Kelvin's ideas were politely ignored by other British physicists.

The Kelvin-Thomson model was one of the hypotheses of atomic structure that Lodge included in his 1906 book *Electrons.* Another was the idea that the atom "consists of a kind of interlocked admixture of positive and negative electricity, indivisible and inseparable into units." This was the picture of the atom favored by Lenard, according to whom the basic structures were rapidly rotating "dynamids," a kind of electrical doublets. Lenard developed the idea in several works between 1903 and 1913, but it failed to attract the interest of other physicists. Yet another of Lodge's candidate hypotheses was that "the bulk of the atom may consist of a multitude of positive and negative electrons, interleaved, as it were, and holding themselves together in a cluster by their mutual attractions" (p. 148). Although it was generally accepted that positive electrons did not exist, in 1901 James Jeans nonetheless proposed such a picture in order to explain the mechanism of line spectra. As a means to circumvent the objection that no equilibrium system exists for a system of charged particles, Jeans suggested that Coulomb's law would break down at very small distances. Undisturbed by the experimentalists' measurements of the electron's charge and mass, Jeans considered an ideal atom in which there was an almost infinite number of (practically massless) electrons concentrated in the atom's outer layer. The positive electrons would then be effectively, and conveniently, hidden within the atom. By means of these and other arbitrary assumptions, he was able to derive spectral series resembling those observed. A somewhat similar model was proposed by Lord Rayleigh five years later, again with the sole purpose of calculating spectral frequencies. In all atomic models before 1913, the emission of light was supposed to result from vibrating electrons. It is interesting to note that Rayleigh, at the end of his paper, considered the possibility that "the frequencies observed in the spectrum may not be frequencies of disturbance or of oscillations in the ordinary sense at all, but rather form an essential part of the original constitution of the atom as determined by conditions of stability" (Conn and Turner 1965, 125). However, he did not develop the suggestion, which later would form a crucial part of Bohr's quantum atom.

Another type of atomic models mentioned by Lodge was the picture of the atom as a kind of solar system, with the electrons revolving ("like asteroids") around a center of concentrated positive electricity. The first suggestion of this kind was enunciated by the French chemist and physicist Jean Perrin in a popular article of 1901. Perrin suggested that the model might explain radioactivity and emission of light, but the suggestion was purely qualitative and he paid no attention to the problem of stability of his planetary atom. The micro-macro analogy between atoms and solar or galactic systems was popular at the time and seems to have been a main reason why planetary atomic models received some attention. The most elaborate attempt was that of Hantaro Nagaoka, whose "Saturnian" model was published in 1904, in the same volume of *Philosophical Magazine* in which Thomson's theory appeared. Nagaoka's model was astronomically inspired, in the sense that it closely relied on Maxwell's 1856 analysis of the stability of Saturn's rings. The Japanese physicist assumed that the electrons were placed uniformly on rings moving around the attractive center of a positive nucleus. Like Thomson and all other model builders, Nagaoka safeguarded his conclusions by adding that "the actual arrangement in a chemical atom may present complexities which are far beyond the reach of mathematical treatment" (Conn and Turner 1965, 113). Nagaoka's calculations led to suggestive spectral formulas and a qualitative explanation of radioactivity. However, they were severely criticized by George A. Schott, a British physicist, who argued that Nagaoka's assumptions were inconsistent and that the model could not lead to the claimed agreement with experimental data. The Saturnian model disappeared from the scene and only reappeared, in an entirely different dressing, with Rutherford's nuclear theory.

In 1911 John W. Nicholson, a mathematical physicist at the Cavendish Laboratory, suggested an atomic model somewhat similar to Nagaoka's. Nicholson's ambitious aim was to derive all the atomic weights of the chemical elements from combinations of proto-atoms, which he supposed existed in the stars only. He considered the positive charge to be of electromagnetic origin, hence much smaller than the electron, and located in the center of the atom. Rotating in spheres around the nucleus (as he called the central charge) were the electrons. Contrary to most other model makers, Nicholson attempted to account for the structure of real atoms, albeit by using hypothetical proto-atoms. Thus, in his scheme, hydrogen contained a ring of three electrons, with the simplest proto-atom being the two-electron system, which he called "coronium." By means of various assumptions, he managed to account for the atomic weights of most elements. Like Haas before him, Nicholson introduced concepts from Planck's quantum theory in order to explain the line spectra. He was led to an atomic explanation of Planck's constant and the conclusion that the angular momentum of the proto-atoms had to be a multiple of this constant. That is, he arrived at the quantization rule $L = nh/2\pi$. This may look very similar to Bohr's reasoning of two years later, but Nicholson's model was not really a quantum theory of the atom. It was founded on classical mechanics and electromagnetism and was much closer to Thomson's than to Bohr's approach. Nicholson continued to develop his theory in papers between 1911 and 1914, and it received some positive response from other British physicists. Yet, by 1915 it was evident that the Nicholson model belonged to the past and not to the future.

RUTHERFORD'S NUCLEAR ATOM

During his fertile years in Montreal from 1898 to 1907, Ernest Rutherford was not particularly interested in atomic models. To the extent he expressed an interest, he was generally in favor of Thomson's theory, which he found useful in understanding radioactive phenomena. It was only in 1910 that Rutherford turned seriously to atomic theory, primarily as a result of his deep interest in the behavior and nature of alpha particles. In 1908 he had definitely shown the alpha particle to be identical with a doubly charged helium ion. In the same year Hans Geiger, a German physicist working with Rutherford in Manchester, reported preliminary results of the scattering of alpha particles on metal foils. Geiger noted an appreciable scattering; the following year, he investigated the matter more thoroughly in collaboration with Ernest Marsden, then a twenty-year-old undergraduate. They found that heavier metals were far more effective as reflectors than light ones and that a thin platinum foil reflected (that is, scattered more than 90°) one of every 8,000 of the alpha particles striking it. When Rutherford learned about the results, he reportedly considered it "the most incredible event that has ever happened to me in my life. . . . almost as incredible as if you fired a fifteen-inch shell at a piece of tissue paper and it came back and hit you." Rutherford made this often-quoted remark in 1936, but it cannot have been the way he reacted in 1909–10. The remark makes sense from the perspective of the almost empty nuclear atom, but Rutherford had no such idea in 1909, when he still thought of the atom as a Thomson-like plenum. At any rate, the experiments induced Rutherford to investigate the scattering of alpha particles and compare the results with Thomson's theory of the scattering of beta particles. This theory, according to which the beta electrons were multiply scattered through small angles by the atomic electrons, seemed to agree nicely with experiments that implied that the number of electrons was about three times the atomic weight.

According to Thomson, the alpha particle was of atomic dimensions and contained about ten electrons. Rutherford, on the other hand, believed that the alpha particle must be considered a point particle, like the electron. Because the alpha particle was a helium atom deprived of two electrons, this view implied, in effect, a nuclear model of the helium atom. Rutherford reached this important conclusion before he developed his scattering theory, which relied on his idea of pointlike alpha particles. The theory that Rutherford presented in 1911 had its experimental basis in the Geiger-Marsden observations of large-angle scattering, which Rutherford found were incompatible with Thomson's theory of multiple electron scattering. In order to produce the observed deflections of more than 90°, scattering had to take place in a single encounter between the alpha particle and a highly charged and concentrated mass. Rutherford therefore suggested that the atom consisted of a massive charge *Ze* surrounded by a cloud of opposite electricity. Since the results of his calculations were independent of the sign of the charge, the nucleus could just as well be a concentration of electrons embedded in a positive fluid, not unlike a special case of the Thomson atom. In Rutherford's words: "Consider an atom which contains a charge $\pm Ne$ at its centre surrounded by a sphere of electrification containing a charge $\mp Ne$ supposed uniformly distributed throughout a sphere of radius *R*. . . . for convenience, the sign [of the central charge] will be assumed to be positive" (Conn and Turner 1965, 138). Based on his nuclear picture of the atom, Rutherford derived the famous scattering formula, which gives the dependence of the scattering probability (cross section) on the scattering angle, energy of the incoming alpha particles, and the thickness and charge of the scattering material.

Rutherford's nuclear atom is justly recognized as a landmark in the history of physics. An observer in 1911 or 1912, however, would never have thought so. When the model was introduced in the spring of 1911, it was met with indifference and scarcely considered to be a theory of the constitution of the atom. The new conception of the atom was not mentioned in the proceedings of the 1911 Solvay Congress (where Rutherford participated), nor was it widely discussed in the physics journals. Not even Rutherford himself seemed to have considered the nuclear atom to be of great importance. For example, in his 1913 textbook on radioactivity, titled *Radioactive Substances and their Radiations*, only 1 percent of the book's 700 pages dealt with the new discovery and its implications. The nucleus was small but not, according to Rutherford, pointlike. On the contrary, Rutherford pictured it as a highly complex body held together by what would become known as nuclear forces: "Practically the whole charge and mass of the atom are concentrated at the centre, and are probably confined within a sphere of radius not greater than 10^{-12} cm. No doubt the positively charged centre of the atom is a complicated system in movement, consisting in part of charged helium and hydrogen atoms. It would appear as if the positively charged atoms of matter attract one another at very small distances for otherwise it is difficult to see how the component parts at the centre are held together" (p. 620).

There were good reasons for the initial lack of interest in the nuclear atom, for Rutherford presented his theory as primarily a scattering theory and only secondarily as an atomic theory. As a scattering theory, it was moderately successful, but its experimental support was limited and indirect; and as an atomic theory, it was incomplete and might even look hopelessly ad hoc. Rutherford had argued from scattering data that the atom's mass was concentrated in a tiny nucleus, but he could offer no suggestions of how the electrons were arranged. For reasons of simplicity, he assumed the negative electricity to form a homogeneous atmosphere around the nucleus, but since the electrons were of no importance in the scattering, this was just an arbitrary picture. "The question of the stability of the atom proposed need not be considered at this stage," he wrote, "for this will obviously depend upon the minute structure of the atom, and on the motion of the constituent charged parts" (Conn and Turner 1965, 138). Rutherford did not suggest a planetary atom in 1911, and his model was therefore completely impotent when it came to chemical questions such as binding and the periodic table. Nor was it any better when it came to physical questions, such as spectral regularities and dispersion. The defining feature of Rutherford's model, the atomic nucleus, was not even new, for nuclear models had already been proposed. Nicholson, who independently suggested his own nuclear model, considered Rutherford's to be merely "a revival of a suggestion of Nagaoka of a simple Saturnian system of the atom, involving only a single positive nucleus" (Heilbron 1968, 303). Incidentally, Rutherford originally wrote of a "central charge"; the word "nucleus" seems to have been used first by Nicholson.

The fate of Rutherford's atomic model changed in 1913, when Geiger and Marsden published new data on the scattering of alpha particles, including a total count of 100,000 scintillations. Their data were in excellent agreement with Rutherford's scattering formula and afforded "strong evidence of the correctness of the underlying assumptions that an atom contains a strong charge at the centre of dimensions, small compared with the diameter of the atom" (Stehle 1994, 221). Still, this was only a confirmation of Rutherford's atomic model seen as a scattering theory, not of other aspects of the model. The Geiger-Marsden results were as irrelevant for the electronic configurations as Rutherford's model was silent about them. An atomic theory would be considered really convincing only if it included the electron system. After all, it was this part of the atom that was responsible for the large majority of the atomic phenomena that could be tested experimentally. This important aspect, absent from the work of the Manchester physicists, was supplied in an unexpected way by a young Danish physicist who turned Rutherford's picture of the nuclear atom into a proper theory of the nuclear atom.

A Quantum Theory of Atomic Structure

Niels Bohr was not originally interested in atomic theory. He wrote his doctoral dissertation on the electron theory of metals and found that this theory, as developed by Lorentz, J. J. Thomson, and others, was unsatisfactory both in its details and in its principles. "The cause of failure is very likely this: that the electromagnetic theory does not agree with the real conditions in matter," he wrote in his 1911 thesis, published only in Danish. Bohr suggested that nonmechanical constraints, or what he called "forces in nature of a kind completely different from the usual mechanical sort," had to be introduced in order to bring the electron theory of metals into agreement with the internal structure of atoms. But in 1911 he did not consider atomic structure and had no clear idea of the kind of constraint or hypothesis needed. He spent the academic year 1911–12 in England, first with J. J. Thomson in Cambridge and then with Rutherford in Manchester, at first continuing his studies of the electron theory of metals but soon concentrating on the new picture of the atom that Rutherford had proposed and that Bohr found highly attractive. He realized that the nuclear atom needed to be completed with an electronic structure and that this would require some nonmechanical hypothesis in order to make the atom stable. Bohr's thinking resulted in the "Manchester memorandum," a document from the summer of 1912 in which he communicated his ideas to Rutherford. In this memorandum Bohr suggested that the atom would be mechanically stabilized if the kinetic energy of the orbiting electrons were confined to be proportional to their frequencies of rotation. As the constant of proportionality, he chose a quantity close to Planck's constant. At this point, Bohr was concerned with mechanical, not electrodynamic, stability.

In the Manchester memorandum, Bohr dealt with electron configurations, molecules, and atomic volumes, but not with spectra. "I do not at all deal with the question of calculation of the frequencies corresponding to the lines in the visible spectrum," he wrote to Rutherford on January 31, 1913, contrasting his theory with that of Nicholson (Bohr 1963, xxxvii). Shortly thereafter, he was asked by a colleague in Copenhagen how his ideas related to Balmer's formula of the hydrogen lines, which Bohr, surprisingly enough, seems to have been unacquainted with or had forgotten about. The question was an eye-opener, and Bohr immediately realized how his ideas could be extended to give an explanation of the discrete spectra. His great paper, "On the Constitution of Atoms and Molecules," appeared in three parts in *Philosophical Magazine* in the summer and fall of 1913. The hydrogen atom was the focus of the first part, where he introduced his famous postulates, namely: (1) the notion of stationary states, where ordinary mechanics is valid but electrodynamics invalid; and (2) the assumption that radiation is emitted or absorbed when the atom passes between different stationary states. The transition process could not be understood classically, Bohr pointed out, "but appears to be necessary in order to account for experimental facts." Most remarkably, inspired by Planck's theory, Bohr assumed that the frequency of light (ν) did not relate directly to the frequencies of the orbiting electrons but was given by the energy difference between two stationary states by the equation $E_i - E_j = h\nu$. From this basic assumption, Bohr was able to derive the Balmer formula for the frequencies of the hydrogen spectrum in a way well known from introductory physics textbooks. The derivation did not merely reproduce a known empirical law, but also resulted in an expression of the Rydberg constant in terms of microphysical constants of nature, namely, the charge and mass of the electron and Planck's constant. Bohr's result was $\nu = Rc(1/n^2 - 1/m^2)$, where $R = 2Z^2\pi^2me^4/h^3$ and n and m are integral quantum numbers characterizing the stationary states; Z is the nuclear charge, 1 for hydrogen. In the summer of 1913, lines in accordance with the formula were known for $n = 2$ (Balmer series) and $n = 3$ (Paschen series); Bohr predicted the existence of further lines corresponding to $n = 1$ and n larger than 4, "series respectively in the extreme ultraviolet and the extreme ultra-red, which are not observed, but the existence of which may be expected." His confidence was justified when Theodore Lyman in 1914 reported lines in agreement with $n = 1$. The ionization potential of a hydrogen atom in its ground state ($n = 1$) followed directly from the Balmer expression. Bohr obtained a value of about 13 volts, which he compared with J. J. Thomson's experimental estimate of 11 volts. (Later and more precise measurements gave complete agreement with Bohr's value.) For the radius of a hydrogen atom in its ground state, later known as the Bohr radius, he obtained 0.55 angstrom, the right order of magnitude for an atom.

Perhaps the most impressive confirmation of Bohr's theory of one-electron atoms was his demonstration that the "Pickering lines" found in stellar spectra and usually attributed to hydrogen were in fact due to singly charged helium ions. These lines satisfied the Balmer-like expression $\nu = R[1/2^2 - 1/(m + 1/2)^2]$ which, if caused by hydrogen, would contradict Bohr's theory, according to which half-quantum states were inadmissible. Bohr turned the threat into a triumph by simply rewriting the expression as $\nu = 4R[1/4^2 - 1/(2m + 1)^2]$ and attributing it to the He^+ ion. His prediction that the Pickering lines must appear in discharge tubes with pure helium was quickly confirmed by spectroscopists. However, the agreement between the measured wavelengths and those predicted by Bohr was not perfect and according to Alfred Fowler, a British spectroscopist, the small discrepancy was large enough to question the validity of the theory. Bohr's reply, published in *Nature* in the fall of 1913, was another brilliant example of what the philosopher Imre Lakatos has called monster-adjustment—turning a counterexample into an example. Bohr pointed out that the quantity m in the expression for R should really be the reduced mass, $mM/(m + M)$, with M being the nuclear mass, and with this correction the discrepancy vanished.

Bohr derived the Balmer formula in more than one way, including a first application of what later became known as the correspondence principle. Bohr noted that for large quantum numbers, there was almost no difference between the frequency of rotation before and after the emission of a quantum; "and according to ordinary electrodynamics we should therefore expect that the ratio between the frequency or radiation and the frequency of revolution also is nearly equal to 1" (Bohr 1963, 13). In an address before the Physical Society in Copenhagen in December 1913, Bohr emphasized that although, in general, there was no connection between classical frequencies of revolution and the frequencies found quantum-theoretically, "[o]n one point, however, we may expect a connection with the ordinary conceptions, namely, that it will be possible to calculate the emission of slow electromagnetic oscillations on the basis of classical electrodynamics" (Jammer 1966, 110). This was the germ of the correspondence principle, which would play a pivotal role in the later development of atomic theory and its transformation into quantum mechanics (see chapter 11).

Bohr's theory was not simply a theory of one-electron atoms, but was planned much more ambitiously. In the second and third part of the trilogy, Bohr applied his theory to chemical atoms larger than hydrogen, and also to molecules. He suggested electron arrangements for the lighter elements and believed that his models provided the periodic system with its first reliable explanation. Moreover, he extended his work to simple molecules, picturing the covalent bond in the hydrogen molecule as two electrons revolving in the same orbit between the two hydrogen nuclei. This part of his work was much less successful and made relatively little impact. Yet it is noteworthy, for it was the first time that definite atomic models were proposed for real atoms. The covalent bond turned out to be outside the reach of Bohr's quantum theory of atoms, but in 1913 he had no reason to think so. On the contrary, there were indications that chemistry would soon be reduced to a branch of Bohr's new physics. For example, Bohr calculated the heat of formation of molecular hydrogen to 60 kcal per mole, in qualitative but certainly not quantitative agreement with the 130 kcal per mole determined experimentally by Irving Langmuir. When Langmuir revised the experimental value to 76 kcal per mole shortly afterward, and Bohr recalculated the theoretical value to 63 kcal per mole, it seemed that the hydrogen molecule was nearly explained. But this was far from from the case.

The strength of Bohr's theory was not its theoretical foundation, which to many seemed unconvincing and even bizarre, but its experimental confirmation over a wide range of phenomena. For example, in 1913–14 the young British physicist Henry Moseley studied the characteristic x-rays emitted by different elements and showed that the square root of the frequencies related proportionally to the atomic number. The Moseley diagram quickly became an important tool for determining an element's place in the periodic table, and it also became an important confirmation of Bohr's theory. Moseley's mechanism of x-ray emission rested on Bohr's theory and in a series of works starting in 1914, Walther Kossel in Munich further explained x-ray spectra in full accordance with the theory. Another important confirmation was the experiments with electron bombardment of mercury vapor that James Franck and Gustav Hertz made in Göttingen between 1913 and 1916. It was soon realized that the experiments brilliantly confirmed Bohr's theory. In 1925, the two physicists were awarded the Nobel prize for having verified Bohr's hypotheses and thereby transformed them into "experimentally proved facts," as Carl Oseen expressed it in his presentation speech in Stockholm. Ironically, Franck and Hertz did not originally relate their experiments to Bohr's theory and, when they first did so, they argued that their measurements could not be explained by the theory. Franck and Hertz measured what they believed was the ionization potential to 4.9 volts, but in 1915 Bohr argued that they had misinterpreted their results and not measured an ionization potential at all, but the energy difference between stationary states in mercury atoms. It was only after Bohr's intervention that Franck and Hertz realized that they unknowingly had provided strong support for Bohr's atomic theory.

The red line of the hydrogen spectrum has a doublet structure, as first shown by Michelson and Edward Morley as early as 1887. Although by 1913 the fine-structure splitting had been measured several times, Bohr seems to have been unaware of this phenomenon, for which there was no room in his theory. But once again an apparent anomaly was turned into a confirmation, although this time it required a major extension of the theory. The extension was made by Arnold Sommerfeld in Munich who, in 1915–16, introduced the special theory of relativity in the mechanics of the Bohr atom. In this way, he was led to a two-quantum atom in which the electronic orbits were described by a principal and an azimuthal quantum number, and an energy expression that depended on both quantum numbers. According to Sommerfeld's more sophisticated theory, there were many more stationary states than in Bohr's 1913 theory, and this made an explanation of the fine structure possible. Sommerfeld derived a value of the fine structure separation, which was completely confirmed by experiments made by Friedrich Paschen at the University of Tübingen in 1916. The remarkable agreement between theory and experiment was considered a great success of the Bohr-Sommerfeld theory and also of the theory of relativity. In fact, the relationship between theory and experiment was rather obscure because the theory allowed calculation only of frequencies but not of intensities; however, although the success was questioned by a few physicists in Germany, to the majority of physicists the work of Sommerfeld and Paschen looked like a striking confirmation of Bohr's quantum theory of atoms.

Chapter 5

THE SLOW RISE OF QUANTUM THEORY

The Law of Blackbody Radiation

QUANTUM THEORY owes its origin to the study of thermal radiation, in particular the "blackbody" radiation that Robert Kirchhoff had first defined in 1859–60. According to Kirchhoff, a perfect blackbody is one that absorbs all the radiation incident upon it; the emitted energy will be independent of the nature of the body and depend only on its temperature. The Austrian physicist Josef Stefan suggested in 1879 that the energy of Kirchhoff's ideal heat radiation varied with the fourth power of the absolute temperature. His suggestion was supplied with a theoretical proof five years later when his compatriot Ludwig Boltzmann combined the second law of thermodynamics with Maxwell's electrodynamics to show that $u = \sigma T^4$, where u is the total energy density and σ is a constant. The first of several laws of blackbody radiation, the Stefan-Boltzmann law helped direct attention to the new area of theoretical and experimental physics. The spectral distribution of the radiation, a question about which the Stefan-Boltzmann law had nothing to say, soon emerged as a major and widely discussed problem. An important step toward the solution of the problem was taken by Wilhelm Wien, who in 1894 showed that if the spectral distribution of blackbody radiation was known at one temperature, it could be deduced at any other temperature. The distribution function $u(\lambda,T)$ would not depend separately on T and the wavelength λ, but on the product λT through some function $\phi(\lambda T)$, namely as $u(\lambda,T) = \lambda^{-5}\phi(\lambda T)$. Wien's displacement law—so called because it implies that the peak of the $u(\lambda,T)$ function will be displaced toward smaller wavelengths when T increases—was found to agree excellently with experiments. The function $\phi(\lambda T)$ was recognized to be of universal significance, but neither the form of the function nor its explanation was known. In 1896 Wien found a possible solution, namely, that $\phi(\lambda T)$ was of the form $\exp(-\alpha/\lambda T)$, with α a universal constant. Wien's radiation law seemed to be correct and was generally accepted, not least after it received confirmation in a series of delicate experiments performed in Berlin between 1897 and 1899. Although Wien's law appeared empirically convincing, however, it rested on theoretical arguments of an unsatisfactory nature and for this reason, a more rigorous derivation was wanted. This was where Max Ludwig Planck, Kirchhoff's successor as professor of physics at the University of Berlin, entered the development.

Planck was a specialist in thermodynamics and was deeply interested in the second law and its applications in physics and chemistry. His main occupation in the early 1890s was not theoretical physics, but rather chemical thermodynamics, to which he tried to give a more rigorous foundation based on the second law. In this work, the concepts of entropy and irreversibility were central. The core of Planck's research program was an attempt to explain irreversible processes on a strict thermodynamic basis, that is, without introducing any statistical or atomistic assumptions in the manner of Boltzmann. Contrary to his senior Austrian colleague, Planck believed firmly in the absolute validity of the second law and denied that there could be any relationship between entropy and probability. In 1895 Planck's reasoning led him to examine the relation between thermodynamics and electrodynamics. He attacked the question of irreversibility from an electrodynamic point of view and argued that the irreversibility of radiation processes was a result of the lack of time symmetry in Maxwell's equations. However, the electrodynamic approach proved unsuccessful. As Boltzmann showed two years later, electrodynamics is no more time-asymmetric—does not provide a "time's arrow"—than mechanics, and so Planck had to find another way to determine the spectrum of blackbody radiation. The result of Planck's renewed efforts was a series of six papers on irreversible radiation processes published in *Annalen der Physik* between 1897 and 1900. In 1899 he found an expression for the entropy of an oscillator by means of which he could derive Wien's radiation law. This was what Planck had hoped for, and had it not been for the experimentalists, he might have stopped there. In the same year as Planck derived Wien's law, experiments proved that the law was not completely correct, contrary to what Planck and most other physicists had assumed.

In the history of blackbody radiation, and hence in the birth of quantum theory, experiment was no less important than theory. Most of the decisive experiments were made at Berlin's Physikalisch-Technische Reichsanstalt (Imperial Institute of Physics and Technology), where the precise spectrum of blackbody radiation was a matter of more than purely academic interest. It was thought that it would lead to knowledge that could be useful to the German lighting and heating industries, which were among the Reichsanstalt's largest customers. Experiments performed by Otto Lummer and Ernst Pringsheim in 1899 indicated that Wien's law was incorrect for long wavelengths. Further experiments by Heinrich Rubens and Ferdinand Kurlbaum, published in the fall of 1900, provided definite proof that the "Wien-Planck law" was only approximately true (see figure 5.1). According to this law, the radiation energy density $u(\nu,T)$ would approach zero for very small values of $\nu/T = c/\lambda T$, while the experiments of Rubens and Kurlbaum showed that $u(\nu,T)$ approached T. As a consequence of the new measurements, several new empirically based laws were proposed, but these were of no importance

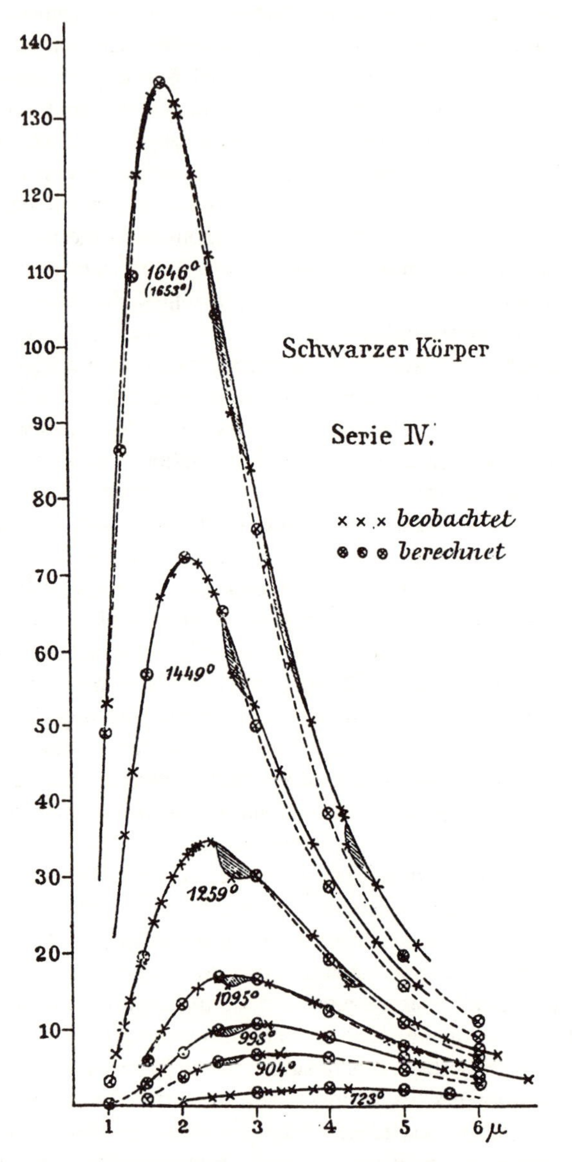

Figure 5.1. Blackbody spectra at different temperatures measured by Lummer and Pringsheim in November 1899. For large values of λT, the measured curve (continuous line) lies above the calculated curve (dashed line), indicating the inadequacy of Wien's radiation law. *Source:* Kangro 1976, 176.

to the theorists. Planck's primary interest was not to find an empirically correct law, but to derive it from first principles. He was now forced to reconsider his work. Something had gone wrong in his derivation of Wien's law, but what? How could he devise from fundamental principles of physics a distribution law that satisfied the Wien expression for large values of ν/T, but led to T for small values? Planck's reconsiderations led him quickly to assume a new expression for the entropy of an oscillator, albeit one that had no clear theoretical justification. About this expression it has been remarked, rightly, "Never in the history of physics was there such an inconspicuous mathematical interpolation with such far-reaching physical and philosophical consequences" (Jammer 1966, 18).

With his new expression for the entropy, Planck was able to derive what he considered to be merely an improved version of Wien's law. The new distribution law—Planck's radiation law, but still with no notion of energy quanta—was announced at a meeting of the Berlin Academy of Sciences on October 19, 1900. According to this first correct version of the law of black-body radiation, the spectral energy density varies as ν^3 divided by the quantity $\exp(\beta\nu/T) - 1$. The law seemed to be in complete agreement with experimental data and was, in this respect, the answer that had been sought for so long. Because the new law rested on an entropy expression that was scarcely more than an inspired guess, however, it was not theoretically satisfactory and so Planck was, once again, led to consider why the formula was so successful. He could not rest content before he understood the new law.

In his attempt to obtain a satisfactory understanding, Planck realized that he had to introduce a new approach, namely, to turn to Boltzmann's idea of entropy as an expression of molecular chaos. This does not mean that Planck surrendered to Boltzmann's probabilistic notion of entropy and irreversibility. Rather than accepting such ideas, Planck reinterpreted Boltzmann's theory in his own nonprobabilistic way. He based his new line of attack on the famous "Boltzmann equation," $S = k \log W$, where k is Boltzmann's constant and W is a combinatorial expression for molecular disorder. The equation was not Boltzmann's, in fact, but appeared in this form only with Planck's work; and it was also Planck who first introduced "Boltzmann's constant" as an important constant of nature. In order to find W, Planck introduced what he called "energy elements," namely, the assumption that the total energy of the oscillators of a black body (E) was divided into finite portions of energy ε. In Planck's words: "I regard E . . . as made up of a completely determinate number of finite equal parts, and for this purpose I use the constant of nature $h = 6.55 \times 10^{-27}$ (erg sec). This constant, once multiplied by the common frequency of the resonators, gives the energy element ε in ergs, and by division of E by ε we get the number P of energy elements to be distributed over the N resonators" (Darrigol 1992, 68). The new derivation was reported at another meeting of the Berlin Academy, on

December 14, 1900, a date that is often referred to as the birthday of quantum theory because it was here that, as seen in retrospect, the quantum hypothesis was first suggested. However, Planck did not really understand the introduction of energy elements as a quantization of energy, i.e., that the energy of the oscillators can attain only discrete values. He did not highlight the quantum discontinuity at all and considered $\varepsilon = h\nu$ to be a mathematical hypothesis with no physical reality behind it. It was, he believed, a temporary feature of the theory that had to be removed in the final formulation.

Here is how Planck, in a letter of 1931 to the American physicist Robert Wood, described his route to the radiation law:

> To summarize, all what happened can be described as simply an act of desperation. . . . [B]y then I had been wrestling unsuccessfully for six years (since 1894) with the problem of equilibrium between radiation and matter and I knew that this problem was of fundamental importance to physics. I also knew the formula that expresses the energy distribution in normal spectra. A theoretical interpretation therefore had to be found at all cost, no matter how high. . . . [The new] approach was opened to me by maintaining the two laws of thermodynamics . . . they must, it seems to me, be upheld in all circumstances. For the rest, I was ready to sacrifice every one of my previous convictions about physical laws. Boltzmann had explained how thermodynamic equilibrium is established by means of a statistical equilibrium, and if such an approach is applied to the equilibrium between matter and radiation, one finds that the continuous loss of energy into radiation can be prevented by assuming that energy is forced, at the onset, to remain together in certain quanta. This was a purely formal assumption and I really did not give it much thought except that no matter what the cost, I must bring about a positive result. (Hermann 1971, 23)

To Planck and his contemporaries, the quantum discontinuity was at first considered a feature that did not merit serious attention. What mattered was rather the impressive accuracy of the new radiation law, confirmed by many later experiments, and the fact that it included the Stefan-Boltzmann law, Wien's displacement law, and, in the limit of large ν/T values, Wien's radiation law. Planck emphasized the constants of nature involved in his law and used it to derive numerical values of k, N (Avogadro's number), and e (the elementary charge). From the blackbody measurements he could find k and since $k = R/N$, where R is the gas constant, N was derived; moreover, from $e = F/N$, where F is Faraday's constant known from electrolysis, he could find e. Planck's numerical determinations were greatly superior to the rather crude estimates obtained by other methods at the time.

In December 1900 Planck did not recognize that the new radiation law necessitated a break with classical physics. Nor, for that matter, did other physicists. It is worth noting that there was no "ultraviolet catastrophe" involved in the formation of Planck's radiation law. The basis of this "catastrophe" is that the equipartition theorem of classical mechanics, when applied to the oscillators of a black body, leads to an energy density of the form $\nu^2 T$ and thus, when integrated over all frequencies, to an infinite total energy. In this sense, there is a conflict between classical physics and the blackbody spectrum, but the conflict played no role at all in the actual events that led to Planck's hypothesis. The mentioned radiation law was obtained by Lord Rayleigh in the summer of 1900, although he added an ad hoc factor of $\exp(-a\nu/T)$ in order to make the formula better agree with data. The $u \sim \nu^2 T$ formula is today known as the Rayleigh-Jeans law because it was rederived by Rayleigh in 1905 and supplied with a numerical correction by James Jeans in the same year. Although Rubens and Kurlbaum included Rayleigh's formula in their paper of 1900, Planck ignored it and the equipartition theorem. The Rayleigh-Jeans law, without the exponential factor, is obviously wrong for high frequencies, but the discrepancy was not seen as a great problem for classical physics. At the time, many physicists, including Rayleigh and Jeans, doubted that the equipartition theorem was generally valid.

Early Discussions of the Quantum Hypothesis

If a revolution occurred in physics in December 1900, nobody seemed to notice it, least of all Planck. During the first five years of the century, there was almost complete silence about the quantum hypothesis, which somewhat obscurely was involved in Planck's derivation of the blackbody radiation law. The law itself, on the other hand, was quickly adopted because of its convincing agreement with experiment. As early as 1902, Planck's radiation formula appeared in the second volume of Heinrich Kayser's authoritative *Handbook of Spectroscopy*, but without any mention of the nature of the quantum assumption. Although criticism was ocassionally raised, by 1908 Planck's result was generally accepted as the correct answer to the question of the blackbody spectrum. Only a handful of theorists found it worthwhile to go into the details of Planck's calculations and ask *why* the formula was correct.

One of the few was Hendrik A. Lorentz, who started concentrating on blackbody theory in 1903 when he independently derived the Rayleigh-Jeans law on the basis of his electron theory. The result puzzled him. Five years later, at a mathematical congress in Rome, he gave a survey of the blackbody problem or, in his terminology, the division of energy between ponderable matter and ether. As Lorentz saw it, the choice was between, on the one hand, the theoretically satisfactory but empirically inadequate Rayleigh-Jeans-Lorentz formula and, on the other, the empirically confirmed but theoretically unsatisfactory Planck formula. Remarkably, he preferred the first choice and suggested vaguely that new experiments were needed in order to decide between the two candidates. The German experimentalists knew better. Convinced that the matter was already firmly settled, they would have nothing of the Rayleigh-Jeans law and protested against Lorentz's suggestion. Consequently, Lorentz was forced to accept Planck's formula and try to understand its true meaning. Recognizing that Planck's theory involved some nonclassical features, Lorentz emerged as one of the leaders of the new quantum theory.

As a result of Lorentz's Rome lecture, the "catastrophical" consequences of the classical Rayleigh-Jeans law became better known in the physics community; it was only from that time that the "ultraviolet catastrophe" came to play a major role in the discussions. (The phrase was introduced by Ehrenfest in 1911 and became a popular theme in physics textbooks.) With electron theory as the dominant and successful microscopic theory of the period, it was tempting to believe that this theory would somehow be able to solve the puzzles of blackbody radiation. This was what Lorentz tried to do in 1903, only to end up with the Rayleigh-Jeans law. For a time Planck pursued the idea that quanta of electricity might lead to quanta of energy. For example, in 1905 he wrote to Paul Ehrenfest, "It seems to me not impossible that this assumption (the existence of an elementary quantum of electricity) offers a bridge to the existence of an elementary energetic quantum h, particularly since h has the same dimension as e^2/c" (Kuhn 1978, 132). Here we have, if only implicitly, the first inkling of the fine structure constant $2\pi e^2/hc$. However, nothing came out of either this idea or other attempts to deduce the quantum of action from existing theory. Yet Planck's deep interest in the universal constants of nature and their possible interrelationships is worth noticing. In a paper of 1899—the first one in which Planck's constant implicitly appeared—he noted that all ordinary systems of units were based on "the special needs of our terrestrial culture" and suggested as an alternative a system based on what corresponded to the constants h, c, and G. Such units, he wrote, would "independently of special bodies and substances, necessarily retain their significance for all times and all cultures, even extraterrestrial and extrahuman ones." The units proposed by Planck had no practical value and were ignored for a long time. However, with the advent of theories of quantum gravity in the 1970s they became widely discussed and today, at the turn of the twentieth century, the Planck mass (10^{-5} gram) and Planck time (10^{-43} seconds) are important quantities in cosmological theory (see also chapter 27).

For most of a decade, Planck believed that his radiation law could be reconciled with classical mechanics and electrodynamics and that the discontinuities were features of the atomic oscillators, not of the energy exchange as such. He realized that some kind of quantization was involved, but not in the sense that the energy values of individual oscillators were limited to a discrete set $h\nu$, $2h\nu$, $3h\nu$, . . . In his early papers, he had written the energy equation as $E = nh\nu$ ($n = 0, 1, 2, \ldots$), but with E meaning the total energy of the oscillators, it did not require the energy of individual oscillators to be similarly restricted. It was only in about 1908, in part as a result of his correspondence with Lorentz, that he converted to the view that the quantum of action was an irreducible phenomenon beyond the understanding of classical physics. Until then he thought of $h\nu$ as the smallest part of the energy continuum rather than something that can exist by itself, an energy quantum in analogy with the electrical quantum, the electron. As he wrote to Lorentz in 1909, he now adopted the hypothesis that "[t]he energy exchange between electrons and free ether occurs only in whole numbers of quanta $h\nu$" (Kuhn 1978, 199).

At the end of the first decade of the twentieth century, quantum theory was still badly understood and studied seriously only by a few theoretical physicists. These included Lorentz, Ehrenfest, Jeans, Einstein, Larmor, and, of course, Planck. Until 1906, Einstein was alone in realizing the radical, nonclassical nature of Planck's theory, but four years later, most specialists recognized that energy quantization was real and necessitated some kind of break with classical physics. During the first decade, quantum theory was largely identical with blackbody radiation theory, and the small field did not make much of an impact on the physics community. This is illustrated in figure 5.2, which gives the number of authors publishing on quantum topics

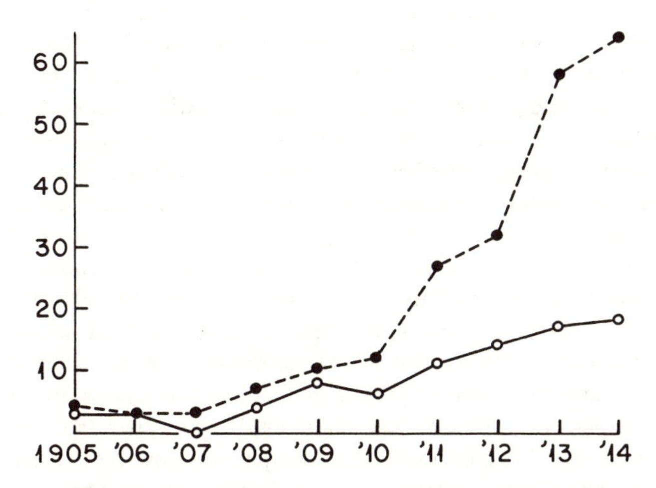

Figure 5.2. The slow rise of quantum theory. The solid circles indicate the number of authors who published on quantum topics. The open circles refer to the number of authors who dealt with blackbody theory, a subset of early quantum physics. *Source:* T. S. Kuhn, *Black-Body theory and Quantum Discontinuity*. Copyright © 1978 by Oxford University Press, Inc. Used by permission of Oxford University Press, Inc.

between 1905 and 1914. Before that time—that is, from 1900 to 1904—the number of authors on quantum theory was either zero or one (the one and only contributor was Planck, in 1900 and 1901). The figure illustrates not only the slow takeoff of quantum theory, but also the dominance of black-body physics until 1910 when publications on specific heats and, from about 1913, atomic and molecular physics began to change both the composition and pace of quantum theory.

Einstein and the Photon

It has been claimed that even if Planck had not found his formula for black-body radiation that initiated the early quantum theory, the theory would nonetheless have arrived within the first years of the twentieth century. The candidate as a hypothetical discoverer of quantum theory is Einstein, who is, of course, best known for his invention of the theory of relativity. Yet young Einstein also made contributions to early quantum theory of such importance—and in 1905, the same year as he introduced relativity—that they may justify the claim. In fact, Einstein seems to have considered his theory of quanta even more important than his slightly later work on special relativity. In a letter of May 1905 to his friend Conrad Habicht, he referred to his forthcoming paper on "radiation and the energetic properties of light" as "very revolutionary." The paper under preparation on special relativity was more modestly referred to as "an electrodynamics of moving bodies with the use of a modification of the ideas of space and time." Whereas Planck, as it has been said, became a revolutionary against his will, Einstein recognized the revolutionary implications of the quantum hypothesis much more clearly and willingly acted as a prophet of the quantum revolution. There is much truth in the claim that quantum theory started in earnest only in 1905, with Einstein's works.

What, then, was the essence of Einstein's self-proclaimed revolutionary work on radiation published in *Annalen der Physik* on June 9, 1905? First, Einstein's approach differed markedly from Planck's and hardly relied at all on Planck's radiation law and its associated quantum of action. Einstein did mention the law, but without using it; instead, he focused on the old Wien law in its experimentally confirmed regime, that is, for high frequencies and low temperatures. Einstein made it clear that this was the interesting and problematic part of the spectrum, the one that required new physical hypotheses. Classical theory would lead to the Rayleigh-Jeans law, Einstein emphasized, and he easily derived the law, including the correct factor in front of ν^2T. From a priority point of view, the law might be called the Einstein or Rayleigh-Einstein radiation law (or, for that matter, if more cumbersomely, the Rayleigh-Lorentz-Einstein-Jeans law). It was the Wien law, however, on which Einstein focused in his 1905 work. Using simple but clever thermodynamic arguments and making full use of Boltzmann's probabilistic theory (the entropy expression $S = k \log W$), he calculated the probability that the entire radiation energy in a container is contained in a small part of the total volume. From this result he reasoned by analogy to classical gas theory that "monochromatic radiation of low density behaves—as long as Wien's radiation formula is valid . . . as if it consisted of mutually independent energy quanta of magnitude $R\beta\nu/N$." That is, according to Einstein, radiation itself had a discrete or atomistic structure, a hypothesis that went far beyond that suggested by Planck. Moreover, according to Einstein, the energy of the oscillators responsible for the emission and absorption of light would change discretely, in multiples of $h\nu$. Note that the symbol β denotes h/k, as used by Planck before he had introduced the quantization hypothesis and, explicitly, the quantum of action, in December 1900. It was no accident that Einstein did not use either Planck's notation or his more mature theory of blackbody radiation. At the time Einstein believed that Planck's theory could not be made to agree with the idea of light quanta, a mistake he corrected in a paper in 1906. With $k = R/N$ and $\beta = h/k$, we have the usual form of the radiation quanta, $E = h\nu$.

Einstein was well aware of the radical nature of his "heuristic viewpoint" of free radiation as consisting of discrete quanta or, as they were later named, photons. (The name was suggested by the American chemist Gilbert Lewis in 1926.) After all, there was impressive evidence for the wave theory of light and for this reason, Einstein emphasized that his concept of light quanta was provisional. Yet he was convinced about the reality of the light quanta and eagerly tried to show that his hypothesis was empirically fruitful. In particular, by considering the photoelectric effect as an energy exchange process in which electrons are liberated from the surface of a metal illuminated by light, he was able to account for experiments made by Philipp Lenard in 1902. Moreover, it followed directly from Einstein's theory that the maximum energy (E) of the photo-generated electrons must be related linearly to the frequency of the incoming light. Einstein's equation was $E = h\nu - P$, where P is a work function depending on the anode metal. At that time, neither Lenard nor others had measured E as a function of ν, and Einstein's photoelectric equation was therefore a truly novel prediction. Einstein's theory was not a response to an experimental anomaly that classical theory could not account for, for in 1905 the photoelectric effect was not considered problematic. It was only some years later that experimentalists took up the question of the relationship between E and ν. And when they did, it was not with the purpose of testing Einstein's theory.

As far as experimental data are concerned, for several years these showed a confusing disparity, ranging from a quadratic over a logarithmic to a linear relationship (i.e., from $E \sim \nu^2$ over $E \sim \log\nu$, to $E \sim \nu$.) Only in about 1914 did evidence accumulate in favor of the linear law and with Robert Millikan's famous series of experiments in 1916, consensus was finally obtained.

It was now established beyond doubt that the maximum energy of the emitted electrons do indeed vary linearly with the frequency of light, just as predicted by Einstein in 1905. One would perhaps believe that this must have been greeted as a great success of Einstein's theory and made most physicists accept the light quantum hypothesis. If so, one would be mistaken. None of the experimentalists concluded in favor of Einstein's "bold, not to say reckless hypothesis," as Millikan called it in 1916. What Millikan had confirmed was Einstein's equation, not his theory, and there was no one-to-one relationship between theory and equation. It was possible to derive the experimentally confirmed equation without the light-quantum hypothesis, and when these more or less classical (and, indeed, more or less ad hoc) alternatives turned out to be untenable, there was always the possibility to declare the photoelectric effect unexplained for the time being. This is what happened. Einstein's theory of light quanta was either ignored or rejected by experimenters and theorists alike. It was just too radical a hypothesis. In 1913, when Einstein was proposed for membership in the prestigious Prussian (or Berlin) Academy of Sciences, the nominators, among them Planck and Walther Nernst, praised Einstein, but they also mentioned that "he may sometimes have missed the target in his speculations, as, for example, in his hypothesis of light-quanta" (Jammer 1966, 44).

Undisturbed by the cool response to the light quantum, Einstein continued to work on the quantum theory, which in the years 1906–11 was his main area of professional occupation and of greater importance to him than the theory of relativity. In a 1909 paper, Einstein derived the energy fluctuations of blackbody radiation. His formula consisted of two terms, one that he traced to the quantum-corpuscular nature of the radiation, and another that he interpreted as a classical wave term. So, in Einstein's view, electromagnetic radiation included both of these features, traditionally seen as contradictory. In 1909 Einstein's fusion of wave and particle theory was highly provisional, but in a series of subsequent works, he developed the idea and after 1925, it would became an integral part of quantum mechanics. One more aspect of Einstein's 1909 work deserves mention, namely, that here he considered momentum fluctuations together with energy fluctuations. A real particle has momentum as well as energy, and in 1909 Einstein clearly thought of the light quantum as a particle in the same sense that electrons and atoms are particles. However, although the momentum of the light quantum ($p = h\nu/c$) follows directly from the theory of relativity, it was only in 1916 that Einstein wrote down the expression.

Specific Heats and the Status of Quantum Theory by 1913

As Einstein was the first to extend the meaning of quantum theory to the radiation field, so was he first to extend it to a problem of what later became known as solid state physics. This he did in 1907, when he applied the quantum theory to calculate the specific heats of solids. It had been known since 1819 that there is a peculiar relationship between the atomic weights of solid elements and their specific heat capacities, namely, that (in modern language) the molar heat capacity is roughly a constant, about 6.4 calories per mole per degree. In 1876 the Dulong-Petit law, named after its French discoverers, was supplied with a solid theoretical explanation by Boltzmann, who showed that it followed from the equipartition theorem of mechanical physics. Although generally considered a great success of the mechanical-atomistic view of matter, the success was not complete. There are a few exceptions to the law of Dulong and Petit, such as carbon (diamond), boron, and silicon. The carbon anomaly had been known since 1841, when experiments showed diamond's molar specific heat to be about 1.8 instead of the 6.4 to be expected from the Dulong-Petit rule. Moreover, experiments made in the 1870s and later on showed that the specific heats increase with the temperature, and in the case of diamond, quite considerably. In 1875 the German physicist Heinrich Weber had established that the specific heat of diamond increases by a factor of 15 over the range from $-100°C$ to $+1000°C$, and later experiments published 1905 by the Scottish chemist James Dewar showed that the specific heats almost vanished at temperatures around 20 K. Attempts to explain the variation failed, and so the problem of specific heats remained an anomaly until Einstein, largely successfully, attacked it in 1907.

Using Planck's distribution law in his work of 1907, Einstein found an expression for the average energy of an atom in a crystal vibrating in three directions with the same frequency. This expression gave the classical value $3kT$ for high temperatures, and thus led to the law of Dulong and Petit, but fell off exponentially for T approaching zero. In this regime, the classical equipartition theorem—the basis of the Dulong-Petit law—could not be applied. By comparing his formula for the heat capacity with the data obtained by Weber, Einstein was able to fit the vibration frequency in such a way that a promising, if not perfect, agreement with experiment was obtained. Einstein's theory was important in attracting interest to the quantum theory. Yet it was only approximate and clearly in need of modification, not least when new experiments showed the low-temperature variation to disagree quantitatively with the theory. A much more sophisticated version of Einstein's theory was developed in 1912 by the Dutch theorist Peter Debye, and this version gave a very close agreement with experiment.

The theory of specific heats helped bring the quantum theory into more traditional areas of physics and make it known to the many physicists who were not interested in, or did not understand, the finer details of the theory of blackbody radiation. It was, in this respect, far more important than the theory of light quanta. However, its impact was not instantaneous. In fact, until 1910–11, Einstein's theory of specific heats remained as unmentioned in the scientific literature as did his theory of light quanta. It was only then that physicists began to take notice of the theory and, rather suddenly, the quantum theory of specific heats became recognized as a major research topic. In 1913 there were more publications on this topic than on blackbody theory. At that time, quantum theory was still a rather esoteric area of physics, but it was now taken seriously by a growing number of physicists. Moreover, it also began to make its presence felt in chemistry. Walther Nernst, the German pioneer of physical chemistry, was instrumental in the rise of interest in quantum theory. Whereas Planck and other students of blackbody radiation focused on the radiation field, and considered it an advantage that this field was independent of the structure of matter, for Nernst quantum theory was important because it might help understanding the structure of matter. His work on chemical thermodynamics in the low-temperature region supported Einstein's theory on specific heats. In 1911 Nernst suggested that the theory should be applicable also to the vibrations of gas molecules, an idea that was taken up by Niels Bjerrum, a young Danish chemist working in Nernst's laboratory in Berlin. In works between 1911 and 1914, Bjerrum applied the quantum theory both to the specific heats of gases and the infrared absorption spectra of molecules. In particular, in 1912 he quantized the rotational energy of a diatomic molecule and used the result to propose a theory of molecular spectra that quickly received experimental confirmation. Bjerrum's work, an early contribution to the area of research later known as chemical physics, was an important success of quantum theory. During most of the 1910s, molecules were more important than atoms to the dissemination of quantum theory.

The growing interest in quantum theory is illustrated by the Solvay Congress in physics held in Brussels in November 1911, which was the first of an important series of international physics meetings. Ernest Solvay, a Belgian industrialist and philanthropist who had earned a fortune by inventing a new method of the production of soda, had a deep if somewhat amateurish interest in theoretical physics. The combination of Solvay's money and Nernst's initiative resulted in the 1911 congress on the problematic relationship among the quantum theory, the kinetic theory of gases, and the theory of radiation. Lorentz presided over the conference; among the 21 invited participants were Europe's finest physicists, including Planck, Nernst, Sommerfeld, Marie Curie, Rutherford, Poincaré, and Einstein. No Americans were invited. Although the discussions in Brussels did not result in any definite answers to the many questions that were raised, they were useful in making the problems of radiation and quantum theory more sharply focused. Einstein found that the congress "had an aspect similar to the wailing at the ruins of Jerusalem," but at least he found it socially stimulating. "It was most interesting in Brussels," he wrote to his friend Heinrich Zangger. "Lorentz is a miracle of intelligence and tact—a living work of art. . . .

Poincaré was altogether simply negative about the relativity theory. . . . Planck is untractable about certain preconceived ideas which are, without any doubt, wrong. . . . The whole thing would have been a delight for the diabolical Jesuit fathers" (Mehra 1975, xiv). Encouraged by the success of the conference, Solvay decided to establish a permanent institution, for which he endowed one million Belgian francs. The International Institute of Physics, founded in 1912, was directed by a board consisting of nine prominent physicists from five countries. The first scientific committee consisted of Lorentz and Kammerlingh Onnes from the Netherlands, Marie Curie and Marcel Brillouin from France, Martin Knudsen from Denmark, Robert Goldschmidt from Belgium, Nernst and Emil Warburg from Germany, and Rutherford from Great Britain. For more than two decades, the Solvay conferences were the most prestigious and scientifically important meetings for elite physicists.

The 1911 meeting was primarily organized by Nernst, who believed that the time was ripe for a conference on problems of matter and radiation in the light of quantum theory. Mainly as a result of Einstein's work on specific heats, Nernst had become convinced about the revolutionary importance of quantum theory, a theory to which he had not given attention earlier. In July 1910 he wrote to Solvay, the benefactor of the planned conference, "It appears that we are currently in the midst of a revolutionary reformulation of the foundation of the hitherto accepted kinetic theory of matter. . . . [T]his conception [of energy quanta] is so foreign to the previously used equations of motion that its acceptance must doubtless be accompanied by a wide-ranging reform of our fundamental intuition" (Kuhn 1978, 215). Planck, with whom Nernst discussed the plan, agreed that quantum theory was a serious challenge to classical concepts of physics. But at first he was skeptical about holding the conference as early as 1911. In June 1910 he wrote to Nernst, "I am of the opinion that hardly half of the participants you have in mind are conscious of a sufficiently active concern for the urgent necessity of a reform [of the theory] to justify their coming to the conference. As for the oldsters (Rayleigh, van der Waals, Schuster, Seeliger) I shall not discuss in detail whether they will be excited by the thing. But even among the younger people the urgency and the importance of these questions has hardly been recognized. Among all those mentioned by you I believe that, other than ourselves, only Einstein, Lorentz, W. Wien and Larmor will be seriously interested in the matter" (Mehra 1975, 5). Although the Brussels conference on "Radiation Theory and the Quanta" included all the key figures of quantum theory, not all of the participants were concerned with quantum problems. Two of the reports, given by Jean Perrin and Knudsen, did not deal with aspects of quantum theory. The titles of the reports delivered at the conference give an impression of which subjects were considered important in the chosen area of theoretical physics:

- Application of the energy equipartition theorem to radiation (by H. A. Lorentz, age 58)
- Kinetic theory of specific heat according to Maxwell and Boltzmann (by J. H. Jeans, age 34)
- The law of blackbody radiation and the hypothesis of the elementary quantum of action (by M. Planck, age 53)
- Kinetic theory and the experimental properties of perfect gases (by M. Knudsen, age 40)
- The proof of molecular reality (by J. Perrin, age 41)
- Application of the quantum theory to physicochemical problems (by W. Nernst, age 47)
- The quantum of action and nonperiodic molecular phenomena (by A. Sommerfeld, age 42)
- The problem of the specific heats (by A. Einstein, age 32)

The Solvay conference did not result in important new insights, but the reports and discussions nonetheless helped to establish a common understanding of what the key problems of quantum theory were. The general attitude was cautious and somewhat skeptical. It was realized that the enigmas of the quantum were far from being solved and that the status of quantum theory continued to be unsatisfactory. "The *h*-disease looks ever more hopeless," wrote Einstein to Lorentz shortly after the congress. In another letter from the same time he concluded that "nobody really knows anything" (Barkan 1993, 68). Yet the few specialists recognized that quantum theory was there to stay and that it marked the beginning of a new chapter in the history of physics. The feeling was well expressed by the conservative Planck in a 1911 lecture to the German Chemical Society. "To be sure, most of the work remains to be done," Planck said, "but the beginning is made: the hypothesis of quanta will never vanish from the world. . . . I do not believe I am going too far if I express the opinion that with this hypothesis the foundation is laid for the construction of a theory which is someday destined to permeate the swift and delicate events of the molecular world with a new light" (Klein 1966, 302).

The undecided state of affairs in quantum theory may be illustrated by Planck's attempts between 1911 and 1914 to revise the theory in order to retain as much as possible of the classical theory of electrodynamics. His new proposal was to abandon the hypothesis that the energy of an oscillator is quantized in the sense that absorption and emission of energy are discrete processes. As an alternative, Planck suggested absorption to be a continuous process, while emission was taken to be discontinuous and regulated by a probabilistic law. On this basis, Planck could derive the blackbody radiation law in what he considered a more satisfactory way. Contrary to the original theory of 1900, in the 1912 version the energy of an oscillator did not vanish at zero temperature. For $T = 0$ the result became $E = h\nu/2$, hence the name zero-point energy. This startling idea aroused much interest and was soon applied to a variety of phenomena, including radioactivity, superconductivity, and x-ray scattering. Nernst even used the idea in cosmological speculations. The existence of a zero-point energy was eventually confirmed and deduced in a natural way from quantum mechanics in the 1920s. But historically, it originated in Planck's incorrect theory of 1912.

Chapter 6

PHYSICS AT LOW TEMPERATURES

The Race toward Zero

CRYOGENICS IS THE study of phenomena and material properties at very low temperatures, and especially of methods to produce such temperatures. Around 1880, cryogenics was a science in its infancy. Throughout the nineteenth century, there had been an interest in the condensation of gases and, in particular, the liquefaction of the constituents of air. The first important result in this area occurred in 1877, when a French mining engineer, Louis Cailletet, announced at a meeting of the Paris Academy of Sciences that he had observed droplets of liquid oxygen. The discovery turned out to be a double one, for two days before Cailletet made his announcement a Swiss physicist, Raoul Pictet, had telegraphed to the Academy a report that he had succeeded in condensing oxygen. It was one of many simultaneous independent discoveries in the history of science. The two researchers used different methods, but in both experiments pure oxygen was cooled under pressure and then suddenly allowed to expand. A few days after having caused the condensation of oxygen, Cailletet repeated his success with the other main constituent of the atmosphere, nitrogen.

A larger amount of liquid oxygen (meaning a few milliliters) was produced in 1883 by the Polish scientists Szygmunt Wrobleski and Karol Olszewski, who modified Cailletet's method into a version in which the sudden expansion of the gas was not needed. In this way they could observe the liquid boiling and not merely droplets formed by the gas. The chemist Wrobleski and the physicist Olszewski worked at Cracow's Jagiellonian University, which was one of the world centers of cryogenics research at the end of the century. (Wrobleski died tragically in 1888, in a fire in his laboratory.) The lowest temperature reported in the early Cracow experiments was about 55 K, some 35 K below the boiling point of oxygen at normal pressure. However, the methods used by the first generation of low-temperature physicists were inefficient and unable to produce larger amounts of liquid gases. The situation changed in the 1890s when new refrigeration technologies were developed, primarily by Carl von Linde in Germany, William Hampson in England, and George Claude in France. Linde was a pioneer in refrigeration engineering and the founder of a successful company that developed refrigerators for industrial uses. In 1895, at the age of 71, he invented an efficient method of gas liquefaction based on the Joule-Thomson effect. The works of Linde, Hampson, and Claude were aimed primarily at the production of liquid air for industrial purposes, but it was also important in the further scientific quest. On the whole, science and technology developed hand in hand in the areas of cryogenics and low-temperature physics. For example, the *Association Internationale du Froid* (International Association of Refrigeration), founded in 1909, was an organization for both scientific and industrial aspects of low temperatures. Another characteristic feature of the early phase of cryogenics was the interdisciplinary nature of the field, which comprised chemists, physicists, and engineers. A third feature characterized the field, namely, that it was very expensive compared with the standards in experimental physics at the time. Only very few physics laboratories could afford to enter low-temperature research.

At the end of the nineteenth century, there was an increasing interest in reaching still lower temperatures, as close to absolute zero as possible. The liquefaction of hydrogen, which was believed to have the lowest boiling point of all gases, became an attractive goal of low-temperature physics and attempts to liquefy the gas soon developed into a race. The main participants in the race were scientists in Britain (London), Poland (Cracow), and the Netherlands (Leiden). As a result of the competitive pressure and the prestige invested in being first to liquefy hydrogen, and later helium, the race took place in an atmosphere of quarrels, priority controversies, and too-hasty claims of success. One of the players in the game was the Scottish chemist James Dewar, who worked at the Royal Institution in London. Equally at home in chemistry and experimental physics, Dewar had investigated low temperatures since 1874, and in 1892 he invented one of the most useful cryogenic devices, the vacuum cryostate or Dewar flask—or, in less scientific language, the thermos bottle. In 1898, using a modification of Linde's method, Dewar succeeded where his competitors had failed. The result of his efforts was 20 milliliters of boiling liquid hydrogen, the temperature of which he estimated to be about 20 K. Having obtained liquid hydrogen he went on to produce the element in its solid state, which he succeeded in doing in 1899. The triple point of hydrogen (where the three phases are in equilibrium) had been reached, but its temperature could not be determined directly. Dewar estimated the temperature to be 16 K, but he had probably reached even lower temperatures in his experiments, perhaps only 12 degrees above absolute zero. Dewar was consciously competing with Olszewski in Cracow and Kammerlingh Onnes in Leiden, and after his victory with hydrogen the race went on to its next goal, the liquefaction of helium.

The element helium was quite new at the time. Although its name and existence were suggested as early as 1868, when Norman Lockyer interpreted an unidentified line in the solar spectrum as evidence of a new element, it was only in 1895 that William Ramsay discovered helium in terrestrial sources. It was originally believed to be "among the rarest of the elements," as the chemist Clemens Winkler put it in 1897. A few years later, in 1903, helium was found to be abundant in American natural gas wells, but it took several years until the technology necessary to extract the helium gas was developed to a commercial level. From a price in 1915 of $2,500 per cubic feet of helium, by 1926 the price had fallen to 3 cents per cubic foot. In the early years of the century, when helium was rare and expensive, scientists suspected that its triple point was below that of hydrogen. Consequently, the low-temperature physicists devoted their efforts to liquefying helium. The competition among Dewar, Olzsewski, and Kammerlingh Onnes was joined by Ramsay and his assistant Morris Travers. Ramsay and Travers were the world's leading specialists in the inert gases, but unfortunately were not on speaking terms with the irascible Dewar. The first attempt to liquefy helium, made by Olszewski, took place as early as 1896. It failed, and so did subsequent attempts of Dewar and Travers. These early experiments were trial-and-error experiments in the sense that the were made without knowing how low a temperature was needed. Only in 1907 was helium's critical temperature estimated reliably, to be between 5 K and 6 K. This was only slightly less than the lowest temperatures obtained at the European centers of cryogenics at the time and so success seemed within reach.

The British and Polish scientists were beaten when Kammerlingh Onnes triumphantly, and to Dewar's great vexation, announced in July 1908 that he had liquefied the gas. The experiment started at 5:45 A.M. on July 10 with 75 liters of liquid air; this was used to condense 20 liters of hydrogen, which was again used to liquefy 60 milliliters of helium under reduced pressure. The first liquefaction of helium was a fact thirteen hours later. Kammerlingh Onnes described the climax as follows: "It was a wonderful moment when the liquid, which looked almost immaterial, was seen for the first time. . . . I was overjoyed when I could show liquefied helium to my friend van der Waals, whose theory had been my guide in the liquefaction up to the end" (Dahl 1984, 2). Kammerlingh Onnes immediately tried to determine if he could solidify the element by evaporation under reduced pressure. He failed, and further attempts were no more successful. Solid helium was eventually achieved in 1924 by Willem Keesom at the Leiden laboratory. Still, Kammerlingh Onnes had produced the first liquid helium ever and in the process reached a new record cold temperature, in 1910 obtaining temperatures as low as 1 K. At this stage, when it proved technically impossible to lower the temperature further, he paused and decided to investigate the physical properties of substances in the newly accessible temperature regime between 1 K and 6 K.

Kammerlingh Onnes and the Leiden Laboratory

Heike Kammerlingh Onnes, director and undisputed leader of the Leiden laboratory from 1882 until his retirement in 1922, had started his career in physics with studies under Kirchhoff in Heidelberg in the early 1870s. His chair in experimental physics at Leiden University was the first of its kind in the Netherlands. Around 1893, he embarked on a large-scale research program in cryogenics and soon most of the work at the laboratory focused on low temperatures. (But not entirely: It was here that Zeeman discovered the magnetic action on spectral lines in 1896.) Kammerlingh Onnes planned his long-term research program systematically and with great managerial skill, as a general or a business executive. His laboratory was one of the first examples of a big science institution in physics. Contrary to the traditional laboratories of a Röntgen, a Rutherford, or a Curie, Kammerlingh Onnes turned his department into an efficiently run and well-funded scientific factory where technical and organizational expertise counted as much as scientific imagination. For example, instead of relying on homemade pieces of apparatus, Kammerlingh Onnes imported foreign technicians and organized a school for instrument makers and glassblowers.

The professional organization of the Leiden laboratory, supported by ample economic resources in part secured through the director's personal connections to Dutch industrialists, made it superior to the competing institutions in Cracow, London, and Paris. By 1906, the laboratory operated an efficient hydrogen liquefier, capable of producing 4 liters per hour, and for more than a decade after 1908 it had a worldwide monopoly on liquid helium. It was only after the end of the war that liquid helium began to be produced elsewhere, first at the University of Toronto, then at the National Bureau of Standards in Washington, and at the Physikalisch-Technische Reichsanstalt in Berlin. Even then, Leiden remained the unchallenged world center of low-temperature physics. Supplies of helium, at the time still a rare and expensive element, were of critical importance to the Leiden laboratory. Kammerlingh Onnes obtained his first helium by heating radioactive monazite sand from North Carolina and subsequently purifying the liberated gas. Other supplies of the valuable gas were derived from thorianite, another radioactive mineral, or received as gifts from foreign chemical companies.

Kammerlingh Onnes was an autocratic leader of the old school, class-conscious and deeply conservative. He ran the laboratory as a "benevolent despot," as Hendrik Casimir, the Dutch physicist and later director of the Philips Research Laboratories, once described him. Yet the despot was well-liked and able to stimulate in his scientific and technical staff a spirit of commitment and cooperation. In order to secure wide circulation of the laboratory's work, in 1885 Kammerlingh Onnes founded the in-house journal *Communications from the Physical Laboratory at Leiden*, whose articles were published in English or, more rarely, French and German. The articles were usually translations or revisions of papers first published in the (Amsterdam) *Proceedings of the Royal Society* and were often reprinted in foreign journals. Whether he had directly participated in the research or not, Kammerlingh Onnes considered it his right that he, as director, should appear as author or coauthor of all publications flowing from the laboratory—a policy that may explain his amazing productivity of scientific papers.

As the Leiden laboratory rose to become the world's leading center of low-temperature physics, it attracted an increasing amount of visitors who wanted to use the unique facilities. Whenever a phenomenon needed to be investigated at very low temperatures, Leiden was the place to do it. And there were many such phenomena. To mention only a few examples, in the early years of the new century Marie Curie performed experiments in Leiden in order to test whether the half-life of radioactive substances was influenced by extreme cold. (Pierre Curie investigated the same problem, but went to London to work with Dewar.) In 1908 Jean Becquerel from Paris examined the behavior of magneto-optical phenomena down to 14 K and found that the experiments in liquid hydrogen supported his controversial hypothesis about positive electrons. Becquerel and, nominally, Kammerlingh Onnes concluded that "the observations in liquid hydrogen seem to furnish strong support to the argument in favour of the existence of positive electrons" (Kragh 1989, 217). This time, Kammerlingh Onnes probably regretted his automatic coauthorship. The "strong support" was not accepted by other physicists who declined to accept the existence of positive electrons.

The identity of the Leiden laboratory was not only made up of buildings, apparatus, and organization, but also of the kind of methodological credo that derived from its director's view of science. The value of quantitative and precise measurements was given great emphasis insofar that these were held to be the essence of science, while theory and qualitative observation were seen as less important. In 1882, in his inaugural lecture at the University of Leiden, Kammerlingh Onnes expressed the credo as follows: "According to my views, aiming at quantitative investigations, that is as establishing relations between measurements of phenomena, should take first place place in the experimental practice of physicists. By measurement to knowledge [*door meten tot weten*], I should like to write as a motto above the entrance to every physics laboratory" (Casimir 1983, 160). This attitude, an important part of the Leiden spirit, was of course shared by many scientists besides Kammerlingh Onnes. The emphasis on quantitative measurements was a characteristic feature of physics at the turn of the century. The German physicist Friedrich Kohlrausch, an important organizer of physics institutes and a writer of influential textbooks, was known as "the master of measuring physics." According to him, measuring was the heart of physics. "Measuring nature is one of the characteristic activities of our age," he proclaimed in 1900 (Cahan 1989, 129). The attitude of Kohlrausch and Kammerlingh Onnes was widespread among experimentalists and perhaps particularly popular among spectroscopists. For example, it was expressed most clearly by the Nobel committee in 1907, when presenting the award to Michelson: "As for physics, it has developed remarkably as a precision science, in such a way that we can justifiably claim that the majority of all the greatest discoveries in physics are very largely based on the high degree of accuracy which can now be obtained in measurements made during the study of physical phenomena. [Accuracy of measurement] is the very root, the essential condition, of our penetration deeper into the laws of physics—our only way to new discoveries" (Holton 1988, 295).

The predilection for high-precision measurement does not necessarily imply a devaluation of either theory or observation. In some cases, experiments may be seen as self-justifying, but in others they are regarded as serving higher purposes, such as discovery and increased understanding of nature. At the turn of the century, it was often argued that precision measurements would lead to qualitatively new phenomena and in this way provide a dialectical transformation of quantity into quality (to use a classical Marxist phrase). This is what the Nobel committee hinted at when presenting the prize to Michelson, and it was certainly not a foreign idea to the American physicist. In a 1902 comment on the fundamental laws of physics and the possibility of future discoveries, he wrote, "It has been found that there are apparent exceptions to most of these laws, and this is particularly true when the observations are pushed to a limit, i.e., whenever the circumstances of experiment are such that extreme cases can be examined." After having mentioned some such examples, Michelson went on, "Many other instances might be cited, but these will suffice to justify the statement that 'our future discoveries must be looked for in the sixth place of decimals.' It follows that every means which facilitates accuracy in measurement is a possible factor in future discovery" (Michelson 1902, 24). This is what has been called "the romance of the sixth decimal," namely, the belief or hope that if scientists know a domain of nature up to some scale, an increase in observational power that gives access to a slightly larger scale may yield dramatic new results. If scientists continue to make their experiments a little more sophisticated—with more accuracy, higher energies, lower temperatures, higher resolution, and (consequently) at higher costs—it will pay off greatly. Naturally, there are no automatics in the process, no guarantee that increased experimental accuracy or range will lead to discoveries. On the other hand, Michelson's view certainly had a basis in fact—that is, it was supported by the history of science. Later physics has continued to provide evidence in favor of the romance of the sixth decimal. Michelson clearly saw his ether drift experiment, according to him leading to the theory of relativity, as belonging to this class. The discovery of superconductivity in 1911 is perhaps an even better example.

The Leiden measurement-to-knowledge philosophy has been termed "sophisticated phenomenalism" and should be distinguished from the crude empiricism in which experiments replace, rather than supplement, theoretical work. In spite of its emphasis on precision experiments, Kammerlingh

Onnes's research program was far from unrelated to theory. What may appear as an obsession with the conquest of new low-temperature ranges was not solely motivated in either simple curiosity or a wish to lay claim to the unkown territory before his competitors (although these factors undoubtedly played a role). Contrary to Michelson and other arch-experimentalists, Kammerlingh Onnes had a solid training in mathematics and was deeply inspired by Dutch theorists such as Lorentz and Johannes van der Waals. Before starting on his great program in cryogenics, he had worked in thermodynamics and molecular physics, and he was especially interested in testing the consequences of the molecular theory of van der Waals. In 1880 van der Waals had formulated the "law of corresponding states," which Kammerlingh Onnes independently developed the following year. According to this law, all substances obey the same equation of state when their pressure, temperature, and volume are expressed as multiples of the values these variables have at the critical point. In 1894 Kammerlingh Onnes stated in clear words the theoretical basis for embarking on a program to liquefy gases: "I was induced to work with condensed gases by the study of van der Waal's law of corresponding states. It seemed highly desirable to me to scrutinize the isothermal lines of the permanent gases especially of hydrogen at very low temperatures" (Gavroglu and Goudaroulis 1989, 51). The same message, the intimate connection between the Leiden low-temperature experiments and the theories of van der Waals, was contained in Kammerlingh Onnes's statement of 1908, quoted above, after he had succeeded in liquefying helium.

Superconductivity

One of the properties investigated at the Leiden laboratory was electrical conductivity in metals. The theory generally accepted at the time was based on a work that the German physicist Paul Drude had published in 1900—largely the same theory that is today presented in college-level textbooks. Drude suggested that metallic conduction was the result of the motion of free electrons under the influence of an external electric field and that the electrons, which he originally assumed to carry both negative and positive charges, had properties like a gas. In a metal, the conduction electrons were supposed to be in thermal equilibrium with the ions and neutral atoms. Assuming for simplicity that all electrons possess the same thermal velocity u, and that u is much larger than the drift velocity, Drude derived for the electric conductivity (the inverse of the resistivity) an expression of the form $\sigma = e^2 n \lambda T^{-\frac{1}{2}}$; here, λ is the mean free path and n the number of free electrons per unit volume. In 1905 Lorentz developed a more sophisticated theory by replacing the unrealistic assumption of equal-speed electrons with electron velocities distributed according to the Maxwell-Boltzmann law. And yet, after lengthy calculations, he arrived at a formula that differed from Drude's only by a numerical factor. The theory of electrical conduction continued to be developed in still more sophisticated versions, by J. J. Thomson, Owen Richardson, Niels Bohr, and others. In these versions, from 1910 to 1915, the electrons in a metallic conductor were conceived as a gas or vapor satisfying the law of ideal gases. It was hoped in this way to find a mechanism for the interaction of electrons and metal atoms that would explain the black-body radiation law. No satisfactory explanation was found, however, and the electron theories were no more successful in accounting precisely for the variation of resistance with temperature.

It was known experimentally that the resistance of pure metals varied proportionally with the absolute temperature, that is, $\sigma \sim T^{-1}$, at least down to 20 K. This posed a problem, for it agreed with the Drude-Lorentz formula only if it was arbitrarily assumed that $n\lambda \sim T^{-\frac{1}{2}}$, and neither Drude's theory nor its successors were able to calculate n and λ as functions of T. Moreover, experiments made at very low temperatures during the first decade of the century disagreed with both the $T^{-\frac{1}{2}}$ dependence and the T^{-1} dependence. Investigations made by Dewar and others of the temperature dependence of resistance about the boiling point of hydrogen indicated a flattening trend of the resistance function for very small temperatures. This was taken to imply one of two possibilities: either that the resistance would approach a nonzero value asymptotically or that it would reach a minimum and then, for even smaller temperatures, increase indefinitely. The latter possibility was widely assumed to agree well with theory. Close to absolute zero, free electrons would supposedly "freeze" and condense onto the atoms; then the density of free electrons would approach zero and, according to the Drude-Lorentz formula, the resistance would increase drastically. Kammerlingh Onnes, among others, found this an attractive hypothesis. In 1904 he described it as follows:

> It seems as if the vapour of electrons which fills the space of the metal at a low temperature condenses more and more on the atoms. Accordingly, the conductivity, as Kelvin has first expressed it, will at a very low temperature reach a maximum and then diminish again till absolute zero is reached, at which point a metal would not conduct at all, any more than glass. The temperature of the maxima of conductivity [lies] probably some times lower than that of liquid hydrogen. At a much lower temperature still, there would not be any free electrons left, the electricity would be congealed, as it were, in the metal. (Dahl 1984, 6)

Armed with his newly produced amounts of liquid helium, Kammerlingh Onnes decided in 1910 to examine the question systematically. The experiments were made in collaboration with Cornelis Dorsman and Gilles Holst, and it was Holst who performed the actual measurements. Yet the report was authored only by Kammerlingh Onnes. The Dutchmen first used a platinum resistance and compared their data with earlier measurements of gold resistances of known purity. The results made Kammerlingh Onnes conclude, "It appears that by descending to helium temperatures the resistance is still further diminished, but when these temperatures are reached the resistance attains a constant value quite independent of the individual temperature to which it has been brought" (ibid.). He realized that even small impurities might affect the result significantly and believed that these masked the real variation of resistance with temperature. Observing that the sample richer in gold showed less resistance than the other sample, he suggested that the resistance of pure metals would vanish asymptotically as the temperature approached zero and become practically zero at 5 K. This was a bold guess. And, like most bold guesses, it was wrong.

New experiments in 1911 made use of mercury, which could be obtained in a highly purified form. An initial experiment, reported in April, seemed to confirm Kammerlingh Onnes's suspicion of an asymptotically vanishing resistance. But when more precise experiments were conducted the following month, they showed a variation that was altogether unexpected, a sudden change to zero resistance at a temperature close to 4.2 K (figure 6.1). In his Nobel lecture of 1913, Kammerlingh Onnes described the discovery as follows: "The experiment left no doubt that, as far as accuracy of measurement went, the resistance disappeared. At the same time, however, something unexpected occured. The disappearance did not take place gradually but *abruptly*. From 1/500 the resistance at 4.2 K drops to a millionth part. At the lowest temperature, 1.5 K, it could be established that the resistance had become less than a thousand-millionth part of that at normal temperature. Thus the mercury at 4.2 K has entered a new state, which, owing to its particular electrical properties, can be called the state of superconductivity." Further experiments made between 1911 and 1913 proved that mercury's superconductivity was there to stay, whereas neither platinum nor gold showed a similar behavior. Yet mercury was no anomaly, for in December 1912 it turned out that tin and lead were superconductors too; the disappearance of resistance was found to occur at 3.78 K for tin and 6.0 K for lead. Contrary to expectations, it was realized that impurities had no effect on the new phenomenon. Moreover, it was definitely confirmed that the vanishing of resistance occurred abruptly and that the "knees" on the resistance curve were experimental artifacts.

The term "superconductivity" first appeared in a paper written by Kammerlingh Onnes in early 1913. Now there was a name for the puzzling phenomenon, but an understanding of what the name covered was completely lacking. For a time, Kammerlingh Onnes did not fully realize the novelty of the phenomenon and he continued to think of it as an extreme case of ordinary electrical conduction, that is, within the framework provided by the

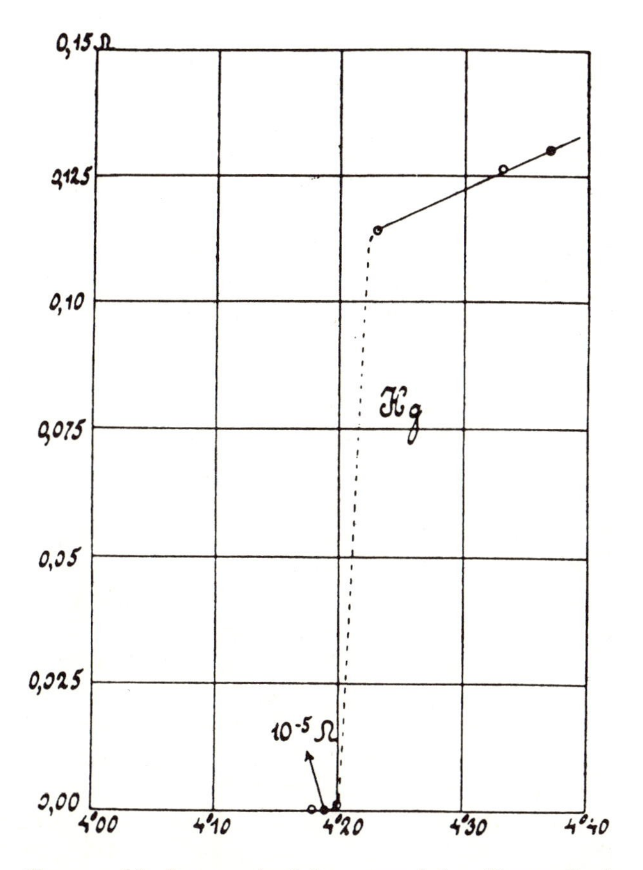

Figure 6.1. Superconductivity revealed: Kammerlingh Onnes's 1911 curve of the resistance of mercury versus temperature.

Drude-Lorentz theory. Perhaps, he thought, it was caused by a sudden increase in the mean free paths of the electrons. This idea, together with the results obtained for lead and tin, made him believe that superconductivity might be a general low-temperature state for all metals. But this was clearly a matter to be decided experimentally, and experiments proved that superconductivity was limited to a few elements. Within the area of experimental superconductivity progress followed quickly, notwithstanding the lack of theoretical understanding. In 1913 the first superconducting magnet was constructed at Leiden, and in 1914 Kammerlingh Onnes and his team began a study of the effect of strong magnetic fields on the superconducting state. A new discontinuity turned up, namely, the existence of a certain critical value of the field above which the zero resistance vanished abruptly and mysteriously. The critical field strength was found to increase with a decreasing temperature. The effect of a supercritical magnetic field thus had the same effect as heating the metal. Yet another novel phenomenon was left to the puzzled theoreticians.

It is sometimes claimed that Kammerlingh Onnes also discovered superfluidity in 1911. The weak basis for this claim is that the Leiden group performed measurements of the variation of the density of liquid helium with the temperature and obtained results that suggested a maximum density near 2.2 K. However, the sharp change in density—a manifestation of helium's superfluidity—was firmly established only several years after World War I, in 1924. Even then, Kammerlingh Onnes did not consider it a particularly interesting phenomenon that deserved a detailed study. Although what he observed in 1911 was indeed a superfluid property, it took many years until it was realized to be a manifestation of a genuinely novel phenomenon. Observation is a necessary but not sufficient precondition for discovery, and it was only in 1938 that superfluidity attained the status of discovery.

Although the discovery of superconductivity was a shining peak in the work of the Leiden laboratory, it was only one part of many in a broadly planned research program. Both before and after 1911, Kammerlingh Onnes and his group spent most of their time investigating other properties at low temperatures, including the Hall effect, piezoelectricity, the Curie law, magneto-optics, and radioactivity. It was for "his investigations of the properties of matter at low temperatures which led, inter alia, to the production of liquid helium" that Kammerlingh Onnes was awarded the Nobel prize in 1913. Superconductivity, in retrospect by far the most important of the discoveries, was not mentioned explicitly in the presentation speech. That superconductivity did not immediately arouse a stir was also reflected in the first Solvay Congress of 1911, which took place half a year after the discovery. In Brussels, Kammerlingh Onnes gave a detailed report on electrical resistance measurements in which he vaguely suggested that the vanishing of resistance might be explained with the help of quantum theory. In his Nobel lecture, he similarly suggested that superconductivity might be connected with "the energy of Planck's vibrators." The brief discussion following the Solvay report, limited to a question from Paul Langevin, indicates that the physicists gathered in Brussels were not particularly interested in the phenomenon.

Attempts to apply quantum theory to develop an improved theory of electrical conduction, and thereby to explain superconductivity, followed quickly after the 1911 discovery. One of the more promising among the theories, proposed by Wilhelm Wien in 1913, was based on the assumption that electrical conduction was essentially determined by the mean-free path of the electrons. At low temperatures, Wien's quantum theory led to a quadratic dependence of the resistance on temperature, but it failed to explain the sudden drop in resistance of superconducting metals. Other applications of quantum theory, suggested by Keesom in 1914 and Frederick Lindemann in 1915, were no more successful. What was the cause of the abrupt change in resistance? Why was the phenomenon restricted to a few of the metals in the periodic table? Theory just could not tell. Yet, in spite of the failure, there was no sense of crisis because of the anomaly.

If superconductivity could not be understood theoretically, perhaps it could be used technologically. At an early stage, the Leiden physicists realized the possibility of constructing powerful superconducting electromagnets, where there would be no heat loss even for very large currents. Such powerful magnets were not merely scientifically interesting, but would also be of great use in the electrotechnical industry. It turned out, though, that strong magnetic fields counteracted the superconducting state; the dream of super-electromagnets therefore had to be shelved, at least temporarily. In early 1914, experiments were made with a superconducting lead ring to which was applied a varying magnetic field of strength up to 10 kilogauss. At small field strengths the resistance was zero, but at a critical value about 600 gauss the resistance rose dramatically, in a manner analogous to the resistance-temperature variation. "The resistance increases . . . as if the introduction of the magnetic field has the same effect as heating the conductor," Kammerlingh Onnes wrote. Further examination of the relationship between resistance and magnetic field had to wait until the 1920s. With the advent of World War I, Leiden was temporarily cut off from its supplies of helium. And without liquid helium, neither superconductivity nor other phenomena at temperatures below 5 K could be studied experimentally.

Low-temperature experiments in Leiden continued after the war ended and new supplies of helium were secured. In 1919 it was established that two more metals, thallium and uranium, are superconducting. The vanishing temperatures were found to be 2.32 K for thallium and about 7.2 K for uranium. On the theoretical side, attempts continued to understand the phenomenon, but progress was notably lacking. The first two Solvay conferences after the war may illustrate the unsatisfactory state of knowledge concerning superconductivity. At the 1921 conference, Kammerlingh Onnes gave an address on "Superconductors and the Rutherford-Bohr Model," in which he reported on the latest Leiden experiments. He suggested that superconductivity was a nonclassical phenomenon that could be understood only in terms of Bohr's quantum atom, but neither Kammerlingh Onnes nor others could tell how. He summarized his report in eight questions, including "Since the Rutherford-Bohr atoms unite to make a metal, what happens to their electrons? Do they lose all or only part of their kinetic energy?"

The subject of the fourth congress in 1924 was "The Electrical Conductivity of Metals" and here, superconductivity was discussed by several of the participants. Lorentz, who spoke on the electron theory of metals, concluded vaguely that the electronic orbits in the superconducting states must be irregular or "particular." Kammerlingh Onnes discussed a possible connection between the electron structures of the few superconducting elements according to Bohr's new theory of the periodic system. Langevin suggested that the discontinuous vanishing of resistance was perhaps a result of a phase change in the material. He was apparently unaware that the suggestion had already been tested experimentally in Leiden, where Keesom's x-ray analyses proved that no change in phase was involved. Owen Richardson proposed a model according to which the electrons could move freely along orbits tangential to each other, and Auguste Piccard wondered whether lightning was perhaps a superconducting phenomenon at normal temperature.

Neither the Solvay discussions nor other contemporary attempts to understand superconductivity brought the subject any closer to an explanation than before the war. As Einstein wrote in 1922, in his only contribution to the literature on superconductivity, "With our wide-ranging ignorance of the quantum mechanics of composite systems we are far from able to compose a theory out of these vague ideas. We can only rely on experiment" (Dahl 1992, 106). Incidentally, this was possibly the first time that the term "quantum mechanics" occurred in a scientific publication.

When the range of ignorance of quantum mechanics was drastically narrowed after 1925, it turned out that a theoretical understanding of superconductivity did not follow in any simple way from the new theory of quanta. Superconductivity was eventually given a satisfactory quantum-mechanical explanation, but it took a long time and many failed attempts until the strange phenomenon was fully understood. A phenomenological theory was developed in 1935 by the brothers Fritz and Heinz London, and in 1957 superconductivity was finally explained on a microscopic basis by the Americans John Bardeen, Leon Cooper, and Robert Schrieffer. We shall look at this later development in chapter 24.

Chapter 7

EINSTEIN'S RELATIVITY, AND OTHERS'

The Lorentz Transformations

THE THEORY OF relativity has its roots in nineteenth-century optics. With the success of Augustin Fresnel's wave theory of light, the question of bodies moving through the ether came into focus. According to a theory Fresnel had suggested in 1818, a moving transparent body would partially drag along the ether. In that case, the velocity of light propagating through a body moving relative to the ether with velocity v would be changed by a fraction of the body's velocity, depending on the quantity v/c, where c is the velocity of light in vacuum. Fresnel's theory accounted for a large number of later optical experiments, which showed that it was impossible to detect, to the first order of v/c, the motion of the earth through the ether. When the elastic theories of light were replaced by Maxwell's electromagnetic theory, the situation was the same: Any theory of electrodynamics of moving bodies had to include the "Fresnel drag." In his first electron version of Maxwell's theory, published in 1892, Lorentz interpreted the Fresnel drag as a result of the interaction between light and the charged particles ("ions," later electrons) in the moving body. However, Lorentz was concerned that his theory was unable to explain an experiment that the American physicist Albert Michelson and his collaborator Edward Morley had performed five years earlier and that Michelson had first made in 1881.

The famous 1887 Michelson-Morley experiment was an attempt to measure the motion of the earth relative to the ether by means of an advanced interferometer technique. The experiment was performed at the Case School for Applied Science in Cleveland, Ohio, where Michelson was a professor of physics. It was to be expected that no first-order effects would be detected, but the Michelson-Morley experiment was precise to the second order, that is, dependent on the tiny quantity $(v/c)^2$. According to Lorentz's theory, the ether drift should be detectable to this order of precision, contrary to the null result of the experiment. The lack of a detectable motion of the earth through the world ether was surprising to both theorists and experimentalists. Rather than accepting the result, for a period Michelson considered the experiment to be a failure. "Since the result of the original experiment was negative, the problem is still demanding a solution," he maintained (Holton 1988, 284). The disagreement between theory and experiment caused Lorentz to modify his theory by assuming what was later called the Lorentz contraction, namely, that the length of a body moving in the direction of the earth's motion will shrink by a factor $\gamma = (1 - \beta^2)^{-\frac{1}{2}}$ or, to the second order of $\beta = v/c$, $1 - \beta^2/2$. The quantity γ is known as the Lorentz factor. Unknown to Lorentz, a similar explanation had been proposed by the Irish physicist George FitzGerald in 1889, although without including the formula. For this reason, it is sometimes referred to as the FitzGerald-Lorentz contraction. Both FitzGerald and Lorentz assumed that the hypothetical contraction was caused by changes in the molecular forces, but at the time none of them could provide an explanation for the assumption.

Lorentz's first explanation of Michelson's result was clearly ad hoc and not even based on his electrodynamic theory. During the following decade he greatly developed the theory, and in 1899 the Dutch theorist was able to derive the length contraction from the more general transformation formulas between the coordinates of a body moving through the ether and those of one at rest with regard to the ether. Lorentz wrote these transformations in a more complete form in 1904, the same form that we know today. He was not, however, the first to put the full "Lorentz transformations" in print. As a purely mathematical transformation, they can be found in a work on the Doppler effect published by Woldemar Voigt as early as 1887. More to the point, in 1900 Larmor derived the equations from his own version of electron theory. By means of the Lorentz-Larmor transformations, the null result of the Michelson-Morley experiment could be explained easily. Indeed, it followed from Lorentz's theory that there could be no detectable effects of uniform motion through the ether, not just to the second order in v/c, but also to all orders.

The Lorentz transformations make up the formal core of the special theory of relativity, and at first glance it might thus seem that Einstein's theory was preceded by the electron theories of Lorentz and Larmor. However, this was not the case at all. In spite of having obtained the same transformations as Einstein in 1905, Lorentz interpreted them in a very different way. First, Lorentz's was a dynamic theory in which the transformations could be ascribed a physical cause, the interaction between the ether and the electrons of the moving body. The length contraction was seen as a compensating effect arising because of the body's motion through the ether. The earth, according to Lorentz, really moved through the ether, only the ether wind was not measurable, in accordance with Michelson's result. Second, Lorentz's ether was an essential part of his theory, in which it functioned as an absolute frame of reference. For example, he maintained (implicitly in 1904 and explicitly in 1906) the existence of absolute simultaneity. That this concept does not agree with the modern interpretation of the time transformation only illustrates the difference between the theories of Lorentz and Einstein. In both theories, the transformation reads $t' = \gamma(t - vx/c^2)$, where t' is the time in the system moving with velocity v relative to the (x,t) system. But Lorentz considered the transformation to be a mathematical device and the "local time" t' to have no real meaning. There was, he thought, only one real time, t (which he called general time). As another aspect of the difference in interpretation, Lorentz did not arrive at the relativistic formula for addition of velocities which, in the relativistic framework, follows straightforwardly from the kinematic transformations.

Although Lorentz, Larmor, and most other physicists stuck to the ether and the associated concepts of absolute space and time, there were also dissenting voices. Ernst Mach had strongly criticized Newton's concept of absolute space and his philosophically-based criticism was well known, not least to the young Einstein. Mach also criticized Newton's notion of absolute time (as had others before him), which he claimed was metaphysical because it did not rest on either experience or intuition. Referring to Mach's criticism of the mechanical world picture in his 1889 book *The Science of Mechanics*, Einstein recalled in his autobiographical notes that "this book exercised a profound influence upon me in this regard while I was a student" (Schilpp 1949, 21).

No sketch of the prehistory of relativity, however brief, can avoid mentioning Henri Poincaré alongside Lorentz. Based on his conventionalist conception of science, around 1900 the French mathematician questioned whether the simultaneity of two events could be given any objective meaning. As early as 1898 he wrote, "Light has a constant speed. . . . This postulate cannot be verified by experience, . . . it furnishes a new rule for the definition of simultaneity" (Cao 1997, 64). Two years later, at the Paris world congress of physics, Poincaré discussed whether the ether really existed. Although he did not answer the question negatively, he was of the opinion that the ether was at most an abstract frame of reference that could not be given physical properties. In his *Science and Hypothesis* of 1902, Poincaré declared the question of the ether to be metaphysical, just a convenient hypothesis that some day would be discarded as useless. In his address to the St. Louis congress in 1904, he examined critically the idea of absolute motion, argued that Lorentz's local time (t') was no more unreal than his general time (t), and formulated what he called the relativity principle, namely, the impossibility of detecting absolute, uniform motion. His formulation of 1904 is worth quoting: "According to the Principle of Relativity the laws of physical phenomena must be the same for a 'fixed' observer as for an observer who has a uniform motion of translation relative to him . . . there must arise an entirely new kind of dynamics, which will be characterized above all by the rule, that no velocity can exceed the velocity of light" (Sopka and Moyer 1986, 293). Up to this point, Poincaré's intervention in the discussion had been mainly programmatic and semiphilosophical. In the summer of 1905, without knowing about Einstein's forthcoming paper, he developed an electrodynamic theory that in some respects went beyond Lorentz's. For example, he proved the relativistic law of addition of velocities, which Lorentz had not done, and also gave the correct transformation formula for the charge density. Apart from restating the principle of relativity as "a general law of nature," Poincaré modified Lorentz's analysis and proved that the Lorentz transformations form a group with the important property that $x^2 + y^2 + z^2 - c^2t^2$ is invariant, that is, remains the same in any frame of reference. He even noticed that the invariant could be written in the symmetric way $x^2 + y^2 + z^2 + \tau^2$ if the imaginary time coordinate $\tau = ict$ was introduced. Poincaré's theory was an important improvement, a relativity theory indeed, but not *the* theory of relativity. Strangely, the French mathematician did not follow up on his important insights, nor did he show any interest in Einstein's simultaneously developed theory of relativity.

Einsteinian Relativity

When twenty-six-year-old Albert Einstein constructed the special theory of relativity in June 1905, he was unknown to the physics community. The paper he submitted to the *Annalen der Physik* was remarkable in several ways, quite apart from its later status as a work that revolutionized physics. For example, it did not include a single reference and thus obscured the sources of the theory, a question that has been scrutinized by later historians of science. Einstein was not well acquainted with the literature and came to his theory wholly independently. He knew about some of Poincaré's nontechnical works and Lorentz's work of 1895, but not about Lorentz's (or Larmor's) derivation of the transformation equations. Another puzzling fact about Einstein's paper is that it did not mention the Michelson-Morley experiment or, for that matter, other optical experiments that failed to detect an ether wind and that were routinely discussed in the literature concerning the electrodynamics of moving bodies. There is, however, convincing evidence not only that Einstein was aware of the Michelson-Morley experiment at the time he wrote his paper, but also that the experiment was of no particular importance to him. He did not develop his theory in order to account for an experimental puzzle, but worked from much more general considerations of simplicity and symmetry. These were primarily related to his deep interest in Maxwell's theory and his belief that there could be no difference in principle between the laws of mechanics and those governing electromagnetic phenomena. In Einstein's route to relativity, thought experiments were more important than real experiments.

Most unusually at the time, the first and crucial part of Einstein's paper was kinematic, not dynamic. He started with two postulates, the first being the principle of relativity formulated as "the same laws of electrodynamics and optics will be valid for all frames of reference for which the equations of mechanics hold good"; the other postulate was "that light is always propagated in empty space with a definite velocity c which is independent of the state of motion of the emitting body." As to the ether, Einstein briefly dismissed it as superfluous. From this axiomatic basis, he proceeded to considerations about apparently elementary concepts, such as length, time, velocity, and simultaneity. His aim was to clarify these fundamental concepts; by very simple arguments, he first showed that simultaneity cannot be defined absolutely but depends on the state of motion of the observers. He next applied this insight to show that there were no consistent notions of absolute time and absolute length of a body. The (Lorentz) transformations between a stationary system and another moving uniformly with respect to it were derived purely kinematically.

Contrary to those of Lorentz and Poincaré, Einstein's formulas related to real, physically measurable space and time coordinates. One system was as real as the other. From the transformation equations followed the formula for addition of velocities, the contraction of moving bodies, and the time dilation, that is, that time intervals are relative to the velocity of the observer. Einstein's transformed time was as real as any time can be and, in this respect, quite different from Lorentz's local time. The addition of two velocities u and v gives the final velocity $V = (u + v)/(1 + uv/c^2)$ and, as Einstein noted, this implies the counterintuitive result that the velocity of light is independent of the velocity of its source.

The foundation of the theory of relativity was given in the kinematic part, and more specifically in its two postulates. It was only in the second part that Einstein justified the title of his paper, "On the Electrodynamics of Moving Bodies." He derived the transformation formulas for electric and magnetic fields which, according to Einstein, were relative quantities in the same way as space and time coordinates. But although the field quantities were relative to the state of motion, the law that governed them was not: The Maxwell-Lorentz equations, Einstein proved, have the same form in any frame of reference. They are relativistically invariant. According to Einstein's theory, many physical quantities are relative to the motion of the observer, but other quantities (such as electrical charge and the velocity of light) and the basic laws of physics remain the same. And it is these invariants that are fundamental. For this reason, Einstein originally would have preferred to call his theory "the invariant theory," a name that might have prevented many misunderstandings. The name "relativity theory" was introduced by Planck in 1906 and quickly became accepted. Ironically, Planck considered the essence of Einstein's theory to be its absolute, not its relative, features.

In 1892 Lorentz had postulated the force that acts on a charge q moving in a magnetic field (the Lorentz force, $\boldsymbol{F} = q\boldsymbol{v} \times \boldsymbol{B}/c$). Einstein kept the law but changed its status. He was able to deduce it in a simple manner from his transformations and thus reduce it to a derived law. At the end of his paper, Einstein considered the mass and energy of moving electrical charges, electrons. His electrons differed from those investigated by contemporary electrodynamicists, for Einstein's were primitive quantities; he was not interested in questions of their shape or internal structure. He predicted that the kinetic energy of an electron would vary with its speed as m_0c^2 $(\gamma - 1)$, where m_0 is the mass of the slowly moving electron, its rest mass. From this result, there is but a small step to the equivalence between mass and energy, a step Einstein took in another paper of 1905. There he derived what is possibly the most famous law of physics, $E = mc^2$. In Einstein's words: "The mass of a body is a measure of its energy-content; if the energy changes by L, the mass changes in the same sense by $L/9 \times 10^{20}$, the energy being measured in ergs, and the mass in grams."

Most readers of Einstein's paper probably considered it to be a contribution to the then-fashionable electron theory and paid less attention to its kinematic part. But Einstein was no electron theorist and his theory was, in accordance with the postulates on which it built, entirely general. The results were claimed to be valid for all kinds of matter, whether electrical or not. Einstein indicated his distance from contemporary electron theory by writing that his results, although derived from the Maxwell-Lorentz theory, were "also valid for ponderable material points, because a ponderable material point can be made into an electron (in our sense of the word) by the addition of an electric charge, *no matter how small*" (Miller 1981, 330; emphasis in original). This kind of "electron" had no place within the electromagnetic worldview. Equivalence between mass and energy was well known in 1905, but in a more narrow, electromagnetic interpretation (see chapter 8). Einstein's $E = mc^2$ was completely general.

Einstein's theory was taken up and discussed fairly quickly, especially in Germany. Its true nature was not recognized immediately, however, and it was often assumed to be an improved version of Lorentz's electron theory. The name "Lorentz-Einstein theory" was commonly used and can be found in the literature as late as the 1920s. The most important of the early relativity advocates was Max Planck, who was instrumental not only in putting his authority behind the theory, but also in developing it technically. Planck was greatly impressed by the theory's logical structure and unifying features. He recognized it to be a fundamental theory that encompassed both mechanics and electromagnetism and was happy when he discovered, in 1906, that the theory of relativity could be presented in the form of a principle of least action. Planck also developed the dynamics of particles according to Einstein's theory and was the first to write down the transformation laws for energy and momentum. Another important advocate was the Göttingen mathematician Hermann Minkowski who, in a 1907 lecture, presented relativity theory in a four-dimensional geometrical framework with a strong metaphysical appeal. Minkowski introduced the notion of a particle's world-line and explained enthusiastically how radical a break with the past the theory of relativity was: "Henceforth space by itself, and time by itself, are doomed to fade away into mere shadows, and only a kind of union of the two will preserve an independent reality" (Galison 1979, 97). However, Minkowski considered Einstein's theory to be a completion of Lorentz's and interpreted it, wrongly, to be within the framework of the electromagnetic worldview.

Thanks to the works by Planck, Minkowski, Ehrenfest, Laue, and others, by 1910 Einstein's theory of relativity had gained firm support and was probably accepted by a majority of elite theoretical physicists. *Annalen der Physik* became the chief journal for a growing number of articles in which the theory was tested, examined conceptually and technically, and applied to new areas, and the old physics recast into relativistic frames. Outside Germany the reception was slower and more hesitant, but whether or not they accepted the theory in a physical sense, by 1910 many physicists used its equations. At that time, the theory was little known outside the physics community. It took some time for relativity to diffuse to the average physicist and, naturally, even longer to enter physics textbooks. The growing familiarity of the special theory of relativity is illustrated by Sommerfeld's famous *Atombau und Spektrallinien*, which was intended to be a book mainly for students and nonexperts in atomic physics. In the first three editions, from 1919 to 1922, Sommerfeld started his chapter on fine structure theory with an eighteen-page introduction to the theory of relativity. In the 1924 edition, he replaced the introduction with the optimistic comment that the theory of relativity was now common knowledge to all scientists.

From Special to General Relativity

> My first thought on the general theory of relativity was conceived two years later, in 1907. The idea occurred suddenly. . . . I came to realize that all the natural laws except the law of gravity could be discussed within the framework of the special theory of relativity. I wanted to find out the reason for this, but I could not attain this goal easily. . . . The breakthrough came suddenly one day. I was sitting on a chair in my patent office in Bern. Suddenly a thought struck me: If a man falls freely, he would not feel his weight. I was taken aback. This simple thought experiment made a deep impression on me. This led me to the theory of gravity. (Einstein 1982, 47)

This was how Einstein, in a 1922 address, described the start of the route that led him to one of the most fundamental theories ever in the history of science. In spite of interesting technical contributions by David Hilbert, Gunnar Nordström, and a few others, general relativity was very much Einstein's work. According to the principle of equivalence, no mechanical experiment can distinguish between a constant (unaccelerated), homogeneous gravitational field and a uniformly accelerated reference frame in which there is no gravitational force. In 1907 Einstein formulated this principle in a generalized way, to be valid for all kinds of experiments, whether mechanical or not. From this point of view, there is no essential difference between inertia and gravitation. Einstein did not immediately follow up on this idea, but in 1911 he developed a first version of a new research program, namely, to find a new theory of gravitation that led both to the equivalence principle and an extended theory of relativity. This first generalization of the relativity principle resulted in two remarkable predictions: First, that the propagation of light was acted on by gravity, and, second, as Einstein had already realized in his 1907 paper, that the rate of a clock is slowed down near a large gravitating mass. As to the first prediction, Einstein found that for a ray grazing the sun, the deflection would be a little less than one arc-second, 0.83″. The clock of the second prediction might be a luminating atom, as measured by a monochromatic spectral line, and here Einstein calculated how the received wavelength increased—became redshifted—with the gravitational field. The result, a direct consequence of the equivalence principle, was $\Delta\lambda/\lambda = \Delta\Phi/c^2$, where $\Delta\Phi$ is the difference between the gravitational potentials at emission and reception of the light.

Einstein realized that his 1911 theory was only a step toward the theory he was looking for. During the next four years, he immersed himself fully in the still more complex search for the new relativistic theory of gravitation. The key to the problem turned out to lie in mathematics. "In all my life I have labored not nearly as hard," he wrote to Sommerfeld in 1912; "I have become imbued with great respect for mathematics, the subtler part of which I had in my simple-mindedness regarded as pure luxury until now" (Pais 1982, 216). Helped by his friend, the mathematician Marcel Grossmann, he recognized that the proper mathematical tool for the theory was the absolute differential calculus (or tensor analysis) originating with the nineteenth-century works of Gauss and Riemann. In collaboration with Grossmann, Einstein developed a tensor theory of gravitation in which space-time was no longer seen as an inert background of physical events, but was itself subject to changes due to the presence of gravitating bodies. He now definitely abandoned the line element of special relativity and replaced it with a more complex tensor expression that in general consisted of ten quadratic terms: $ds^2 = \Sigma g_{mn} dx^m dx^n$. Similarly, he no longer based his theory on the Lorentz transformations, which he wanted to replace with a more general invariance group. In 1913, Einstein discussed the requirement of general covariance—that is, that the field equations must have the same form in every frame of reference. He considered physical laws satisfying this requirement to be preferable, as this would minimize the arbitrariness and maximize the simplicity of the world picture. However, having formulated the principle of general covariance, he abandoned it and instead proposed a set of field equations that did not have this property. Einstein's main reason for this retrograde step was that his generally covariant equations failed to reduce to the Newtonian limit for weak static gravitational fields. He developed an argument, later known as the "hole argument," which convinced him that a theory based on generally covariant equations could not be the correct answer. The reason was that such a theory, Einstein mistakenly thought, would violate determinism and causality. It took him two years of hard thinking before he was forced to realize that general covariance was indeed the key to all his problems. This phase culminated during the summer and fall of 1915, and in November 1915 he presented to the Berlin Academy of Sciences the final form of his generally covariant field equations for gravitation. It was, he wrote to Sommerfeld, "the most valuable discovery I have made in my life."

Einstein's paper of November 18, 1915 did not only include a new set of gravitational field equations that were generally covariant and logically satisfactory. He also used his new theory to conclude that his earlier prediction of gravitational deflection of light was wrong by a factor of 2. According to the improved theory, a light ray passing the sun should be deflected an angle of 1.7″. Apart from the theory's appealing logical structure, it was another prediction that really made Einstein feel that his theory was correct, and this time it was a prediction of a known effect, namely, the anomalous precession of Mercury's perihelion. It had been known since 1859 that Mercury does not move around the sun exactly as it should according to Newtonian mechanics. Its perihelion precesses slowly around the sun, as explained by celestial mechanics, but with a slightly different speed of rotation. The anomaly was only 43″ per century (about 8 percent of the observed precession), yet was enough to constitute a problem for Newton's theory of gravitation. Einstein was not the first who sought to explain the Mercury anomaly, but he was the first to give a quantitative explanation based on a fundamental theory not tailored to solve the problem. His calculated value of the precession agreed almost perfectly with the observed value. Einstein had long known that the Mercury perihelion problem would be a litmus test of any new theory of gravitation. As early as Christmas day 1907, he wrote to his friend Conrad Habicht: "I hope to clear up the so-far unexplained secular changes of the perihelion length of Mercury . . . [but] so far it does not seem to work."

In early 1916, while Europe was bleeding in World War I, Einstein had his general theory of relativity ready. Few physicists were able to understand (or even have access to) the theory because of its complexity and unfamiliar mathematical formulation. But the theory resulted in three predictions, which could be tested in order to judge whether the theory was really correct physically and not just the dream of an imaginative mathematician. The prediction of Mercury's perihelion advance was a great success, but far from enough to convince skeptics about the truth of Einstein's theory. After all, Einstein had known about the anomaly all along, so perhaps he had somehow built the result into his theory. And couldn't the correct result be obtained without the doubtful theory of relativity? This was what a few German antirelativists claimed, referring to a theory that Paul Gerber had published in 1902 and in which he obtained the same expression for the perihelion advance of a planet as Einstein did in 1915. Ernst Gehrcke, a leading German antirelativist, had the paper republished in 1917 in order to use it against the theory of relativity and Einstein's priority. The result was a minor controversy, but few physicists, even among the antirelativists, were fooled. Gerber's theory was simply wrong and his correct perihelion expression purely coincidental.

The gravitational redshift prediction, the same in 1915 as in 1911, was extremely difficult to test, primarily because the measured spectral lines came from the sun's hot atmosphere, where a gravitational redshift could not be easily distinguished from other effects such as the Doppler effect. Between 1915 and 1919, several researchers tried to verify the "Einstein effect," but in all cases they failed to find an effect of the size predicted by Einstein. On the other hand, in 1920 two physicists from Bonn University, Albert Bachem and Leonhard Grebe, reported results in essential agreement with the prediction. Unsurprisingly, Einstein quickly endorsed the Bachem-Grebe results. Although the German claim was criticized by other experimentalists and the whole question not clarified until much later, from about 1920 many physicists came to believe that Einstein's prediction agreed reasonably well with observations, or, at least, that there was no clear disagreement. For example, the American astronomer Charles E. St. John at the Mount Wilson Solar Observatory believed, from 1917 to 1922, that his careful observations were at variance with Einstein's theory, but in 1923 he "converted" to relativity and decided that he had confirmed Einstein's prediction. There are reasons to assume that St. John's change in attitude, as well as that of many other researchers, were to a large extent the result of the third test, the 1919 eclipse measurements that so spectacularly confirmed the light-bending prediction.

Shortly after Einstein's first (and incorrect) prediction of 1911, a few astronomers sought to test the prediction by measuring the position of stars near the rim of the sun during an eclipse. Eclipse expeditions in 1912 in Brazil and in 1914 in southern Russia failed to produce any results, in the first case because of persistent rain and in the second case because of the outbreak of the war. In 1918 American astronomers at the Lick Observatory succeeded in taking photographs of an eclipse passing the United States. The report, which included data conflicting with the 1915 prediction, was not published, however. It was instead the British expedition headed by Frank Dyson and Arthur Eddington and planned by 1917 that produced the confirming results. The total solar eclipse of 1919 was studied at two locations, at Principe Island off the West African coast (by Dyson) and at Sobral in Brazil (by Eddington). Photographs were taken by both expeditions; when they were analyzed, they showed a deflection of light in excellent (but not perfect) agreement with Einstein's theory. Dyson, who reported the results at a joint meeting of the Royal Society and the Royal Astronomical Society on November 6, 1919, concluded, "A very definite result has been obtained that light is deflected in accordance with Einstein's law of gravitation" (Earman and Glymour 1980a, 77). This was, in fact, a too-optimistic conclusion that could be obtained only by a treatment of the available data that came close to manipulation, including the rejection of data that did not agree with Einstein's prediction. Eddington, the British authority in and prophet of relativity, was fully convinced of the truth of the general theory of relativity and his preconceived view colored the conclusion. At any rate, the theory was accepted by the majority of physicists and astronomers, if not by all. As Gehrcke had invoked Gerber's old theory as an alternative to Einstein's Mercury perihelion calculations, so did conservatives like Wiechert and Larmor suggest elaborate electromagnetic explanations of light bending in the early 1920s. But compared with the enormous interest in Einstein's theory, these and other nonrelativistic attempts attracted little attention. The 1919 eclipse expedition became a turning point in the history of relativity, if more from a social than from a scientific point of view.

According to the recollections of Ilse Rosenthal-Schneider, who was a student of philosophy and physics at Berlin University in 1919, when Einstein received the news about the Dyson-Eddington confirmation, he was "quite unperturbed." He said to her, "I knew that the theory is correct. Did you doubt it?" To Rosenthal-Schneider's question of what he would have said if the observations had disagreed with the theory, Einstein replied, "I would have had to pity our dear God. The theory is correct all the same" (Rosenthal-Schneider 1980, 74).

The picture of Einstein as an all-knowing and somewhat arrogant rationalist who did not care about experiments is undoubtedly widespread and part of the Einstein myth. But it is basically wrong, at least as far as the younger Einstein is concerned. On the contrary, Einstein was deeply interested in experimental tests of his theories and often tried to arrange for such tests. For example, when Einstein predicted that light would be deflected in gravitational fields in 1911, it was he who tried to interest the astronomers in testing the prediction. Einstein emphasized the closed and logical structure of his general theory of relativity, which to him implied not that it must therefore be correct, but that it could not be modified in order to accommodate some experimental refutation. In a 1919 letter to Eddington, he wrote, "I am convinced that the redshift of spectral lines is an absolutely inevitable consequence of relativity theory. If it were proven that this effect does not exist in nature, the whole theory would have to be abandoned" (Hentschel 1992, 600). He expressed himself similarly with regard to the predicted deflection of light rays. Far from being "unperturbed," he expressed great joy when he heard about the results. Rosenthal-Schneider's story is not reliable. That Einstein later came to adopt a Platonic, rationalist attitude with regard to experiments versus mathematical theories is another matter.

Reception

Einstein's theory of relativity shared with Darwin's evolutionary biology, Röntgen's invisible rays, and Freud's psychoanalysis the fact that it was met with an enormous interest outside as well as within the scientific community. It became one of the symbols of the modernism of the interwar period and, as such its importance extended far beyond physics. Einstein's theory was labeled "revolutionary," a term commonly associated with the passage from Newtonian to Einsteinian physics. The theory of relativity was indeed a kind of conceptual revolution and in the early 1920s the revolution metaphor, freely associating to political revolutions, became a trademark of Einstein's theory. It was a trademark that Einstein did not want to sanction. Einstein did not consider himself a revolutionary; in papers and addresses, he repeatedly stressed the evolutionary nature of the development of science. The theory of relativity, he often said, was the natural outcome of the foundations of physics laid by Newton and Maxwell. Thus, in a 1921 paper, Einstein noted, "There is someting attractive in presenting the evolution of a sequence of ideas in as brief a form as possible, and yet with a completeness sufficient to preserve throughout the continuity of development. We shall endeavour to do this for the theory of relativity and to show that the whole ascent is composed of small, almost self-evident steps of thought" (Hentschel 1990, 107).

The general public—including many scientists and most philosophers—discovered relativity only after the end of World War I, in part as a result of the much-publicized eclipse expedition announcing the confirmation of Einstein's theory. A large part of the literature on relativity in the 1920s was written by nonscientists who, more often than not, thoroughly misunderstood the theory and discussed its implications in areas where it could not be legitimately applied. Some authors "applied" relativity to art theory, some to psychological theories, and still others drew wide-ranging philosophical and ethical consequences from Einstein's theory. Not uncommonly, ethical relativism was claimed to follow from the theory of relativity, which for this reason was declared unwanted in some quarters. Didn't Einstein claim that everything is relative and that no point of view is better than any other point of view?

The eminent Spanish philosopher José Ortega y Gasset was one of those who misused Einstein's theory to argue his own pet philosophy, which he called "perspectivism." Here is a sample: "The theory of Einstein is a marvelous proof of the harmonious multiplicity of all possible points of view. If the idea is extended to morals and aesthetics, we shall come to experience history and life in a new way. . . . Instead of regarding non-European cultures as barbarous, we shall now begin to respect them, as methods of confronting the cosmos which are equivalent to our own. There is a Chinese perspective which is fully as justified as the Western" (Williams 1968, 152). Incidentally, Ortega y Gasset's perspectivism would later be accepted as politically correct by many Westeners and used in criticizing the scientific worldview. Whatever the merits of perspectivism or relativism, these ideas have nothing to do with the theory of relativity.

Figure 7.1 shows the annual distribution of German books on relativity between 1908 and 1944, with a sharp peak in 1921 when the public debate culminated. The figure refers to both textbooks in physics and to popular, philosophical, and antirelativist books and booklets, of which the latter category made up about three-quarters of the total number of titles published during the peak years 1920–22. Of course, physicists had "discovered" relativity many years before 1921, in the shape of the special theory, although, as mentioned, until about 1913 many physicists did not clearly distinguish between Einsteinian relativity and the equations of the electron theories. The fine structure of early relativity publications is displayed in figure 7.2, curi-

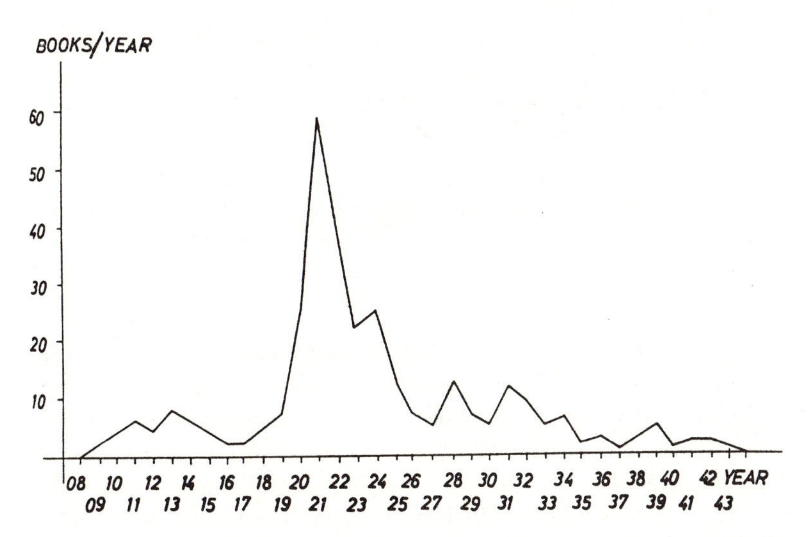

Figure 7.1. German books on relativity, 1908–43. *Source:* Goenner 1992. Reprinted with the permission of Einstein Studies, series editors Don Howard and John Stachel.

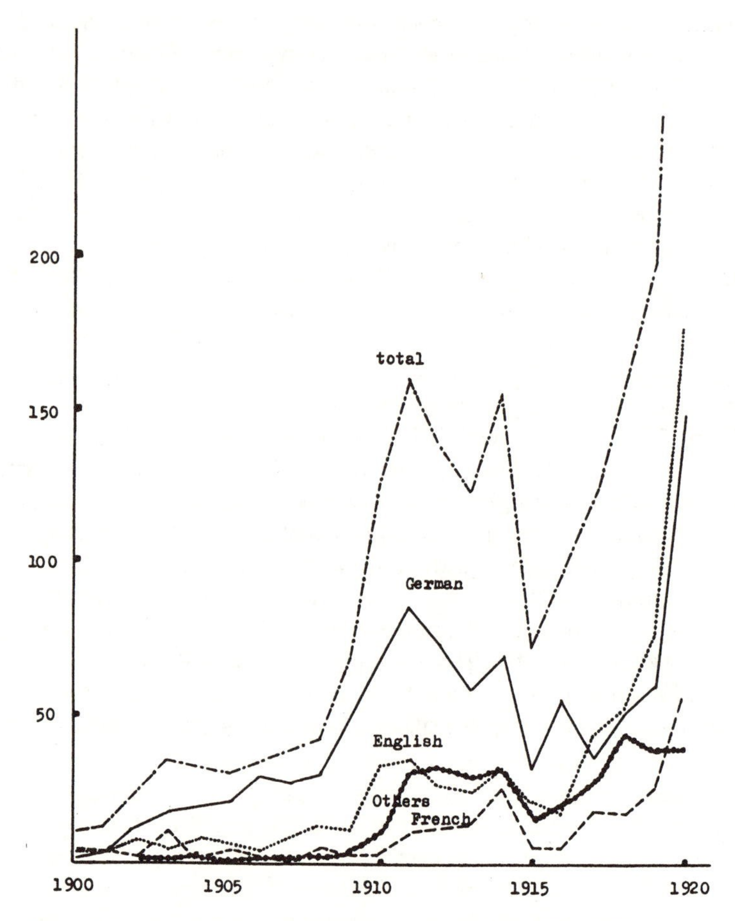

Figure 7.2. Distribution of publications on relativity, 1900–1920. *Source:* Reprinted from J. Illy, "Revolutions in a revolution," *Studies in the History and Philosophy of Science* 12, 1981, 173–210, with permission from Elsevier Science.

ously starting in 1900. Note the German dominance, the lack of French literature until about 1912, and the decrease in the total number of publications between 1910 and 1915 (see also table 7.1). The latter feature probably reflects that, by 1911, the special theory of relativity was widely accepted among physicists and was no longer considered to be at the cutting edge of physics. It was only after the extended theory appeared in 1915 that momentum was regained.

The receptions of relativity depended on particular national and cultural environments. In France, Einstein's theory was originally met with silence, a

TABLE 7.1
National Distributions of Scientific Publications on Relativity until 1924

Country	*No. of authors*	*No. of publications*
Germany	350	1435
England	185	1150
France	150	690
United States	128	—
Italy	65	215
Netherlands	50	126
Austria(-Hungary)	49	—
Switzerland	37	—
Russia (USSR)	29	38

Note: Data from Hentschel 1990, p. 67.

fact undoubtedly related to the rigid, centralized French system of education and research and, possibly, also to the influence of Poincaré. Except for Paul Langevin and a few of his students, French physicists discovered relativity only when the 1919 eclipse observation made international headlines. And even then, relativity was widely received with some suspicion; in the aftermath of the Great War, it did not ease the perception that relativity was a German theory.

The situation in the United States was not much different, if for different reasons. Significantly, the first serious study was a 1909 paper written by two physical chemists, Gilbert N. Lewis and Richard C. Tolman. They, and almost all other American scientists aware of the theory of relativity, tended to present the theory in accordance with positivistic norms. For them, relativity theory was an inductive generalization from experimental facts, an attitude very different from that held by Einstein and his German colleagues. Millikan's view, that "the special theory of relativity may be looked upon as starting essentially in a generalization from Michelson's experiment," was shared by almost all American physicists (Holton 1988, 280). In the United States, in particular, it was not uncommon to criticize Einstein's theory for being "undemocratic" and against people's common sense. William Magie, a professor of physics at Princeton University, was a spokesman for this kind of quasipolitical opposition. In a 1912 address, he argued that fundamental physical theories "must be intelligible to everybody, to the common man as well as to the trained scholar." Holding, somewhat unrealistically, that "all previous physical theories have been thus intelligible," Magie suggested that the democratic foundation of science lay in "the primary concepts of force, space and time, as they are understood by the whole race of man." But Einstein's theory did not rest on these natural concepts, which led Magie to ask, rhetorically, "Have we not the right to ask of those leaders of thought to whom we owe the development of the theory of relativity, that they . . . pursue their brilliant course until they succeed in explaining the principle of relativity by reducing it to a mode of action expressed in terms of the primary concepts of physics?" (Hentschel 1990, 75).

Relativity came to Russia a little earlier than to the United States, and the subject was discussed at Paul Ehrenfest's theoretical seminar at St. Petersburg from about 1907. Shortly after the civil war, when Russia had become the Soviet Union, courses on the extended theory of relativity were offered in St. Petersburg, where Alexander Friedmann established an important school of theoretical physics and made fundamental contributions to relativistic cosmology. Contrary to the situation in the 1930s and 1940s, relativity was not seen as an ideologically controversial theory; although some defenders of the true Marxist faith rejected the theory of relativity in the name of dialectical materialism, philosophical and political objections did not interfere with the scientific study of relativity that took place in St. Petersburg (or, from 1924, Leningrad) and elsewhere in the vast communist empire (see also chapter 16).

The theory of relativity was very much a German theory and perhaps, as some argued, a German-Jewish theory. Compared with the lively discussions in Germany and the important contributions from that country, British physicists were slow to respond constructively to the new theory. Part of the problem was Einstein's rejection of the ether, a concept that most British physicists adhered to and hesitated to abandon. The first textbook on relativity written by a Briton was Ebenezer Cunningham's *The Principle of Relativity*, published in 1914. This book, from which young Paul Dirac learned relativity, still included vestiges of the ether worldview. For example, the principle of relativity was formulated as "the general hypothesis, suggested by experience, that whatever be the nature of the aethereal medium we are unable by any conceivable experiment to obtain an estimate of the velocities of bodies relative to it" (Sànchez-Ron 1987, 52). In England, as elsewhere, relativity really broke through only after 1919, now with Eddington as the fountainhead of the theory. Several monographs on relativity theory appeared in 1920–23, of which Eddington's *Space, Time and Gravitation* (1920) and *The Mathematical Theory of Relativity* (1923) became international best-sellers. The sudden rise in interest in relativity may, in some way, have reflected the general cultural situation following the Great War. This was what Dirac recalled:

> It is easy to see the reason for this tremendous impact. We had just been living through a terrible and very serious war. . . . Everyone wanted to forget it. And then relativity came along as a wonderful idea leading to a new domain of thought. It was an escape from the war. . . . Relativity was a subject that everybody felt himself competent to write about in a general philosophical way. The philosophers just

put forward the view that everything had to be considered relative to something else, and they rather claimed that they had known about relativity all along. (Kragh 1990, 5)

Dirac's recollection of 1977 is supported by primary sources, such as an editorial in the British newspaper *The Morning Post* of September 27, 1922, which offers another possible explanation: "One of the results of the war, in which the scientific brains of this country were mobilized to such good purpose, is an appreciable increase of public interest in achievements of science, whether theoretical or practical. In prewar days, neither the man in the street nor the man at the club window could have been persuaded to read articles about the Einstein versus Newton controversy" (Sànchez-Ron 1992, 58).

But it was in German-speaking Europe that things really happened. Not only was this Einstein's world, but it was also the world of Planck, Laue, Weyl, Hilbert, Minkowski, Mie, Born, Lorentz, and Pauli. It was only in Germany that a constructive mathematically and conceptually informed debate occurred; and it was also here that the destructive antirelativist discussion flourished first and lasted longest. The theory of relativity was attacked by a loose group of German right-wing physicists who were not only politically on the extreme right, but also endorsed conservative scientific virtues, such as classical mechanicism, causality, visualizability, and physical insight based on experiments rather than on theory (see also chapter 10). The most articulate and radical members of the antirelativity and also, sometimes, antiquantum camp were Philipp Lenard, Johannes Stark, and Ernst Gehrcke, all physicists of renown. Lenard and Stark were Nobel laureates, and Gehrcke, since 1921 ordinary professor at the Physikalisch-Technische Reichsanstalt, was a recognized expert in experimental optics. The antirelativist physicists rejected Einstein's theories, which they found were absurd, spiritually dangerous, and lacking experimental confirmation. If some of the relativistic equations were confirmed, they argued that the equations could equally well be derived on a classical foundation. Four-dimensional space-time, curved space, and the twin paradox were among the concepts that were declared mathematical abstractions devoid of physical meaning. From 1920, the German crusade against relativity accelerated with an anti-Einstein meeting in Berlin organized by Paul Weyland, a political activist, and with Gehrcke as speaker. The meeting caused a sharp public reply by Einstein, in which he indicated that anti-Semitism was part of the agenda of the antirelativist campaign. During the next few years, Lenard, Stark, Gehrcke, and their allies continued to attack Einstein and the theory of relativity in a large number of articles, pamphlets, and books. According to them, relativity was not a scientific theory at all, but an ideology, a mass illusion cleverly camouflaged by Einstein and his friends in power in German physics. One of Gehrcke's antirelativity booklets was titled, characteristically, *The Theory of Relativity:*

A Scientific Mass Suggestion (1920). Although noisy and fairly numerous, the antirelativists failed to impede the progress of German theoretical physics in the 1920s and, instead, marginalized themselves. It was only after 1933, when a new political system had seized power, that antirelativism became of some, if still only limited, importance in German science.

Chapter 8

A REVOLUTION THAT FAILED

IN PREVIOUS CHAPTERS we have often mentioned the electromagnetic worldview as the main contender to the mechanical view of physics. It may seem strange to deal with this incorrect worldview after having described the advent of the true theory of relativity, for after 1905 physicists must surely have recognized that the electron theory (in its broad meaning) was inferior to relativity. Perhaps they should have to, but they did not. In fact, the electromagnetic worldview experienced its zenith just after 1905, and it took at least five more years until it was recognized that this grand attempt to formulate a new basis for physics was probably not tenable.

The Concept of Electromagnetic Mass

As mentioned in chapter 1, at the turn of the century the mechanical worldview was under attack and on its way to be replaced by a view based on electromagnetic fields. The more radical and elaborate version of the new conception, known as the electromagnetic worldview, emerged about 1900 and flourished for about a decade. Its core program was the full reduction of mechanics to electromagnetism, a new physics in which matter had disappeared as a substance and been replaced by effects of electromagnetic fields—electrons. The program was based on developments in electrodynamics in the 1880s and 1890s, during which period a less complete electromagnetic view of nature emerged and resulted in the electron theories of Larmor, Lorentz, and Wiechert. In this process, two traditions played a role. One, of importance only to Lorentz and Wiechert, was the corpuscular electrodynamical tradition of Weber and his followers, which operated with electrical particles in instantaneous interaction, but without a field concept; the other tradition, on the contrary, was firmly rooted in Maxwellian field electrodynamics. Already in 1894, Wiechert had clearly expressed the conviction that material mass might be an epiphenomenon and that the only true mass was of electromagnetic origin, consisting of the hypothetical electrical particles he depicted as excitations in the ether. Wiechert's suggestion anticipated the electromagnetic worldview, but his work in this area was somewhat overshadowed by the works of Lorentz and German theorists such as Max Abraham and Adolf Bucherer.

In 1881, young J. J. Thomson showed that when a charged sphere moves through the ether, it will acquire a kind of apparent mass, analogous to a sphere moving through an incompressible fluid. For a sphere with charge e and radius R, he found the electromagnetically induced mass to be $m' = 4/15\ e^2/R^2c$. This first introduction of electromagnetic mass was later improved by the excentric British engineer-physicist Oliver Heaviside, who in 1889 derived the expression $m' = 2/3\ e^2/R^2c$. Contrary to Thomson, who did not think of the "apparent mass" as real, Heaviside considered it as real as the material mass. It was part of the sphere's measurable or "effective" mass. A further improvement was obtained by Wilhelm Wien in 1900, in a paper significantly titled "On the Possibility of an Electromagnetic Foundation of Mechanics." Wien confirmed Heaviside's expression in the limit of small velocities and added the important result that the electromagnetic mass would depend on the velocity and differ from Heaviside's if the velocity approached the velocity of light. Exactly how the mass of an electron, or any charged body, would depend on the velocity quickly became a problem of crucial importance in the new electron physics. Wien's paper of 1900 has been seen as the first clear pronouncement of the electromagnetic worldview, which (in spite of Wiechert's earlier formulation) is justified insofar that Wien assumed that all mass was of electromagnetic nature. Matter, Wien argued, consisted of electrons, and electrons were particles *of* electricity, not tiny spheres on which electricity resided. Moreover, he contended that Newton's laws of mechanics had to be understood electromagnetically and, if a complete correspondence could not be achieved, electron theory was the more profound and fundamental theory of the two.

The first detailed electron model was constructed by Max Abraham, a Göttingen physicist who had graduated under Planck in 1897. According to Abraham, writing in 1902, the most important question in physics was this: "Can the inertia of electrons be completely accounted for by the dynamical action of its field without using the help of mass which is independent of the electric charge?" (Goldberg 1970, 12). It was, in part, a rhetorical question. Abraham believed it *had* to be answered affirmatively. He made a detailed study of the dynamics of the electron in a paper of 1903, which in more than one way contrasted with Einstein's relativity paper two years later. Whereas Einstein's paper was mathematically simple, Abraham's was a mathematical tour de force; and whereas Einstein's filled 31 pages, Abraham's was no less than 75 pages. Both works appeared in *Annalen der Physik*, a journal that often included papers of a length that would be unheard of in modern physics journals. In his works of 1902–1903, Abraham argued that the only kind of electron that could be understood fully in terms of electromagnetism was a rigid sphere with a uniform surface or volume charge distribution. For such an electron, he calculated its mass—arising purely from the electron's own fields—and found a velocity variation which, to the first order of $\beta^2 = (v/c)^2$, can be written $m = m_0(1 + 2/5\ \beta^2)$. Here, m_0 denotes the electromagnetic rest mass, $2/3\ e^2/R^2c$. We shall shortly return to Abraham's important work, which was not the only electron theory of the period.

The idea of mass varying with velocity was part of Lorentz's research program as he developed it from 1899. Lorentz's approach differed from that of Wien and Abraham, and he was reluctant to follow them in their belief that all mass is electromagnetic. In 1904, however, he overcame his natural caution and in his theory of the electron from that year he supported the electromagnetic worldview. Not only was the mass of his electron of electromagnetic origin, but he also argued that all moving matter (whether consisting of electrons or not) must obey the mass-variation characteristic of electrons. In 1906, in a lecture given at Columbia University, he stated, "I for one should be quite willing to adopt an electromagnetic theory of matter and of the forces betwen material particles." He continued: "As regards matter many arguments point to the conclusion that its ultimate particles always carry electric charges and that these are not merely accessory but very essential. We should introduce what seems to me an unnecessary dualism, if we considered these charges and what else there might be in the particles as wholly distinct from each other" (Lorentz 1952, 45). Lorentz's electron was, however, different from Abraham's rigid particle. It was a deformable electron, meaning that it would contract in the direction of motion and thus acquire an ellipsoidal shape instead of the spherical shape it had at rest. Abraham objected to this feature of Lorentz's theory because the stability of the electron would require some nonelectromagnetic force; this, Abraham claimed, went against the spirit of the electromagnetic worldview. Exactly what this spirit was could, however, be debated; not all physicists saw in Abraham's theory the essence of the electromagnetic worldview. Thus, in 1908 Minkowski considered the rigid electron to be "a monster in relation to Maxwell's equations" and wittily remarked that "approaching Maxwell's equations with the concept of the rigid electron seems to me the same thing as going to a concert with your ears stopped up with cotton wool" (Miller 1981, 350).

For the variation in mass, Lorentz deduced the famous inverse-square-root expression that is today exclusively associated with Einstein's theory of relativity: $m = m_0(1 - \beta^2)^{-\frac{1}{2}}$ or, approximately, $m = m_0(1 + 1/2\ \beta^2)$. It is readily seen that the practical difference between Lorentz's and Abraham's formulae is small and that it requires electrons at very high speeds to distinguish experimentally between them.

Abraham's rigid and Lorentz's deformable electrons were the most important models, but not the only ones. The German physicist Adolf Bucherer, and independently Paul Langevin in Paris, proposed in 1904 yet another model, characterized by an invariant volume under motion so that the electron's contraction in one dimension would be compensated by inflation in the other. The Bucherer-Langevin theory led to a mass-velocity relationship different from both Abraham's and Lorentz's. In order to decide among the different theories—including Einstein's, which was usually seen as a variant of Lorentz's—appeal had to be made to the precision experiments. Before we turn to the experiment-theory issue, we shall consider some of the wider aspects of the electromagnetic world picture.

Electron Theory as a Worldview

By 1904 the electromagnetic view of the world had taken off and emerged as a highly attractive substitute for the mechanical view that was widely seen as outdated, materialistic, and primitive. As an indication of the strength of the new theory, it was not only discussed in specialized journals, but also began to appear in physics textbooks. For example, Bucherer introduced his electron theory in a 1904 textbook. Of more importance was the textbook in electrodynamics that Abraham published the same year and which, in its several editions, became widely used both in Germany and abroad during more than twenty years. The work was a revision of a textbook on Maxwell's theory written by August Föppl in 1894 (used by, among others, the young Einstein), but whereas Föppl had given a mechanical derivation of Maxwell's equations, Abraham used his revised version to reverse Föppl's priority between mechanics and electromagnetism. In a companion volume of 1905, Abraham came out without much reserve as a missionary for the electromagnetic worldview.

In commemoration of the centenary of the United States' purchase of the Louisiana Territory, a Congress of Arts and Sciences was held in St. Louis in September 1904. Among the physics delegates were several international leaders of physics, including Rutherford, Poincaré, and Boltzmann. The general message of many of the addresses was that physics was at a turning point and that electron theory was on its way to establishing a new paradigm in physics. In his sweeping survey of problems in mathematical physics, Poincaré spoke of the "general ruin of the principles" that characterized the period. Poincaré was himself an important contributor to electron theory and he was now willing to conclude that "the mass of the electrons, or, at least, of the negative electrons, is of exclusively electro-dynamic origin . . . [T]here is no mass other than electro-dynamic inertia" (Sopka and Moyer 1986, 292). The address of another French physicist, thirty-two-year-old Paul Langevin, was more detailed, but no less grand, no less eloquent, and no less in favor of the electromagnetic world picture. Langevin argued for his own (and Bucherer's) model of the electron, but the detailed structure of the electron was not what really mattered. The important thing was the coming of a new era of physics. As Langevin explained in his closing words:

> The rapid perspective which I have just sketched is full of promises, and I believe that rarely in the history of physics has one had the opportunity of looking either so far into the past or so far into the future. The relative importance of parts of this immense and scarcely explored domain appears different to-day from what it did in the preceding century: from the new point of view the various plans arrange themselves in a new order. The electrical idea, the last discovered, appears to-day to dominate the whole, as the place of choice where the explorer feels he can found a city before advancing into new territories. . . . The actual tendency, of making the electromagnetic ideas to occupy the preponderating place, is justified, as I have sought to show, by the solidity of the double base on which rests the idea of the electron [the Maxwell equations and the empirical electron]. . . . Although still very recent, the conceptions of which I have sought to give a collected idea are about to penetrate to the very heart of the entire physics, and to act as a fertile germ in order to crystallize around it, in a new order, facts very far removed from one another. . . . This idea has taken an immense development in the last few years, which causes it to break the framework of the old physics to pieces, and to overturn the established order of ideas and laws in order to branch out again in an organization which one foresees to be simple, harmonious, and fruitful. (ibid., 230)

Evaluations similar to Langevin's, and often using the same code words and imagery, can be found abundantly in the literature around 1905. They rarely included references to quantum theory or the new theory of relativity.

The methodology behind the electromagnetic research program was markedly reductionistic. Its aim was to establish a unitary theory of all matter and forces that exist in the world. The basis of the theory was Maxwellian electrodynamics, possibly in some modified or generalized version. It was an enormously ambitious program. When it was completed, nothing would be left unexplained—at least in principle. In this sense, it was clearly an example of a "theory of everything." Elementary particles, atomic and quantum phenomena, and even gravitation were held to be manifestations of that fundamental substratum of the world, the electromagnetic field. Attempts to explain gravitation in terms of electromagnetic interactions went back to the 1830s, when such a theory was proposed by the Italian physicist Ottaviano Mossotti. Later in the century the idea was developed in mathematical details by the Germans Wilhelm Weber and Friedrich Zöllner, who based their theories on the notion of electrical particles in instantaneous interaction. Electrogravitational theories continued to be discussed after field electrodynamics became the dominant framework of electricity and magnetism. For example, in 1900 Lorentz derived, on the basis of his electron theory, a gravitational law that he considered a possible generalization of Newton's. Attempts to unify the two basic forces of the universe, usually by reducing gravitation to electromagnetism, was part of the electromagnetic program, but in spite of much work, no satisfactory solution was found.

The electromagnetic worldview was also a matter of interest outside physics and was ocassionally discussed by philosophers. It even entered politics, as illustrated by Lenin's *Materialism and Empiriocriticism*, a political-philosophical work written while Lenin was an emigré in Geneva and London in 1908. In his attempt to formulate a dialectical-materialistic conception of nature, Lenin quoted passages from Poincaré, Righi, Lodge, and other physicists to the effect that physics was in a state of crisis. "The electron," wrote the future leader of the Soviet Union, "is as *inexhaustible* as the atom; nature is infinite, but it *exists* infinitely." What Lenin meant by this is not quite clear, but then perhaps it was not intended to be clear.

The revolutionary atmosphere in theoretical physics is further illustrated by a meeting of the German Association of Scientists and Physicians that took place in Stuttgart in 1906, that is, after the introduction of Einstein's relativity. Most of the leading electron theorists participated and the general opinion was in favor of the electromagnetic worldview in its pure form as developed by Abraham, for instance. Lorentz's theory was criticized for its reliance on a nonelectromagnetic stabilizing force; only Planck defended Lorentz's (and Einstein's) theory against this theoretical objection. The rigid electron was incompatible with the relativity postulate, Planck argued, and this spoke in favor of the Lorentz theory. To Abraham and his allies, it spoke against it. Among the allies was thirty-seven-year-old Arnold Sommerfeld, who would later become a leader of quantum and atomic physics, but at that time had specialized in the fashionable electron theory. Sommerfeld made it clear that he considered the Lorentz-Einstein theory hopelessly conservative, an attempt to save what little could be saved of the old and dying mechanistic worldview. Although Planck admitted that the electromagnetic program was "very beautiful," it was still only a program, he countered, and one that could hardly be worked out satisfactorily. To Sommerfeld, such an attitude was unwarranted "pessimism."

Generational revolts are often parts of revolutionary movements and, according to Sommerfeld, the new paradigm in physics held special appeal to the younger generation: "On the question of principles formulated by Mr. Planck," Sommerfeld said, "I would suspect that the gentlemen under forty will prefer the electrodynamic postulate, those over forty the mechanical-relativistic postulate" (Jungnickel and McCormmach 1986, 250). The generalization was probably fair enough, although there were exceptions. One of them was Einstein, Sommerfeld's junior by ten years. Another was Lorentz, well above the forty-year-old limit. Less than a decade after the Stuttgart meeting, the situation between revolutionaries and conservatives had reversed. Now, Abraham wrote in 1914, "physicists of the old school must shake their heads in doubt on this revolution in the conception of mass. . . . The young mathematical physicists who filled the lecture halls in the epoch of its influence were enthusiastic for the theory of relativity. The physicists of the older generation, . . . mainly regarded with scepticism the bold young men who undertook to overthrow the trusted foundations of all physical measurement on the basis of a few experiments which were still under discussion by the experts" (Goldberg 1970, 23).

One of the important issues in the worldview discussion was the status of the ether. There was a wide and confusing variety of views about the ether and its relationship to the electron theory, but only a minority of physicists wanted to do without the ether. The majority view around 1905 seems to have been that the ether was an indispensable part of the new electron physics—indeed, another expression of the electromagnetic field. For example, in his Columbia University lectures of 1906, Lorentz spoke of the ether as "the receptacle of electromagnetic energy and the vehicle for many and perhaps all the forces acting on ponderable matter . . . we have no reason to speak of its mass or of forces that are applied to it" (Lorentz 1952, 31). The ether survived the attack on the "old physics," but it was a highly abstract ether, devoid of material attributes. Thus, in 1909 Planck wrote that "Instead of the so-called free ether, there is absolute vacuum . . . I regard the view that does not ascribe any physical properties to the absolute vacuum as the only consistent one" (Vizgin 1994, 17). Planck was not the only one to use the term "vacuum" synonymously with "ether." From this position, there is but a small step to declare the ether nonexistent. This was precisely the conclusion of the German physicist Emil Cohn, a specialist in electrodynamics who developed his own ether-free version of electron theory between 1900 and 1904. One could consistently deny the relativity principle and the ether, as Cohn did; or deny the relativity principle and accept the ether, as Abraham did; or accept both the relativity principle and the ether, as Lorentz did; or accept the relativity principle and deny the ether, as Einstein did. No wonder many physicists were confused.

Mass Variation Experiments

The answer to the confusion was obviously experiments. In principle, at least, it should be possible to test the predictions of the various theories and, in this way, decide which came nearest to the truth. In fact, the first experiments made in order to determine the distribution between mechanical and electromagnetic mass had already been performed at the time Abraham proposed his electron model. Walter Kaufmann, the Göttingen physicist who in 1897 had measured the charge-to-mass ratio of cathode rays simultaneously with Thomson, started a series of experiments in 1900 in which he deflected beams of electrons in electric and magnetic fields. In order to find the velocity-dependent part of the electron's mass (that is, the electromagnetic mass), electrons of extreme speeds were necessary; for this purpose, Kaufmann used beta rays, which had recently been identified with electrons moving with speeds up to 90 percent or more of the velocity of light.

Kaufmann, who was not only an excellent experimenter but also an able theorist, was an early advocate of electron theory and the electromagnetic world picture. There is little doubt that his enthusiasm for the new physics colored his interpretations of his experimental data. In 1901 he concluded that about one-third of the electron's mass was of electromagnetic origin. When he was confronted with Abraham's new theory, he quickly reinterpreted his data and managed to come to a quite different conclusion, namely, that the entire mass of the electron was electromagnetic, in agreement with Abraham's view. Kaufmann and Abraham were colleagues at Göttingen University and this personal relationship, as well as Kaufmann's deep sympathy for the electromagnetic worldview, led Kaufmann to claims that were not justified by his data alone. In 1903 he concluded that not only beta ray electrons, but also cathode ray electrons, behaved in accordance with Abraham's theory.

After 1905, when Lorentz's theory (or the "Lorentz-Einstein theory") appeared as a competitor to Abraham's, Kaufmann performed new experiments in order to settle the question of the variation of mass with velocity. The complex experiments led to results that apparently refuted Lorentz's theory, but agreed reasonably well with Abraham's and also with the predictions of Bucherer and Langevin. In light of the new data, Kaufmann concluded that the attempt to base physics on the relativity postulate "would be considered as a failure." His experiments were eagerly discussed at the 1906 Stuttgart meeting, where most of the participants were on Kauffman's side, for the rigid electron and against the deformable electron and the relativity principle. It was realized, however, that the correct interpretation of the experiments was far from straightforward, and Planck warned that there were many uncertainties and questionable points in Kaufmann's analysis. He therefore suggested that the experiments were unable to decide between the theories of Abraham and Lorentz and that new experiments were needed.

The reactions of the theorists differed. Abraham was happy to accept Kauffman's conclusion, and Lorentz, much less happily, accepted it too. Lorentz felt that his entire theory was threatened—indeed, proved wrong. As he wrote to Poincaré, "Unfortunately my hypothesis of the flattening of electrons is in contradiction with Kaufmann's new results, and I must abandon it. I have, therefore, no idea of what to do" (Miller 1981, 337). Philosophically speaking, Lorentz acted as a "falsificationist," in accordance with Karl Popper's philosophy of science. It was not so with Einstein, whose response was rather in agreement with the recommendations made by the philosopher Imre Lakatos. Einstein at first ignored Kaufmann's alleged refutation, but suspected that errors had entered the reduction of data and their interpretation. Yes, the experiments disagreed with the theory of relativity but, no, that did not imply that the theory was wrong: It must mean that the experiments were wrong. "In my opinion," Einstein wrote in 1907, "both theories [Abraham's and Bucherer-Langevin's] have a rather small probability, because their fundamental assumptions concerning the mass of moving electrons are not explainable in terms of theoretical systems which embrace a greater complex of phenomena" (ibid., 345). Einstein's attitude (and Planck's, too) was not simply to deny the validity of Kaufmann's experiment because it disagreed with a favored theory. Einstein and Planck went deeply into the experimental details, analyzed the entire situation carefully, and then concluded that there were good reasons to suspect systematic errors.

In any case, Einstein's youthful confidence in his theory soon turned out to be warranted. In experiments of 1908, Bucherer, as competent as Kaufman in theory and experiment, measured the electric and magnetic deflection of beta rays in a manner different from Kaufmann's. His results were different too, leading to a confirmation of the Lorentz-Einstein theory. In regard of the fact that Bucherer had himself proposed a rival theory of electrons, which had received some support from Kaufmann's experiments, it is noteworthy that he did not hesitate in criticizing these experiments and conclude from his own that only Lorentz's theory was viable. At that time, Bucherer had lost confidence in his own theory of the constant-volume electron because it was contradicted by dispersion phenomena. He therefore considered his experiment to be a test between only two alternatives, Abraham's and Lorentz's. Bucherer's experiments were far more transparent than Kaufmann's and more difficult to criticize. But of course they were not beyond criticism, and did not make up a crucial experiment in favor of the Lorentz-Einstein theory. During the following years experiments continued, mostly in Germany, and it took some time until the experimental situation stabilized with the result that it became generally accepted that the Lorentz-Einstein mass variation was experimentally confirmed. By 1914, the question was largely settled. It was not a complete stabilization, though, and discussions continued many years after Einstein's theory had become accepted and the electromagnetic view of nature fallen into oblivion.

We shall not carry on with this story except to mention that the question of the mass variation acquired a political dimension after 1920, when antirelativism began to flourish in parts of German cultural life. Many conservative physicists longed for the day when Einstein's theory would be replaced by a theory based on the ether or electromagnetic concepts. Ironically, among these conservatives were some of the former advocates of the electromagnetic worldview who, years earlier, had considered themselves bold progressives in their fight against old-fashioned mechanicism. Abraham, once an arch-revolutionary, was one of those who never accepted relativity and protested against an ether-free physics. Bucherer, another of the electromagnetic revolutionaries and unwillingly a contributor to the victory of relativity, became in the 1920s an ardent antirelativist on the right wing of German physics. In this situation, new experiments were made on the electron's variation of mass, sometimes with the clear intention of refuting the theory of relativity. And indeed, some of the antirelativists concluded that their experiments confirmed Abraham's old theory, while refuting Einstein's. Other physicists, including Bucherer, accepted the Lorentz-Einstein formula but were careful to distinguish between the formula and Einstein's theory. "Today," Bucherer wrote in 1926, "the confirmation of the Lorentz formula can no longer be adduced as proving the Einsteinian theory of relativity" (Kragh 1985, 99).

Decline of a Worldview

The electron experiments were one factor in the decline of the electromagnetic worldview, but they formed only one factor among many. Whereas experiments are subject to experimental testing, worldviews are not. In fact, the possibility of a fully electromagnetic physics, comprising all aspects of physical reality, was never disproved, but rather melted away as it gradually lost its early appeal. It is difficult to be a revolutionary for a longer period, especially if the high hopes of a better future show no sign of becoming reality. By 1914, at the latest, the electromagnetic worldview had lost its magic and the number of its advocates diminished to a small crowd at the periphery of mainstream physics. Shortly before the outbreak of the World War I, Emil Warburg edited a book in a series called *Contemporary Culture* (*Kultur der Gegenwart*), with thirty-six summary articles on different fields of physics. Among the authors of this semiofficial work were leading physicists within the German culture, including Wiechert, Lorentz, Rubens, Wien, Einstein, Kaufmann, Zeeman, and Planck. The content of the book may give some indication of the composition of physics at the time:

Mechanics	79 pages
Acoustics	22 pages
Heat	163 pages
Electricity	249 pages
Optics	135 pages
General Principles	85 pages

The chapters on heat included not only blackbody radiation, but also Einstein's article on "Theoretical Atomistics," that is, the kinetic theory of matter. Among the thirteen articles in the category of electricity, there was one about wireless telegraphy, one about x-rays, and two dealing with radioactivity. The article on the theory of relativity, written by Einstein, was placed in the General Principles category. The electromagnetic worldview was barely visible and entered only indirectly in Lorentz's article on "The Maxwell Theory and the Electron Theory."

Why did the electromagnetic program run out of power? The process of decline was a complex one that involved both scientific reasons and reasons related to changes in the period's cultural climate. As mentioned, experimental predictions, such as those arising from Abraham's theory, did not agree with the results of actual experiments. On the other hand, this was hardly a major cause for the decline, for the rigid electron was not a necessary ingredient of the electromagnetic worldview. More important was the competition from other theories that were either opposed to the electromagnetic view or threatened to make it superfluous. Although the theory of relativity was sometimes confused with Lorentz's electron theory or claimed to be compatible with the electromagnetic worldview, about 1912 it was evident that Einstein's theory was of a very different kind. It merely had nothing to say about the structure of electrons and with the increasing recognition of the relativistic point of view, this question—a few years earlier considered to be essential—greatly changed in status. To many physicists, it became a pseudo-question. As the rise of relativity theory made life difficult for electromagnetic enthusiasts, so did the rise of quantum theory. Around 1908, Planck reached the conclusion that there was a fundamental conflict between quantum theory and the electron theory, and he was cautiously supported by Lorentz and other experts. It seemed that there was no way to derive the blackbody spectrum on a purely electromagnetic basis. As quantum theory became more and more important, electron theory became less and less important. The worst thing that can happen to a proclaimed revolution is that it is not needed.

In general, electron theory had to compete with other developments in physics that did not depend on this theory, and after 1910 new developments in physics attracted interest away from the electron theory. So many new and interesting events occurred, so why bother with the complicated and overambitious attempt to found all of physics on electromagnetic fields? Rutherford's nuclear atom, isotopes, Bohr's atomic theory, the diffraction of x-rays by crystals, Stark's discovery of the electric splitting of spectral lines, Moseley's x-ray-based understanding of the periodic system, Einstein's extension of relativity to gravitation, and other innovations absorbed the physicists' intellectual energy and left the electromagnetic worldview behind. It was a beautiful dream indeed, but was it physics? Much progress took place in atomic physics and as the structure of the atom became better understood, it became increasingly difficult to uphold the electromagnetic view. The positive charge had been a problem in atomic theory since 1896, when the Zeeman effect indicated that electrons carry a negative charge. A theory of matter in harmony with the electromagnetic worldview needed positive electrons, but none were found. In 1904 a British physicist, Harold A. Wilson, suggested that the alpha particle was the positive electron ("exactly similar in character to an ordinary negative electron") and that its mass was of purely electromagnetic origin, that is, it satisfied the expression $2/3\ e^2/R^2c$. The price to be paid was that the alpha particle would then have a radius several thousand times less than that of the negative electron. With the identification of alpha particles with helium ions, the price turned out to be too high. Other suggestions to introduce positive electromagnetic electrons were no more successful, and with Rutherford's discovery of the true positive particles—atomic nuclei of mass many thousand times the electron's—the attractive symmetry between positive and negative elementary charges was lost.

On the other hand, advances in atomic physics could be, and in fact sometimes were, interpreted to be in qualitative agreement with the electromagnetic view. Rutherford himself was inclined to this view, that all matter was electrical in nature, and in 1914 he argued that the smallness of the hydrogen nucleus was just what was to be expected if it was entirely of electromagnetic origin. Other early researchers in nuclear physics followed him, and it was only in the 1920s that electromagnetic mass disappeared from this area of physics. This does not mean, however, that Rutherford and his colleagues were adherents of the electromagnetic worldview or electron theory in the Continental sense. It is no accident that most of the contributors to electron theory were Germans and very few, if any, were British. There were subtle differences between the British and the German attitudes. It was one thing to believe in the electric theory of matter, as many British physicists did, but another to eliminate all mechanical concepts and laws in favor of electromagnetic ones such as those Abraham and his allies aimed at. In any case, the hypothesis that atomic nuclei were of electromagnetic origin rested on faith, and not on experimental evidence. Contrary to electrons, nuclei or ions could not be accelerated to the extreme speeds that were necessary in order to test what part of their mass was electromagnetic and what part mechanical.

Unified Field Theories

By 1910, the original electromagnetic program of Wien, Abraham, and their allies had run into serious troubles, not least because of its opposition to the progressive theory of relativity. There was not necessarily any contradiction between the electromagnetic view and special relativity, however, and in the second decade of the century a few physicists revitalized the electromagnetic program by incorporating into it Einstein's theory of relativity. Einstein himself was greatly interested in this kind of unified electromagnetic theory, which also attracted the interest and respect of several other mathematical physicists, including Hilbert, Pauli, Sommerfeld, Born, and Weyl. The German physicist Gustav Mie, at the University of Greifswald, was the most productive and eminent of the contributors to the later phase of the electromagnetic program. Contrary to Abraham and most other supporters of the program's earlier phase, Mie accepted the theory of relativity and made full use of its concepts and mathematical methods. For example, Lorentz invariance and the notion of a four-dimensional space-time were important parts of Mie's theory. Yet, Mie fully agreed with the physical view of his predecessors—that ultimately, the world consists of structures in an electromagnetic ether. The following description from 1911 gives the essence of the electromagnetic worldview and shows that Mie's view was basically the same as that of earlier physicists such as Larmor, Wien, Lorentz, and Abraham:

> Elementary material particles . . . are simply singular places in the ether at which lines of electric stress of the ether converge; briefly, they are "knots" of the electric field in the ether. It is very noteworthy that these knots are always confined within close limits, namely, at places filled with elementary particles. . . . The entire diversity of the sensible world, at first glance only a brightly colored and disordered show, evidently reduces to processes that take place in a single world substance—the ether. And the processes themselves, for all their incredible complexity, satisfy a harmonious system of a few simple and mathematically transparent laws. (Vizgin 1994, 18 and 27)

Mie's theory, which he developed in three long papers in 1912–13 (comprising a total of 132 pages), was primarily a theory of elementary particles. He wanted to include gravitation in his theory as well, but did not succeed in accounting for the gravitational field by means of the same electromagnetic equations that he used in his theory of matter.

Mie believed that the electron was a tiny portion of the ether in "a particular singular state" and pictured it as consisting of "a core that goes over continuously into an atmosphere of electric charge which extends to infinity" (Vizgin 1994, 28). That is, strictly speaking, the electron did not have a definite radius. Near the center of the electron, the strength of the electromagnetic fields would be enormous and Mie thought that under such conditions, the Maxwell equations would no longer be valid. He therefore developed a set of generalized nonlinear electromagnetic equations that, at relatively large distances from the electron's core, corresponded to the ordinary equations of electromagnetism. From Mie's fundamental equations, it was possible to calculate the charges and masses of the elementary particles as expressed by a certain "world function." This was a notable advance and the first time that a field model of particles had been developed in a mathematically precise way. Interestingly, in developing his formalism Mie made use of matrix methods, a branch of applied mathematics that would later enter quantum mechanics. The advance was limited to the mathematical program, however, and the grandiose theory was conspicously barren when it came to real physics. By 1913, two elementary particles were known—the electron and the proton—and their properties could be derived from the theory *in principle*. Alas, as yet this was in principle only, for the form of the world function that entered the equations was unknown. On the other hand, neither could it be proved that there were no world functions compatible with the existence of electrons and protons, and believers in Mie's program could therefore argue that future developments might lead to success.

Although Mie's theory never led to the results hoped for, it was not without consequences. It influenced some of the works made by Einstein, Hilbert, Weyl, and others, and as late as the 1930s, Max Born reconsidered Mie's theory as a candidate for a classical framework of developing a consistent quantum electrodynamics. What appealed to some mathematical physicists around 1920 was not so much the details of Mie's theory as its spirit and methods. As Weyl expressed it in 1923: "These [Mie's] physical laws, then, enable us to calculate the mass and charge of the electrons, and the atomic weights and atomic charges of the individual elements whereas, hitherto, we have always accepted these ultimate constituents of matter as things given with their numerical properties" (Vizgin 1994, 33).

The aim of a unified theory is to understand the richness and diversity of the world in terms of a single theoretical scheme. The mass and charge of the electron, for example, are usually considered to be contingent properties, that is, quantities that just happen to be what they are ("things given"). They do not follow uniquely from any law of physics and for this reason, they could presumably be different from what they are. According to the view of the unificationists, the mass and charge of the electron (and, generally, the properties of all elementary particles) must ultimately follow from theory—they must be turned from contingent into law-governed quantities. Not only that, but the number and kinds of elementary particles must follow from theory too; not merely those particles that happen to be known at any given time, but also particles not yet discovered. In other words, a truly successful unified theory should be able to predict the existence of elementary particles—no more than exist in nature, and no less. This is a formidable task, especially because physical theories cannot avoid relying on what is known empirically and thus must reflect the state of art of experimental physics. In 1913 the electron and the proton were known, and thus Mie and his contemporaries designed their unified theories in accordance with the existence of these particles. But the impressive theories of the electromagnetic program had no real predictive power. For all its grandeur and advanced mathematical machinery, Mie's theory was a child of its age and totally unprepared for the avalanche of particle discoveries that occurred in the 1930s. When Gustav Mie died in 1957, the world of particles and fields was radically different from what it had been in 1912. It was much more complex and much less inviting to the kind of grand unified theories that he had pioneered in the early part of the century. A couple of decades after Mie's death, on the other hand, grand unified theories would come to the forefront of physics and promise a unified worldview which, in a general sense, shared some of the features of the old electromagnetic program. For more on these modern unified theories, see chapter 27.

Chapter 9

PHYSICS IN INDUSTRY AND WAR

INDUSTRIAL PHYSICS

PHYSICS IS NOT only academic physics taught and cultivated at universities and similar institutions with the purpose of penetrating still deeper into nature's secrets. A large and important part of physics consists of applying physical knowledge for technological purposes, or rather developing new kinds of concepts, theories, and apparatus that can be used for such purposes. Applied or engineering physics has often been essential for technological progress, but it is only in rare cases that new technologies arise directly from scientific knowledge. In most cases of so-called science-based technologies, the connection between basic science and technology is complex and indirect.

The usefulness of physics is not, of course, a modern issue. Quite the contrary, the idea that physics may or will lead to technological innovations beneficial to humankind is an integral part of the history of physics since the days of Bacon and Galileo. Yet it was only in the second half of the nineteenth century that the rhetoric was followed by practice to an appreciable extent. As the century neared its close, it became still more evident that physics might have technological potentials of the same magnitude as chemistry, the science that had earliest and most visibly proved its character as a productive force. This was an important element in the increased support of physical research at the time, and it was a view shared by many physicists. One of them was Emil Warburg, the director of a new institute of physics at the University of Freiburg. On the occasion of the inauguration of the institute in 1891, he argued: "As far as physics is concerned, the so-called rise of the natural sciences, which characterizes modern times, lies not in the number and significance of discoveries or principles of research. It is due much more to the greatly increased effect which this science exerts on civil life and on branches of technology dependent on it. And, as we must add, to the countereffects which thereby result" (Cahan 1985, 57). From that time onward, an increasing number of physicists began to work with applied, technical, or industrial physics, either in private industry, in polytechnical institutes, or in nonprofit research institutions. "Industrial physics" took off in the 1920s, but even before the outbreak of World War I, this and other kinds of nonacademic physics were firmly established in the large countries.

The areas of physics that were most suited for industrial application were classical fields such as thermodynamics, electromagnetism, optics, and materials science; the "new physics"—fields like radioactivity, relativity, and quantum and atomic theory—did not immediately result in practical applications on a large scale. But even a new and exotic field like radioactivity was of industrial interest, and work with radiation standards became an important part of the early laboratories specializing in radioactive measurements. Marie Curie is traditionally pictured as the heroine of pure physics, but the purity of her work did not prevent her from being quite concerned with establishing an international radium standard and securing links with the French radium industry. For Curie, industrially oriented work was a natural and integrated part of her scientific work. From about 1905, there were close ties between the laboratory of the Curies and the company of Armet de Lisle, a chemical industrialist. The industrial dimension extended to the laboratory and the courses on radioactivity offered by the Institut du Radium in Paris, which attracted many scientists working in industry. "It was not uncommon for Curie's co-workers to lead a double life in science and industry," a recent study observed (Roqué 1997b, 272). The other major center of early radioactive research, the Institute of Radium Research in Vienna founded in 1908, served industrial and metrological functions similar to those of the Paris institute. The Viennese physicists cooperated with the Auer Company, which processed uranium minerals in order to obtain the valuable radium (in the form of radium chloride) that was of such great interest to both science and industry.

The Physikalisch-Technische Reichsanstalt in Berlin, founded in 1887 on the initiative of Werner von Siemens and Hermann von Helmholtz, was perhaps the most important of the early applied physics institutions. The aim of the Berlin institute was to apply physics to technical innovations useful to German industry and thus, by implication, the country. For, as Siemens wrote to the Prussian government, "In the competition between nations, presently waged so actively, the country that first sets foot on new [scientific] paths and first develops them into established branches of industry has a decisive upper hand" (Hoddeson et al. 1992, 13). Helmholtz, the first president of the Reichsanstalt, was eager to make it a research institution and not merely a place where established knowledge was passively applied to solve industrial problems. His successor, Friedrich Kohlrausch, who served from 1895 to 1905, greatly expanded the institution and turned it into the world's foremost center for applied physics and precision measurements. In 1903 the Reichsanstalt consisted of ten buildings and employed 110 individuals, of whom 41 were occupied with scientific work. The importance of the Reichsanstalt in promoting and assisting high-tech industries was also recognized abroad and other nations established similar institutions. The National Physical Laboratory in England (1898), the National Bureau of Standards in the United States (1901), and the Institute of Physical and Chemical Research in

Japan (1917) were all founded as imitations of the Reichsanstalt. Although much original research, both basic and applied, was done at the institute, practical tasks such as testing were of more economic importance and took up a large part of the resources. An enormous number of tests were made, especially of thermometers and electric lamps (for example, between 1887 and 1905, almost 250,000 thermometers were tested). Such routine work contributed to the scientific decline of the Reichsanstalt and the departure of talented scientists for the universities, which was a problem especially during the presidency of Emil Warburg (1905–22).

The application of physics in private industries was most advanced in the United States, where several industrial research laboratories were founded in the early part of the century, primarily in the electrical industry. Contrary to the traditional laboratories for testing and quality control, scientific research had a prominent place in the new laboratories, which were manned with university-trained personnel, often with Ph.Ds. Also contrary to the situation in many European countries, there were close contacts between American academic physicists and those employed by the large industries. Academic physicists frequently worked as consultants for industry, and applied physicists frequently contributed to the scientific-academic literature. The difference between pure and applied physics could sometimes be difficult to distinguish. In 1913 physicists from industrial laboratories made up about 10 percent of the membership of the American Physical Society; seven years later, at a time when total membership had doubled, they made up about 25 percent of the Society's membership. The scientific importance of the industrial laboratories is further illustrated by the distribution of papers in *Physical Review*. In 1910, only 2 percent of the papers had their origin in industrial laboratories; five years later, the proportion had risen drastically to 14 percent, and in 1920 the figure was 22 percent. By far the most research-intensive companies were the two electrotechnical giants, General Electric and American Telephone and Telegraph Company (AT&T), which contributed more to pure physics than many universities. The scientific strength of the two companies is illustrated by figures from the years 1925–28, when 27 contributions to *Physical Review* came from General Electric and 26 from the Bell Telephone Laboratories, the research branch of AT&T. In comparison, Columbia University contributed 25 papers and Yale 21 (California Institute of Technology ranked first, with 75 papers). At that time, Bell Laboratories was the world's largest and richest institution for industrial research. With 3,390 technical personnel and a research staff of about 600 scientists and engineers, Bell Laboratories had initiated the era of big science. AT&T expenditures for research and development, measured in 1948 dollars, grew from $6.1 million in 1916 to $10.8 million in 1920; in 1926 the expenditures were $16.6 million, and in 1930 the figure had increased to $31.7 million. (Large as these figures were, they were only a prelude to the even more explosive development that occurred after 1945. In 1981, shortly before AT&T was ordered by law to break up into several smaller, independent companies, the number of Bell Labs employees was 24,078, of whom 3,328 had Ph.D. degrees. In that year, AT&T supplied Bell Labs with no less than $1.63 billion.)

Although AT&T was exceptional, it was only one of a growing number of American companies that employed scientists. By 1931, more than 1,600 companies reported to the National Research Council about laboratories employing nearly 33,000 people. Nine years later, more than 2,000 corporations had research laboratories, employing a total of some 70,000 people.

European industries followed the trend set by the Americans, but not with the same pace and intensity. The relatively few physicists from industrial laboratories in Europe did not contribute much to the pure physics literature. The largest European electrotechnical company, the Siemens Corporation in Germany, recognized the value of applied scientific research but did not have a central research department comparable to those of its American rivals. It was only in 1920 that such a department was formed under the Göttingen-trained physicist Hans Gerdien. The new research laboratory issued its own journal, called *Wisenschaftlichen Veröffentlichungen aus dem Siemens-Konzern*, which was similar in scope and content to *The Bell System Technical Journal*. However, the Siemens journal did not have the same scientific importance as its American counterpart. We shall now consider two important cases of applied physics, both of them related to aspects of electronics.

Electrons at Work, I: Long-Distance Telephony

Alexander Graham Bell's telephone was not a product of science. During the first two decades of telephony, the new system of telecommunications developed in an empirical way, guided—and often misguided—only by the telegraph theory that William Thomson had pioneered in 1855. When it turned out that good voice transmission over distances of more than a few hundred kilometers was difficult to achieve, engineers suggested reducing the resistance and capacitance of the line system. Self-inductance was generally believed to be a harmful quantity, which could be reduced by changing from iron to copper wires. However, these ways of extending the speaking distance had a rather limited effect, and by the mid-1890s it became increasingly clear that the empirical methods of the telegraph engineers would never produce good-quality telephony over long distances. It was time to apply more scientific procedures based on physics.

In fact, such procedures had been applied as early as 1887, when Aimé Vaschy in France and Oliver Heaviside in England analyzed theoretically the transmission of telephone currents by means of fundamental electrical theory. They obtained formulas for the variation of the attenuation and distortion with the line's electrical parameters and concluded that self-inductance was beneficial: The more self-inductance in a line, the better. The works of Vaschy and Heaviside did not result in practical methods until much later, and in England, Heaviside's scientifically based recommendations were resisted by engineers within the powerful Post Office. It was in the United States and, on a smaller scale, in Denmark that the Vaschy-Heaviside theory was first transformed into a practical technology.

Scientific research was not given high priority in the early years of the Bell Telephone Company, but Hammond Hayes, a university-trained physicist employed by Bell, urged the company to place more emphasis on research. In 1897 he hired George Campbell to develop the kind of high-inductance cables that Heaviside had suggested. In his approach and training, Campbell differed from the engineers and technicians who used to work in telegraphy and telephony, and this difference was of crucial importance to his success. Campbell had a strong background in theoretical physics, obtained at MIT, Harvard, and three years in Europe, where he had studied under luminaries such as Poincaré, Boltzmann, and Felix Klein. Thoroughly familiar with the works of Vaschy and Heaviside, Campbell used his mathematical skills to develop a theory of cables (or aerial wires) "loaded" with discretely spaced inductance coils. The result, completed in the summer of 1899, was a mathematical analysis of a coil-loaded line, which predicted where to place the coils in the circuit, how heavily to load them, and the most efficient distribution of copper between the cable and the coils. Campbell realized that the technological solution could not be obtained through empirical methods and that it would be in the company's interest to attack the problem by means of theoretical physics. As he wrote in a memorandum of 1899, "It is economy to carry theory as far as possible and leave as little as possible for cut and try methods" (Kragh 1994, 155). Campbell's innovation had a practical aim, but it was no less scientific for that. He published his loading theory in 1903, not in an engineering journal, but in the distinguished physics journal *Philosophical Magazine*.

Results very similar to Campbell's were obtained independently at the same time by Michael Pupin, a Serbian-born professor of electrical engineering at Columbia University. Like Campbell, Pupin had a strong background in theoretical physics (he had studied under Kirchhoff and Helmholtz) and had no confidence in the trial-and-error methods that still dominated much of the engineering profession. He had the advantage over Campbell that he was a free university researcher, not a company man, and for this reason, among others, in 1900 he managed to obtain a patent for his system of coil loading (or "pupinization," as it became known). After an extended legal dispute, AT&T bought the rights to exploit the patent and Pupin's German patent was likewise bought by Siemens & Halske.

The first coil-loaded line was built in 1901; during the following decade, the loading system became the backbone of a large number of long-distance lines, in both America and Europe. The innovation based on the work of Campbell and Pupin was a great commercial success and great propaganda for the usefulness of theoretical physics. By the end of 1907, the group of Bell Telephone companies had equipped 138,000 km of cable circuit with about 60,000 loading coils. Four years later, when the number of coils had increased to 130,000, the company estimated that it had saved about $100 million from the invention of inductively loaded circuits. As physicists and chemists played a leading role in the invention and development of the loading system in the United States, so they did in Europe, where German companies were leaders in the field. The first European telephone line equipped with Pupin coils was made by Siemens & Halske in 1902 and based on experiments made by August Ebeling and Friedrich Dolezalek. Their careers indicate the increased contacts between industry and academic physicists: Ebeling had studied under Helmholtz, worked with Werner von Siemens, and spent five years at the Physikalisch-Technische Reichsanstalt until he became a leading engineer in Siemens & Halske; Dolezalek had studied under Nernst and after a period within Siemens & Halske, he returned to academia to replace his former professor as director of the institute for physical chemistry at Göttingen University.

In the case of Germany, and most other European countries, there was another important actor missing on the American scene, namely, government institutions. German telecommunications technology was supported by the Reichpost, which had its own staff of highly qualified researchers. Its Imperial Telegraph Test Department, founded in 1888, included Franz Breisig, a former student of Heinrich Hertz and possibly Europe's foremost expert in electrical communications theory.

Pupinization, or coil loading, was the most important method of extending the range of telephony, but was not the only one. For cables, and especially submarine cables, the alternative was to wind the copper wires tightly with soft iron. The idea of continuous loading went back to Heaviside, but it took fifteen years until it was implemented into a practical technology. This was done by a Danish engineer, Carl E. Krarup, who designed the first loaded submarine cable, which was laid in 1902 between Denmark and Sweden. Krarup, who had been a research student under Wilhelm Wien at the university of Würzburg, was another of the new generation of scientifically trained engineers. Continuous cables of Krarup's type were used extensively for three decades—for example, in 1921 on the 190-km Key West-to-Havana cable.

With the success of the loading method, it became important to design the Pupin coils as efficiently as possible, that is, with a maximum magnetic permeability and a minimum loss of energy due to eddy currents. Materials science became of crucial importance to the telephone enterprises. As early as 1902, Dolezalek had obtained a patent on an inductance core made of finely divided iron powder mixed with an insulating binding substance, but the idea was transformed into an innovation only more than a decade later. In 1911, AT&T had organized a branch within its Western Electric Engineering Department with the purpose of applying basic science systematically to problems of telephony. One of the early fruits of the new research program was the 1916 invention of an iron-powder core. These were quickly put into commercial production at Western Electric's manufacturing plants at Hawthorne, Illinois, which by 1921 had a weekly output of 25,000 pounds of iron powder. At the same time, Western Electric researchers worked to find a substitute for iron with superior magnetic properties as a loading material. Swedish-born Gustaf Elmen developed a method of heating and cooling nickel-iron alloys in such a way that their permeabilities were large and their hysteresis loss low; the result was "permalloy," consisting of roughly 20 percent iron and 80 percent nickel. In the 1920s this alloy replaced iron in Pupin cores and by around 1930, AT&T companies produced more than one million permalloy core coils per year.

No less important was the application of permalloy as a loading material for transatlantic telegraphy. A reliable theory for this purpose was developed at Western Electric, primarily by Oliver Buckley, a Cornell-educated physicist who had joined the company in 1914 to work with vacuum tubes and magnetic materials. After much theoretical and experimental work, a realistic theory was obtained that could guide the construction of high-speed loaded telegraph cables. When the 3,730-km-long cable between New York and the Azores, operated by Western Union Telegraph Company and manufactured by a British cable company, was tested in 1924, it fulfilled all expectations. The operating speed of 1,920 letters per minute was four times as high as the record of conventional cables of the Kelvin type. The phenomenal speed outstripped the capacity of existing terminal equipment and necessitated the construction of special high-speed recorders to follow it. The success of this and other permalloy-loaded cables revitalized transoceanic telegraphy and proved, once again, the practical and economic value of physical research. In 1924, reflecting on his long career as a researcher within the Bell system, Campbell concluded that "electricity is now preeminently a field for mathematics, and all advances in it are primarily through mathematics."

Electrons at Work, II: Vacuum Tubes

The electron tube (or radio tube, or valve) is one of the most important inventions of the twentieth century. Its history goes back to 1880, when

Edison noticed that if a plate was melted into one of his newly invented light bulbs, a small current would flow from the filament to the plate. The Edison effect attracted the interest of electrical engineers. One of them, the Englishman John Ambrose Fleming, demonstrated in 1889 that the particles emitted from the filament were negatively charged. A decade later, it was realized that these particles were electrons. In 1904 Fleming, by then professor of electrical engineering at University College in London and technical consultant to Marconi's Wireless Telegraph Company, discovered that lamps using the Edison effect could be used as detectors for high-frequency electromagnetic waves. Fleming constructed the first vacuum tube, a diode that was patented by the Marconi Company in 1905, but the invention was not a success. The diode turned out to be too insensitive to make it practical as a detector in wireless telegraphy; instead, it was the American Lee De Forest who made the first practical vacuum tube. Although initially not a great commercial success, De Forest's invention in 1906 of the triode (or "audion") was the starting point of the electronic age. The Austrian Robert von Lieben, another of the tube pioneers, constructed diodes, and later triodes, to be used as amplifiers in telephony. At that time, the term "vacuum tube" was a misnomer, for neither De Forest, Lieben, nor others recognized the importance of a high vacuum at first. De Forest's original triode was indeed a vacuum tube, but was not highly evacuated; he believed that gas remnants were essential to the proper working of the tube. Lieben's tubes of 1910 were only partially exhausted and operated with a rarefied mercury vapor.

Vacuum tubes were turned into technological wonders in 1912, primarily through the research of private companies. AT&T's expert in the area was Harold Arnold, a young physicist with a fresh Ph.D. taken with Millikan in Chicago. When De Forest demonstrated his triode to the Bell Telephone Company, Arnold realized that the crude device could be developed into a powerful amplifier or relay. He quickly made his own improved version, in particular by exhausting it thoroughly by means of one of the new high-vacuum pumps invented by Wolfgang Gaede in Germany. (Tube electronics, as well as other parts of applied and experimental physics, depended crucially on progress in vacuum technology.) Whereas earlier types of tubes had been "soft"—not very highly exhausted—the work at the Bell laboratories resulted in "hard" triodes, in which the current passed purely thermionically and not by ionization. Other improvements made by Arnold and his team included the replacement of the hot filament by a cathode coated with calcium or barium oxide; such cathodes, first constructed by the German physicist Arthur Wehnelt in 1904, could operate at a lower temperature, and hence extended the lifetime of the tubes. It took less than a year to develop operable tube repeaters. These were tested on the New York-Washington telephone line in 1913, and two years later they proved their commercial worth by securing telephone conversation over the first transcontinental line, between New York and San Francisco. Robert Millikan, who participated in the opening ceremony, wrote in his autobiography: "From that night on, the electron—up to that time largely the plaything of the scientist—had clearly entered the field as a potent agent in the supplying of man's commercial and industrial needs. . . . The electronic amplifier tube now underlies the whole art of communications, and this in turn is at least in part what has made possible its application to a dozen other arts. It was a great day for both science and industry when they became wedded through the development of the electronic amplifier tube" (Millikan 1951, 136). Millikan did not mention that the real breakthrough in vacuum tube technology came as a result of the demand during World War I. When the United States entered the war in 1917, the country's weekly production did not exceed 400 tubes; two years later, the production was about 80,000 per week.

Developments in Germany ran parallel to those in the United States, but less quickly and less smoothly. Four of the largest electrotechnical companies (Siemens & Halske, Allgemeine Elektrizitäts-Gesellschaft, Felten und Guilleaume, and Telefunken) formed a consortium in 1912 that acquired Lieben's patents in order to develop them commercially. The focus on the gas-filled Lieben tubes was a mistake. It took two years for the Germans to realize that the future belonged to high-vacuum tubes; then, Telefunken began production of vacuum triodes. During the war, German efforts to perfect vacuum tube technology accelerated (as they did in other belligerent countries). One of the physicists occupied with tube research was Walter Schottky, a student of Planck who shared his talent between general relativity and electronics. Working part-time for Siemens & Halske, Schottky developed an improved vacuum tube by introducing an extra grid between the anode and the normal control grid. Although Schottky's work in this area did not immediately lead to a practical tube, it was important because it included a study of the fundamental physics of vacuum tubes and led to the discovery of discrete noise effects, which Schottky explained in a now-classic contribution to information theory. Schottky's insight was developed to a great extent by later researchers, particularly Harry Nyquist (in 1928) and Claude Shannon (in 1948), both at Bell Laboratories.

In the United States, tube technology was the concern not only of AT&T, but also of General Electric. In 1912 William Coolidge, a university-trained physicist working at the General Electric Research Laboratory, had developed the first incandescent lamps with tungsten as a filament and had begun work on improving x-ray tubes. His assistant, Irving Langmuir, had studied physical chemistry under Nernst in Göttingen, where he also followed Felix Klein's mathematical lectures. After a few years as instructor at an American polytechnic institute, he came to General Electric in 1909 and stayed there for more than forty years. He was the first corporate scientist to win a Nobel prize (in chemistry, 1932) and probably the only one to have a mountain named after him—Mount Langmuir in Alaska. At General Electric, Langmuir started a research program in electrical discharges in gases and vacuums that led him to suggest that introduction of inert gases into the incandescent bulb would extend the bulb's lifetime. The discovery was quickly turned into a highly profitable innovation. At the same time, he studied the emission of electrons in a vacuum as a mechanism of the vacuum tube. Langmuir realized that the essential process in diodes and triodes was electron emission, which required a high vacuum. Until that time, engineers working with the new tubes had little understanding of the physical mechanisms, although a general theory of electron emission from hot bodies had been published by the British physicist Owen Richardson in works from 1901 to 1903.

Richardson, a student of J. J. Thomson and a specialist in electron theory, coined the term "thermionic" in 1909 and formulated an empirical law relating the thermionic emission rate to the temperature of the metal; during the following decade, he struggled to test and perfect "Richardson's law." Richardson's phenomenological theory came to be recognized as the scientific explanation of vacuum tubes, but during the formative years of the tubes, 1904–1908, its role was insignificant. It was only later that the theory became of fundamental importance in tube electronics. When Richardson was awarded the Nobel prize in 1929, Carl Oseen emphasized, in his presentation speech, the close connection between Richardson's work in pure physics and the amazing progress in communications technology:

> Among the great problems that scientists conducting research in electrotechnique are today trying to solve, is that of enabling men to converse in whatever part of the world each may be. In 1928 things had reached the state when we could begin to establish telephonic communication between Sweden and North America. . . . Every owner of a valve receiving-set knows the importance of the valve in the apparatus—the valve, the essential part of which is the glowing filament. . . . The most important fact was that Mr. Richardson's opinion about the thermion-phenomenon with fixed laws was totally confirmed. Through this fact a solid basis was obtained for the practical application of the phenomenon. Mr. Richardson's work has been the starting-point and the prop of the technical activity which has led to the progress of which I have just spoken.

Oseen's presentation of Richardson's work as the scientific basis from which technological marvels followed was exaggerated. Until about 1913, tube technology relied very little on scientific theory, and Richardson himself showed little interest in technological questions. In his Nobel lecture, he did not mention either the vacuum tube or the electronic technology of which he was supposed to be the scientific father. On the other hand, Langmuir was well acquainted with Richardson's theory, which he used to obtain a thorough understanding of the working of vacuum tubes. In particular, Langmuir confirmed Richardson's expression for the thermionic current and proved conclusively that electron emission did not require a residual gas. Langmuir's scientific understanding of the triode was an important factor in General Electric's development of the tube. The General Electric researchers, led by Langmuir and his colleague Saul Dushman, had their improved high-vacuum tube ready in 1913. One of the results was a protracted patent litigation with AT&T.

The Nobel prize-winning discovery of electron diffraction in 1927 had its starting point in research connected with the Arnold-Langmuir patent suit. Clinton J. Davisson, a Ph.D. student of Richardson's, had joined the Western Electric engineering department in 1917. Two years later he and his assistant, Lester Germer, began work on the thermionic emission of Wehnelt cathodes. Although the work did not influence the outcome of the suit, as Arnold and Davisson had initially believed, it eventually developed in a highly satisfactory and completely unexpected way. In their extended research program, Davisson and Germer were led to study electron emissions from metal surfaces under electron bombardment. In 1924 these electron-electron scattering experiments resulted in patterns that Davisson interpreted as arising from the crystal structure of the metal, but nothing further seemed to follow from the experiments. The Bell physicists did not think of electrons as waves. It was only in the spring of 1926, when Davisson spent a holiday in England, that he happened to learn about Louis de Broglie's ideas of 1924 that electrons might be diffracted by crystals. At that time, Davisson did not know of the new quantum mechanics, but he was told by European colleagues that Schrödinger's theory might be the key to understanding his experiments. On his way back to New York, he studied the new wave mechanics. In a letter to Richardson, he wrote, "I am still working at Schrödinger and others and believe that I am beginning to get some idea of what it is all about. In particular I think that I know the sort of experiment we should make with our scattering apparatus to test the theory" (Russo 1981, 145).

The new direction of the old research program resulted in the famous Davisson-Germer experiment, proving experimentally de Broglie's formula that electrons moving at speed v can be ascribed a wavelength $\lambda = h/mv$. Appropriately, the first full report of the discovery was published in the April 1927 issue of the *Bell Laboratories Record.* Ten years later, Davisson, together with George P. Thomson, received the Nobel prize for the discovery.

Physics in the Chemists' War

Until the outbreak of World War I in August 1914, most scientists considered themselves members of a supranational class, a republic of learning where nationality was less important than scientific accomplishment. When the ideology of supranationalism clashed with the reality of war, however, it broke down almost immediately and was quickly replaced with a chauvinism no less primitive than that hailed by other groups in the European nations. Before the end of 1914, a propaganda war within the real war had started—a war fought on paper by scientists and other scholars. Physicists were no longer just physicists—they were now German physicists, French physicists, Austrian physicists, or British physicists. An important factor in the *Krieg der Geister* (war of the learned) was a manifesto issued in October 1914 in which ninety-three German scientists, artists, and scholars tried to justify the actions of their army, including its attack on neutral Belgium and the destruction of Louvain. In the *Aufruf* or manifesto "To the Civilized World," the self-proclaimed German "heralds of truth" denied that Germany had caused the war and that German soldiers acted without the discipline and honor to be expected from the army of a civilized nation. All such accusations were claimed to be infamous lies. To the German scientists, militarism and culture were inseparably connected: "Were it not for German militarism, German civilization would long since have been extirpated. . . . Have faith in us! Believe, that we shall carry on this war to the end as a civilized nation, to whom the legacy of a Goethe, a Beethoven, and a Kant is just as sacred as its own hearths and homes" (Nicolai 1918, xii). Among the signers of the manifesto were several leading physicists and chemists, including Planck, Nernst, Ostwald, Haber, Lenard, and Röntgen. Einstein, one of the few physicists who opposed the war and the general climate of chauvinism, did not sign. On the contrary, he signed a counter-manifesto in favor of peace and cooperation that a physiologist, Georg Nicolai, had drafted. However, the "Appeal to the Europeans" was not a success: Only four persons signed it.

The German manifesto provoked sharp replies from scientists in England and France. For example, foreign membership of the Paris Academy of Sciences was terminated for German scientists who had signed the manifesto. Soon, French scholars attacked their colleagues across the Rhine and all that they stood for, or what the French thought they stood for. Although the Germans were admittedly good organizers of science, according to the French propaganda they lacked originality and tended to appropriate ideas that originated elsewhere. In some cases, French scientists argued that German science was intrinsically distinct from, and of course inferior to, the science of civilized nations such as France. According to the physicist and chemist Pierre Duhem, *la science allemande* had its own character and was marked by the German race's deplorable mental characteristics. German physicists were just unable to think intuitively and they lacked the common sense that was necessary to tie abstract physical theories to the real world. As a typical example of abstract German theorizing, Duhem mentioned the theory of relativity, with its absurd postulate of the velocity of light as an upper limit of velocity. Another French author, a biologist, singled out the quantum theory as an example of German "mathematical-metaphysical delirium." In a 1916 book, he wrote: "The principle of relativity is the basis of a scientific evolution which can best be compared with futurism and cubism in the arts. . . . We find a good example of this mathematical-metaphysical delirium in the theory of quanta of Max Planck, a professor of physics in Berlin and one of the 93 intellectuals on the other side of the Rhine. Planck . . . introduces . . . atoms of heat, of light, mechanical energy (!), indeed of energy in general; as a result of the theory of relativity these atoms even possess a mass endowed with inertia (!!)" (Kleinert 1978, 520). It is ironic that what French patriotic scientists in 1916 considered to be theories typical of the German mind were the very same theories that Nazi German scientists, twenty years later, would consider the prime examples of a non-German, Jewish mind! The reaction against German science was somewhat more tempered in Britain, but there was no understanding and no attempts at reconciliation were made. In an article in *Nature* in October 1914, William Ramsay, the Nobel laureate in chemistry, wrote that "German ideals are infinitely far removed from the conception of the true man of science." Although Ramsay recognized the brilliance of individual German scientists, as a whole world science could easily do without the Germans: "The greatest advances in scientific thought have not been made by members of the German race," he wrote. "So far as we can see at present, the restriction of the Teutons will relieve the world from a deluge of mediocrity."

When war was declared, many scientists in the belligerent countries offered suggestions to the military as to how they could employ their qualifications in the service of the nation. Political and military authorities were reluctant to accept the offers and, in many cases, chose to ignore them. Traditionally, military officers were skeptical about the usefulness of scientific research, especially if the research was done by civilian scientists. In Germany and the Austro-Hungarian empire, younger scientists were considered ordinary soldiers, who served on the battlefields like any other group of conscripts. In the early phase of the war, there were no plans to employ physicists or other scientists in war projects or otherwise take advantage of their expertise. This only came later and although the recognition of the importance of science increased during the war, it remained at a limited level. The largest and best known of the German war projects, the chemical gas warfare project under the direction of Fritz Haber, employed several physicists. For shorter or longer periods of time, James Franck, Otto Hahn, Erwin Madelung, and Gustav Hertz worked on the project. Hahn experimented with poison gases in the Kaiser Wilhelm Institute in Berlin and developed new gases at the Bayer Chemical Works in Leverkusen. Another military institution where physicists found employment was the Artillery

Testing Commission (Artillerie-Prüfungs-Kommission, or APK) in Berlin. Here, various methods of ranging were investigated—optical, acoustical, seismometric, and electromagnetic. Max Born spent part of the war working at the APK, and so did Alfred Landé, Fritz Reiche, Ferdinand Kurlbaum, and Rudolf Ladenburg. The idea of using sound ranging as a method of determining the position of enemy artillery came from Ladenburg who, as a cavalry captain, was able to persuade the military to set up a group of scientists to work with the problem. Among other things, Born studied the influence of the change of wind with altitude on the propagation of sound by means of methods of Hamiltonian dynamics. "I think it was a good piece of applied mathematics," Born recalled (Born 1978, 171).

Still other German physicists worked with problems of telecommunications, a field that originally was neglected by the German army. Max Wien, the physics professor at the University of Jena, organized a unit of physicists and engineers to work on the use of wireless in airplanes. Sommerfeld, for a while, did research on listening to enemy telephone communication by detecting weak "earth currents" propagated through the ground. Tube-amplified listening devices based on Sommerfeld's research were able to detect signals more than one kilometer from a line and proved to be of considerable value to the German army. (Technical physics was not a foreign field to the theorist Sommerfeld, who had earlier published on electrotechnical problems and the hydrodynamic theory of lubrication.)

In England, the mobilization of science and technology took its start in the summer of 1915 with the Committee of the Privy Council for Scientific and Industrial Research which, the following year, was changed to the Department for Scientific and Industrial Research (DSIR). The aim of the DSIR was to promote, fund, and coordinate scientific and technical research of industrial and military significance. Although it was aimed at the production of useful goods, such as chemicals and optical glass, the DSIR recognized that the goal required investments in pure science. As *Nature* put it in 1916, "The neglect of pure science might be compared with the ploughing and manuring of a piece of land, followed by an omission to sow any seed." The National Physical Laboratory, which since 1902 had been run under the supervision by the Royal Society, was transferred to the new department in 1918. At the outbreak of the war, the laboratory had 187 employees. During the next four years, militarily oriented work was given high priority and by the end of 1918, the number of people employed at the National Physical Laboratory had increased to about 550. The DSIR was the most important of the early British bodies of science policy and was to form the backbone of support for physics and the other sciences for the next three or four decades. Like their German counterparts, British physicists were involved in testing and developing methods of ranging. William H. Bragg and other physicists investigated sound-ranging methods for pinpointing German guns behind the trench lines. At the Munitions Inventions Department, Ralph Fowler and the young Edward Milne worked with optical instruments designed to guide antiaircraft fire.

A problem of even greater importance to the British was the threat of German submarines. Work in this area took place within the Royal Navy's Board of Inventions and Research (BIR), which was created in the summer of 1915. Rutherford was one of several physicists who were engaged in attempts to develop means of detecting the feared submarines. He had part of his Manchester laboratory turned into a large water tank, where he performed numerous experiments. According to Rutherford's biographer, not only did his systematic reports to the admiralty mark the birth of underwater warfare science, but also, "What Rutherford said about submarine-hunting in 1915 remains true in the 1980s" (Wilson 1983, 348). Rutherford advised that acoustic detection was the only practical way of locating submarines, and he and other physicists worked hard to develop efficient "hydrophones" and other listening apparatus. His work also included research on the use of piezoelectric quartz for the production of high-frequency sound for the detection of submarines. This line of work, which was independently reproduced by Paul Langevin in France, later developed into the technology of sonar. The invention of sonar is sometimes traced back to the war work of Langevin and Rutherford.

As in the case of England, the war induced the American government to go into science planning. In June 1916 a National Research Council (NRC) was established with the purpose of organizing and promoting research for "national security and welfare." Also as in England, submarine detection was given high priority, with Millikan appointed as chairman of an NRC committee against submarines. The Americans' work in this area was, to some extent, coordinated with the considerably more advanced work of their allies in France and England. In June 1917 a French-British mission, which included Rutherford, visited its American colleagues. Whereas American research in both submarine detection and artillery ranging was inferior to that of their allies, it was superior in the field of communications. When war was declared in April 1917, the U.S. Army commissioned John Carty of AT&T and a select number of other scientists and engineers as officers in the Signal Corps. Carty organized a Science and Research Division, which was headed by Millikan and included scientists and engineers from universities and private corporations. Within a few months, AT&T supplied 4,500 engineers and operators organized into 14 battalions, thereby tripling the staff of the Signal Corps. The efficiency of the U.S. Army's communications systems, both wired and wireless, depended on developments in vacuum tube technology made by Oliver Buckley, the AT&T physicist who was in charge of the Signal Corps' laboratory in Paris.

Compared with World War II—"the physicists' war"—physicists played a relatively minor role in World War I, rightly called "the chemists' war." Their contributions were in no way of decisive importance, and the kind of physics that entered the war effort had little connection with frontier research in the physical sciences. On the other hand, it was the first time in history that physicists made themselves visible and useful in a major military conflict. When the war ended, it was clear that physics could be of considerable importance to the military and that government support was necessary if physics should be applied for security purposes. Although "military physics" had thus become established by 1918, no one would have dreamt of the crucial role that the physicists would play in warfare a little more than two decades later. And even less would they have dreamt of the militarization of physics that became such a pronounced feature in the American-dominated world of physics in the 1950s (see chapter 20).

PART TWO

FROM REVOLUTION TO CONSOLIDATION

Chapter 10

SCIENCE AND POLITICS IN THE WEIMAR REPUBLIC

GERMANY WAS the world's leading scientific nation in the early part of the twentieth century and, in many ways, served as a model for other countries. It was in Germany that many of the great innovations in physics and the other exact sciences had their origin. The year 1918 was a watershed in both German and world history. With the lost war and the humiliating Versailles treaty came internal unrest, lack of food, political murders, a drastic fall in the country's economy, and a hyperinflation that lasted until 1923, when a loaf of rye bread sold for half a trillion marks. Germany was in a severe state of crisis, economically, politically, and spiritually. Yet, against all odds, German physics did remarkably well during the difficult years and succeeded in maintaining its high international position. In some of the new and exciting areas, such as atomic and quantum theory, German physicists set the international agenda. The seeds of quantum mechanics were sown during the early Weimar republic, despite all its difficulties and miserable conditions of living.

Science Policy and Financial Support

The German scientific community remained intact after the war, but it was a poor community in desperate search of money. Not only was Germany a relatively poor nation in the immediate postwar years, but its scientists were also excluded from international collaboration. The lack of foreign currency added to the Germans' problems, because it made the purchase of foreign literature and instruments almost impossible. The universities witnessed a drastic increase in the number of students, many of them ex-soldiers, for whom they had neither space, teachers, nor money. Institute budgets lagged far behind their prewar levels, and so did salaries for scientific personnel; because of the inflation, savings might lose their value from one day to the next. Otto Hahn recalled how he and Max von Laue went for a week's holiday to southern Germany in 1923: "At the end of that short trip Max von Laue found himself a million Marks short of his fare home. I was able to lend him the money. In later years, when money had regained its value in Germany, I used now and then to remind him that he still owed me a million. Finally he handed me a fifty milliard [billion] note, saying that there it was, together with interest both simple and compound" (Hahn 1970, 137).

The leading German scientists realized that they needed to legitimize their sciences in a new way, both to satisfy themselves and to appeal more convincingly to potential government sponsors. Germany had lost the war, the Kaiser had fled, the good old days had gone. What was left to carry the nation on to new honor and dignity? According to many scientists, the answer was (not surprisingly) science. In an address to the Prussian Academy of Sciences in November 1918, Max Planck made the message clear: "If the enemy has taken from our fatherland all defense and power, if severe domestic crises have broken in upon us and perhaps still more severe crises stand before us, there is one thing which no foreign or domestic enemy has yet taken from us: that is the position which German science occupies in the world. Moreover, it is the mission of our academy above all, as the most distinguished scientific agency of the state, to maintain this position and, if the need should arise, to defend it with every available means" (Forman 1973, 163). It was a theme often heard in the young Weimar republic. Planck, who at the time emerged as an unofficial spokesman for German science, repeated the message in a newspaper article in 1919: "As long as German science can continue in the old way, it is unthinkable that Germany will be driven from the ranks of civilized nations" (Heilbron 1986, 88). Science should be supported, said Planck, not primarily because it would lead to technological and economic progress (although this was an argument too), but because it was Germany's premier cultural resource. Science should be seen as a bearer of culture, a *Kulturträger* in German. It was something the country could be proud of, which could act as a surrogate for the political and military power that, alas, no longer existed. Science was seen as a means for the restoration of national dignity, and Germany's famous scientists became instruments of a national and international cultural policy on par with the country's poets, composers, and artists.

The appeal to the cultural and political values of science was not only the rhetoric of a few scientists in search of support. Surprisingly, in view of the traditional utilitarian legitimation of science, it was consonant with the view of many politicians and humanist scholars who favored an antiutilitarian and antimaterialistic attitude toward science. The most valuable asset in the struggle for international cultural recognition was perhaps Einstein, who, as a German (of a kind), could be used for propagandistic purposes as long as he stayed in Germany. Planck, among others, feared that Einstein might leave Germany, which would mean a loss not only to German science, but also to German cultural politics. The issue was spelled out with clarity in a report of September 2, 1920 from the German chargé d'affaires in London to the Foreign Ministry in Berlin. After having mentioned rumors in British newspapers that Einstein might leave Germany and go to America, the chargé d'affaires continued, "Particularly at this time, Professor Einstein counts as a cultural factor of the first rank for Germany, since his name is known far and wide. We should not drive such a man out of Germany; we could use him for effective *Kultur* propaganda. If Professor Einstein should really be intending to leave Germany, I should deem it desirable in the interests of Germany's reputation abroad, if the famous savant could be persuaded to remain in Germany." This report was just one of many. Whenever Einstein traveled abroad to lecture, an official from the German embassy or consulate would secretly send reports to Berlin. The reports, from Paris, Copenhagen, Tokyo, Madrid, Oslo, Chicago, and elsewhere, paid particular attention to how the foreign press connected the famous physicist with Germany. A German diplomat in Montevideo reported with satisfaction on June 4, 1925 that because Einstein "was all over celebrated as a 'sabio aleman' [German scholar] (the Swiss citizenship, which he also carries, was scarcely mentioned), his visit has been most valuable for the German cause" (Kirsten and Treder 1979, 207 and 234).

In Wilhelmian Germany, academic science was funded mostly by the German states (to which to universities belonged) and not by the federal government in Berlin. Under the dire conditions after 1918, German science organizations had to find new sources for funding in order to avoid starvation. The most important of the new central scientific-political agencies was the *Notgemeinschaft der deutschen Wissenschaft* (Emergency Society for German Science and Scholarship), founded in 1920. This organization represented several German science institutions, including the Kaiser-Wilhelm Gesellschaft, the universities and technical colleges, the academies of science, and the Society of German Scientists and Physicians. The main activity of the Notgemeinschaft was to raise and allocate money for research of all kinds—in the natural sciences, engineering, the social sciences, and the humanities. The Notgemeinschaft supported individual scientists and research projects on the basis of merit and the grants were allocated independent of the recipient's university. Applications were judged by panels of experts, without interference from the central government. From 1922 to 1934, Max von Laue was chairman of the physics panel, an influential position in German science policy. By far, most of the Notgemeinschaft's resources came from the government in Berlin, but there were also substantial donations from abroad, including major contributions from General Electric and the Rockefeller Foundation in the United States. Only a minor part came from German industry, which preferred to channel its support to the competing Helmholtz Society or to individual scientific projects of a more technical nature.

The Notgemeinschaft was controlled by Berlin scientists and, as far as physics was concerned, it favored the kind of pure theoretical physics that Planck thought was culturally (and hence, to his mind, politically) important. As a member of the Notgemeinschaft's executive committee, Planck arranged that the atomic theorists in Göttingen and Munich received sufficient funds. In general, however, a large portion of the research grants went to Berlin physicists. For example, during 1924–26, Berlin physicists received about half the grants, a fact that angered many conservative physicists outside the capital, who accused the Notgemeinschaft of Berlin favoritism. Throughout the 1920s, Johannes Stark and other conservative physicists sought to win influence in the Notgemeinschaft, but they were resolutely kept out by Laue and Planck. In spite of the Notgemeinschaft's predilection for modern theoretical physics, it had enough money and magnanimity also to support antirelativists like Stark, Lenard, and Rudolf Tomaschek.

The Electrophysics Committee, a subcommittee of the Notgemeinschaft based on General Electric donations, was essential to German atomic physics. Of the 140 grant applications it received between 1923 and 1925, it approved 71, most of which were in atomic and quantum physics. The money were ostensibly donated to help technical and experimental physics, but the members of the committee (which included Planck as chairman) often wanted to use them for atomic theory. "To be sure, the money is designated 'especially for experimental research,'" Planck wrote to Sommerfeld in 1923, "but your project can be presented as the working out of experimental research. The main thing of course is your name" (Heilbron 1986, 92). In physics as elsewhere, it is nice to have a name and to have friends. Quantum mechanics would undoubtedly have come into being irrespective of the support of the Notgemeinschaft. Yet Heisenberg and Born were supported by Notgemeinschaft money, and it is understandable that the Electrophysics Committee congratulated itself in 1926 when quantum mechanics had proved its worth—in more than one sense. The money and the philosophy of the Notgemeinschaft had paid off handsomely: "As is well known, quantum mechanics stands at the center of attention among physics circles of all nations. The work of Heisenberg and Born, which the Electrophysics Committee has supported and without which the work would very probably not have been done in Germany but elsewhere, has shown the usefulness of the Electrophysics Committee in the development of physics in Germany" (Cassidy 1992, 160).

During 1921–25, the Notgemeinschaft supported projects in the physical sciences with annual grants of about 100,000 gold marks, which was a very substantial sum; in addition, it sponsored a small number of fellowships. The organization's importance for German physics is reflected in the research publications. It has been estimated that at least one-quarter of the 8,800 publications appearing in the three principal German physics journals between 1923 and 1938 were based on research wholly or partially supported by the Notgemeinschaft. On a less statistical note, consider the case of Werner Heisenberg, in 1923 a promising twenty-one-year-old student of physics. Heisenberg was the kind of physicist that the Notgemeinschaft wanted to keep in the field and he received support from the organization in 1923–25, although not quite enough to live on. His professor in Göttingen, Max Born, provided additional means. Like several other professors, Born befriended German industrialists and foreign philanthropists, some of whom helped fund his institute. Born received generous contributions from a New York financier, Henry Goldman, and he used the money to help his students and assistants, among them Heisenberg. There were many ways to survive the difficult years, some more regular than others.

The Physikalisch-Technische Reichsanstalt was one of the central physics organizations that survived the war. But it did not live up to its former glory and position as one of the world's leading institutes of pure physics. On the contrary, as the Reichsanstalt became ever more oriented toward testing and technical physics, it deteriorated scientifically. Shortly after Friedrich Paschen became director of the Reichsanstalt in 1924, he complained to Sommerfeld that "proper scientific work has ceased here." He continued, "The institute has become more and more technical. Purely scientific research is retreating. The very attitude of the science officers is very technical. On most of them modern physics has left no trace" (Kragh 1985, 110).

International Relations

At a medical congress in Copenhagen in 1884, Louis Pasteur confirmed the neutrality and internationality of science. "Science has no native country," he said, but then went on, "Even if science does not have any native country the scientist should particularly occupy himself with that which brings honor to his country. In every great scientist you will always find a great patriot" (Kragh 1980a, 293). The problematic ambivalence between patriotism and scientific internationalism was fully displayed during and after World War I, when it turned out that patriotism was the stronger of the two ideals (see also chapter 9). In a letter to George Hale in 1917, the American physicist and inventor Michael Pupin wrote, "Science is the highest expression of a Civilization. Allied Science is, therefore, radically different from Teutonic Science. . . . we see today more clearly than we have ever seen before, that Science cannot be dissociated from the various moods and sentiments of man. . . . I feel that scientific men are men first and scientists after that" (Forman 1973, 158).

During the last phase of the war, French, British, and American scientists discussed the structure of a new international science organization to replace the International Academy of Sciences. The leading figures in these negotiations were the American astronomer George Hale, the British (German-born) physicist Arthur Schuster, and the French mathematician Charles Émile Picard. The result was the formation, in the summer of 1919, of the International Research Council (IRC), with membership at first restricted to the Allied powers and those nations that had been associated with them or otherwise opposed the Germans. The member countries were France, England, the United States, Belgium, Italy, Australia, Canada, New Zealand, South Africa, Japan, Brazil, Greece, Poland, Portugal, Romania, and Serbia—clearly a political selection, rather than one based on scientific excellence. Membership in the scientific unions under the IRC, such as the planned International Union of Pure and Applied Physics (IUPAP), was likewise restricted to these countries. IUPAP was formed in 1922 by representatives from thirteen nations, with William H. Bragg as its first president. He was followed by Millikan and, during the difficult war years, Manne Siegbahn from Sweden. IUPAP was not very active until 1931, when it was reorganized in connection with the change from the IRC to the ICSU, the International Council of Scientific Unions. At that time, IUPAP was one of eight scientific unions within the council, the others being those for astronomy, geodesy and geophysics, chemistry, mathematics, radio science, biology, and geography.

Because of the hatred and suspicion resulting from the terrible war, Germany, Austria, Hungary, and Bulgaria were kept out of the IRC, and even the neutral countries were not allowed in. Hardliners feared that the neutral countries might change the balance of power in the IRC and vote for the admission of the former Central powers. Only in 1922 were some of the neutral countries admitted, among them the Netherlands and the Scandinavian countries. French suspicion seemed justified, for it was these countries that first pleaded for acceptance of the Central powers. In 1925 Lorentz suggested, on behalf of the Dutch, Norwegian, and Danish delegations, that the exclusion policy be annulled. Although the resolution was supported by Britain and the United States, and at the General Assembly received ten votes out of sixteen, it did not receive the neccesary majority of two-thirds of all member nations. (At the time, the IRC included twenty-nine member nations, of which only sixteen were present, so even a unanimous vote would not have changed the statutes.) Winds changed the following year, after the Locarno Treaty had secured a milder political atmosphere in Europe. Then, leading scientists in the IRC, including Schuster, Hale, and Picard (who served as president), accepted German admission and invitations were sent to Germany and the other former enemy countries. However, when the invitation finally came, it was rejected by the German and Austrian scientists. The result was that the two countries remained outside the official body of international science until after World War II. Hungary joined the IRC in 1927, and Bulgaria in 1934.

The exclusion of Germany from the IRC was only one of the ways in which the victors of the war sought to isolate and diminish the importance of German science. From 1919 to about 1928, German science was subject to an international boycott, in the sense that German scientists were not allowed to attend many international conferences. During the early years the boycott was fairly effective, with exclusion of Germans from most international conferences; among 275 international scientific conferences between 1919 and 1925, 165 were without German participation (see figure 10.1). Many international institutes, bureaus, and agencies were removed from German soil. In 1914, there were 60 such organizations, of which 14 were located in Germany and 18 in France; in 1923, the number had increased to 85, now with 37 in France and only 6 in Germany. These international organizations felt that not only should German science be boycotted, but so should German language and scientific publications. At most international conferences, German was not allowed as an official language. After all, as it was stated in *Science* in 1920, "German is, without doubt, a barbarous language only just emerging from the stage of the primitive Gothic character, and . . . it should be to the advantage of science to treat it as such from the date August 1, 1914."

As intended, the boycott did cause some inconvenience to German science, although the harm was more psychological than substantial. It probably did more harm to the boycotting nations which had to do without Germans at conferences devoted to subjects in which German scientists were the

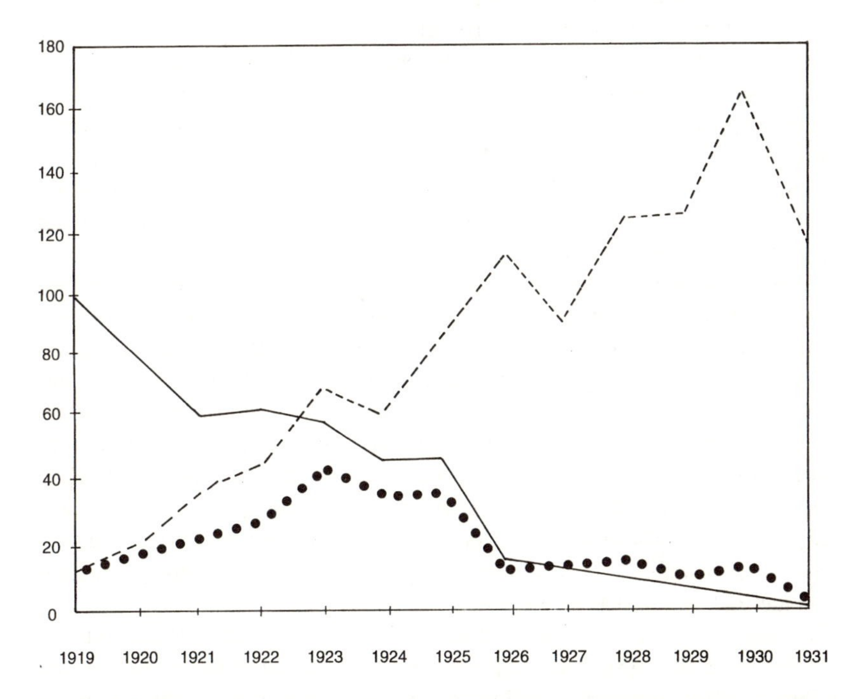

Figure 10.1. The number of international scientific congresses (dashed line) and those from which Germany was excluded (dotted line). The solid line gives the percentage of congresses that excluded German participation. *Source:* Redrawn from Schröder-Gudehus 1966.

undisputed experts. The Solvay congresses of 1921 and 1923, on "Atoms and Electrons" and "Electrical Conductivity of Metals," respectively, included no German physicists, which clearly deflated the value of these meetings. The boycott was never total, and it was not followed by either the Netherlands or Scandinavia. In 1919 the Swedish Academy of Sciences decided to award the Nobel prize in physics to Planck and Stark, and the prize in chemistry to Fritz Haber. It was widely considered offensive that the prizes went to three German scientists; the award to Haber was particularly controversial because of his involvement in German gas warfare. From the perspective of France, the Swedes had made a political choice in an effort to rehabilitate German science. Haber, who shared the science-as-*Macht-Ersatz* view of Planck and many other German scientists, seems to have agreed. He saw the award "as a tribute to the accomplishments of German science from professional colleagues in Sweden," as he wrote to the Swedish Academy of Sciences (Widmalm 1995, 351).

Einstein was one of the few German scientists who was welcome at international conferences. Not only was he the world's most famous scientist, but he was also a pacifist and democrat, and was not considered a "true" German. (At that time, Einstein held double citizenship, Swiss and German.) But even a pacifist Swiss-German Jew could be too controversial in the early 1920s. When Einstein accepted an invitation from Paul Langevin to speak in Paris in the spring of 1922, he was brought to the French capital secretly in order to avoid demonstrations. Moreover, he had to cancel a lecture before the dignified Academy of Sciences because thirty members had planned to walk out of the auditorium as soon as Einstein entered. To them, he was enough of a German to be unwanted.

The attempts to use Einstein politically against his German colleagues were no more successful than German attempts to use him for their own political purposes. For example, the organizers of the 1923 Solvay Congress wanted Einstein to participate, but Einstein refused when he realized that other German scientists would not be invited. As he wrote to Lorentz, "[Sommerfeld] is of the opinion that it is not right for me to take part in the Solvay Congress because my German colleagues are excluded. In my opinion it is not right to bring politics into scientific matters, nor should individuals be held responsible for the government of the country to which they happen to belong. If I took part in the Congress I would by implication become an accomplice to an action which I consider most strongly to be distressingly unjust" (Mehra 1975, xxiii). In an article titled "All About Relativity" from the same time, the popular English journal *Vanity Fair* wrote, lightheartedly: "And, anyway, Einstein is a German, and the whole thing is without doubt a German plot to regain control of the aniline dye trade. . . . He answers one question by putting another. But there is where the German of it comes in. If only this discovery could have been made by a representative of one of the friendly Allies (question for collateral reading: name five

Allies who are friendly at the present time.)" (Hentschel 1990, 125). *Vanity Fair*'s comment was ironic, but the irony pinpointed a real problem in the political context of science at the time.

Another case illustrates the same context in a different manner. In January 1923, the Hungarian George Hevesy and the Dutchman Dirk Coster announced that they had discovered element number 72 while working at Niels Bohr's institute in Copenhagen. The announcement of the discovery of hafnium (Hf), as they called the element, gave rise to a major priority controversy with French scientists, who argued that they had already discovered the element—which they called celtium (Ct). The fact that the controversy started in early 1923, at a time when German-French relationships were strained to the breaking point as a result of the French-Belgian occupation of the Ruhr district, gave it a particular flavor. The discovery of hafnium, and especially the discredit thereby brought upon French science, was regarded by militant Allied scientists as a conspiracy against the French *gloire*, a sinister attempt to provide Germany with intellectual revenge for its military defeat. When Hevesy, at Rutherford's advice, submitted a paper on hafnium to the pro-French and pro-celtium *Chemical News*, its editor, W. P. Wynne, responded, "We adhere to the original word celtium given to it by Urbain as a representative of the great French nation which was loyal to us throughout the war. We do not accept the name which was given it by the Danes who only pocketed the spoil after the war" (Kragh 1980a, 294). Although no German scientists had contributed to the discovery of hafnium, it nevertheless was associated with "Teutonic science." Hevesy was an Hungarian, hence a former enemy, and Coster was a Dutchman who had spent most of his scientific career in Sweden and Denmark, countries that were accused of being pro-German. Although Niels Bohr's status in the scientific world prevented any direct criticism of his activities, many radical advocates of the boycott line found him too soft on the Germans. During, as well as after, the war Bohr had kept normal and friendly relations with his German colleagues and during the zenith of the boycott, visits between Copenhagen and Germany were frequent. Bohr had lectured in Germany, participated in unauthorized conferences, and been the host to many German physicists; he was always careful to acknowledge the results of his German colleagues; and he published many of his most important papers in German journals, thereby ignoring the efforts to isolate German as an unworthy language for scientific communication. All this made Bohr and Copenhagen physics look pro-German in some quarters and increased the vague feeling of Teutonism connected with hafnium. The matter was delicate to the International Committee on Chemical Elements, a branch of the IRC with no representatives from the neutral countries. Celtium had no scientific credibility, but for political reasons the committee found it impossible to sanction hafnium. Only in 1930 was hafnium officially accepted.

Germany was not the only country where science suffered because of the war. Material conditions were even worse in the young Soviet Union, which was isolated from other countries much more effectively than Germany. Suffering and death because of malnutrition, typhus, and the civil war did not distinguish between scientists and nonscientists. The conditions of physical hardship during the terrible years 1918–22 made it very difficult for the physicists to resume their research activities and build up institutional frameworks. The Russian Association of Physicists was founded in 1919, primarily through the efforts of Orest Khvol'son and Abram Ioffe, but the organization proved inadequate to the task of normalizing physics. For a period, lack of paper prevented or greatly delayed scientific publications. For example, the first Russian textbook on general relativity, completed in 1922 by Alexander Friedmann and Vsevelod Frederiks, first appeared in 1924 because of the shortage of paper. The most important of the Soviet physics journals (*ZhRFKhO*) appeared infrequently between 1918 and 1923 and not at all in 1920 and 1921. Attempts by Russian physicists to rebuild their discipline and break their isolation from Western countries met with only partial success from about 1922. In that year, the Soviet Union established diplomatic relations with Germany and as a result, contacts between German and Russian physicists increased. Ioffe, founder of the Leningrad Physico-Technical Institute, went to Berlin to order much-needed scientific books, journal subscriptions, and instruments. He and other Russian physicists were received with great sympathy by their German colleagues, who were eager to establish professional connections between the two pariah countries in spite of their political differences. In 1922 the German Physical Society elected the Leningrad physicist Khvol'son as their only honorary member. From about 1925, contacts with the West increased further and many Soviet physicists visited Germany and other Western countries, including the United States. The Soviet physicists were eager to publish in Western journals and chose the German *Zeitschrift für Physik* as their favorite journal. In the late 1920s, more than 12 percent of the papers in the *Zeitschrift* were written by Russians, who contributed with a total of 592 articles between 1920 and 1936. One-third of the articles were authored by physicists at the Leningrad Physico-Technical Institute, the most important of the physics institutes in the Soviet Union.

The Physics Community

As indicated previously, in the early 1920s the German physics community was split up in questions of science, politics, and ideology. The large right-wing group included fascists like Stark and Lenard, as well as less ideological physicists like Wilhelm Wien and Otto Lummer. The right-wing physicists largely shared the same political views, including chauvinism, ultraconservatism, and opposition to the Weimar republic. Anti-Semitism, too, was common to most of them. In 1922, when Paschen proposed Alfred Landé for an associate professorship at Tübingen, a conservative stronghold, Paschen felt that his colleagues were negatively prejudiced not only because Landé was a "progressive" atomic theorist, but also because he was a Jew. (Such incidents, and anti-Semitism in general, were not exclusive to Germany; discrimination against Jews was common also at universities in the United States and some other countries.) Whereas there was an identifiable right wing in German physics, there was no left wing with socialist sympathies. To the extent that they showed an interest in politics at all, Paschen, Planck, Laue, Sommerfeld, and most other established progressive physicists held generally conservative, antigovernment views. Einstein and Born were among the few exceptions to the rule. Only if measured against the extreme standards of the right wing would the bourgeois conservatism of German mainstream physicists appear as liberalism. The young generation of physicists who had not served under the war, including Heisenberg, Jordan, and Pauli, was largely apolitical.

The scientific views of the right-wing physicists were, to a considerable extent parallel with their political views, both being conservative. They kept loyal to the standards of Wilhelmian science and opposed the Zeitgeist of the Weimar period. Most right-wing physicists stuck to the worldview of classical mechanicism and electrodynamicism, including such notions as the ether, determinism, causality, and objectivity that became unfashionable in the early 1920s. Yet they were not in complete disharmony with the Weimar Zeitgeist. The predeliction for visualizability (*Anschaulichkeit*) and the drive to substitute physical insight for abstract mathematical reasoning had equal value for the conservatives and the "progressives" who adapted their views to the Zeitgeist. The standards of the right wing manifested themselves in a more or less direct dissociation from quantum and relativity theories and a preoccupation with experiments at the expense of theory. Not all experimentalists were conservatives, and not all theorists were "progressives," but there was a clear connection nonetheless.

To some extent, the division between "progressives" and "reactionaries" reflected the tension between the powerful Berlin physicists and the physics institutes at the provincial universities. The division was not sharp, however, as illustrated by the strong theoretical schools in Munich and Göttingen. It was a stereotype that did not fit at all with quantum mechanics, which was of non-Berlin (Göttingen) origin, and where most Berlin physicists were in favor of Schrödinger's conservative version rather than Heisenberg's radical version. Yet, fairly or not, to many physicists "Berlin" came to signify abstract theory, Jewish intellectualism, arrogance, and bad taste. The German Physical Society was dominated by Berlin physicists and, for this reason, was seen as suspect in some quarters. Einstein had served as president in

1916–18 and he was followed by Sommerfeld after Max Wien, an experimentalist from the University of Jena, had declined the position. In 1920 Stark organized a rival organization, the Association of German Academic Physics Teachers (*Fachgemeinschaft Deutscher Hochschullehrer der Physik*), but the importance of the new association was limited; with the Munich physicist Wilhelm Wien as new president of the Physical Society, the accusations of "Berlinerei" were less convincing.

Another question that divided German physicists concerned the application of physics to technical and industrial needs. Such applications were out of tune with the Weimar Zeitgeist's antiutilitarianism and, indeed, most mainstream physicists had little respect for technical physics. On the other hand, many right-wing physicists were eager to apply their science for technical purposes. A large percentage of German physicists had been occupied with industrial or military physics during the war and they felt that their work was not sufficiently respected in the Physical Society. When the German Society for Technical Physics was founded in 1919, it was partly as a reaction against the perceived predominance of theory in the Berlin-dominated German Physical Society. Within a year, membership in the new society soared to 500, and between 1925 and 1935 it could boast of having as many members as the Physical Society, about 1,300 (see figure 10.2). Gehrcke, the antirelativist, was one of the founders of the Society for Technical Physics and published frequently in its journal, the *Zeitschrift für technische Physik*.

Among the prestigious journals for academic publications, the *Annalen der Physik* was preferred by the right-wing physicists. If they could not publish in the *Annalen*, they would rather publish in less prestigious or even obscure journals than in the *Zeitschrift für Physik*, the other important physics journal in Germany. The *Zeitschrift*, founded in 1920 by the Physical Society and associated with liberal and avant-garde views, was the favorite journal of the young generation of quantum physicists but was almost boycotted by the "reactionaries," to whom it signified the decadence and dogmatism of modern physics. It was scarcely a coincidence that Schrödinger's papers on wave mechanics appeared in the *Annalen* and not in the *Zeitschrift*. When the *Zeitschrift* was launched in 1920, it was planned to be limited to three annual volumes with a maximum of 1,440 pages. But its success made it expand beyond the most optimistic expectations. In 1924 it included ten volumes with a total of 4,015 pages, and in 1929 its seven thick volumes included 6,094 pages. The language of the journal was German and its success was a strong argument for the importance of German as an international language of physics. By 1925, it was evident that French and British efforts to limit German as a scientific language had failed. When a paper in English, written by the Indian physicist R. N. Ghosh, appeared in the *Zeitschrift* in 1925, it caused vehement protests from many German physi-

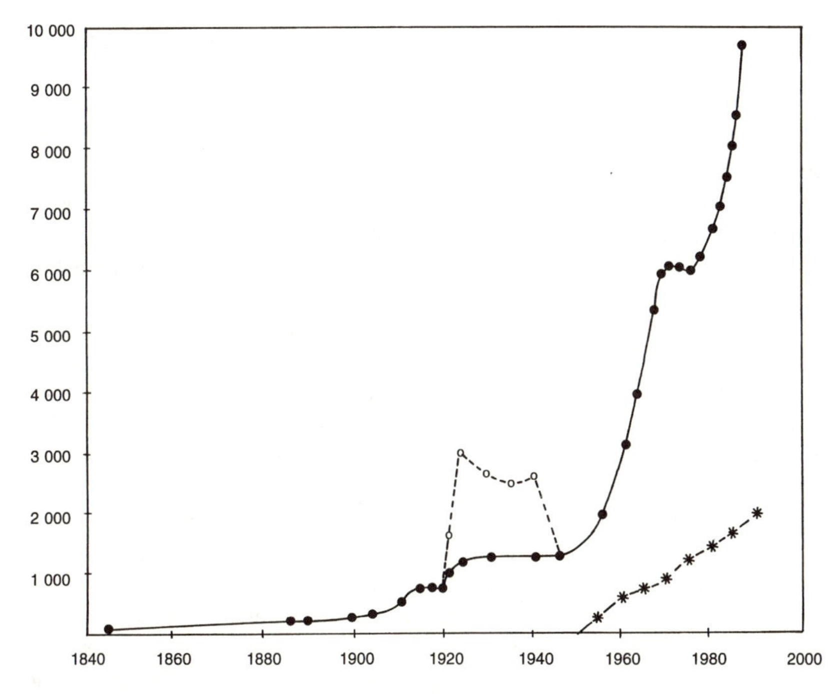

Figure 10.2. Membership of the German Physical Society from its origin in 1845 to 1986. *Source:* Redrawn from Mayer-Kuckuk 1995, with permission from Wiley-VCH Verlag GmbH.

cists. The journal's editor, the able Karl Scheel, was forced to promise that all papers had to be in German language. To Lenard, who detested everything British, the promise came too late. He left the society in protest and posted in his Heidelberg physics institute a sign with the message, "Entrance is forbidden to members of the so-called German Physical Society" (Beyerchen 1977, 98).

Zeitgeist and the Physical Worldview

During the decade following 1918, physics in Germany faced not only economic difficulties, but also a changed intellectual environment that in many ways was hostile to the traditional values of physics. Physics, and science in general, was now increasingly being accused of being soulless, mechanistic, and contrary to human values. Such charges were old, in Germany and elsewhere, but they became more frequent and were stated with more authority in the Weimar republic. Non- or antiscientific attitudes were popular in philosophy, psychology, and sociology, and astrology, cabalism, and other brands of mysticism flourished. In 1927 Sommerfeld wrote, "The belief in a rational world order was shaken by the way the war ended and the peace dictated; consequently one seeks salvation in an irrational world order. . . . We are thus evidently confronted once again with a wave of irrationality and romanticism like that which a hundred years ago spread over Europe as a reaction against the rationalism of the eighteenth century and its tendency to make the solution of the riddle of the universe a little too easy" (Forman 1971, 13). Many physicists felt that the Zeitgeist of the period was basically antagonistic to their science and that what counted to the educated public were ideas foreign to the scientific mind. Authors and philosophers emphasized *Lebensphilosophie* and *Weltanschauung* rather than the results of science, which they identified with an outdated materialism and unwarranted belief in causality and objective knowledge. Holism, intuition, relativism, and existentialism were on the agenda of the time, not mathematical analysis, controlled experiments, and causal explanations.

The new-age program was most influentially presented by Oswald Spengler in his gloomy, yet enormously popular *Der Untergang des Abendlandes* (*The Decline of the West*) which, from 1918 to 1926, was printed in about 100,000 copies. According to Spengler's sweeping analysis of the ups and downs of civilizations, modern science was in a state of deep crisis, but only because it had not managed to adapt to the new culture. Dismissing old-fashioned notions such as causality and objectivity, Spengler argued that science was anthropomorphic and merely an expression of a particular culture at a particular time; physics was relative to culture and history, and there was no physics apart from or independent of local cultural environments—the reader may sense an association with views popular in the last decades of the century. Spengler, whose insight in contemporary physics was less than satisfactory (but few of his readers would know), believed that time was ripe for a new physics that unified thought and spirit. However, the new physics could not build on the old; it would need a whole new civilization: "The vast and ever more meaningless and threadbare fabric woven by natural science falls apart. It was, after all, nothing but the inner structure of the mind. . . . But today, in the sunset of the scientific epoch, in the stage of victorious skepsis, the clouds dissolve and the quiet landscape of the morning reappears in all distinctness . . . weary after its striving, the Western science returns to its spiritual home" (Forman 1971, 37). Spengler's philosophy was perhaps confused and badly argued, but it was representative of the Zeitgeist and highly influential in the cultural discussions of the time. It was known by practically all educated Germans, including the physicists.

Given that this antirationalistic and antipositivistic climate dominated a large part of the Weimar culture, and given that it questioned the very legitimacy of traditional science, it was only natural that physicists felt forced to respond to the new ideas. And they did, usually not by defending the traditional values of science but, in many cases, by bowing to the new Zeitgeist. One result of the adaptation was, as mentioned, that many physicists abstained from justifying their science by utility and instead stressed that physics is essentially culture. Some theoretical physicists "capitulated" to the new Spenglerian views in the sense that they unknowingly assimilated their values to what they perceived was the new Zeitgeist. For example, in the early 1920s several German physicists addressed the question of crisis in physics and argued that the principle of causality could no longer be considered a foundation of physical theories. This repudiation of causality was not rooted in specific experimental or theoretical developments in physics.

The quantum mechanics that appeared in 1925 was a German theory, which broke with microphysical causality. It is therefore natural to ask if there was a causal connection between the general ideas about acausality before 1925 and the formation and interpretation of quantum mechanics. The historian Paul Forman has argued that German physicists, because of the influence of the Weimar Zeitgeist, were predisposed toward an acausal quantum mechanics and craved for a crisis in the existing semimechanical atomic theory. According to Forman, the very possibility of the crisis of the old quantum theory depended on the Weimar intellectual milieu. However, there are good reasons to reject the suggestion of a strong connection between the socio-ideological circumstances of the young Weimar republic and the introduction of an acausal quantum mechanics. Suffice to mention a few of the reasons:

1. Whereas the physicists often discussed the (a)causality question and other Zeitgeist-related problems in talks and articles addressed to general audiences, these topics were almost never mentioned in either scientific papers or addresses before scientific audiences.
2. To the extent that physicists adapted to the Zeitgeist, the adaption was concerned with the values of science, not with its content.
3. Many of the physicists had good scientific reasons to reject detailed causality and did not need to be "converted." At any rate, only a very small proportion of the German physicists seem to have rejected causality before 1925–26.
4. Sommerfeld, Einstein, Born, Planck, and other leading physicists did not bow to the Zeitgeist, but criticized it explicitly.
5. The recognition of some kind of crisis in atomic physics was widespread around 1924, primarily because of anomalies that the existing atomic theory could not explain. Bohr and a few other physicists suggested vaguely that energy conservation and space-time description might have to be abandoned (see chapter 11).
6. The first acausal theory in atomic physics, the 1924 Bohr-Kramers-Slater radiation theory, was not received uniformly positively among German physicists, contrary to what one would expect according to the Zeitgeist thesis. And those who

did accept the theory were more impressed by its scientific promises than by its ideological correctness. The theory's element of acausality was not seen as its most interesting feature. Moreover, the theory had its origin in Copenhagen, with a cultural climate very different from that of Weimar Germany, and was proposed by a Dane, a Dutchman, and an American.

7. Among the pioneers of acausal quantum mechanics were Bohr, Pauli, and Dirac, none of whom was influenced by the Weimar Zeitgeist. The young German physicists who created quantum mechanics were more interested in their scientific careers than in cultural trends and sought deliberately to isolate themselves from what went on in society.

In conclusion, there were good reasons—internal as well as external—for why quantum mechanics originated in Germany. As far as I can judge, adaptation to the Weimar Zeitgeist was of no particular importance.

Chapter 11

QUANTUM JUMPS

QUANTUM ANOMALIES

AS BOHR WAS well aware, his 1913 quantum theory of atoms was merely the beginning of a research program that would carry him and his colleagues into unknown territory; it was in no way considered a finished theory. During the war years, the theory was extended and modified, mainly by Bohr and a group of German physicists, but also—in spite of the divides caused by the war—by British, Dutch, and Japanese colleagues. The most important of the early generalizations was the result of work done by Sommerfeld in Munich, who quickly established himself as an authority of atomic theory and obtained a status surpassed only by Bohr's. Sommerfeld's masterful and comprehensive textbook in atomic theory, *Atombau und Spektrallinien*, based on courses given at the University of Munich in 1916–17, was first published in 1919. During subsequent years, it ran through several new editions and became the "bible" of atomic theory to the postwar generation of physicists. In 1923 the third German edition was translated into English as *Atomic Structure and Spectral Lines*.

Sommerfeld, trained in pure mathematics and an expert in applying advanced mathematical methods to physical problems, developed Bohr's simple theory by basing it on an extended use of action integrals. According to Sommerfeld, the dynamical description of atomic systems with several degrees of freedom was based on quantum conditions of the type $\int p_i dq_i = n_i h$, where the q's and p's are generalized position and momenta coordinates, respectively, and the n's are integral quantum numbers. Sommerfeld's 1916 atom was characterized by two quantum numbers, the principal n and the azimuthal k, and was soon extended with a "magnetic" quantum number m. By using this technique, Sommerfeld was able to reproduce Bohr's formula for the energy levels of the hydrogen atom and to extend it to the relativistic fine structure case, a major success of the old quantum theory. The Munich formalism also included the use of the action-angle variables methods known from celestial mechanics but, until then, rarely used in physics. By means of these methods, in 1916 the Russian physicist Paul Epstein (staying in Munich, where he was interned as an enemy alien) and the astronomer Karl Schwarzschild gave independently detailed calculations of the Stark effect in agreement with experiments. The splitting of spectral lines in strong electric fields, discovered by Johannes Stark in 1913, defied classical explanation but its general features could be understood on the basis of Bohr's theory. The earliest works on the Stark effect, including a theory Bohr proposed in 1914, resulted in qualitative agreement only. Epstein concluded in 1916 that his new results "prove the correctness of Bohr's atomic model with such striking evidence that even our conservative colleagues cannot deny its cogency" (Jammer 1966, 108). In the same year, Sommerfeld and Peter Debye, also independently, explained the simple Zeeman splitting into triplets.

Bohr's favorite tool in atomic theory was not action integrals à la Sommerfeld, but the correspondence principle that he had used loosely in 1913 and which he sharpened in an important memoir of 1918. This principle became the hallmark of the early Bohr school in quantum theory and an important conceptual guide to the construction of the new quantum mechanics. The essence of the correspondence principle, as Bohr understood it around 1920, was the following: In the limit of large quantum numbers ($n \rightarrow n - m$, $m \ll n$), transitions to stationary states not very different from the initial one will result in frequencies almost identical with those to be expected classically, that is, from Maxwellian electrodynamics. This simply followed from Bohr's quantum theory of the atom and was not the content of the correspondence principle in its wider sense.

Bohr realized that the original quantum theory was incomplete in the sense that although it prescribed frequencies, it had nothing to say about intensities and polarizations. This was a serious deficiency, for in order to compare a theoretically derived spectrum with one obtained experimentally, the intensities must be known. Bohr therefore extended the principle to cover a correspondence with the classical intensities as well, namely, by requiring a correspondence between the squares of the Fourier coefficients of the dipole moment (the classical measure of the intensity) and the transition probability coefficients introduced by Einstein in his radiation theory of 1916–17. In 1918, Bohr wrote, "We may expect that also for small values of n the amplitude of the harmonic vibrations corresponding to a given value of τ [where $n \rightarrow n - \tau$] will in some way give a measure for the probability of a transition between two states for which $n' - n''$ is equal to τ" (Darrigol 1992, 126).

As an important consequence of the correspondence principle, if a particular harmonic component is zero, the transition probability will be zero as well; that is, the transition will be "forbidden." In this way, Bohr and his assistants applied the correspondence principle in order to estimate intensities of spectral lines and to derive selection rules. For example, whereas the Polish physicist Adalbert Rubinowicz, an assistant of Sommerfeld's, had derived the selection rule $\Delta k = 0, \pm 1$ for the azimuthal quantum number, Bohr reasoned from his correspondence principle that transitions with $\Delta k = 0$ could not occur. Spectroscopic evidence showed Bohr's result to be correct. One of the first and most impressive applications of the correspondence principle was made by Hendrik A. Kramers, a young Dutchman studying under Bohr in Copenhagen. In his 1919 dissertation, Kramers calculated in detail the intensities and polarizations of the hydrogen spectral lines, including Zeeman and Stark effects. His results were in good, if not perfect, agreement with experimental data.

The correspondence principle was very much Bohr's invention and it was received rather skeptically in Germany, where Sommerfeld's more formal quantum principles and deductive procedures were seen as more promising. To Sommerfeld, the correspondence principle had too much of the character of a "magic wand," as he called it in the first edition of his *Atombau*. In 1935, Kramers recalled how many physicists perceived Bohr's semi-intuitive use of the principle: "In the beginning the correspondence principle appeared to the physicists as a somewhat mystical magic wand, which did not act outside Copenhagen" (Kragh 1979, 156). Bohr's magic was particularly strong, and particularly confusing, in his use of the correspondence principle in his "second atomic theory" of 1921–23, in which he attempted to account for the atomic structure of all the elements in the periodic table. Building eclectically on a mixture of x-ray spectroscopy, chemical data, correspondence arguments, and vague symmetry principles, Bohr constructed two-quantum (n, k) atomic models all the way from hydrogen to uranium. He even derived the electron configuration for the hypothetical element of atomic number 118, which he predicted would belong to the group of inert gases. Contrary to his 1913 picture, the electrons now moved in Keplerian ellipses and during their orbits, they penetrated the region of internal electrons, thereby causing a coupling of the revolving electrons. Bohr gave a full exposition of his theory during a meeting in Göttingen in 1922. His style and way of thinking—perhaps more than his theory itself—deeply impressed the participants, among whom were most of the German atomic physics élite. Several of the young physicists left their first meeting with the master of atomic wizardry convinced of the great value of the correspondence principle.

Although Bohr's theory of the periodic system was soon superseded by better theories, for the first time it gave a reasonably satisfactory explanation of the atomic structure of all the chemical elements. In particular, Bohr argued that the rare earth elements were characterized by a gradual building up in the $n = 4$ level, from 3×6 electrons in lanthanum to 4×8 electrons in lutetium, implying that the unknown element of atomic number 72 would be chemically related to zirconium and not a rare earth. Bohr's prediction was verified in late 1922 when element 72 (hafnium) was discovered in zirconium minerals. Although the prediction did not follow unambiguously from Bohr's theory, the discovery was generally seen as a brilliant confirmation of this theory in particular and of the quantum theory of atoms in general. As mentioned in chapter 10, the discovery led to an extended priority conflict.

For every success of the Bohr-Sommerfeld theory, there was a failure or an anomaly. The hydrogen atom (and other one-electron atomic systems) was brilliantly explained by the theory, but even the next-simplest atom, helium, proved a problem. Between 1918 and 1922, Bohr and Kramers in Copenhagen, Edwin Kemble and John Van Vleck at Harvard, Alfred Landé in Frankfurt am Main, and James Franck and Fritz Reiche in Berlin investigated the helium spectrum with the tools available from quantum theory. They used different methods and approaches, but the overall result was uniformly disappointing: Although the predictions of quantum theory were not entirely at variance with the measurements, the discrepancies were blatant enough to make most experts agree with Van Vleck's conclusion of 1922 that "some radical modification of the conventional quantum theory of atomic structure appears necessary" (Kragh 1979, 132). This became even more clear in 1923, when Born and Heisenberg gave a detailed, systematic analysis of the helium atom and concluded that the derived spectrum differed from the one observed. This was, as Born wrote to Bohr, a "catastrophe" (Darrigol 1992, 177). The anomalous Zeeman effect in many-electron atoms was no less catastrophical. During 1920–24, many physicists attacked the problem, including Landé, who was able to give a phenomenological explanation of the observed splitting of spectral lines. However, neither Landé, Sommerfeld, Pauli, Heisenberg, nor other physicists occupied with the problem could justify their results in terms of quantum theory. "It's a great misery with the theory of anomalous Zeeman effect," Pauli wrote to Sommerfeld on July 19, 1923. He added that the misery included "[all] atoms containing more than one electron" (Mehra and Rechenberg 1922:1, 502).

In fact, the misery was not limited to atoms with more than one electron. Even one-electron systems caused trouble, although this was only reluctantly accepted by the physicists who wanted to believe that the theory of hydrogenic atoms was in perfect shape. In 1921–23 the hydrogen molecule ion (H_2^+) was investigated by Pauli and, independently, Karl F. Niessen in the Netherlands. The calculated ionization energy did not agree with experiments made shortly afterward. Another anomaly was the discovery in 1922 by Otto Oldenberg, at the University of Munich, that the hydrogen lines exhibit an anomalous Zeeman effect in weak magnetic fields (contrary to the normal Zeeman effect in stronger fields). According to the Bohr-Sommerfeld theory, this so-called Paschen-Back effect—named after Friedrich Paschen and Ernst Back, who discovered it in 1912—ought not to occur in hydrogen. In 1924 it was realized that quantum theory could not account for the case of an electron moving in crossed electric and magnetic fields. Also, the covalent bond between, for example, two hydrogen atoms remained unexplained—in general, the physicists' quantum theory had little to offer to the chemists. In late 1924 Robert Mulliken concluded from his study of molecular spectra that the lowest state of the harmonic oscillator was not zero, but given by

$\frac{1}{2}$ = $h\nu$. The existence of a zero-point energy agreed with Planck's ill-fated theory of 1911, but not with later knowledge. The problem was that according to the Bohr-Sommerfeld theory, there should be no zero-point energy, contrary to Mulliken's discovery. Finally, between 1921 and 1924 it was established experimentally that slow electrons penetrate freely through an argon gas despite the strong interatomic fields of force. This so-called Ramsauer effect, discovered (among others) by the Heidelberg physicist Carl Ramsauer, was recognized to be a quantum phenomenon of some kind, but it defied explanation. There were thus many relevant experiments and facts that the Bohr-Sommerfeld theory was unable to explain and, in this sense, were anomalies. However, most of them were not considered to be very serious problems and were of limited importance in the process that created the quantum crisis in about 1924. For example, most physicists ignored or explained away the theory's inability to account for valence, and even the hydrogen molecule ion anomaly was not seen as a major problem of the same serious nature as the helium atom and the anomalous Zeeman effect.

By 1924, the accumulation of experimental anomalies, together with a widespread dissatisfaction with the conceptual and logical structure of the existing quantum theory, created a situation of crisis in the small community of atomic physicists. Several physicists concluded that the Bohr-Sommerfeld quantum theory was irremediably wrong and had to be replaced by some other theory. On the other hand, given its many successes, the "old" theory could hardly be completely wrong and was generally expected to relate to the new quantum theory in some correspondence-like way. Max Born believed in 1923 that "the whole system of concepts of physics must be reconstructed from the ground up" (Forman 1968, 159). It was also Born who coined the term "quantum mechanics" in a 1924 paper, in which he dealt with the problematic translation of classical formulas into their quantum-theoretic analogues by means of the correspondence principle. The inadequacy of the Bohr-Sommerfeld theory was recognized and there existed a name—quantum mechanics—for its successor. Unfortunately, no one knew what this quantum mechanics of the future would look like.

The early quantum theory was cultivated at three research centers in particular. In Munich, Sommerfeld established his important school with scientists such as Epstein, Rubinowicz, Gregor Wentzel, and Wilhelm Lenz among his collaborators. Born turned to atomic theory relatively late, only in 1921 after having become professor at Göttingen; with assistants such as Pauli, Friedrich Hund, and Pascual Jordan, he made Göttingen a world center of quantum theory. There were close connections between the two German centers, as exemplified by Sommerfeld's students Pauli and Heisenberg, who came to Göttingen from Munich. It was Bohr in Copenhagen, however, who was the dominating force in the early phase of atomic theory. A new university institute of theoretical physics, generally known as "Bohr's Institute," was founded in 1921 and attracted a large number of visitors from all around the world. Both before and after quantum mechanics, Bohr's institute was recognized to be the mecca of atomic theory (table 11.1).

TABLE 11.1
Foreign Visiting Physicists in Copenhagen, 1916–30

Name	*Years*	*Age*	*Country*
H. Casimir	1929, 1930	20	Netherlands
C. Darwin	1927	40	UK
D. Dennison	1924–26, 1927	24	USA
P. Dirac	1926–27	24	UK
R. Fowler	1925	36	UK
J. Franck	1921	39	Germany
E. Fues	1927	34	Germany
G. Gamow	1928–29, 1930	24	USSR
S. Goudsmit	1926, 1927	24	Netherlands
D. Hartree	1928	31	UK
W. Heisenberg	1924–25, 1926–27	22	Germany
W. Heitler	1926	22	Germany
G. Hevesy	1920–26	35	Hungary
E. Hückel	1929	33	Germany
F. Hund	1926–27	30	Germany
P. Jordan	1927	25	Germany
O. Klein	1918–22, 1926–31	24	Sweden
H. Kramers	1916–26	22	Netherlands
L. Landau	1930	22	USSR
A. Landé	1920	32	Germany
N. Mott	1928	23	UK
Y. Nishina	1923–28	33	Japan
W. Pauli	1922–23	22	Austria
L. Pauling	1927	26	USA
S. Rosseland	1920–24, 1926–27	26	Norway
A. Rubinowicz	1920, 1922	31	Poland
J. Slater	1923–24	23	USA
L. Thomas	1925–26	22	UK
G. Uhlenbeck	1927	30	Netherlands
H. Urey	1923–24	30	USA
I. Waller	1925–26, 1927, 1928	27	Sweden

Note: This list is not complete. The age refers to the physicist's age at his first visit. All the visiting physicists worked with Bohr at the University of Copenhagen, from 1921 at Bohr's Institute for Theoretical Physics. The 63 physicists visiting for at least one month between 1920 and 1930 were from 17 countries, with the most visitors from the United States (14), Germany (10), Japan (7), England (6), and the Netherlands (6).

Source: Robertson 1979.

Heisenberg's Quantum Mechanics

The path to quantum mechanics went over radiation theory, involving as an important component attempts to construct quantum-theoretical dispersion theories on the basis of difference rather than differential equations. This work was pursued in Göttingen by Born and his collaborators, and also in Copenhagen, where Kramers was deeply occupied with the dispersion problem. In the fall of 1924, Kramers and Heisenberg published an important theory of dispersion which, in retrospect, can be seen as the first decisive step toward the new quantum mechanics. The radiation problem had by then become acute, especially as a result of Arthur Compton's discovery in 1923 that pulses of monochromatic x-rays act as particles with momentum and energy in accordance with Einstein's old light quantum hypothesis ($p = h\nu/c$ and $E = h\nu$, where ν is the frequency). A similar conclusion was obtained independently by Debye in Zurich. Compton's important discovery caused great concern in Copenhagen, where Bohr firmly resisted the light quantum or photon interpretation. As an alternative, Bohr and Kramers developed an idea of John Slater's into a nonphoton radiation theory. The 1924 Bohr-Kramers-Slater theory was based on the notion of "virtual oscillators" and the assumption that the energy was only statistically conserved in interactions between atoms and radiation. Bohr and his collaborators not only abandoned strict energy conservation, but also argued that radiative processes could not be described causally in space and time. Although the controversial Bohr-Kramers-Slater theory was short-lived—it was abandoned when Walther Bothe and Hans Geiger showed in the spring of 1925 that it disagreed with experiments—it was very influential and guided Heisenberg in his further thinking. Bohr accepted the experimental refutation, but not the photon.

The growing crisis in quantum theory was discussed by Bohr, Kramers, Heisenberg, and Pauli at a meeting in Copenhagen in March 1925. A few months later, back in Göttingen, Heisenberg had found a way to formulate an abstract quantum mechanics that promised to be fundamental, logically consistent, and not plagued by the difficulties of the Bohr-Sommerfeld theory. The leading conceptual theme in the new theory was, as Heisenberg wrote in the brief abstract to his seminal paper in *Zeitschrift für Physik* of September 18, 1925, the search for "a basis for theoretical quantum mechanics founded exclusively upon relationships between quantities which in principle are observable." Later in the article, he elaborated: "[I]t seems sensible to discard all hope of observing hitherto unobservable quantities, such as the position and period of the electron. . . . Instead it seems more reasonable to try to establish a theoretical quantum mechanics, analogous to classical mechanics, but in which only relations between observable quantities occur." The positivistic observability criterion was the philosophical basis of the theory, but not a particularly new or controversial basis. The idea that quantum theory ought to build on observable quantities, and that electronic orbits were therefore to be dispensed with, was widely discussed at the time, especially by Pauli and Born. As early as 1919, in connection with a criticism of Weyl's unified theory of gravitation and electromagnetism, Pauli had emphasized that only quantities that are observable ought to enter physical theories. In the spring of 1925, Born and Jordan repeated the same message in a quantum-theoretical context, calling it "a fundamental principle of great importance and fertility." However, there was no royal road from the observability principle to quantum mechanics, the roots of which are rather to be found in Heisenberg's sophisticated use of Bohr's correspondence principle and his intense intellectual interaction with Pauli.

To give just a glimpse of Heisenberg's reasoning, consider the position coordinate of an electron in an atom, $x(n,t)$, where n may denote the energy. The electron may perform some periodic motion with frequency $\omega(n)$. Although $x(n,t)$ is not an observable quantity, it can be written as a Fourier series, the terms of which can be related to observables, namely, as the sum $\Sigma a_\alpha(n) \exp[i\alpha\omega(n)t]$, where summation is over the integral α values. The expression is characterized by a double set of indices, n and α, and Heisenberg suggested that the classical term $a_\alpha(n) \exp[i\alpha\omega(n)t]$ corresponded to the quantum term $a(n, n - \alpha) \exp[i\omega(n, n - \alpha)t]$. The new quantity was an array or table of symbols that depended on the transition between the two quantum states n and $n - \alpha$. Heisenberg found that the multiplication of two such quantum tables, x and y, did not satisfy the commutative law, that is, that xy differed from yx. This was a mysterious result, and Heisenberg at first considered it more disturbing than important.

Heisenberg's new "reinterpretation" (*Umdeutung*) of mechanics was highly abstract and not easily understood, not even by Heisenberg himself. The theory led to some physical results, although at first these were not very impressive. Heisenberg initially had sought to apply his theory to the hydrogen atom, only to find out that even this simple case was too complicated to solve. He therefore turned to the less realistic, but still nontrivial, case of the anharmonic oscillator, which he was able to treat satisfactorily. At the same time, he could offer a justification of Bohr's frequency condition ($E_n - E_m = h\nu_{nm}$) and the Bohr-Sommerfeld quantization conditions. For the harmonic oscillator, he found the energy spectrum $E_n = (n + \frac{1}{2})h\nu$ and thus derived the zero-point energy that had recently received support from molecular spectroscopy. In Göttingen, Born quickly realized the importance of Heisenberg's theory, which he examined and extended in a paper written with Jordan. Born realized that Heisenberg's symbolic noncommutative multiplication could be written in terms of matrix calculus and that the quantum mechanical variables were matrices. On this basis, Born and Jordan proved the fundamental commutation relation between momentum and position, that

pq − **qp** = $(h/2\pi i)$ **1** where **1** is the unit matrix. With the insight that matrix calculus was as designed for quantum mechanics, things went quickly. Heisenberg's theory was established on a firm basis in November 1925 with the famous "three-man paper" (*Dreimännerarbeit*) of Born, Heisenberg, and Jordan. Without using (or knowing about) matrices, the theory was also developed by twenty-three-year-old Paul Dirac at Cambridge University. In the fall of 1925, Dirac had ready his own version of quantum mechanics, which built on a translation of Heisenberg's products of two quantum quantities $(xy - yx)$ into the Poisson brackets known from classical dynamics. Many of the results found by Born, Heisenberg, and Jordan were independently obtained by Dirac, whose algebraic version of quantum mechanics became known as q-number algebra.

In whatever version, the new quantum mechanics was more impressive from a mathematical than from an empirical point of view. Many physicists were skeptical because of the theory's lack of visualizability and its unfamiliar mathematical formalism. They wanted to see if the theory was also empirically fruitful and could deal with simple physical systems actually occurring in nature. A minimum requirement would be that it could reproduce the energy spectrum of hydrogen in accordance with Bohr's old theory. Pauli and Dirac independently treated the nonrelativistic hydrogen atom according to quantum mechanics and showed in early 1926 that it gave the correct results. At that time, the hypothesis of the spinning electron had been introduced from spectroscopic evidence by the Dutch physicists Samuel Goudsmit and George Uhlenbeck. This extremely important discovery was not welcomed by all the creators of quantum mechanics. Bohr and Pauli at first rejected the idea of electron spin, among other reasons because it seemed inconsistent with hydrogen's fine structure. A closer look proved that there was no conflict between spin and quantum mechanics, however, and in the spring of 1926 Heisenberg and Jordan used a simple form of spin quantum mechanics to derive hydrogen's fine structure in approximate agreement with Sommerfeld's formula. This was satisfactory, and it was even more satisfactory that they also succeeded in explaining the anomalous Zeeman effect, the old puzzle that had haunted the earlier quantum theory. Yet, these were only reproductions of already known results. So far, quantum mechanics had not produced a single prediction of a novel phenomenon.

Schrödinger's Equation

The Austrian Erwin Schrödinger, professor of physics at the university of Zurich, did not belong to the Copenhagen-Göttingen-Munich tradition. He had worked in a variety of fields, including radioactivity, general relativity, thermodynamics, and gas theory, but had not shown much interest in either spectroscopy or atomic theory. When he started developing wave mechanics, he was aware of Heisenberg's new theory, but it was a theory that did not appeal to him and failed to inspire him. On the contrary, as he wrote in one of his 1926 papers on wave mechanics, "I was absolutely unaware of any genetic relationship with Heisenberg. I naturally knew about this theory, but because of the to me very difficult-appearing methods of transcendental algebra and because of the lack of visualizability [*Anschaulichkeit*], I felt deterred by it, if not to say repelled" (Moore 1989, 205). Compared with Heisenberg, Dirac, Jordan, and Pauli, Erwin Schrödinger was not only conservative, but at age 39, he was also "old." But, it turned out, not hopelessly so.

In 1925, while working on gas theory, Schrödinger studied the work of a relatively unknown French physicist, Louis de Broglie, who in his thesis of 1924 had suggested a deep-lying duality between matter and waves. In an attempt to bridge quantum theory and relativity, de Broglie proposed to combine Einstein's two 1905 formulas for the energy of light quanta and matter by the simple but speculative relationship $h\nu = mc^2$. That is, according to de Broglie, a particle of mass m could be ascribed a frequency and be characterized by a phase wave. De Broglie used his idea to propose a wave interpretation of Sommerfeld's quantization condition and predicted that a beam of moving electrons with momentum p should exhibit wave nature with a wavelength given by $\lambda = h/p$. This result, valid for all kinds of particles and subsequently derived from quantum mechanics, states the famous de Broglie wavelength. The theory was not received favorably and was ignored by most physicists outside Paris. French theoretical physics had a low reputation among atomic physicists, and it was not expected that anything interesting could come from Paris. The official French boycott of German physics did not make it easier to appreciate a glimpse of Parisian genius. However, Einstein found some of de Broglie's ideas to be valuable in his own work on gas quantum theory (Bose-Einstein statistics) and, through Einstein, they were taken up by Schrödinger. He, too, was primarily interested in de Broglie's theory in connection with gas theory, but in late 1925 he concentrated on a new wave theory of atoms inspired by de Broglie's wave-particle dualism. The result was a wave equation for the hydrogen atom which, when solved, would give the energy eigenvalues—that is, the spectrum. This first Schrödinger equation was, in accordance with de Broglie's theory, fully relativistic and therefore expected to yield Sommerfeld's fine structure formula. When Schrödinger finally succeeded in solving the equation, however, he found that although it indeed gave a fine structure formula, it did not reproduce the correct spectrum. Something had gone wrong, and Schrödinger decided to publish only the nonrelativistic approximation and the resulting Bohr formula.

Schrödinger's work on wave mechanics appeared in four long papers in

Annalen der Physik in the spring and summer of 1926, under the common title "Quantization as an Eigenvalue Problem." The Austrian physicist introduced his fundamental wave equation in several different ways, none of which reflected the route that had originally led him to the equation. The important thing was the energy eigenvalue equation itself, which he wrote as $\Delta\psi + 8\pi^2 m/h^2(E - V) = 0$ for a particle subject to a force with potential energy V. For the hydrogen atom, $V = -e^2/r$. Schrödinger noted that the equation would follow from the classical equation of motion $E = p^2/2m + V$ if the momentum was replaced by the differential operator $h/2\pi i\ \partial/\partial x$ acting on the wave function ψ. Quantization was not introduced axiomatically but, in a sense, explained, namely by requiring that the wave function satisfying the equation be single-valued. Along the same line, Schrödinger believed to have explained Bohr's discontinuous quantum jumps in terms of wave theory, a notion he disliked profoundly. "It is hardly necessary to point out," he reminded the readers of *Annalen*, "how much more gratifying it would be to conceive a quantum transition as an energy change from one vibrational mode to another than to regard it as a jumping of electrons. The variation of vibrational modes may be treated as a process continuous in space and time and enduring as long as the emission process persists" (Jammer 1966, 261). Similarly, in a letter to Lorentz, Schrödinger described Bohr's model of light emittance as "monstrous" and "really almost inconceivable." He was evidently excited to be rid of such monstrosities: "I was so extremely happy, first of all, to have arrived at a picture in which at least something or other really takes place with that frequency which we observe in the emitted light that, with the rushing breath of a hunted fugitive, I fell upon this something in the form in which it immediately offered itself, namely as the amplitudes periodically rising and falling with the beat frequencies" (MacKinnon 1982, 234). In the summer of 1926, Schrödinger introduced the energy operator $E = -h/2\pi i\ \partial/\partial t$ and formulated the time-dependent wave equation as $ih/2\pi\ \partial\psi/\partial t = H\psi$ or just $H\psi = E\psi$, where ψ is a function of both space and time coordinates and H is the Hamilton operator. At the same time, he realized that the wave function had to be a complex function (in the mathematical sense) and not, as he had earlier believed, real.

Schrödinger's wave mechanics had great advantages over the competing systems of quantum mechanics. In particular, it built on mathematical concepts and operations well known from other areas of theoretical physics and was therefore much easier to use in practical calculations. Physicists who were not well acquainted with the mathematical methods could look them up in Richard Courant and David Hilbert's *Methods of Mathematical Physics* that appeared in 1924 and happened to cover just the methods needed in the new quantum mechanics. In addition to facilitating calculations, wave mechanics was also less abstract than matrix mechanics and, according to many physicists, preferably from a conceptual point of view. Most of the results derived by Schrödinger were duplications of results already obtained, usually in a more cumbersome way, from quantum mechanics. For example, Schrödinger found the eigenvalues of the harmonic oscillator to be $E_n = (n + \frac{1}{2})h\nu$, the same result as Heisenberg had found. How did it happen that quantum mechanics and wave mechanics, two theories based on completely different concepts of nature and using very different mathematical tools, gave the same results when applied to simple physical systems? That the two theories are mathematically (but not physically) equivalent was suspected at an early date and in the spring of 1926, Schrödinger proved that any wave-mechanical equation could be translated into a corresponding equation in quantum or matrix mechanics, and vice versa. Similar equivalence proofs were independently provided by Carl Eckart in the United States and, without bothering to publish it, by Pauli.

With regard to the question of the nature and significance of the wave function, Schrödinger was not very clear. For a time, he wanted to understand particles as consisting of waves, that is, to construct particles purely of concentrated wave packets. However, this wave model of matter ran into difficulties and later in 1926, Schrödinger suggested that the wave function, through the product $\psi\psi^*$ (where ψ^* is the complex conjugate of ψ), was a kind of electrical weight function, the charge density being represented by $e\psi\psi^*$. According to this picture, the electron was thus not a sharply localized particle, but was smeared out in space. Shortly afterward Born, in a study of collision processes, suggested his famous probability interpretation, according to which $\psi\psi^* dV$ is the probability that the particle is in the ψ-state in the volume element dV. Born's interpretation, quickly adopted and developed by Pauli, Jordan, Dirac, and others, was of great importance because it introduced explicitly into microphysics an irreducible element of probability. This implied a change in the meaning of natural laws, but not that causal laws were no longer fundamental in physics. As Born formulated it in his paper from the summer of 1926, "The motion of particles conforms to the laws of probability, but the probability itself is propagated in accordance with the law of causality" (Jammer 1966, 285).

Schrödinger's wave mechanics was initially received with some scepticism, and sometimes even hostility, by the quantum theorists in Göttingen and Copenhagen. They tended to consider the emphasis on classical virtues such as spatio-temporal continuity and visualizability a retrograde step. Heisenberg reported to Pauli that he found Schrödinger's theory to be "disgusting." On the other hand, they recognized the force in Schrödinger's system and after the equivalence proofs, most of them adopted a pragmatic attitude toward the two competing formulations of quantum mechanics. The majority of physicists used the language and mathematics of wave mechanics, but interpreted Schrödinger's theory in accordance with the ideas of Bohr, Heisenberg, and Born. After Born's probabilistic interpretation of wave mechanics, the question of how to generalize the interpretation and relate it to matrix mechanics came to the forefront. The essential step in this process, leading to a completely general and unified formalism of quantum mechanics, was the transformation theory developed independently by Dirac and Jordan at the end of 1926. With this theory, quantum mechanics had obtained an elegant mathematical structure and the difference between Schrödinger's and Heisenberg's formulations lost most of its former significance.

Even with this satisfactory state of affairs, and the many successful applications of quantum mechanics to spectroscopy and other areas of physics, the theory was not without problems. There was, for example, the question of the relationship between relativity and quantum mechanics. If quantum mechanics was really a fundamental theory of the microcosmos, it ought to be consistent with the fundamental theory of macroscopic bodies, the (special) theory of relativity. Yet it was obvious from the very beginning that this was not the case. The Schrödinger equation is of the second order in the space derivative and of the first in the time derivative, in contradiction to the theory of relativity. It was not too difficult to construct a relativistic quantum wave equation, such as Schrödinger had already done privately and as Oskar Klein, Walter Gordon, and several other physicists did in 1926–27. Unfortunately, this equation, known as the Klein-Gordon equation, did not result in the correct fine structure of hydrogen and it proved impossible to combine it with the spin theory that Pauli had proposed in 1927. The solution appeared in January 1928, when Dirac published his classical paper on "The Quantum Theory of the Electron," which included a relativistic wave equation that automatically incorporated the correct spin. Dirac's equation was of the same general form as the Schrödinger equation, $H\Psi = \mathrm{i}h/2\pi\ \partial\Psi/\partial t$, but the Hamilton function was of the first order in $\partial\Psi/\partial x$ and included matrices with four rows and four columns; correspondingly, the Dirac wave function had four components. Most remarkably, without introducing the spinning electron in advance, the equation contained the correct spin. In a certain, unhistorical sense, had spin not been discovered empirically, it would have turned up deductively in Dirac's theory. The new theory was quickly accepted when it turned out that the Dirac eigenvalue equation for a hydrogen atom resulted in exactly the same energy equation that Sommerfeld had derived in 1916. Dirac's relativistic wave equation marked the end of the pioneering and heroic phase of quantum mechanics, and also marked the beginning of a new phase. It soon turned out that the equation contained surprises, subtleties, and problems undreamed of by Dirac when he derived the equation.

Dissemination and Receptions

Quantum mechanics spread very rapidly from Göttingen and the few other centers where the theory had originally been constructed. The transmission of new ideas and results took place formally, through the scientific journals, as well as informally, through conferences and exchanges of letters and manuscripts. It was only a small group of the European physicists who was part of the informal network organized around Göttingen and Copenhagen, and they had the advantage of speedy communications that outsiders had not. From the summer of 1925 to the spring of 1927, publications on quantum mechanics exploded, initially with a doubling time of about two months. Between July 1925 and March 1927, more than 200 papers on the new theory were submitted to the scientific journals; the time between a manuscript being received by the editors and the paper appearing in print was often short. To give an example, Dirac's paper on the hydrogen spectrum was received by the *Proceedings of the Royal Society* on January 22, 1926 and was published on March 1 of that year. Even speedier was the publication of Born's paper on collision problems, in which he introduced the probability interpretation. *Zeitschrift für Physik* received the paper on June 25, 1926 and it was published on July 10. The speeds of publication has an explanation: According to the policy of the *Zeitschrift*, any "reputable" physicist could have his or her paper published without refereeing. Max Born was definitely a reputable physicist. In the case of the *Proceedings*, a Fellow of the Royal Society could communicate papers by someone else and in this way obviate the need for referees. Dirac's paper was communicated by Ralph Fowler, F.R.S.

Just keeping abreast with the literature was a difficult task. Edward Condon recalled the pace with which quantum mechanics developed at the time: "Great ideas were coming out so fast during that period (1926–1927) that one got an altogether wrong impression of the normal rate of progress in theoretical physics. One had intellectual indigestion most of the time that year, and it was most discouraging" (Sopka 1988, 159). The climate was highly competitive and frequently, results were published independently by different physicists or physicists had to give up publishing their results altogether because they were beaten by competing colleagues. In this publication game, the German physicists and their allies in Copenhagen had the advantage of easy and quick access to both published and unpublished results. Americans, on the other hand, normally had to wait at least one additional month until they could read the articles in the German physics journals. The competitive spirit in quantum mechanics at the time was expressed by John Slater who, in a letter to Bohr in May 1926, wrote somewhat bitterly of his frustration at being beaten in the publication race: "It is very difficult to work here in America on things that are changing so fast as this [quantum mechanics] is, because it takes us longer to hear what is being done, and by the time we can get at it, probably somebody in Europe has already done the same thing" (Kragh 1990, 21).

Pauli once referred to quantum mechanics as *Knabenphysik*—boys' physics—because so many of the main contributors were still in their twenties. For example, in September 1925 Heisenberg was 23 years old, Pauli 25, Jordan 22, and Dirac had just turned 22. More than half of the early generation of quantum physicists, that is, the eighty or so contributors to quantum mechanics in this period, were born after 1895, and they wrote about 65 percent of all the papers. Many of these bright young physicists felt, arrogantly, that quantum mechanics belonged to them and that most elder physicists were just incapable of mastering the theory. "It was very difficult not to be senile after having lived thirty years," recalled Friedrich von Weizsäcker, in 1932 a twenty-year-old member of the new-generation physics gang. "I feel that the general attitude was just an attitude of . . . an immense 'Hochmut,' an immense feeling of superiority, as compared to old professors of theoretical physics, to every experimental physicist, to every philosopher, to politicians, and to whatever sorts of people you might find in the world, because we had understood the thing and they didn't know what we were speaking about" (Kragh 1996b, 89). As it appears from table 11.2, Germany and its neighboring countries dominated the early phase of quantum mechanics. Whether the quantum physicists were Germans or not, the principal language of quantum mechanics was German. The most important of the journals was the *Zeitschrift für Physik*, in which many non-Germans published their results and which included sixty-eight papers on quantum mechanics between July 1925 and March 1927. Also worth noticing is the weak position of France, where quantum mechanics only disseminated slowly and with no contributions of great importance. The only French contributors to the literature of quantum mechanics in the period were Louis de Broglie and Léon Brillouin.

Compared to the theory of relativity, quantum mechanics developed rapidly, disseminated very quickly, and met almost no resistance. Also contrary to relativity, quantum mechanics attracted little public interest. Eddington was one of the few scientists who wrote about the theory for a nonscientific readership. Although quantum mechanics was no less counterintuitive than relativity, there was no quantum counterpart to the antirelativistic literature that flourished in the 1920s. The rapid scientific dissemination took place not only by publication of papers, but also through lecture visits, review articles, textbooks, and lecture courses on quantum mechanics. The first textbook devoted specifically to quantum mechanics seems to have been *The New Quantum Mechanics* by the Cambridge physicist George Birtwistle. Published in 1928, the book gave a detailed review of the main contributions to matrix and wave mechanics, from de Broglie's 1923 theory on matter waves

TABLE 11.2
Publications and Authors in Quantum Mechanics, July 1925 to March 1927

Country	*Number of authors*	*Papers written by them*	*Papers written in the country*	*Papers published in the country*
Germany	19	59.5	54	120
Switzerland	5	17	21	0
Austria	5	7	6	0
Denmark	4	7	17	1
Netherlands	2	4	5	1
Central Europe, total	35	94.5	103	122
France	2	12	12	14
Britain	6	15	18	30
USA	19	34.5	26	27
USSR	11	11	11	9
Italy	3	4	4	1
Sweden	2	6	5	0
Other	3	5	3	0
Total	81	182	182	203

Note: The first two columns do not refer to the authors' nationalities, but to the countries where they mainly stayed in the period. The numbers of papers in column 2 include original papers and reviews, but not translations and preliminary notes, which are included in column 4. Column 3 shows the effect of foreign visitors.

Source: Based on data in Kojevnikov and Novik 1989.

to Bohr's complementarity principle. Other early monographs included Arthur Haas's *Materiewellen und Quantenmechanik* (1928), Weyl's *Gruppentheorie und Quantenmechanik* (1928), Sommerfeld's wave mechanical supplement to his *Atombau und Spektrallinien* (1929), Edward Condon's and Philip Morse's *Quantum Mechanics* (1929), and Born and Jordan's *Elementare Quantenmechanik* (1930). The last book, which was based on abstract matrix mechanical methods rather than the more easily applicable wave mechanics, was not a success. The most influential of the early books on quantum mechanics was undoubtedly Dirac's *The Principles of Quantum Mechanics* of 1930 which, in spite of being abstract and generally unpedagogical, was highly successful. It went through several editions and translations and was the standard work on quantum mechanics in the 1930s.

Whereas the old quantum theory had not been cultivated much by American physicists, the new quantum mechanics was eagerly and positively received by a physics community in strong growth. During the last half of the 1920s, American physics matured and rose to a leading position in world physics. For example, the number of members and fellows of the American Physical Society increased rapidly, from about 1,100 in 1920 to 1,800 in 1926 and 2,400 in 1930. *Physical Review* correspondingly increased in both size and importance. In 1929 the journal's 2,700 pages were distributed among 281 papers, of which approximately 45 dealt with aspects of quantum mechanics. At that time there was felt a need for broader reviews, for example of quantum mechanics, and the result was the launching in 1929 of *Reviews of Modern Physics*, the first two issues of which appeared as *Physical Review Supplement*. Among the first review articles was Edwin Kemble's and Edward Hill's "The General Principles of Quantum Mechanics," which covered both matrix and wave mechanics in more than 100 pages. An important reason for the quick and smooth reception of quantum mechanics in the United States was the trips to Europe made by young American physicists and the visits to American universities made by European quantum physicists. In the late 1920s, more than thirty Americans studied at the European centers of theoretical physics, of which Göttingen, Munich, Zurich, Copenhagen, and Leipzig were the most popular places. Visits and lecture tours of European physicists became common and included pioneers such as Sommerfeld, Born, Heisenberg, Dirac, Kramers, Hund, and Brillouin. Born's lectures at the Massachusetts Institute of Technology and other American universities during 1925–26 were of particular importance, because they introduced quantum mechanics to American physicists at a time when the theory was still under completion.

The last half of the 1920s witnessed a remarkable shift in the internal structure of physics, from experimental to theoretical work. Whereas in 1910 only about 20 percent of the world physics literature consisted of papers that were mainly theoretical, in 1930 the percentage was close to 50. The shift was a worldwide trend, not least caused by the new quantum mechanics, but it was particularly important in the United States, where theory had traditionally been less cultivated than in Europe. Although no Americans participated in the creative phase of quantum mechanics, they quickly caught up with the subject and contributed importantly to the second phase. Among the early generation of American quantum physicists were Carl Eckart, John Slater, John Van Vleck, David Dennison, Robert Oppenheimer, and the chemist Linus Pauling. Like most other members of the first generation of quantum physicists, they worked primarily with theoretical aspects of quantum mechanics.

Many European physicists were deeply occupied with the philosophical implications of the new mechanics and devoted much time to discussing the broader meaning of the theory's strange nonclassical features. Do physical properties come into existence only as a result of measurements? If so, is the observed world real and objective? Can the object and subject be distinguished or do they form an indissoluble whole? Can the lessons of quantum mechanics be extrapolated to society and culture? For Bohr, Einstein, Heisenberg, Jordan, and others, it was as important to understand these features as it was to calculate physical problems with the new technique. The Americans' attitude was markedly different. Although there was considerable interest among the Americans in foundational issues, for example in the correct formulation of the uncertainty principle, they did not care much about the larger philosophical problems associated with quantum mechanics. These simply did not enter the papers and books published by American physicists, whose attitude was pragmatic and inspired by the operationalism laid out in Bridgman's *Logic of Modern Physics*. According to this attitude, experimental results were all that mattered and could be meaningfully discussed; the job of the quantum physicist was therefore to make calculations that could be checked experimentally. Slater's philosophical (or perhaps antiphilosophical) views, as expressed in 1937, were accepted by the large majority of American physicists:

> A theoretical physicist in these days asks just one thing of his theories: if he uses them to calculate the outcome of an experiment, the theoretical prediction must agree, within limits, with the result of the experiment. He does not ordinarily argue about philosophical implications of his theory. . . . Questions about a theory which do not affect its ability to predict experimental results correctly seem to me quibbles about words, rather than anything more substantial, and I am quite content to leave such questions to those who derive some satisfaction from them. (Schweber 1990, 391).

The Copenhagen spirit—that is, Bohr's complementarity principle and related interpretations of quantum processes—left the Americans as cold as it excited many Continental physicists. Of course, the lack of interest in the complementarity principle was not exclusive to the Americans, but the extent to which American physicists—including those who had stayed with Bohr in Copenhagen—ignored Bohr's philosophy is nonetheless noticeable. "I didn't altogether like it," Dirac recalled in 1963 about the complementarity principle, and explained that "it doesn't provide you with any equations which you didn't have before" (Kragh 1990, 84). For Dirac, this was reason enough to dislike the idea, and his colleagues in the United States tended to agree.

When quantum mechanics arrived, theoretical physics was essentially limited to Europe and North America. Modern physics came late to Japan, where it took off only in the 1930s. The Japanese pioneer was Yoshio Nishima, a physicist who had spent most of the 1920s in Europe on an extended stay that included six years in Copenhagen with Bohr. Quantum mechanics was in part introduced through lectures given by invited Westerners, including Otto Laporte, Sommerfeld, Dirac, and Heisenberg. From the spring of 1931, Nishima lectured on quantum mechanics at the University of Kyoto, using Heisenberg's new *Die physikalische Prinzipien der Quantentheorie* and later Dirac's *Principle of Quantum Mechanics*. Nishima's efforts were instrumental in the creation of the strong school of theoretical quantum physics that emerged in Japan later in the 1930s.

Chapter 12

THE RISE OF NUCLEAR PHYSICS

THE ELECTRON-PROTON MODEL

SHORTLY AFTER the nuclear model of atomic structure had been accepted, several physicists started speculating about the structure of the atom's tiny nucleus. The general view was that proposed by Rutherford, namely, that the nucleus was made up of electrons and positive unit particles, the latter identical with hydrogen nuclei and often termed "positive electrons" or H-particles; or, from 1920 on, protons. It appeared evident that the nucleus included electrons, for the positive particles were clearly in need of some negative electricity to prevent the nucleus from exploding. Moreover, it was known since 1913 that the beta electrons had their origin in the nucleus and not in the outer layers of electrons. This had been argued by Bohr, among others, who in his 1913 work pointed out that "the fact that two apparently identical elements [isotopes] emit β-particles of different velocities, shows that the β-rays as well as the α-rays have their origin in the nucleus" (Bohr 1963, 53).

According to the nuclear atomic model, the mass number A and the atomic number Z would depend on the number of protons (p) and electrons (e) as $A = p$ and $Z = p - e$. On the other hand, nuclei of radioactive bodies also gave rise to alpha particles, which therefore were often assumed to be additional nuclear constituents. For a nucleus with a alpha particles, the equations would then be $A = 4a + p$ and $Z = 2a + p - e$. This hypothesis enjoyed general acceptance between 1915 and 1932. In fact, not a single physicist seems to have doubted the nuclear electron hypothesis and what has rightly been called the two-particle paradigm, namely, that all matter consisted of electrons and protons (although some of these might exist in bound forms, as alpha particles or other combinations). How, then, were the two or three nuclear species arranged in the nucleus? Given the almost complete lack of experimental evidence, it was a hopeless task to construct reliable nuclear models in the 1910s and 1920s. Nonetheless, a surprisingly large number of physicists (and chemists, too) were undeterred by the difficulties and speculated more or less freely on nuclear structures. Most of these models were pure speculations, often based on loose numerological arguments, but a few were of a more serious nature. For example, in 1918 the Munich physicist Wilhelm Lenz constructed a model of the alpha particle in accordance with the rules of quantum theory, namely, four protons revolving in an equatorial plane with one electron at each pole. Sommerfeld referred approvingly to this model in his *Atombau*.

The most eminent of the earliest generation of nuclear physicists was also the most prolific of the model makers, and not the least speculative. Inspired by his earlier alpha scattering experiments, Rutherford suggested in his 1920 Bakerian lecture that other particles in the nucleus than electrons, alphas, and protons might exist. Rutherford argued that there was evidence for a light helium nucleus, consisting of three protons bound by one electron (X_3^{++} in Rutherford's notation), and perhaps also a heavy hydrogen isotope consisting of two protons and one electron. And—why not?—the nucleus also contained a neutral particle made up of one electron and one proton, a "neutron," according to Rutherford. It was also at this occasion that Rutherford introduced the name "proton." Rutherford was particularly fascinated by the possibility of neutrons because these would "enter readily the structure of atoms, and may either unite with the nucleus or be disintegrated by its intense field, resulting possibly in the escape of a charged H atom [proton] or an electron or both" (Badash 1983, 886). Rutherford may not have been aware that his "neutrons" had been proposed as early as 1899 by the Australian physicist William Sutherland, who suggested that the ether consisted of doublets of positive and negative electrons. Sutherland's suggestion was taken over by Nernst in his authoritative textbook *Theoretical Chemistry*, where it appeared in all editions between 1903 and 1926. Whereas the idea of light helium nuclei was based on weak experimental evidence and abandoned in 1924, the neutron hypothesis had a long life and was taken quite seriously in Cambridge. James Chadwick believed in its existence as much as Rutherford, and tried on several occasions in the 1920s to detect the hypothetical particle. He did not succeed until 1932, and then it turned out that the observed neutral particle was not Rutherford's neutron after all. Rutherford continued to develop his ideas of nuclear structure, and in 1925 he reached the conclusion that the nucleus consisted of a massive core surrounded by positive and negative satellites (protons and electrons). He found his satellite model important enough to include it in *Radiations from Radioactive Substances*, written with Chadwick and Charles D. Ellis and published in 1930.

Most of Rutherford's hypotheses were based on interpretations of experiments in which substances were bombarded with alpha particles. In December 1917 he wrote to Bohr, "I am detecting & counting the lighter atoms set in motion by α particles & the results, I think, throw a good deal of light on the character & distribution of forces near the nucleus. I am also trying to break up the atom by this method" (Stuewer 1986a, 322). The most important of these experiments was made in Manchester in 1919, shortly before Rutherford left for Cambridge to become director of the Cavendish Laboratory. In a reinvestigation of experiments made earlier by Ernest Marsden,

Rutherford studied the action of alpha particles on various gases by detecting the scintillations produced by long-range particles formed by the action. With pure nitrogen, he observed what he called an anomalous effect, namely, the production of long-range particles similar to those obtained with hydrogen. "It is difficult to avoid the conclusion," he wrote, "that these long-range atoms arising from the collision of alpha particles with nitrogen are not nitrogen atoms but probably charged atoms of hydrogen, or atoms of mass 2. If this be the case, we must conclude that the nitrogen atom is disintegrated under the intense forces developed in a close collision with a swift α particle, and that the hydrogen atom which is liberated formed a constituent part of the nitrogen nucleus" (Beyer 1949, 136). Further work done in the Cavendish proved that Rutherford's conclusion was largely correct. He had achieved the first artificial disintegration of an atomic nucleus and thus opened up a new stage in the history of modern alchemy. The process was $^{14}N + {}^{4}He \rightarrow {}^{17}O + {}^{1}H$, although Rutherford originally interpreted it as $^{14}N + {}^{4}Hc \rightarrow {}^{13}C + {}^{4}He + {}^{1}H$. It was only in 1924, when cloud chamber photographs failed to show the track of any alpha particles from the recoil atoms, that the error was corrected.

During the following years, Rutherford and his Cavendish group continued this kind of experiments in the hope of transforming still more elements. Similar work was done at the Vienna Radium Institute, but not with the same results. Whereas Rutherford and Chadwick did not find evidence for disintegration of elements heavier than potassium, nor for beryllium and lithium, the Vienna physicists Gerhard Kirsch and Hans Petterson (the latter of whom was Swedish) claimed to be much more successful in disintegrating elements. Not only did they report results widely different from those produced in Cambridge, but they also attacked Rutherford's satellite model of the nucleus. The disagreement evolved into a protracted controversy with some of the same components that characterized the notorious N-ray episode at the beginning of the century. As a result of a visit Chadwick made to the Vienna Institute in 1927, he found that the Austrian-Swedish team did not control their results and that the counting of the scintillations was systematically biased toward the (too) high values they wanted to find. According to an historian: "The counting was done by women, the reasoning being that they could concentrate on the task more intensely than men, having little on their minds anyway, and by Slavic women because their large, round eyes were best suited for counting. The women were told what sort of counting rate was anticipated, and, being anxious to please, they provided it" (Badash 1983, 887).

Nuclear physics in the 1920s was intimately linked with radioactivity, the only source of high-energy projectiles being the alpha and beta particles emitted by naturally occurring radioactive substances. In order to measure the intensity and direction of particles, either scattered or produced by disintegration, simple scintillation devices were normally used. Visual counting of scintillations went back to 1908, when Erich Regener of Berlin University concluded that each alpha particle hitting a phosphorescent screen produced a scintillation. The simple method was used extensively until the early 1930s. Rutherford's experiment of 1919 did not use apparatus more advanced than that used in Geiger and Marsden's alpha scattering experiments some ten years earlier. Lack of money and available technology made the Cavendish experiments simple and in agreement with Rutherford's own predilection toward sealing wax and string methods. Yet it was not all sealing wax and string. One of the most important instruments in the infancy of nuclear physics was the electromagnetic mass spectrograph, developed by Francis Aston from its first version in 1919. By the late 1920s, the mass spectrograph had become a complicated and expensive instrument.

For detection purposes, the cloud chamber began to play in important role in the 1920s. The principle of making tracks of ionized droplets visible by means of sudden expansion was discovered by Charles T. R. Wilson in the Cavendish Laboratory in the late 1890s in connection with meteorological studies. By 1911, Wilson had constructed the first cloud chamber to study the paths of ionizing particles and taken the first cloud chamber photograph. In 1921 T. Shimizu, a Japanese physicist working at the Cavendish, found a means for working the chamber automatically, and the technique was further improved by Patrick M. S. Blackett. Practically all the innovative work on scintillation and cloud chamber techniques was done in the Cavendish Laboratory, which was also the birthplace of the gas ionization chambers or counters. The most effective version of the early ionization counters was designed by Hans Geiger in 1913. With the starting of the war, Geiger returned to Germany to serve in the artillery; after 1918, he continued his work on ionization counters. The modern, highly sensitive Geiger-Müller counter was a German invention. It was developed in 1928 by Geiger and his collaborator at the University of Kiel, Walther Müller. The development in detection methods in the 1920s and later owed much to electronics engineering, especially the use of vacuum tube circuits.

Quantum Mechanics and the Nucleus

Quantum mechanics was a general theory of atoms and electrons. It was assumed to hold for the atomic nucleus as well, but during the first phase of quantum mechanics, there were no attempts to apply the new theory to nuclear physics. The situation changed in the summer of 1928, when it was shown that alpha radioactivity could be understood in terms of quantum mechanics. The important quantum-mechanical theory of alpha decay was proposed independently by George Gamow in Göttingen and Ronald Gurney and Edward Condon in Princeton. Gamow, a twenty-four-year-old Russian physicist, argued (as Rutherford had done earlier) that the nuclear potential must be strongly attractive for very small distances and attain a maximum height before it merged with the repulsive Coulomb potential. He pictured the alpha particles as preexisting in the nucleus, vibrating or orbiting within the potential well. Classically, the alpha particle would not be able to penetrate the potential, but according to Born's interpretation of Schrödinger's wave mechanics, there would be a finite probability that a particle escaped the nucleus with an energy smaller than the maximum height of the potential. This is the famous case of a particle or matter wave penetrating or "tunneling" a potential barrier, a case that today enters all introductory textbooks in quantum mechanics but in 1928 was a new and exciting phenomenon.

Gamow's use of quantum mechanics enabled him to find the penetration probability and, after translating it into the decay constant, derive a linear relationship between the logarithm of the decay constant and the energy of the emitted alpha particles. This was just the Geiger-Nuttall relationship, which had been known empirically since the 1912 work of Geiger and the English physicist John Nuttall. There had earlier been derivations of the Geiger-Nuttall law—for example, by Frederick Lindemann in 1915—but these were pseudo-explanations based on ad hoc assumptions. Gamow's explanation, as well as the largely identical one proposed by Gurney and Condon, was much more satisfactory because it was based on fundamental theory. The Gamow-Gurney-Condon theory was extremely important, both because it provided a convincing demonstration that quantum mechanics applies to the atomic nucleus and because it formed the basis of other applications of quantum mechanics to nuclear physics.

The statistical nature of radioactivity had been a puzzle ever since it was recognized in the early part of the century. Numerous attempts had been made to provide a causal explanation of the origin of radioactivity, but it was only with quantum mechanics that it was realized that such attempts to make sense of radioactivity's statistical nature were futile. As Gurney and Condon put it in 1929, referring to these attempts: "This has been very puzzling so long as we have accepted a dynamics by which the behaviour of particles is definitely fixed by the conditions. We have had to consider the disintegration as due to the extraordinary conjunction of scores of independent events in the orbital motions of nuclear particles. Now, however, we throw the whole responsibility on to the laws of quantum mechanics, recognizing that the behaviour of particles everywhere is equally governed by probability" (Kragh 1997a, 357).

During the years following 1928, quantum mechanics was successfully applied to several other problems involving atomic nuclei, among which collision problems were particularly important. For example, in 1928 Nevill

Mott, a twenty-three-year-old Cambridge physicist, reproduced Rutherford's 1911 expression for the scattering of charged particles from a pointlike nucleus. More interestingly, for collisions between two identical particles (such as alpha particles scattered by a helium gas) he predicted a result that differed from Rutherford's, namely, that for low velocities scattering at 45° would occur about twice as frequently as classically expected. The prediction, confirmed by cloud chamber experiments by Patrick Blackett and Frank Champion in 1931, was one of the first novel predictions of quantum mechanics in the nuclear regime.

The works of Gamow, Mott, and Gurney and Condon did not shed any new light on the structure of the nucleus, which was still assumed to consist of electrons and protons. These were the only known elementary particles and, besides, radioactive nuclei emitted electrons in the form of beta particles. On the other hand, during the late 1920s it was gradually realized that somehow, electrons ought to have no place in the atomic nucleus. They were necessary, but unwelcome. One of the problems of the electron-proton model was that it did not agree with the experimentally determined statistics of nuclei. Studies of the rotational spectrum of the N_2^+ molecule showed that the spin of the nitrogen nucleus must be one. But if the nucleus consisted of 14 protons and 7 electrons, an odd number of particles with spin one-half, it should itself have spin one-half. The discrepancy between measurements and theoretical expectations was pointed out by Ralph Kronig, who suggested in 1928 that "probably one is therefore forced to assume that protons and electrons do not retain their identity to the extent they do outside the nucleus" (Pais 1986, 301). Neither Kronig nor others could be more concrete at that time. The following year, studies of Raman spectra confirmed the result that the nitrogen nucleus obeyed Bose-Einstein statistics, that is, possessed an integral spin. In Göttingen, Walter Heitler and Gerhard Herzberg amplified Kronig's conclusion: "[I]t seems as if the electron in the nucleus loses, along with its spin, also its right of participation in the statistics of the nucleus" (ibid., 302).

Even more important than the nitrogen anomaly was the problem of understanding the beta spectrum. In 1914 Chadwick had found that the spectrum of beta radioactivity was continuous, although mixed with a line spectrum. According to Chadwick and the Cavendish physicists, the continuous spectrum was the real one, whereas the discrete lines had their origin in, for example, an inner photoelectric effect in the electron system such as proposed by Charles Ellis in 1922. However, it was possible to account for the spectrum without assuming the beta electrons to be emitted with energies in a continuous range. In Berlin, Lise Meitner suggested that the electrons started out with the same energy but that some of the energy was converted to gamma radiation, which would produce secondary beta rays. Meitner's alternative led to a protracted controversy with the Cavendish scientists. The controversy was solved only in the late 1920s, when experiments proved incompatible with Meitner's theory. It was now firmly established that the continuous beta spectrum had its origin in the nucleus. This conclusion, however, was most discomforting from a theoretical point of view. According to quantum mechanics, a nucleus can exist only in discrete energy states; assuming energy conservation, a two-particle decay into a daughter nucleus and a beta electron cannot, therefore, reproduce the continuous spectrum.

Together with the spin-statistics problem and problems of relativistic quantum mechanics, the continuous beta spectrum led to a kind of crisis in parts of the physics community in 1929–31. Niels Bohr's answer to the crisis was radical, namely, that energy conservation fails in beta decay. In an unpublished note of June 1929, he emphasized "how little basis we possess at present for a theoretical treatment of the problem of β-ray disintegrations." He continued: "Indeed, the behaviour of electrons bound within an atomic nucleus would seem to fall entirely outside the field of consistent application of the ordinary mechanical concepts, even in their quantum theoretical modification. Remembering that the principles of conservation of energy and momentum are of a purely classical origin the suggestion of their failure in accounting for β-ray emission can on the present state of quantum theory hardly be rejected beforehand." Bohr continued to advocate violation of energy conservation for at least three years; he was supported by some of the younger physicists, including Gamow and Landau. Gamow was the author of the first textbook ever in nuclear physics, in its modern sense: a book titled *Constitution of Atomic Nuclei and Radioactivity* and prefaced on May 1, 1931. At that time, nuclear physics was still a nascent field. The book, aimed to give "as complete an account as possible of our present experimental and theoretical knowledge of the nature of atomic nuclei," filled 114 pages. Gamow referred approvingly to Bohr's idea of energy nonconservation and wrote, in complete accord with his master in Copenhagen, "The usual ideas of quantum mechanics absolutely fail in describing the behaviour of nuclear electrons; it seems that they may not even be treated as individual particles, and also the concept of energy seems to lose its meaning" (p. 5).

Pauli was no less worried than Bohr and Gamow, but he would have nothing to do with energy nonconservation. In December 1930 he proposed, in an "open letter" to Meitner and Geiger, that the beta puzzle as well as the N-14 problem might be solved by introducing into the nucleus a new, neutral particle: "[There is] the possibility that there could exist in the nuclei electrically neutral particles that I wish to call neutrons, which have spin 1/2 and obey the exclusion principle, and additionally differ from light quanta in that they do not travel with the velocity of light: The mass of the neutron must be of the same order of magnitude as the electron mass and, in any case, not larger than 0.01 proton mass. The continuous β-spectrum would then become understandable by the assumption that in β decay a neutron is emitted together with the electron, in such a way that the sum of the energies of neutron and electron is constant" (Brown 1978, 27). Pauli hesitated in publishing his idea, but it was nonetheless well known in the physics community. Only in 1933, in a discussion at the seventh Solvay Congress, did he publically defend the hypothesis, which appeared in print in the proceedings published in 1934. At that time, the "heavy neutron" had been discovered and Enrico Fermi proposed to call Pauli's particle a "neutrino." As shown by his letter to Meitner and Geiger, Pauli believed originally that the neutrino was a nuclear constituent with a small but nonzero mass and also with a magnetic moment. According to this picture, the nucleus consisted of protons, electrons, and neutrinos. Pauli's hypothetical neutrino broke with the two-particle paradigm and was, in this ontological respect, revolutionary. From a methodological point of view, however, it was a conservative theory, because its entire rationale was to preserve the well-tested conservation laws.

The neutrino was met with resistance or indifference at first. Among the antagonists was Bohr, among the protagonists Fermi. It was only after Fermi's successful 1934 theory of beta decay that the neutrino acquired some measure of respectability. But it was still a hypothetical particle, and was widely assumed to be beyond detection. In an article in *Nature* in April 1934, Hans Bethe and Rudolf Peierls concluded that "there is no practically possible way of observing the neutrino." As late as 1936, Dirac rejected the neutrino and found energy nonconservation to be a preferable alternative. The neutrino played a role not only in beta theory, but also in attempts to understand electromagnetic phenomena in terms of nuclear theory. In 1934 Louis de Broglie suggested that the photon might be conceived as a pair of a neutrino and an antineutrino. His idea aroused much interest among theoretical physicists. During 1934–38, it was developed in various directions by, among others, Jordan, Kronig, Gregor Wentzel, and Ernst Stueckelberg. However, the many papers on this topic failed to make connections with experiments and did not lead to a satisfactory neutrino theory of light. By 1940 the theory was largely abandoned, not because it was proven false but because it had proven fruitless.

Pauli's proton-electron-neutrino model was only one of several speculative attempts to understand the atomic nucleus in a new way. In 1930 the Russian physicists Dmitri Iwanenko and Victor Ambarzumian suggested a theory of beta decay based on Dirac's new theory of the electron. According to the Russians, the beta electron did not preexist in the nucleus, but was born there together with a proton originating from an electron in a negative energy state.

Yet another ephemeral hypothesis was suggested by Heisenberg at about the same time, in a letter to Bohr. By conceiving the world as a lattice (*Gitterwelt*) with cell length h/Mc^3, where M is the proton's mass, Heisenberg argued that he could understand the nucleus as consisting of protons and what he called "heavy light quanta." The advantage of the model was that it avoided nuclear electrons, the disadvantage (among others) that it violated most of the conservation laws, including charge conservation. "I do not know if you find this radical attempt completely mad," he wrote to Bohr. "But I have the feeling that nuclear physics is not to be had much more cheaply" (Carazza and Kragh 1995, 597). Bohr agreed that the paradoxes of nuclear physics required drastic changes in theory, but he found the lattice world hypothesis to be, if not completely mad, then too mad. After discussions in Copenhagen, Heisenberg shelved the idea, only to return to the hypothesis of a fundamental length later in the 1930s. In 1932 he found another and more acceptable candidate for heavy neutral nuclear particles—the neutron.

Astrophysical Applications

Nuclear physics, or rather nuclear speculations, entered astronomy at an early date. The new insight into the atomic constitution was first used in attempts to understand one of the classical mysteries of physics, the production of energy in the sun and the other stars. As early as 1917, Arthur Eddington speculated that the source of energy might be the annihilation of electrons and protons into radiation energy. This hypothetical process was widely discussed in astronomy for more than a decade. After all, although there was no experimental evidence for the process, neither were there any good reasons why it should not occur in the interior of stars. As a possible alternative, Eddington suggested in 1920 that the energy might come from the formation of helium from four hydrogen atoms—that is, a fusion process. "What is possible in the Cavendish Laboratory may not be too difficult in the sun," was his sound argument, referring to Rutherford's recent nuclear experiments. Eddington knew that according to Aston's mass-spectrographic measurements, the mass of a helium nucleus was nearly 1 percent less than that of four hydrogen nuclei, and that the formation reaction would therefore be followed by a considerable release of energy. Millikan was another scientist who applied the new nuclear physics in his astrophysical work, although in his case it was not the stellar energy but the cosmic radiation that was the target. In a series of works between 1926 and 1930, the American experimentalist argued that the cosmic rays consisted of distinct bands of high-energy photons originating from the nuclear buildup processes in the depths of the universe. The rays were, as he expressed it, "the birth cries of the elements" or "the signals sent out through the ether announcing the continuous creation of the heavier elements out of the lighter" (Kragh 1996b, 147). According to Millikan, the cosmic atom-building processes did not take place step by step, but in a single act, the heavier elements being formed directly from protons, electrons, and alpha particles. This may seem fantastic, but it was scarcely more fantastic than the alternative proposed by James Jeans, namely, that stars consisted mainly of transuranic elements that would transform spontaneously into radiation. Jeans suggested that the transformation would occur not only by ordinary radioactive decay, but also by annihilation of entire atomic nuclei. Very little was known about cosmic nuclear processes in the 1920s, and the lack of experimental knowledge invited speculations.

With Gamow's quantum mechanical theory of alpha decay, a new and less speculative chapter in nuclear astrophysics began. The Austrian physicist Fritz Houtermans and the British astrophysicist Robert d'Escourt Atkinson were both doing postdoctoral work in Germany. They realized that Gamow's tunneling process might be inverted and thus possibly explain the building up of elements by nuclear reactions. Calculations made in 1929 indicated that the probability of alpha particles entering even a light nucleus was vanishingly small under the conditions assumed to exist in the interior of stars. Proton-nucleus reactions were found to be more promising and Houterman and Atkinson, collaborating with Gamow, derived a general expression relating the capture cross section to the temperature and atomic number of the target nucleus. The theory suggested that the source of stellar energy might be a transmutation of four protons into an alpha particle—that is, the process originally suggested by Eddington. According to Houterman and Atkinson, however, the process was not Eddington's improbable four-particle collision, but the consecutive capture of protons by a light nucleus and the subsequent expulsion of an alpha particle. And whereas Eddington's suggestion had no basis in physical theory, the Houterman-Atkinson theory was a quantitative theory based on the most recent quantum mechanical knowledge. The theory counts among the pioneering contributions to modern astrophysics, but at first it attracted little attention.

The theory of Houterman and Atkinson presupposed that hydrogen existed abundantly in the stars, an assumption that won general acceptance among astronomers only about 1930. With the new knowledge of hydrogen's predominant role in stars, Atkinson, who had meanwhile moved to the United States, gave a greatly expanded version of the theory in 1931. Without the still unknown neutrons and deuterons, helium could not be built up directly from protons, but Atkinson devised a cyclic model in which helium was formed by disintegration of unstable nuclei. In this way, he attempted to explain the abundance of the entire range of elements by proton capture processes. It was only after 1932, and especially after the introduction of the neutron, however, that nuclear astrophysics began to deliver really promising results. Atkinson, Gamow, Bethe, and T. E. Sterne in the United States, Harold Walke in England, and von Weizsäcker, Ladislaus Farkas, and Paul Harteck in Germany were among those who pioneered nuclear astrophysics in the 1930s. Characteristically, the four Americans originally came from Europe: Atkinson and Sterne from England, Gamow from Russia, and Bethe from Germany. The idea that stellar element formation was to be based on neutron capture was developed independently by Walke and Gamow in 1935. Both physicists were inspired by current laboratory experiments, such as Fermi's in Rome and Cockcroft's and Walton's in Cambridge. Walke expressed the analogy between stellar processes and laboratory processes in this way: "The atomic physicist, with his sources of high potential and his discharge-tubes, is synthesizing elements in the same way as is occurring in stellar interiors, and the processes observed, which result in the liberation of such large amounts of energy of the order of million of volts indicate how the intense radiation of stars is maintained and why their temperatures are so high" (Kragh 1996b, 92).

The early ideas of neutron-induced nuclear processes did not lead to a satisfactory explanation of either element formation or stellar energy production. The breakthrough with regard to the latter problem came in 1938 and was very much a result of progress in nuclear theory. Hans Bethe, a former student of Sommerfeld's and an expert in quantum and nuclear physics, had fled Germany in 1933 and settled at Cornell University. In 1938 he participated in a conference on "Problems of Stellar Energy Sources" in Washington, D.C., which was attended by both astronomers and nuclear physicists. Although he had no previous knowledge of astrophysics, Bethe used his superb knowledge of nuclear physics to devise a detailed theory of solar energy production, which was quickly recognized to be the foundation of all later work in the area. A less detailed theory along the same lines was proposed by von Weizsäcker in 1938. The essence of Bethe's theory was that four protons fused into a helium nucleus through a cyclic process, wherein carbon nuclei acted catalytically. Contrary to earlier theories, it was based on detailed calculations supplied with values of cross sections determined experimentally. The nuclear-physical calculations made Bethe conclude that in order to give the energy produced by the sun, the cycle would require a central temperature of 18.5 million K, a value in excellent agreement with the one based on astrophysical models of the sun. Bethe's theory was widely applauded by astronomers and physicists alike. It took a little longer for the Nobel physics committee in Stockholm to recognize its value—twenty-eight years, to be exact.

1932, Annus Mirabilis

In 1932, the neutron was a well-known but missing particle. The physics literature of 1929–31 contains a dozen references or more to the neutron, but these were all to electron-proton composites in Rutherford's sense. The real neutron was found in the spring of 1932 and about a year later, it was recognized to be an elementary particle. The course of events that led to Chadwick's celebrated discovery, and for which he was awarded the Nobel prize only three years later, started with experiments made in 1930 by Walther Bothe and Herbert Becker at the Physikalisch-Technische Reichsanstalt in Berlin. The two physicists found that beryllium exposed to alpha particles produced what they believed were energetic gamma rays. In Paris, Irène Curie and Frédéric Joliot examined the "beryllium radiation" and reported in early 1932 that this radiation was able to knock out protons from hydrogen-rich paraffin. They thought that the mechanism might be some kind of Compton effect. On the other side of the English Channel, in Cambridge, Chadwick thought otherwise. Rutherford's neutron was still very much alive at Cavendish and Chadwick realized that it might explain the beryllium radiation. He therefore repeated and modified the Paris experiments and soon reached the conclusion that what was produced were neutrons, not gamma rays. According to Chadwick, the process was $^4He + ^9Be \rightarrow ^{12}C + n$, where the symbol n stands for a neutron of mass number 1. In his note on "Possible Existence of a Neutron," Chadwick discussed the possibility that the observed effects might be due to "a quantum of high energy" rather than a neutron. "Up to the present," he concluded, "all the evidence is in favour of the neutron, while the quantum hypothesis can only be upheld if the conservation of energy and momentum be relinquished at some point." Chadwick also found his neutrons when bombarding boron with alpha particles and from this process, he inferred the neutron's mass to be close to 1.007 proton masses.

Now that the neutron had been discovered, one might think that there was no longer any need for electrons in the nuclei and that everything was fine with the atomic nucleus. But this was far from the situation in 1932. Chadwick interpreted his neutron to be Rutherford's long-sought proton-electron composite, and supposed "that the proton and electron form a small dipole, or we may take the more attractive picture of a proton embedded in an electron." As to the possibility that the neutron might be elementary, Chadwick commented that "[this view] has little to recommend it at present, except the possibility of explaining the statistics of such nuclei as N^{14}" (Beyer 1949, 15 and 19). For a long time he, and most other physicists, hesitated in admitting that the neutron was elementary. The first one to propose the neutron as a new elementary particle of spin one-half was the Leningrad physicist Dmitri Iwanenko. In the summer of 1932, Iwanenko emphasized that the proposal would solve the nitrogen-14 puzzle. Yet it took at least one more year until the majority of physicists, who were used to the two-particle paradigm for so long, accepted the elementary nature of the neutron. The ambiguous attitude was clearly exhibited in Heisenberg's important theory of nuclear structure of 1932–33, in which he introduced exchange forces between the protons and neutrons and treated the nucleus quantum-mechanically. In spite of considering the protons and neutrons to be the constituents of the nucleus, Heisenberg at first made use of nuclear electrons and treated the neutron as a proton-electron compound. The elementary particles were still protons and electrons. Heisenberg's theory of nuclear structure marked the beginning of a new chapter of nuclear theory—or rather the beginning of the field as such—and was quickly followed by important contributions from Eugene Wigner, Ettore Majorana, and others.

During the 1933 Solvay Congress on "The Structure and Properties of Atomic Nuclei," the neutron was at center stage. Dirac suggested that the nucleus was made up of three kinds of particles: protons, neutrons, and electrons. The suggestion was not considered particularly strange at the time. Chadwick still wavered between the two views, those of the complex and the elementary neutron. Whatever the correct view, the neutron was a useful particle, for, as Chadwick explained to the Solvay participants, it could be used as a projectile in nuclear processes. As an example, he reported that he had observed alpha particles from neutrons reacting with oxygen by the process $^{1}n + {}^{16}O \rightarrow {}^{13}C + {}^{4}He$. What convinced the physicists of the elementary nature of the neutron was, in part, new developments within nuclear theory, and in part, more precise measurements of the neutron's mass. Chadwick had originally obtained for the neutron's mass 1.0067 proton masses—that is, slightly less than the mass of a proton plus an electron (1.0078). Subsequent experiments seemed to confirm that it would require energy to split a neutron into its constituents and that the particle was therefore a bound proton-electron system. However, new and more accurate measurements showed that mass of the neutron was a little larger than the proton-plus-electron system. This was considered an established fact by October 1934, when physicists convened at a conference on nuclei and cosmic rays in London. It was then realized that the neutron was unstable and must decay spontaneously into a proton and an electron, a suggestion first made by Chadwick and Maurice Goldhaber in 1935. After that date, the composite neutron ceased to be discussed and electrons were finally excluded from the nucleus. It took a long time, however, until the neutron decay was actually observed. This required intense neutron sources from nuclear reactors and was first reported in 1948 by A. H. Snell and his collaborators at Oak Ridge. Two years later, J. M. Robson at the Chalk River reactor in Canada determined the neutron's half-life to be about 13 minutes.

The neutron was perhaps the most dramatic of the actors in what is routinely referred to as the *annus mirabilis* of nuclear and particle physics, 1932, but what more appropriately might be called the *anni mirabiles* 1931–33. It was not the only actor, nor was it the first. In late December 1931, Harold Urey, a chemist at Columbia University who in 1923–24 had spent a year with Bohr in Copenhagen, reported the discovery of deuterium. Together with his collaborators, Ferdinand Brickwedde and George Murphy, Urey isolated the heavy hydrogen isotope by evaporating four liters of liquid hydrogen. They identified the isotope spectroscopically by the slight change in wavelength caused by the heavier nucleus. The subsequent preparation of heavy water in 1933 was first made by another American chemist, Gilbert Lewis. It was quickly realized that artificially accelerated nuclei of the heavy isotope were excellently suited for projectiles in nuclear reactions. For some time, a confusing variety of names were used for the particles—among them di-proton, deuton, and diplon—but eventually Urey's name "deuteron" won acceptance, together with deuterium for the corresponding atom. (The name "protium" for the ordinary hydrogen isotope never became popular.) Yet another very important development in the early 1930s was the discovery of artificial radioactivity, made in early 1934 by the Joliot-Curies in connection with irradiation of aluminum by alpha particles. The two French scientists detected the production of the recently discovered positrons, which they first suggested were decay products of the emitted protons, that is, $p \rightarrow n + e^+$. However, they soon realized that the positron activity continued after the alpha source was removed and that they had, in fact, discovered positive beta radioactivity: $^4\mathrm{He} + {}^{27}\mathrm{Al} \rightarrow {}^{30}\mathrm{P} + {}^1n$, followed by $^{30}\mathrm{P} \rightarrow {}^{30}\mathrm{Si} + e^+$. The importance of the discovery of artificial radioactivity was immediately recognized and resulted in a Nobel prize in chemistry to the Joliot-Curies in 1935. The new phenomenon immediately became widely employed in nuclear physics, chemistry, biology, and medicine.

The years 1932–33 were truly exciting, not only as seen in retrospect but also as experienced by physicists at the time. In May 1932, Bohr wrote to Rutherford, "Progress in the field of nuclear constitution is at the moment really so rapid, that one wonders what the next post will bring . . . One sees a broad new avenue opened, and it should soon be able to predict the behaviour of any nucleus under given circumstances" (Weiner 1972, 41). Two and a half years later, Frank Spedding, an American physicist, reported in a letter about the London conference: "There was also a symposium on nuclear physics. This field is moving so rapidly that one becomes dizzy contemplating it. With talk of the experimental properties of H, He, the new artificial radioactive elements, the neutron and positron, and the predicted properties of the neutrino and proton of minus charge, one who has been brought up on the old naive picture of protons and electrons in the nucleus feels bewildered" (ibid.).

These fascinating new theories and discoveries were only part of the miraculous years. New instrument technologies were no less important. Until about 1930 the only way to make a nuclear reaction occur was to use the projectiles that nature happened to provide in the form of alpha rays; the alternative was to make use of the even less controllable cosmic radiation, but this method was still in its infancy. The first successful nuclear disintegration brought about by purely artificial means was obtained in the spring of 1932 by John D. Cockcroft and Ernest Walton at the Cavendish Laboratory. The two physicists applied a voltage multiplier system, supplied in part by the Metropolitan-Vickers Electrical Company, where Cockcroft had worked as an electrical engineer trainee before he switched to physics and joined Rutherford in 1924. With this apparatus, they obtained proton energies up to 380 keV in 1929 and, three years later, 700 keV. Cockroft and Walton studied lithium bombarded with high-energy protons and, using both visual scintillation methods and cloud chamber photographs as detectors, they concluded that "the lithium isotope of mass 7 captures a proton and that the resulting nucleus of mass 8 breaks up into two α-particles" (Beyer 1949, 30). Moreover, they noted that the process occurred at energies and rates in qualitative agreement with Gamow's quantum-mechanical calculations. In fact, these calculations entered directly into Cockcroft's design of the high-tension apparatus. Cockcroft knew about Gamow's theory and realized that it allowed 300 keV protons to be fairly effective nuclear projectiles against targets like boron and lithium.

At about the same time as Cockcroft and Walton performed their pioneering experimental work, other types of accelerators were in progress in the United States. In 1931 the American engineer Robert Van de Graaf constructed an electrostatic accelerator with a maximum voltage of 1.5 million volts. In the same year, Ernest Lawrence and his student David Sloan at the University of California, Berkeley, constructed the first practical linear accelerator with which they obtained mercury ions with 1.3 MeV of energy. It was another of Lawrence's machines, however, that revolutionized nuclear physics and ushered in the era of "big science." The first experimental cyclotron, using strong magnetic fields to make nuclear particles spiral with increasing radius as their speeds increased, was constructed in 1931. The 1932 version, with magnet pole faces of 11 inches in diameter, produced a current of 10^{-9} ampere with 1.2 MeV protons. Lawrence and his collaborator, M. Stanley Livingston, predicted confidently that beams of 10 MeV protons would be generated "in the not distant future," a prediction that was soon confirmed. Originally the machine was not given its own name, but Lawrence and his group used the word "cyclotron" in what they described as "a sort of laboratory slang." By 1936, the name was in general use. The cyclotron proved eminently useful in nuclear physics in a broad range of areas, extending from pure research of nuclear reactions to industrial and medical applications. It was very much an American technology and in the 1930s, it was mastered fully only by Lawrence and the cyclotroneers trained by him. By 1934 the machines began to multiply in the United States, first with a small cyclotron at Cornell University built by Livingston; five years later, there were ten more machines in operation or under construction. Outside the United States, the first cyclotron was installed at Riken (the Institute for

Physical and Chemical Research) in Tokyo in 1935 to become operative in the spring of 1937. In Europe, cyclotrons were introduced at about the same time, but more hesitatingly than in the United States. Money was among the reasons for the relatively slow introduction, but conservatism and lack of familiarity with the new technology also played a role. In mid-1939 there were five cyclotrons operating in Europe, located in Cambridge, Liverpool, Paris, Stockholm, and Copenhagen.

Experimental nuclear physics in general, and accelerator physics in particular, helped change the geographical distribution of world physics. It is remarkable that the strong German physics community did not join the development as leaders but, on the contrary, fell behind the development not only in the United States and Britain, but also in France. And it is also notable that this occurred even before the Nazi regime changed the conditions of physics in Germany, and at a time when German physicists were still leaders in many other areas of physics.

Since about 1910, France had not been able to keep its former position as one of the leaders of world physics. Compared with what happened in the other large nations, French physics had little to boast of. For example, French journals were no longer among the leading physics journals. In 1934 the most quoted of the French journals, the *Comptes Rendus*, ranked number 7 of all physics journals (number one was *Zeitschrift für Physik*); *Journal de Physique* ranked number 11 and *Annales de Physique* was only number 34 on the list. As another indication of the low international reputation of French physics, while 27 percent of all references in American physics journals were to articles in German, only 3 percent were to articles in French. The rise of nuclear physics, however, was one of the factors that helped revitalize French physics in the late 1930s and reinstate Paris as an important city of world physics. The laboratory of the Joliot-Curies became one of the leading centers of nuclear physics, attracting many foreign physicists.

Notwithstanding the important work done in Britain and France, it was in the United States that nuclear physics first experienced its remarkable growth. The share of papers on nuclear physics in *Physical Review* was 8 percent in 1932; in 1933 it had increased to 18 percent, in 1935 to 22 percent, and in 1937 to no less than 32 percent. Nuclear physics was not only growing, but it was also becoming increasingly expensive. In order to be in the nuclear research front, external funding was often a necessity, as illustrated by the fact that 46 percent of the total number of funded papers in *Physical Review* of 1935 were on nuclear physics. By 1939, one-third of the *Physical Review* papers on nuclear physics were funded by external agencies.

Chapter 13

FROM TWO TO MANY PARTICLES

ANTIPARTICLES

AT THE MEETING of the British Association for the Advancement of Science in Bristol in September 1930, Dirac gave a lecture in which he said, "It has always been the dream of philosophers to have all matter built up from one fundamental kind of particle, so that it is not altogether satisfactory to have two in our theory, the electron and the proton" (Kragh 1990, 97). At thc time, Dirac believed that he had succeeded where the philosophers had failed and that he had reduced all matter to be manifestations of the electron only. (The reader will recognize the vague similarity with J. J. Thomson's atomic theory and, more generally, the electrodynamic worldview.) Dirac was deeply fascinated by the unitary view of matter, and it is ironic that his unitary considerations led him to introduce, less than a year after the Bristol address, three or four new elementary particles in addition to the electron. With Dirac's 1931 theory and Pauli's neutrino hypothesis, the first break was made with the two-particle paradigm.

It was clear to Dirac and several of his colleagues that the relativistic electron theory of 1928 led to strange consequences. The problem, sometimes referred to as the "± difficulty," had its basis in the Dirac equation, which formally included solutions with negative energy. Contrary to the situation in classical physics, these could not be dismissed as nonphysical but had to be taken seriously—that is, somehow to be related to some objects of nature. In November 1929 Dirac believed he had found the solution to the problem. "There is a simple way of avoiding the difficulty of electrons having negative kinetic energy," he wrote to Bohr, and continued: "[I]f the electron is started off with a +ve energy, there will be a finite probability of its suddenly changing into a state of negative energy and emitting the surplus energy in the form of high-frequency radiation. . . . [I]f all states of −ve energy are occupied and also few of +ve energy, those electrons with +ve energy will be unable to make transitions to states of −ve energy and will therefore have to behave quite properly. . . . It seems reasonable to assume that not all states of negative energy are occupied, but that there are a few vacancies of 'holes.' . . . [O]ne can easily see that such a hole would move in an electromagnetic field as though it had a +ve charge. These holes I believe to be protons" (Kragh 1990, 91).

Dirac's proton-as-electron theory, published in 1930, assumed a world of negative-energy states occupied by an infinite number of electrons governed by Pauli's exclusion principle. Only the few unoccupied states, the "holes," would appear as observable physical entities. But why would they appear as protons, two thousand times as heavy as electrons? There were two reasons for Dirac's choice: For one thing, if protons and electrons were the only elementary particles—as almost all physicists believed at the time—there seemed to be no other possibility; for another thing, the hypothesis would be the realization of the age-old and, to Dirac, highly attractive "dream of philosophers."

Attractive or not, the hypothesis was universally met with skepticism and immediately ran into serious problems. For example, if the proton were the electron's antiparticle (a name not yet introduced), it would supposedly annihilate according to $p^+ + e^- \rightarrow 2\gamma$, and calculations indicated that the mean lifetime of matter would, in that case, be absurdly low, about 10^{-9} seconds. This argument alone was not sufficient to convince Dirac that his theory was wrong, but in the spring of 1931 he realized (as others had done) that the hole had to have the same mass as the electron. In the new version, as it appeared in a remarkable paper in the *Proceedings of the Royal Society*, the antielectron was introduced for the first time as "a new kind of particle, unknown to experimental physics, having the same mass and opposite charge to an electron (Kragh 1990, 103)." Moreover, because the proton was now a separate species of particle, it would, according to Dirac, probably have its own antiparticle. A few years later, in his 1933 Nobel lecture, Dirac went a step further, speculating about matter made up entirely of antiparticles: "We must regard it rather as an accident that the Earth (and, presumably, the whole solar system), contains a preponderance of negative electrons and positive protons. It is quite probable that for some of the stars it is the other way about, these stars being built up mainly of positrons and negative protons. In fact, there may be half the stars of each kind. The two kinds of stars would both show exactly the same spectra, and there would be no way of distinguishing them by present astronomical methods." However, in 1931 the antielectron was a purely hypothetical particle, and most physicists declined to take Dirac's theory seriously. It was only later that it became recognized as "perhaps the biggest jump of all big jumps in physics of our century," as Heisenberg generously called it in 1973.

Dirac's 1931 paper did not deal primarily with antielectrons, but was an ambitious, and failed, attempt to explain the reason for the existence of a smallest electric charge. During the course of this work, Dirac was led to introduce a nonintegrable phase factor into the wave function; he showed that this was equivalent to introducing a magnetic field with a magnetic charge as its source—that is, a magnetic monopole as a magnetic analog to the electron. The Dirac monopole was a hypothetical particle, justified only in the sense that it was not forbidden by quantum mechanics. Dirac realized that this did not secure the actual existence of monopoles in nature, but since there was no theoretical reason to bar the existence of monopoles, then "one would be surprised if Nature had made no use of it [the possibility]." In the history of ideas, this kind of argument—that entities that can exist must exist—goes back to Leibniz, and is known as the principle of plenitude. Contrary to the positron, the magnetic monopole aroused no interest among physicists, most of whom ignored the suggested particle. There were, however, a few speculations that the neutron might consist of two oppositely charged magnetic poles or that the monopole might otherwise play a role in the atomic nucleus. It was only in the 1970s that the monopole theory, as well as experimental searches for the particle, became a major research area. Since then, there have been several claims to have detected magnetic monopoles, but none have been confirmed. Monopoles may exist, or they may once have existed. But the magnetic monopole of the late 1990s still has the status it had in 1931: It is hypothetical (see also chapter 21).

The status of the antielectron, on the other hand, changed during 1932–33. At the California Institute of Technology, Carl Anderson, a former student of Millikan, noted in cloud chamber photographs from the cosmic radiation some tracks that he first thought might be due to protons. In a later paper in March 1933, he suggested that he had discovered a positively charged electron, or a "positron," as he called it. He also suggested "negatron" for the ordinary electron, but the name, although occasionally used, did not catch on. Anderson's detection of positrons might look like a nice case of a discovery inspired by theory, but the discovery was actually not indebted to Dirac's theory at all. Rather than interpreting the positive electron to be a result of pair production à la Dirac, Anderson believed that it was ejected from an atomic nucleus split by an incoming cosmic ray photon. His tentative explanation did not refer to either Dirac or quantum mechanics. It had an old-fashioned air and relied on a visualizable concept of the nucleus that agreed with Millikan's ideas, but had almost nothing to do with the sophisticated quantum models that Heisenberg and others were establishing at the time. Anderson wrote, "If we retain the view that a nucleus consists of protons and neutrons (and α-particles) and that a neutron represents a close combination of a proton and electron, then from the electromagnetic theory as to the origin of mass the simplest assumption would seem to be that an encounter between the incoming primary ray and a proton may take place in such a way as to expand the diameter of the proton to the same value as that possessed by the negatron" (Beyer 1949, 4).

It was only after Blackett and the Italian physicist Guiseppe Occhialini reported new cosmic ray experiments, and explicitly referred to Dirac's theory, that it was realized that Anderson had, in fact, discovered the Dirac particle. Within a year, the positron was generally accepted and entered as an important particle in both theory and experiment. However, the positive reception of the positron did not imply an equally positive reception of Dirac's hole theory, which continued to be criticized. As late as 1934, Millikan and Anderson stuck to the view that the cosmic ray positrons preexisted or were formed in atomic nuclei, and therefore were not to be identified with antielectrons. Some theorists, including Pauli, suggested that the positron might satisfy Bose-Einstein statistics and that the neutrino, which was thought to be perhaps a nuclear particle, consisted of a positron-electron pair. Whereas the positron—alias the antielectron—was accepted in the mid-1930s, other antiparticle hypotheses had a more quiet existence. The negative proton was occasionally discussed, but not necessarily as the antiparticle of the proton. For example, Gamow suggested in several papers between 1934 and 1937 that the atomic nucleus included negative protons different from Dirac's antiprotons. As to the antineutron, it was introduced for the first time in 1935, by the Italian (later Brazilian) physicist Gleb Wataghin.

Surprises from the Cosmic Radiation

At the time that Dirac proposed his hole theory, cosmic radiation was still considered a most mysterious area of nature and its composition was a subject of controversy. Some of the cosmic ray particles possessed very large energies—much larger than those the new accelerators could provide—and for this reason, the radiation was of interest to nuclear and particle physicists. Cosmic ray physics was highly relevant to fundamental theories of physics and at the same time, with its balloon flights and mountain climbing, it had some of the charm of natural explorations. As the American physicist Karl Darrow wrote in 1932, the new field was "unique in modern physics for the minuteness of the phenomena, the delicacy of the observations, the adventurous excursions of the observers, the subtlety of the analysis and the grandeur of the inferences" (Cassidy 1981, 2). The discovery of the positron did much to make cosmic ray research a central field of physics. This can be seen, for example, from the number of papers (including letters) published on the subject in *Physical Review*. In 1928 two papers on cosmic rays appeared, and in 1929 only a single one; in 1930 the number rose to 4, and in 1931 to 9. The following year the number jumped to 30 papers, and in 1933 there were no fewer than 43 papers on cosmic rays or the positron.

The cosmic radiation was known as the poor man's laboratory because nature freely delivered particles with energies unheard of in real laboratories. The disadvantage, of course, was that the projectiles were completely uncontrollable and, in many cases, even unknown. The poor-man comparison with accelerator experiments was justified insofar as cost and organization were concerned. There was little that separated the cosmic ray research of the 1930s from the classical low-cost experiments at the Cavendish in the 1920s.

Rutherford would have had no objections to cosmic ray research. In 1933 Jabez Street, a leading American cosmic ray physicist and later a codiscoverer of the muon, applied to Theodore Lyman, his director at Harvard, for a grant of $800 for a year's research. The entire application was a one-page handwritten sheet arguing that "the discordant character of the results obtained by Profs. Compton and Millikan makes it clear that the accumulation of data on the nature of Cosmic Rays is much to be desired" (Galison 1987, 78). Street received his $800, and he used it wisely.

In order to understand what happened in the cosmic ray detectors, whether balloon-borne or placed on high mountain peaks, the physicists needed to know the nature of the cosmic rays themselves. Millikan, and his students with him, still advocated the "birth cry" theory, according to which the primary cosmic rays were high-energy photons. This implied that there should be no geomagnetic latitude effect, that is, no east-west variation as a result of the particles' deflection in the magnetic field of the earth. And, indeed, Millikan and his coworkers found no evidence at all for such an effect. On the other hand, American physicists on the East Coast, and Arthur Compton in particular, argued that there was incontestable evidence for a latitude effect and that the primary cosmic rays therefore had to consist of charged particles. The disagreement evolved into a major controversy, which was well covered by American newspapers. The outcome of the controversy was, by and large, that Millikan lost and Compton won. As early as 1929, Walther Bothe and Werner Kohlhörster in Berlin claimed to have demonstrated that cosmic rays included penetrating charged particles rather than the "ultra gamma rays" that they had thought of until then. Their conclusion was confirmed by the young Italian physicist Bruno Rossi of the University of Florence. Rossi developed an important new technique by connecting Geiger-Müller counters in a "coincidence circuit" in such a way that only particles traversing all the counters would be registered. About 1933, experimental evidence, essentially measurements at different geographical latitudes, mounted against Millikan's photon theory. Most of the penetrating particles of the primary cosmic radiation turned out to be charged and, for some unknown reason, positively charged. For some time Millikan, Anderson, and other Californians resisted this conclusion, but from about 1935, the controversy came to an end. British experiments using coincidence circuits constructed of cloud chambers controlled by counters confirmed the conclusions of Bothe, Kohlhörster, and Rossi. In 1935 Street showed that the large majority of events emerging from the lead plates were individual, hence highly penetrating, charged particles and not shower or cascade particles originating from photons. This more or less ended the controversy, but it also raised a pertinent question: What were the particles?

The penetrating cosmic ray particles were at first believed to be electrons, but the way they lost their energy indicated that they might not be ordinary electrons. They acted strangely, and Anderson and his group at Caltech referred informally to them as "green" electrons to distinguish them from the ordinary, absorbable or "red" electrons. In 1936 the nature of these green electrons became a matter of much discussion, not least because one possible solution to the problem was theoretical, namely, that current quantum electrodynamics might be unable to account for the phenomenon. There seemed to be two alternatives, either that quantum theory broke down at high energies or that a new particle, intermediate in mass between the proton and the electron, was in play. Almost all physicists chose the first alternative, but the solution of the enigma turned out to lie with the second.

Could the particles be protons? Not according to Anderson and Seth Neddermeyer, who concluded at the 1934 London conference that "most of the high energy cosmic ray particles at sea level have electronic mass" (Galison 1983, 287). If these particles of electronic mass were not electrons, what were they? On December 12, 1936, Anderson received the Nobel prize for his discovery of the positron. It could have been an occasion for speculations about the mysterious green electrons, but Anderson resisted the temptation to speculate. In his address in Stockholm, he mentioned the new cosmic ray data and observed that "these highly penetrating particles, although not free positive and negative electrons, will provide interesting material for future study." They did: In the spring of 1937, Anderson and Neddermeyer reached the conclusion that the most reasonable hypothesis was that "there exist particles of unit charge, but with a mass (which may not have a unique value) larger than that of a normal free electron and much smaller than that of a proton" (ibid., 298). Slightly later, on the other side of the North American continent, Street and his group reached a similar conclusion, although by very different arguments, and so did a Japanese group led by Yoshio Nishina. The discovery of what today is known as the muon was thus a triplet, although there is no question about the priority: It belonged to the Californians. A particle with mass intermediate between the electron and the proton, a meson, had been discovered. Or had it? Exactly when the heavy electron, μ-meson or muon was discovered is hardly a question worth contemplation. But if it is worth contemplation, it is a complex question. At any rate, it is a question related to other developments in cosmic ray and particle physics that took place at about the same time and can therefore not be answered in isolation from these developments.

After the Anderson-Neddermeyer-Street particle had been recognized to be real, the next question was obviously to determine its mass and other characteristics, such as its spin and decay modes. A less important question, but still one that had to be decided, related to the particle's name. Among the names suggested were meson, mesotron, barytron, heavy electron, and yukon; the first two derived from the mass of the particle, the last from Yukawa's recent theory of "heavy quanta." For a while, the "mesotron" suggested by Millikan, Anderson, and Neddermeyer was in general use, but in the 1940s, the abbreviated "meson," suggested by Homi Bhabha, became the more popular name and was officially approved by the Cosmic Ray Commission of the International Union of Physics in 1947. In 1937 Street and Edward C. Stevenson estimated, from the specific ionization and length of the cloud chamber track, a mass value for the mesotron of about 130 electron masses, not widely different from the 200-electron-mass estimate in Yukawa's theory of nuclear forces. The identification of the Yukawa particle and the 1937 mesotron was first suggested by Oppenheimer and Robert Serber in the same year. It was a minor disaster, as we will see below. The fact that Yukawa's prediction was published before the discovery of the mesotron, and the physicists' belief that the prediction was about the mesotron, does not mean that the experimental discovery was causally related to the meson theory. As in the case of the positron, there was no connection between theory and experiment. Anderson and Neddermeyer did not know about Yukawa's prediction until the summer of 1937. As Neddermeyer later recalled: "The muon, like the positron, was a purely experimental discovery in the sense that it was made entirely independently of any theoretical considerations of what particles should or should not exist" (Brown 1981, 132).

More precise mass values for the mesotron had to await the end of World War II. By 1950 the mass value had been determined to be 215 $\pm$ 6 electron masses, distinctly different from what was then known about the mass of the Yukawa meson. With respect to the decay, the identification of the mesotron with the Yukawa particle indicated the decay schemes $\mu^+ \rightarrow e^+ + \nu$ or $\mu^- \rightarrow e^- + \bar{\nu}$ in analogy with beta decay. Very little was known about the muons, as the particles were eventually called, in the 1930s. It was only in 1941 that Franco Rasetti from the University of Rome (but at the time working in Canada) determined a mean lifetime of $(1.5 \pm 0.3) \times 10^{-6}$ seconds, a value not very different from what is currently accepted. At that time there was no firm indications of the mesotron's spin. Generally speaking, although the discovery of the mesotron was helpful in the understanding of the particle content of the cosmic radiation, and then of the range of validity of quantum electrodynamics, it did not really clarify the situation in elementary particle physics (see table 13.1). In fact, although this was happily unknown at the time, it made the situation even more complicated.

Crisis in Quantum Theory

A quantum theory of the electromagnetic field was first developed by Dirac and, independently, by Jordan in 1927. Jordan found it "very probable" that the theory would soon be developed into "the natural formulation of the quantum theoretical electron theory by describing both light and matter as

TABLE 13.1
Some Particle Discoveries, 1897–1956

Current name	*Older name(s)*	*Prediction*	*Discovery*
Electron	corpuscle, negatron	1894; J. Larmor	1897; J.J. Thomson
Proton	H-particle	——	ca. 1913 (no discoverer)
Neutrino	neutron	1929; W. Pauli	1956; F. Reines, C. Cowan
Positron	positive electron	1931; P. Dirac	1932; C. Anderson
Neutron	——	1920; E. Rutherford	1932; J. Chadwick
Antiproton	negative proton	1931, P. Dirac	1955; O. Chamberlain, E. Segré, C. Wiegand, T. Ypsilantis
Antineutron	——	1935; G. Wataghin	1956; B. Cork, G. Lambertson, O. Piccioni, W. Wenzel
Muon	mesotron, μ-meson	——	1937; C. Anderson, S. Neddermeyer
Pion, charged	π-meson	1935; H. Yukawa	1947, C. Powell, G. Occhialini, C. Lattes
Pion, neutral	π-meson	1938; N. Kemmer	1950; R. Bjorklund, W. Crandall, B. Moyer, H. York
Λ^o baryon	V-particle	——	1947; C. Butler, G. Rochester
$K\pi_3$ baryon	τ-meson	——	1949; C. Powell et al.

interacting waves in three-dimensional space" (Rueger 1992, 312). However, the development turned out to be much more frustrating than expected in the year of 1927. Two years later, Pauli and Heisenberg proposed an ambitious theory of quantum electrodynamics (QED) that was relativistically invariant and included quantization of radiation as well as of matter waves. The Heisenberg-Pauli theory was a masterpiece of mathematical physics and the foundation of later theories of QED, but it was also complicated and indigestible for the majority of physicists. In spite of its promising features, it was infected with paradoxes and divergent quantities. In particular, the self-energy of the electron (the energy of an electron in its own electromagnetic field) turned out to be infinite, which was, of course, an unacceptable result.

Yet, many applications of the Heisenberg-Pauli formalism were independent of the theoretical deficiencies.

After the successful and rapid development of nonrelativistic quantum mechanics from 1925 to 1927, a period of grave doubts concerning the foundation of quantum mechanics followed with the attempts to establish a relativistically invariant theory of electromagnetic interactions. Many physicists believed that a new quantum revolution, based on some radically new concept, was near. The problems were in part of a logical and conceptual nature, and in part rooted in the inability of the existing theory to account for new empirical facts. Within the first group of problems, the archetypical case was the self-energy of the point electron, which turned out to be infinite. And this was only one of the severe problems facing the quantum electrodynamics based on the 1929 theory of Heisenberg and Pauli. Whereas the infinite self-energy had its counterpart in classical theory, new divergencies of a non-classical nature turned up in relativistic quantum field theory. For example, J. Robert Oppenheimer proved in 1930 that in addition to the classical electrostatic self-energy, a new quantum effect would contribute with a quadratically diverging term to the self-energy. As Oppenheimer pointed out, the divergencies would cause an infinite displacement of spectral lines. When the positron was incorporated into quantum field theory, the infinities remained: In positron theory, the contribution to the electric charge density from vacuum polarization proved to be divergent. Still another kind of divergence, the "infrared catastrophe," arose in the late 1930s in connection with attempts to take into account the emission of soft electrons during the scattering of charged particles. The logical consistency of relativistic quantum theory was also questioned in connection with the legitimacy of the electromagnetic quantum field. In 1931, Landau and Peierls argued that field measurements could not be performed unambiguously and hence that current quantum electrodynamics was inconsistent. The Landau-Peierls criticism caused concern until 1933, when Bohr and Léon Rosenfeld showed that the consequences of quantum electrodynamics were consistent with the best possible measurements of electromagnetic field quantities. The Bohr-Rosenfeld work was widely interpreted as if the failures of QED could be avoided if only questions about averaged field quantities at non-pointlike regions of space-time were asked.

The theoretical and conceptual situation in QED was a source of much worry to the physicists. In 1930 Bohr wrote to Dirac: "I . . . believe firmly that the solution of the present troubles will not be reached without a revision of our general physical ideas still deeper than that contemplated in the present quantum mechanics" (Cassidy 1981, 9). Three years later, the problems had become even more acute. Robert Oppenheimer summarized them to his brother Frank: "As you undoubtedly know, theoretical physics—what with the haunting ghosts of neutrinos, the Copenhagen conviction, against all evidence, that cosmic rays are protons, Born's absolutely unquantizable field theory, the divergence difficulties with the positron, and the utter impossibility of making a rigorous calculation at all—is in a hell of a way" (Kragh 1990, 165). In 1936 Dirac reached the conclusion that QED had to be abandoned because it was an ugly and complicated theory that did not really explain anything. Einstein, who found QED to be "awful," was pleased with Dirac's conclusion.

Within the second group of problems, the experimental anomalies, the challenges arising from understanding the atomic nucleus and the high-energy part of the cosmic radiation were the most serious. The theory of stopping of fast charged particles in matter did not agree with experiments, as Walter Heitler, an expert in quantum field theoretical calculations, concluded in 1933. When he joined Bethe in developing a more rigorous stopping theory, the disagreements remained. "The theoretical energy loss by radiation for high initial energy is far too large to be in any way reconcilable with the [cosmic ray] experiments of Anderson," Bethe and Heitler wrote in 1934. "It is very interesting that the energy loss of fast electrons . . . provides the first instance in which quantum mechanics apparently breaks down for a phenomenon outside the nucleus" (Galison 1983, 285). The data that Anderson and Neddermeyer presented at the London conference in October 1934 showed good agreement with theory for low energies, but a total disagreement for energies larger than about 150 times the rest energy of the electron ($mc^2 = 0.51$ MeV). In order to account for the discrepancy between theory and experiments, it was suggested that the penetrating particles were protons and not electrons; and when this hypothesis turned out to be untenable, it was concluded that QED had failed at high energies. According to Bethe's report to the London conference, "The experiments of Anderson and Neddermeyer on the passage of cosmic-ray electrons through lead . . . [show that] the quantum theory apparently goes wrong for energies of about 10^8 volts" (ibid., 288). Bethe, Heitler, Oppenheimer, and other quantum theorists were faced with a choice between introducing a new particle and accepting the QED breakdown. Between the two possibilities, they chose the second one. With the discovery of the mesotron (muon) in 1937, the situation eased considerably and many physicists concluded that there was no need for a quantum revolution after all. But the feeling of crisis continued, for QED was still plagued by infinities, and the new meson field theories that emerged in the late 1930s had their own divergence problems.

The responses to what was perceived as a continued crisis varied. Many prominent physicists, including Bohr, Dirac, Heisenberg, Pauli, and Landau, believed in a revolutionary approach, namely, that the problems could not be solved within existing theory but should be exploited in constructing a theory of the future that might differ as much from existing quantum theory as it differed from classical theory. Other, more pragmatically disposed physicists (including Bethe, Heitler, Fermi, and Oppenheimer) held that the problems could be avoided by technical improvements or that some appropriate reformulation of the existing theory might at least lead to sensible answers to all problems turning up empirically. Although these two attitudes can be identified as historical trends, the difference between "revolutionaries" and "conservatives" was neither absolute nor permanent. For example, although Heisenberg and Dirac in general favored a revolutionary view—that existing theory would have to be replaced by an entirely new one—this did not prevent them from exploring ways out of the problems based on modifications of existing theory.

The attitudes of the physicists in the 1930s to the problems in quantum theory can be divided conveniently into four classes.

1. Some of the revolutionary-minded physicists rather welcomed the series of crises, which they saw as genuine manifestations of the limitations of the existing theory and hence a clue to the future theory they dreamed of. Heisenberg believed that the infinite terms should not be discarded and that they, in a suitably interpreted version, would occur also in the correct theory of the future.

2. Other physicists concentrated on avoiding the divergencies without changing the framework of existing theory. One way of doing this was to cut off high-frequency contributions. Cutoff procedures were used frequently and proved helpful in practical calculations, but since they lacked theoretical justification and destroyed relativistic invariance, they were widely seen as pragmatic pseudosolutions. Another way of extracting reliable information from a presumably unreliable theory was to omit or subtract unwanted terms by suitable calculation techniques. This approach, pioneered by Dirac and Kramers, contained the germ of the renormalization procedures that were developed after 1945 and to which we shall return in chapter 22.

3. Related to this approach were attempts to eliminate the divergencies either directly in quantum theory or by way of changing the classical foundation. The first approach was followed by Gregor Wentzel in his so-called λ-limiting method (1933) and by Dirac in his introduction of negative probabilities and an indefinite metric for the Hilbert space (1941). The approach of reforming classical theory was adopted, in different ways, by Born and Leopold Infeld in their 1934 nonlinear field theory and by Dirac in his 1938 electron theory.

4. A last approach was to abandon quantum electrodynamics and replace it, at least temporarily, with a more modest theory, or set of rules, based on correspondence arguments. This alternative was for a period followed by Christian Møller, Oppenheimer, and Bethe, but was recognized to be less fundamental than quantum electrodynamics.

Heisenberg was among the most active contributors to the foundational discussion of QED, both in its purely theoretical contexts and regarding its application to cosmic ray phenomena. His favorite way of solving the divergence problems was by introducing a smallest length, as he did in 1938, when he argued that the smallest or fundamental length should be derived from Yukawa's new meson theory. Heisenberg hoped on this basis to construct a new, relativistically invariant, quantum theory that would contain the old one as a limiting case and differ from it only where the fundamental length could not be considered an infinitesimal quantity. However, his theory was criticized by most other physicists, including Pauli, who had initially cooperated with Heisenberg in his research program.

Yukawa's Heavy Quantum

"At the present stage of the quantum theory little is known about the nature of interaction of elementary particles." Thus began an article by Hideki Yukawa, a twenty-eight-year-old Japanese physicist from Osaka Imperial University, in the first issue of the 1935 volume of the *Proceedings of the Physico-Mathematical Society of Japan*, a journal not widely known outside Japan. The article was based on a ten-minute talk given to the meeting of the Physico-Mathematical Society in Tokyo on November 17, 1934. Yukawa was concerned with nuclear forces and was inspired by, on the one hand, Heisenberg's 1932 nuclear theory and, on the other, Fermi's 1934 theory of beta radioactivity. Basing his reasoning on ideas from these two sources, he sought to develop a unified picture of what later would be called weak and strong interactions. During this attempt, he was led to postulate "a new sort of quantum" that mediated the exchange forces in the atomic nuclei in analogy with the photon in electromagnetic fields. Yukawa suggested that the new nuclear potential would decrease very rapidly with the distance and that the range of the potential would be about $\lambda = 2 \times 10^{-15}$ m, the characteristic size of atomic nuclei. He further suggested the range parameter to be related to the mass of the charged "*U*-quantum" by $\lambda = 2\pi mc/h$, and thus predicted for its mass a value of about 200 electron masses.

But did the heavy quanta exist or were they just mathematical artifacts? Yuakawa cautiously mentioned that "as such a quantum with large mass and positive or negative charge has never ben found by the experiment, the above theory seems to be on a wrong line" (Beyer 1949, 144). Yet there is little doubt that Yukawa believed his quanta to exist and that he did not share his Western colleagues' fears of introducing new particles. Although the heavy quanta would not turn up in ordinary experiments, he argued that they would be observable in high-energy interactions, such as those occurring in cosmic radiation.

Yukawa's prediction of a new elementary particle intermediate between the electron and the proton was received with silence. It went unnoticed for more than two years, not only in Europe and America, but also in Japan.

Even Yukawa himself left the matter for more than a year. It was only when the anomalous measurements of Anderson and Neddermeyer began to attract attention that Yukawa, on January 18, 1937, sent a note to a Western journal about his theory, suggesting that "it is not altogether impossible that the anomalous tracks discovered by Anderson and Neddermeyer, which are likely to belong to unknown rays with *e/m* larger than that of the proton, are really due to such [*U*-]quanta" (Brown and Rechenberg 1996, 123). The editor of *Nature* denied publication because the suggestion was judged speculative. (Four years earlier, the same journal had also rejected Fermi's paper on beta decay.) However, with the announcement in May 1937 that the mesotron had been discovered in the cosmic radiation, response to Yukawa's theory changed dramatically. It was first referred to in a Western journal by Oppenheimer and Serber in June 1937, when they judged Yukawa's theory to be artificial and incorrect. Yet a negative response may sometimes be better than no response, and Oppenheimer and Serber's critical evaluation undoubtedly acted as an effective advertisement for the Japanese theory. In the fall of 1937, physicists in Europe and the United States were busy studying the paper of the unknown Japanese physicist. The reason for the sudden change of attitude from indifference to enthusiasm was clearly that the physicists found in Yukawa's prediction the very particle that had recently been discovered experimentally by Anderson and Neddermeyer. In other words, the mesotron was the *U*-quantum. The identification was generally assumed, but soon turned out to be problematical. During the years following 1937, the meson theory of nuclear forces attracted much interest and was developed by a large number of physicists, especially in Japan and Europe. Among the most important Japanese contributors were Yukawa, Sin-Itiro Tomonaga, Shoichi Sakata, and Mituo Taketani; the Europeans included Heitler, Nicholas Kemmer, Herbert Fröhlich, Pauli, and Homi Bhabha (who was from India, but worked in England).

The nuclear meson was unstable, with a mean lifetime at rest that Yukawa in 1938 estimated to be 10^{-7} seconds. In the same year, Heisenberg and Hans Euler found, from an analysis of experimental data, 2.7×10^{-6} seconds for the lifetime of the cosmic ray mesotron. The value did not change much when the first direct measurements were made during the war. Heisenberg and Euler were not worried by the discrepancy and found the agreement to be be "quite satisfactory." Neither did problems with the mass of the mesotron cause physicists to raise the red flag. Experiments in the late 1930s failed to give a precise value of the mesotron mass and instead resulted in values that varied considerably, from 120 electron masses to 350 electron masses or more. Yet physicists did not find the divergence alarming and for a while, they managed to convince themselves that mesotrons had a unique value of about 200 electron masses. Again, if the cosmic ray mesotron was the same as Yukawa's heavy quantum, it should interact strongly with nuclei.

Experiments did not show the expected high probability of capture in matter, but neither did this anomaly cause the physicists to question the one-meson assumption.

The outbreak of World War II naturally meant that work in pure physics, such as meson theory and cosmic ray research, was greatly reduced. But it did not stop completely and in several of the belligerent countries, including Italy, Japan, and the United States, important work continued throughout the war years. According to the accepted theory of mesotrons, proposed by Tomonaga and Toshima Araki in 1940, the negative particles would be captured and absorbed in the nucleus, leaving only the positive ones to decay into electrons and neutrinos. Since the probability of absorption of a negative meson (as I shall henceforth call the particle) was much larger than the decay probability, according to this theory, negative mesons should be almost completely absorbed and no decay electrons turn up. The Tomonaga-Araki effect was confirmed by the three Italian physicists Marcello Conversi, Oreste Piccioni, and Ettore Pancini in 1945, using iron as an absorber. However, when graphite was used in subsequent experiments the effect mysteriously disappeared, and in other light elements, the negative mesons turned out to decay with about the same rate as the positive particles. This result, obtained in early 1947 and quickly confirmed in the United States, was a genuine anomaly, since it contradicted the Tomonaga-Araki prediction. At this time, the accumulated difficulties with the one-meson theory could no longer be ignored and things suddenly progressed quickly. Although the Italian physicists did not suggest any explanation for the anomaly, their colleagues in the United States did.

During the first Shelter Island conference in June 1947 (which will be discussed further in chapter 22), the Italian experiment was among the subjects of discussion. Robert Marshak of the University of Rochester suggested to resolve the anomaly by means of a two-meson hypothesis. According to this hypothesis, there existed two different kinds of mesons, with different masses and lifetimes; the penetrating, weakly interacting particles arose from the decay of strongly interacting particles in the upper atmosphere. Shortly afterward, he became aware of the Bristol evidence for two mesons and realized that it fitted well with his idea. Together with Bethe, Marshak developed the idea into a proper two-meson theory that included a calculation of the decay time of the heavy meson as about 10^{-7} seconds. A proposal somewhat similar to the Bethe-Marshak theory was independently made by Sakata and Takesi Inoue in Japan, building on an even earlier idea of Sakata and Yasutaka Tanikawa.

Another response to the experiment of Conversi, Piccioni, and Pancini came from the Italian physicist Bruno Pontecorvo, at the time working in Canada (he would later emigrate to the Soviet Union). In the summer of 1947, Pontecorvo suggested that the Anderson-Neddermeyer meson was a heavy electron and thus belonged to what would later be called the lepton family. Pontecorvo realized that the capture of mesons into nuclei resembled nuclear electron capture and suggested the process $\mu^- + p \rightarrow n + \nu$ in analogy with the inverse beta process $e^- + p \rightarrow n + \nu$. The important insight that the electron and the (μ) meson are both "weak" particles was also reached slightly later by Oskar Klein in Sweden and Giovanni Puppi in Italy.

During the spring of 1947, the decade-long identification of Yukawa's particle with the Anderson-Neddermeyer particle was quickly crumbling. The earlier conservative atmosphere gave way for a "severe antidogmatism," as Pontecorvo recalled. The only missing piece in the puzzle was now the identification of the strongly interacting Yukawa particle.

The first observation of a nuclear disintegration by a meson was made in January 1947, when Donald Perkins, a London physicist, suggested identifying a cosmic-ray event as a "sigma" particle of intermediate mass reacting with a light nucleus. A few weeks later, Cecil Powell and his group at Bristol University reported other similar tracks in photographic plates exposed to the cosmic radiation. In their communication to *Nature* of May 1947, the Bristol team reported the observation of "double mesons," that is, what appeared to be a meson originating from another stopped meson. It was however only in the fall of 1947 that Powell, together with Occhialini and the Brazilian physicist Cesare Lattes, concluded that a heavy "π-meson" decayed into a lighter "μ-meson" (the electron track was not visible in the 1947 emulsions.) Originally the three physicists concluded that $m_\pi/m_\mu = 2$, which implied that the π-to-μ decay was accompanied by a heavy neutral particle. It was only when Lattes went to Berkeley and produced the π-mesons artificially that the mass was determined accurately. The result, $m_\pi/m_\mu = 1.33$, contradicted the Bristol value. Consequently, Powell and his group were forced to conclude that they were wrong and that there was no heavy neutral particle involved in the decay. All the same, the important thing was that the π-meson (pion) had been discovered and that it was found to be different from the lighter μ-meson (muon).

As the discoveries of the positron and the muon were unrelated to theory, so was the discovery of the pion. The physicists welcomed the true meson of the nuclear forces, the mass of which fitted well with Yukawa's theory. But with the discovery of the pion, the status of the muon changed and there seemed to be no room for the Anderson-Neddermeyer particle in the physical theories. "Who ordered that?" Isidore Rabi is said to have asked about the muon. The year 1947 marked the end of the first phase, and the beginning of a new phase, in the still-young discipline of elementary particle physics. The π-meson was predicted by a Japanese physicist, and Yukawa's theory was primarily developed by theoretical physicists in Europe and Japan; experiments on mesons were for a time an Italian speciality; the 1947 discovery was made in England by an international team, including an Italian and a Brazilian; and the discovery, as well as all earlier experiments, used as a meson source the cosmic radiation. After 1947, particle physics would change in many ways. The field would become dominated by American physicists and the role of the cosmic radiation would be challenged by new high-energy accelerators. We shall follow up on the story in chapter 21.

Chapter 14

PHILOSOPHICAL IMPLICATIONS OF QUANTUM MECHANICS

UNCERTAINTY AND COMPLEMENTARITY

DURING THE YEARS immediately following 1925, the small population of quantum physicists was intensely occupied not only with developing the theory and applying it to new areas, but also with understanding the conceptual foundation of the theory. Implicitly, and in a few cases also explicitly, the physicists acted as philosophers. Schrödinger, as we have seen, originally tried to interpret wave mechanics electrodynamically, but had to admit that the interpretation was untenable. Other physicists suggested hydrodynamics rather than electrodynamics as the appropriate continuum theory on which a semiclassical understanding of wave mechanics might be obtained. For example, in 1926 the German physicist Erwin Madelung developed a hydrodynamic model that reproduced some, but far from all, of the basic features of Schrödinger's theory. Such models or analogies continued to be suggested both before and after World War II, but they were not very successful and the majority of quantum physicists paid no attention to them. From the fall of 1926, Born's probabilistic interpretation was accepted by most physicists, although the precise understanding and implications of this interpretation were a matter of discussion. De Broglie, whose work had been the starting point of wave mechanics, was reluctant to join the majority view. In 1927 he proposed as an alternative a "theory of double solution," which built on a double system of solutions to the Schrödinger equation. According to de Broglie's theory, a particle could be described as a concentrated packet of energy, corresponding to a singular solution, and the particle would be guided by a continuous ψ-wave (a "pilot wave") interpreted in accordance with Born's probabilistic view. In this way, de Broglie was able to formulate a deterministic theory of microphysics without abandoning Born's insight into the probabilistic nature of quantum processes entirely. De Broglie's theory was severely criticized by Pauli at the 1927 Solvay Congress, and neither Schrödinger nor Einstein supported it. Disappointed and unable to counter Pauli's objections, de Broglie abandoned his theory. By 1928, he had already embraced the Copenhagen interpretation favored by Born, Heisenberg, Bohr, and others, and for more than two decades he remained a loyal Copenhagener.

In 1952 de Broglie returned to a modified version of his double solution theory and from that time onward, he went his own way. In the same year, an approach similar to de Broglie's was adopted by David Bohm, a young American physicist who, until then, had followed an orthodox path in his view on quantum theory. By introducing what he called a "quantum potential," Bohm was able to formulate a quantum theory that, although nonclassical, retained some classical features, such as particles moving along specific trajectories in accordance with the principle of causality. Bohm's theory was either ignored or criticized for being unnecessary, because it merely reproduced the results known from ordinary quantum theory. According to Heisenberg it was "ideological," and Pauli regarded it as "artificial metaphysics." It took about twenty-five years until Bohm's theory became widely discussed, and even then it was considered only by a minority of physicists.

After this brief detour to the postwar period, we now return to the golden 1920s. What was the deeper meaning of the lack of commutativity between canonically conjugate quantities, such as a particle's position and its momentum? This was, among other things, what Heisenberg answered with his famous uncertainty principle in the spring of 1927. (We will use the terms "uncertainty" and "indeterminacy" indiscriminately, although they are sometimes taken to have somewhat different meanings.) The general idea of this fundamental principle had been in the air for some time and was, for example, discussed in a letter that Pauli wrote to Heisenberg in October 1926. "The first question is, why may only the p's, and in any case not both the p's and also the q's be described with any accuracy," Pauli stated. "One can see the world with p-eyes and one can see it with q-eyes, but if one opens both eyes together then one goes astray" (Hendry 1984a, 99). Heisenberg agreed, and replied that "it is meaningless to talk of the position of a particle of fixed velocity. But if one accepts a less accurate position and velocity, that does indeed have a meaning" (ibid., 111). The same theme entered Dirac's transformation theory of December 1926, which was an important source for the later uncertainty principle. Dirac concluded, "One cannot answer any question on the quantum theory which refers to numerical values for both the q_{ro} [position] and the p_{ro} [momentum]. . . . [If] one describes the state of the system at an arbitrary time by giving numerical values to the co-ordinates and momenta, then one cannot actually set up a one-one correspondence between the values of these co-ordinates and momenta initially and their values at a subsequent time" (Kragh 1990, 42).

Heisenberg was indebted to discussions with Dirac, Jordan, and Pauli, but it was his discussions with Bohr on the foundation of quantum mechanics, most of all, that led him to the formulation of the uncertainty principle. When Heisenberg presented his first version of the uncertainty paper to Bohr, the Danish physicist did not like it. "He pointed out to me that certain statements in the first version were still incorrectly founded, and as he always insisted on relentless clarity in every detail, these points offended him deeply," Heisenberg recalled. "After several weeks of discussion, which were not devoid of stress, we soon concluded, not least thanks to Oskar Klein's participation, that we really meant the same, and that the uncertainty relations were just a special case of the more general complementarity principle" (Wheeler and Zurek 1983, 57).

Heisenberg's paper was characterized by the same kind of positivistic arguments that had served as a motivation for his 1925 paper, in which quantum mechanics was first introduced. His starting point was clearly philosophical: "If one wants to make clear what is meant by the words 'position of an object,' for example of an electron . . . then one has to describe definite experiments by means of which the 'position of an electron' can be measured; otherwise this term has no meaning at all." It is important to realize that Heisenberg did not state the uncertainty relations as a philosophical doctrine, but that he derived them from quantum mechanics and illustrated their significance by means of thought experiments. They were (and are) consequences of quantum mechanics, not the conceptual foundation of the theory. Heisenberg showed that the minimum indeterminacy in the position of a particle is related to the particle's indeterminacy in momentum by the expression $\Delta q \Delta p = h/4\pi$. He also showed that a corresponding relationship exists between the uncertainty in measuring the energy of some quantity and the corresponding uncertainty in the time measurement: $\Delta E \Delta t > h$.

The Heisenberg relations were quickly taken up, discussed, and sought to be extended or modified by many physicists, including Schrödinger, Edward Condon, and Howard Robertson. Robertson, a Princeton physicist better known for his work in cosmology, proved a more general version of the uncertainty relations, valid for any pair of conjugate variables, in 1929. According to the view of Heisenberg and most other physicists, the uncertainty relations involved necessarily the product of two uncertainties; one of the quantities (say q) could well be precisely determined, but then the other (p) would be wholly undetermined. During the attempts to sharpen and generalize the uncertainty relations, this view was questioned by several physicists, who argued that independent of the uncertainty of one of the variables, the other could not possibly be smaller than a certain value. This view was advocated in 1928 by Arthur Ruark in the United States and Henry Flint in England, both of whom suggested that $\Delta q = h/mc$. This relation and the corresponding minimal time interval $\Delta t = h/mc^2$ enjoyed considerable reputation between 1928 and 1936, when they were supported by authorities such as Pauli, de Broglie, and Schrödinger, as well as a number of less eminent physicists. However, the idea failed to lead to interesting physical applications and by the late 1930s, most physicists had abandoned it and returned to the conventional view.

Because the uncertainty relations followed from quantum mechanics, they were accepted by practically all physicists. But it was one thing to accept the mathematics, and quite another to agree on the meaning and philosophical implications. What did the innocent looking $\Delta q \Delta p = h/4\pi$ really mean? As Heisenberg made clear in his 1927 paper, for one thing it meant that the classical concept of causality had to be abandoned—not because it was illegitimate to infer from a present cause to a future effect, but because a physical system could never be defined precisely. Because we can know the present only within the limitations given by quantum mechanics, we can also know the future only inaccurately. "Since all experiments obey the quantum laws and, consequently, the uncertainty relations, the incorrectness of the law of causality is a definitely established consequence of quantum mechanics itself," Heisenberg argued. "Even in principle we cannot know the present in all detail. For that reason everything observed is a selection from a plenitude of possibilities and a limitation on what is possible in the future" (Wheeler and Zurek 1983, 83). Of course, one might imagine that the world was causal on some deeper level and that the acausality was confined to the phenomenal world only. But from Heisenberg's positivistic point of view, this objection made no difference: "Such speculations seem to us, to say it explicitly, to be without value and meaningless, for physics must confine itself to the description of the correlations between perceptions." Yet the uncertainty relations do not necessarily rule out strict determinism and causality. During the 1930s, the question was discussed by many physicists and philosophers, and it is a question that is still a subject of discussion more than seventy years after Heisenberg proposed his principle.

If Heisenberg's uncertainty principle is a consequence of quantum mechanics, Bohr's complementarity principle is not. It is a considerably broader and considerably less well-defined doctrine, which is primarily of a philosophical nature. There is little doubt that the formulation of the principle was indebted to Heisenberg's work with quantum uncertainties, but the idea of complementarity was not merely a philosophical generalization of Heisenberg's principle. It grew out of reflections about quantum theory that Bohr had entertained before Heisenberg started his work. Bohr presented his ideas on complementarity for the first time at an international congress of physics in Como in the fall of 1927, commemorating the centenary of Volta's death. On this occasion, he stressed that in the quantum world, contrary to the classical world, an observation of a system can never be made without disturbing the system. But how can we then know the state of the system? The quantum postulate would seem to imply that the classical distinction between the observer and the observed was no longer tenable. How then would it be possible to obtain objective knowledge? Bohr's reflections on these and related questions led him to introduce the notion of complementarity as denoting the use of complementary but mutually exclusive viewpoints in the description of nature. Two years later, he defined the complementarity principle as "a new mode of description . . . in the sense that any given application of classical concepts precludes the simultaneous use of other classical concepts which in a different connection are equally necessary for the elucidation of phenomena" (Jammer 1974, 95). This was about the clearest formulation of the complementarity principle, a doctrine that is notoriously vague and ambiguous. The wave description and the particle description are complementary, and thus in conflict. But Bohr argued that the physicist is still able to account unambiguously for his experiments, for it is he who chooses what to measure and thereby destroys the possibility of the realization of the conflicting aspect. In agreement with Heisenberg, Bohr emphasized that the aim of physics was to predict and coordinate experimental results, not to discover the reality behind the phenomenal world. "In our description of nature," he wrote in 1929, "the purpose is not to disclose the real essence of the phenomena, but only to track down, as far as it is possible, the relations between the manifold aspects of our experience" (Heilbron 1985, 219).

Although the wave-particle dualism is the standard example of complementarity, to Bohr and his disciples the principle had a much wider significance. Bohr soon applied it to other areas of physics, to biological questions, to psychology, and to cultural questions in general. For example, at the International Congress of Anthropological and Ethnological Sciences in 1938, Bohr explained that emotions and perceptions of them stand in a complementary relationship analogous to situations of measurements in atomic physics. Other physicists associated with the Copenhagen program went even further. Jordan, in particular, extrapolated complementarity to areas of psychology, philosophy, and biology in such an exaggerated way that an embarrassed Bohr had to emphasize that the concept had nothing to do with vitalism and could not be taken as a defense of either antirationalism or solipsism. Jordan's extreme interpretation of the measurement process included that observations not only disturb the measured quantity but literally produce it. "We ourselves produce the results of measurement," he emphasized in 1934 (Jammer 1974, 161).

The complementarity principle became the cornerstone of what was later referred to as the Copenhagen interpretation of quantum physics. Pauli even stated that quantum mechanics might be called "complementarity theory," in an analogy with "relativity theory." And Peierls later claimed that "when you refer to the Copenhagen interpretation of the mechanics what you really mean is quantum mechanics" (Whitaker 1996, 160). What the Copenhagen interpretation is, exactly, is no clearer than the nature of the complementarity principle itself, however, which means that it is not very clear. It is a matter still discussed by philosophers and a few philosophically minded physicists. In fact, the term "Copenhagen interpretation" was not used in the 1930s but first entered the physicists' vocabulary in 1955, when Heisenberg used it in criticizing certain unorthodox interpretations of quantum mechanics.

Many of the important physicists of the 1930s, including Pauli, Heisenberg, Jordan, and Rosenfeld, became enthusiastic supporters of Bohr's complementarity philosophy, which they saw as the true conceptual kernel of quantum mechanics. It is noteworthy that almost all the physicists who explicitly adopted Bohr's viewpoint had personal contacts to Bohr and had been visitors to his institute. Outside the Copenhagen circle, the reception of the complementarity philosophy was considerably cooler, either politely indifferent or, in a few cases, hostile. Dirac, who had close connections to the Copenhageners and great respect for Bohr, did not see any point in all the talk about complementarity. It did not result in new equations and could not be used for the calculations that Dirac tended to identify with physics (see also chapter 11). Nor were all the students at Bohr's institute converted to the complementarity philosophy. Consider the case of Christian Møller, who studied at the institute between 1926 and 1932 and remained there throughout his active life. Although Møller was a typical product of the Copenhagen school and deeply influenced by its working spirit, complementarity arguments left no trace in his published works. He was familiar with these arguments, of course, but not particularly interested in the broad conceptual problems highlighted by Bohr. As he recalled in an 1963 interview, "Although we listened to hundreds and hundreds of talks about these things [complementarity and mesurements problems], and we were interested in it, I don't think, except Rosenfeld perhaps, that any of us were spending so much time with this thing. . . . When you are young it is more interesting to attack definite problems. I mean this was so general, nearly philosophical" (Kragh 1992, 304).

This was also the attitude of many young quantum physicists, especially in the United States, where Bohr's reputation as a quantum sage was much more limited than in Europe. Problems that were "nearly philosophical" were not considered attractive. American physicists had a more pragmatic and less philosophical attitude to physics than many of Bohr's associates. They concentrated on experiments and specific calculations and, for these purposes they had no use for the complementarity principle. This is not to say that there was no interest in foundational problems among the Americans, only that it went in different directions and was on a less grand scale than in Denmark and Germany. The uncertainty principle was eagerly taken up by several American physicists, including Kennard, Ruark, Van Vleck, Condon, and Robertson, but they showed almost no interest in Bohrian complementarity. That the contemporary importance of the complementarity principle was relatively modest is also seen from the textbooks from which students were taught quantum theory. Most textbook authors, even if sympathetic to Bohr's ideas, found it difficult to include and justify a section on complementarity. Among forty-three textbooks on quantum mechanics published between 1928 and 1937, forty included a treatment of the uncertainty principle; only eight of them mentioned the complementarity principle. In spite of the fact that a large part of the world's physicists did not endorse the Copenhagen interpretation, or rather did not care about it, the opposition to it was weak and scattered. Whatever the reasons, by the mid-1930s Bohr had been remarkably successful in establishing the Copenhagen view as the dominant philosophy of quantum mechanics.

Against the Copenhagen Interpretation

Perhaps the most famous, and most romanticized, episode in the history of twentieth-century physics is the debate between Einstein and Bohr concerning the interpretation of quantum mechanics. This series of Socratic discussions between two profound and legendary scientist-philosophers has become part of the physics folklore and, indeed, of the general intellectual folklore. Whatever the details of their discussions, they have a place in Western intellectual history comparable to, say, the controversy between Newton and Leibniz about three hundred years ago. One sometimes gets the impression that Bohr and Einstein hold continual discussions during most of two decades, often face to face. In reality, the two physicists met personally only a few times, and the importance of their discussions has been exaggerated and romanticized in many accounts from the last quarter of the century.

Although quantum mechanics was indebted to Einstein's fundamental contributions to quantum theory from 1905 to 1925, Einstein did not take much interest in the new theory at first. His general attitude was skeptical and he denied, more on philosophical than on scientific grounds, that the microworld could only be described statistically. In a famous letter to Born in December 1926, he wrote about his "inner voice" that told him that quantum mechanics "hardly brings us nearer to the secret of the Old One. . . . I am convinced that he does not throw dice" (Jammer 1974, 155). Einstein's dissatisfaction with the statistical interpretation gave rise to a paper presented orally to the Prussian Academy of Sciences in early 1927. The manuscript, titled "Does Schrödinger's Wave Mechanics Determine the Motion of a System Completely or Only in the Sense of Statistics?" sketched a kind of hidden-variables theory somewhat similar to Madelung's hydrodynamic theory. But Einstein must have realized that his alternative was unsatisfactory for he never submitted the manuscript for publication.

Einstein did not participate in the Volta congress, but was among the participants of the fifth Solvay Congress in October 1927, where Bohr, Dirac, Heisenberg, Pauli, Schrödinger, and other leading physicists discussed the foundation of quantum mechanics. Bohr lectured on his new ideas of complementarity, which Einstein heard for the first time. Einstein was not convinced, and argued that the Bohr-Heisenberg interpretation, according to which quantum mechanics is a complete theory of individual processes, contradicted the theory of relativity. He discussed various thought experiments in the hope of demonstrating that the uncertainty relations were not necessarily valid and that atomic phenomena could be analyzed in more detail than specified by Heisenberg's relations. When Bohr showed Einstein's arguments to be untenable, Einstein came up with a new thought experiment, which Bohr again countered. According to Bohr, quantum mechanics (including the uncertainty relations) was a complete theory that exhausted all possibilities of accounting for observable phenomena. There is no doubt that Bohr came out as the "winner" of the 1927 discussions and that most of the participants recognized the force of his arguments. "BOHR [was] towering over everybody," Ehrenfest wrote after the congress. "It was delightful for me to be present during the conversation between Bohr and Einstein. Like a game of chess, Einstein all the time with new examples. . . . Bohr from out of philosophical smoke clouds constantly searching for the tools to crush one example after the other. Einstein like a jack-in-the-box; jumping out fresh every morning. Oh, that was priceless. But I am almost without reservation pro Bohr and contra Einstein" (Whitaker 1996, 210).

Einstein admitted that Bohr was a clever discussant, but not that his views were correct. In a letter to Schrödinger half a year after the Solvay Congress, Einstein described the Copenhagen interpretation sarcastically as "the Heisenberg-Bohr tranquilizing philosophy—or religion?" And he added that "it provides a gentle pillow for the true believer from which he cannot very easily be aroused" (Jammer 1974, 130). The second round of the celebrated Bohr-Einstein debate took place during the following Solvay congress, in October 1930, at a time when Bohr's complementarity idea was gaining strength among European physicists. This time, Einstein focused on the energy-time uncertainty relation ($\Delta E \Delta t > h$), which he attempted to refute. The means of refutation was the same as that of three years earlier, a thought experiment. In his new thought experiment, later known as the photon box experiment, Einstein made use of the mass-energy relation of special relativity, $E = mc^2$, and argued that the energy of a photon and its time of arrival at a screen could be predicted with unlimited accuracy, in contradiction to the uncertainty relation. But Bohr brilliantly answered the challenge, this time by invoking the redshift formula of Einstein's general theory of relativity. The outcome of the second round of debate was the same as that of the first round: Bohr's conception of quantum mechanics was strengthened and Einstein's skepticism seemed to be unwarranted. Until then, Einstein had hoped to refute quantum mechanics by showing the uncertainty relations to be wrong; his belief in ultimate causality was unshaken and in the 1930s, he shifted the focus of his objections from inconsistency to incompleteness.

The statistical meaning of the wave function does not necessarily exclude the possibility that individual atomic events are determined by parameters not yet discovered. The general hypothesis of such well-defined sublevel parameters has a long history in the physical sciences and goes back long before quantum mechanics. To mention but one example, early twentieth-century attempts to explain radioactivity causally made use of a version of the hypothesis (see chapter 4). The possibility of "hidden variables" was recognized at an early stage of quantum mechanics, but as long as the hypothetical parameters had no physical significance, they were not given much thought. Yet, it was a possibility, and an attractive one, to those who disliked the Copenhagen interpretation. If a quantum mechanics could be formulated with hidden variables, and if it reproduced all the results of the standard theory, there would seem to be no compelling reason for physicists to accept the Copenhagen picture of the atomic world.

The question of hidden variables was among the problems examined by the Hungarian-American mathematician John von Neumann in a 1932 book titled *Mathematische Grundlagen der Quantenmechanik* (*Mathematical Foundations of Quantum Mechanics*). Von Neumann gave a mathematically precise formulation of the foundation of quantum mechanics, basing the theory on the use of Hilbert spaces. In a work of 1933, the French physicist Jacques Solomon was led independently to the same conclusion, that hidden parameters are inconsistent with the accepted formalism of quantum mechanics.

In a small part of his important book, von Neumann proved that a causal understanding of quantum mechanics based on hidden variables is not possible. Imagine two systems described by the same ψ-function. The same measurements will in general lead to different results, which, according to the standard interpretation, is ascribed to quantum mechanics being acausal. But might it not be explained if the two systems differed in some hidden parameters that determine the outcome of the measurements? This was what von Neumann proved could not be the case. According to von Neumann, "It is therefore not, as is often assumed, a question of a re-interpretation of quantum mechanics,—the present system of quantum mechanics would have to be objectively false, in order that another description of the elementary processes than the statistical one be possible" (Pinch 1977, 185). On the other hand, von Neumann admitted the slight possibility that quantum mechanics might be wrong. Being a physical theory, it could not be proved mathematically: "Of course it would be an exaggeration to maintain that causality has thereby been done away with. . . . In spite of the fact that quantum mechanics agrees well with experiment, and that it has opened up for us a qualitatively new side of the world, one can never say of the theory that it has been proved by experience, but only that it is the best known summarization of experience" (Jammer 1974, 270).

In spite of von Neumann's words of caution, his mathematical proof was accepted widely and sometimes taken to be a proof of the Copenhagen interpretation. There were, in fact, considerable differences between Bohr's position and von Neumann's interpretation, but the distinctions were rarely noted. For example, "the measurement problem" was not the same for Bohr and von Neumann. Bohr tended to see it as a problem of generalizing the classical framework in order to avoid contradictions between two mutually incompatible classical concepts, both necessary in the description of experiments. His solution was complementarity. To von Neumann, on the other hand, the problem of measurement meant the mathematical problem of proving that the formalism gave the same predictions for different locations of the "cut" between observer and object. The role of human consciousness in the measuring process was part of the quantum-philosophical discussion in the 1930s. Von Neumann argued that the element of consciousness could not be excluded and in a monograph of 1939, Fritz London and Edmond Bauer claimed explicitly that the reduction of the wave function was the result of a conscious activity of the human mind. "It looks as if the result of a measurement is intimately linked to the consciousness of the person making it, and as if quantum mechanics thus drives us toward complete solipsism," they wrote, only to argue that the new role of the observing consciousness did not, after all, undermine objectivity. In the spirit of positivism, they noted with satisfaction that nothing in the measuring situation would "keep us from predicting or interpreting experimental results" (Wheeler and Zurek 1983, 258).

It is debatable how great a role von Neumann's impossibility proof played in the process that led to the Copenhagen hegemony, for most physicists already believed that there were good empirical reasons to support the position of Bohr and his allies. On the other hand, von Neumann's mathematical authority helped the process greatly and his proof was often referred to as the final word on the matter. Almost no physicists undertook a critical study of the proof, and many of the physicists who referred to the proof probably had only glanced through it (or just heard about it—until 1955, von Neumann's book existed in German only). The few philosophers who had the competence and courage to criticize it were not taken very seriously. It was only in the 1950s, when the hidden variable debate was revived, that von Neumann's argument became the subject of critical examination. It then turned out, as shown by the British physicist John Bell in the mid-1960s, that von Neumann's alleged proof did not, in fact, rule out all theories operating with hidden parameters. Bell, who played a leading role in the debate over the interpretation of quantum mechanics, was inspired by Bohm's theory and in general in favor of hidden variable theories.

Is Quantum Mechanics Complete?

After his "defeat" in 1930, Einstein continued to think deeply about the epistemological situation in quantum mechanics, convinced that an exact and causal description of natural phenomena must be possible. In the spring of

1935, now settled in the United States, Einstein published, together with his young Princeton colleagues Boris Podolsky and Nathan Rosen, a brief but famous paper titled "Can Quantum-Mechanical Description of Physical Reality Be Considered Complete?" The final version of the paper was written by Podolsky and formulated in a way that Einstein did not entirely approve of. The three authors started by stating that physical concepts must correspond to aspects of physical reality. Their reality criterion was this: "If, without in any way disturbing a system, we can predict with certainty (i.e., with probability equal to unity) the value of a physical quantity, then there exists an element of physical reality corresponding to this physical quantity." The correspondence led to a necessary condition for the completeness of a physical theory, namely, "Every element of the physical reality must have a counterpart in the physical theory." Einstein, Podolsky, and Rosen (EPR) now argued that quantum mechanics combined with the reality criterion led to a contradiction and that the only alternative was to accept that a quantum mechanical description of reality is not complete. The argument of the EPR paper was essentially negative, in the sense that it aimed at undermining the standard view of quantum mechanics without proposing an alternative. In their conclusion, Einstein and his collaborators "left open the question of whether or not such a [complete] description exists," adding, "We believe, however, that such a theory is possible" (Wheeler and Zurek 1983, 138–41).

Bohr was greatly disturbed by the EPR argument and started at once to develop a counterargument, which he had ready after a period of about five months (in the case of Bohr, a methodical thinker, this was fast). His main line of argument was a rejection of the criterion of physical reality proposed by Einstein, Podolsky, and Rosen. He found this criterion to be invalid because it presupposed that the object and the measuring apparatus could be analyzed in distinct parts; this was not possible according to the Copenhagen view, in which they formed a single system. In Bohr's careful but convoluted style:

> A criterion of reality like that proposed by the named authors contains . . . an essential ambiguity . . . regards the meaning of the expression "without in any way disturbing a system." Of course there is in a case like that just considered no question of a mechanical disturbance of the system under investigation during the last critical stage of the measuring procedure. But even at this stage there is essentially the question of *an influence on the very conditions which define the possible types of predictions regarding the future behavior of the system.* Since these conditions constitute an inherent element of the description of any phenomenon to which the term "physical reality" can be properly attached, we see that the argumentation of the mentioned authors does not justify their conclusion that quantum-mechanical description is essentially incomplete. (ibid., 148; emphasis in original).

Whereas the EPR argument became extremely famous in the 1960s and later, in the 1930s this third round of the Bohr-Einstein debate did not arouse much of interest among physicists. The EPR paper did not succeed in convincing physicists to abandon the Copenhagen interpretation, and the general impression was that Bohr had once again countered Einstein's objections satisfactorily. It merely confirmed to mainstream quantum physicists what they had always thought, namely, that Einstein and his allies—"the conservative, old gentlemen," as Pauli described them in a letter to Schrödinger—were hopelessly out of tune with the development. The large majority of physicists seems to have been simply uninterested. They could easily find better things to do than trying to understand philosophical arguments with no relevance for their daily work as physicists.

More philosophically inclined physicists, however, among them Schrödinger, found the EPR discussion highly interesting. In contributions of 1935, the father of wave mechanics supported Einstein's view and developed his own objections to Bohr's position on quantum theory. In one of these contributions, he proposed an argument, different from that of EPR, against the completeness of quantum mechanics. He famously illustrated his point by means of a thought experiment involving a poor cat confined in a chamber with an amount of radioactive material and a diabolical device which, when triggered by a disintegration, would release deadly vapors of hydrocyanic acid. Schrödinger's paradoxical conclusion was this: "If one has left this entire system to itself for an hour, one would say that the cat still lives *if* meanwhile no atom has decayed. The first atomic decay would have poisoned it. The ψ-function of the entire system would express this by having in it the living and the dead cat (pardon the expression) mixed or smeared out in equal parts" (Whitaker 1996, 234). If it followed from the Copenhagen interpretation that a cat could be "half dead" and ascribed a wave function $\Psi_{cat} = 1/2(\Psi_{alive} + \Psi_{dead})$, would that not indicate that some other interpretation must be preferred? Schrödinger's cat paradox became immensely popular from about 1970, finding its way to schoolchildren's T-shirts and much else, but in the 1930s it did not arouse much debate. Bohr did not respond to it, perhaps because he found its premises so evidently flawed. After all, according to his view, a macroscopic body such as a cat, or a bottle with sodium cyanide, could not be assigned a wave function. Within the framework of the Copenhagen interpretation, there was no paradox to solve. Neither did Schrödinger seem to have considered the cat paradox very paradoxical at all. He described it as a "quite ridiculous case" and considered its lesson to be a warning against the naive acceptance of a "blurred model" as representing reality. As he pointed out: "In itself it [the cat example] would not embody anything unclear or contradictory. There is a difference between a shaky or out-of-focus photograph and a snapshot of clouds and fog banks" (ibid.).

Chapter 15

EDDINGTON'S DREAM AND OTHER HETERODOXIES

PHYSICS IN THE 1930s was not an unbroken string of new discoveries and theoretical developments leading to a fuller and more correct understanding of the quantum world. At the same time as nuclear physics flourished and progress finally followed the many problems in quantum theory, much work was being invested in alternative research programs. Some of these were highly ambitious, aiming at formulating what would in effect have been a new foundation of physics. The attempts were clearly unorthodox, and were recognized as such at the time, but they nonetheless attracted a good deal of interest. Diverse as they were, several of them had in common that physics was intimately connected with cosmology. On the philosophical level, they had a clear orientation toward rationalism and a priori reasoning. The trend was primarily a British phenomenon; although it inspired several physicists in Europe and North America, it was only in Britain that it was of enduring importance. The attempts to establish a new cosmophysics failed, and today have been largely forgotten. So why pay any attention to them? For one thing, such attention provides a healthy antidote to a simplistic, linear conception of scientific progress. During every period in the history of science, mainstream physics has been challenged by heterodox views, and often it is only in retrospect that we can see what belonged to the progressive highways and what to the blind alleys of mistakes. For another thing, some of the approaches and aims of the failed cosmophysical revolution of the 1930s continued to play a role in the postwar period, and are still of interest to some modern physicists and astronomers.

EDDINGTON'S FUNDAMENTALISM

Arthur Eddington was not only one of the world's most eminent theoretical astronomers, but he was also a pioneer in cosmology and an authority on the general theory of relativity (and, in addition, a successful author of popular expositions of science and philosophy). His fascination with Einstein's theory had convinced him that the method of tensor calculus—the mathematical method of general relativity—was the only possible means of studying physics at its most fundamental level. He was therefore greatly disturbed when he realized that Dirac's 1928 equation for the electron was not expressed in tensor form, and he decided to generalize and reinterpret the equation. Attempts to formulate the Dirac equation within the mathematical framework of general relativity were common among mathematical physicists around 1930, but Eddington's was a starting point for a grander theory of the entire world of physics. It inspired him to develop a radically new kind of physical theory and to embark on a program that would last until his death in 1944.

According to Eddington, an electron could not be considered an individual particle, but had to be described in connection with all the other electrons in the universe, a view that opened up what he believed was a deep connection between microphysics and cosmology. Following the path of earlier unificationists, one of Eddington's aims was to reduce the contingencies in the description of nature—for example, by explaining the fundamental constants of physics rather than accepting them as merely experimental data. One of these constants was the fine-structure constant $\hbar c/e^2$, which entered prominently in Dirac's theory and was known to be about 1/137. Another dimensionless constant believed to be of fundamental importance in about 1930 was the mass ratio between the two building blocks of matter, the proton and the electron ($M/m = 1838$). Eddington argued that the value of the inverse fine-structure constant could be deduced to be the whole number 136. In 1936, after experiments had shown the value to be close to 137, he provided arguments that the number must be increased by one. As to the proton-to-electron mass ratio, he suggested that this number is the ratio between the solutions to the quadratic mass equation $10m^2 - 136m + 1 = 0$. Whereas these numbers are small and not directly connected with the universe at large, Eddington suggested that the constants of nature may also relate to cosmological quantities, of which he considered the "cosmical number" the most important. This number, N, is the number of electrons (or protons) in the observable part of the universe, approximately 10^{79}. The number can be estimated from the mean density of the universe, but Eddington claimed that he was able to deduce it from his theory, namely, as $N = 2 \times 136 \times 2^{256} = 3.15 \times 10^{79}$. Moreover, he argued that it was related to other constants of nature by formulas such as $\pi\, e^2/GmM = (3N)^{\frac{1}{2}}$, where G is Newton's constant of gravitation.

Eddington's method for deriving the relationship between cosmic and atomic constants changed as his research program developed, and so did the results he obtained, but not materially. In 1936 he collected and elaborated his results in *Relativity Theory of Electrons and Protons*, a work that was as remarkably ambitious as it was remarkably unsuccessful. According to Eddington, a proper understanding of the universe—both in its microscopic and cosmic aspects—necessarily involved the extraction of the meaning of the fundamental constants, such as the elementary charge, Planck's constant, the masses of the electron and the proton, the gravitational constant, the velocity of light, and the cosmical number. Deliberately associating his view with Pythagorean and Keplerian reasoning, he wrote in 1935, "We must thus look on the universe as a symphony played on seven primitive constants as music played on the seven notes of a scale" (Kragh 1982a, 82). The point of Eddington's program was that he strived to deduce the numerical values of dimensionless combinations of natural constants from epistemological considerations and to connect them with the ultimate large number, the cosmical number.

It is impossible to describe Eddington's methods in a few lines. Not only were they complex, but they also changed as his research progressed. Yet his epistemological foundation remained largely the same. This foundation was that the Dirac equation describes the structural relation of the electron (or the proton) to the entire universe in the sense that the wave equation of the electron and the general-relativistic field equations of the expanding universe hold symmetrically. Structure, rather than substance, was the essence of physics. More generally, Eddington believed that it was possible to obtain knowledge of the fundamental laws of nature from the peculiarities of the human mind. "All the laws of nature that are usually classed as fundamental," he wrote, "can be foreseen wholly from epistemological considerations. They correspond to a priori knowledge, and are therefore *wholly subjective*" (ibid., 84; emphasis in original).

Eddington's view has been charaterized as "selective subjectivism," namely, the idea that our sensory and intellectual equipment has a selective effect in the sense that it largely determines our knowledge of the natural world. He did not deny the existence of an objective world, but he identified it with the conscious and spiritual world, not with the phenomenal world studied experimentally by physicists. If the laws of nature are essentially the subjective constructions of the physicists, little or no room for the empirical-inductive method is left in fundamental physics. According to Eddington, his theory "does not rest on . . . observable tests. It is even more purely epistemological than macroscopic theory. . . . It should be possible to judge whether the mathematical treatment and solutions are correct, without turning up the answer in the book of nature" (ibid.).

Agreement with experiments played no important role to Eddington, but of course he could not totally ignore the development in empirical knowledge that took place in the 1930s. For example, he had originally assumed elementary particles to include electrons and protons only, and when new particles were discovered (such as positrons and mesons), he had somehow to incorporate them in his system. This he did, but in a way that was clearly ad hoc and appeared artificial to most physicists. Likewise, his use of quantum mechanical principles differed from the ordinary understanding of quantum mechanics. Eddington relied on Heisenberg's uncertainty principle to argue that quantum theory implies considerations of the universe as a whole and that isolated physical objects can have no measurable properties; but his use of the uncertainty principle was grossly at odds with the standard interpretation of quantum mechanics. The same was the case with Eddington's idiosyncratic use of Pauli's exclusion principle, on the basis of which he attempted to construct a theory of all fundamental interactions.

Eddington developed his cosmophysical system in a large number of articles and books, but he did not succeed in convincing many other physicists about its soundness. With few exceptions, his theory was met by skepticism or indifference. Toward the end of his life, he busied himself with writing a comprehensive and revised account of his theory, but he died before the work was completed. The book, titled *Fundamental Theory*, was published posthumously in 1946. Although it was studied by some physicists and philosophers, and was sought to be developed further by a few mathematical physicists, its impact was limited. Like earlier versions of Eddington's theory, it had a reputation for being difficult, idiosyncratic, and obscure, if also fascinating. The two books of 1936 and 1946 became famous, but more for their obscurity than for their scientific merits.

In spite of Eddington's failure, his grand project was not without implications, and it did enjoy some support. Schrödinger found Eddington's approach attractive and in the late 1930s, he took it up eagerly in order to construct a unified theory of quantum mechanics and relativistic cosmology. Schrödinger's aim was to explain quantization in terms of the relativistic field equations governing the universe and, in this sense, to give priority to classical field theory over quantum discontinuities. As he wrote to Sommerfeld in 1937, "The world is finite and therefore it is atomistic—because a finite system possesses discrete proper frequencies. And in this way the general theory of relativity gives birth to quantum theory" (Rueger 1988, 395). Although Schrödinger was unable to carry out his unification scheme, his attempt led him to pioneering studies of quantum waves in closed-world models, both static and expanding. Among other results, he found that in an expanding universe, there was a possibility that particles would be formed out of the vacuum. At that time, this and other results went unappreciated, but in hindsight, they mark the beginning of a line of research dealing with the interaction of quantum fields and gravitational fields. It was only much later, in the 1970s, that this kind of study became recognized as an interesting area among astrophysicists and specialists in quantum theories of gravity (see chapter 27).

Cosmonumerology and Other Speculations

Eddington was not the first to call attention to the possible significance of dimensionless combinations of natural constants, but it was his work in particular that stimulated many physicists to take up similar investigations. As an example, consider the reputed German physicist Reinhold Fürth, who in

1929 used his own version of Eddington's theory to deduce, purely theoretically, that the proton-to-electron mass ratio was 1838.2. Moreover, he argued that the mass of the hypothetical neutron—a composite proton-electron system—could be expressed by the formula $M_n = 16^{16}(hc/\pi G)^{\frac{1}{2}}$. Fürth was only one among many physicists in the period who took such numerological arguments seriously. Some of the Eddington epigones concentrated on microphysics and small numbers, others on cosmology and very large numbers, and many attempted to establish numerical relationships between atomic and cosmological quantities. Arthur Haas, the Austrian-American physicist who, back in 1910, had introduced the quantum of action in atomic theory (chapter 4), was another prolific Eddingtonian. In 1934 he published *Die Kosmologischen Probleme der Physik* (The Cosmological Problems of Physics) and two years later, after he had left Vienna for the United States, he deduced the mass of the universe from basic, but arbitrary, assumptions.

The kind of numerological speculations exemplified by the works of Fürth and Haas enjoyed a certain popularity in the decade, but most mainstream physicists realized that the coincidences had no explanatory force and consequently ignored them. Or they ridiculed them, as did Hans Bethe, Guido Beck, and Wolfgang Riezler in a satirical article in the German periodical *Die Naturwissenschaften* in 1931. The three young physicists pretended to derive from Eddington's theory the zero-point temperature as $T_0 = -(2/\alpha - 1)$ degrees, where α is the fine-structure constant. As they joyfully remarked, "Putting $T_0 = -273°$, we obtain for $1/\alpha$ the value 137, in perfect agreement within the limits of accuracy with the value obtained by totally independent methods." When it was discovered that the paper was a joke, the editor of the journal apologized for the three physicists' bad manners and explained that the parody was "intended to characterize a certain class of papers in theoretical physics of recent years which are purely speculative and based on spurious numerical agreements."

Yet, the step between speculation and fruitful imagination is often small, and numerology à la Eddington attracted not only second-rate physicists. As we shall see shortly, Dirac believed that numerical coincidences were deeply significant. In a paper of 1929, Sommerfeld, not a physicist attracted by loose speculations, judged Eddington's derivation of the fine structure constant to be "extremely beautiful and satisfactory." Other eminent physicists, including Lewis, Landau, Gamow, Jordan, and Chandrasekhar, admitted a weakness for the kind of numerological reasoning cultivated by Eddington, although none of them accepted his theory.

Numerology was not the only form of speculative theory that flourished in the 1930s. For example, there was a minor tradition, going back to the mid-1920s (and with roots much farther back), that examined the hypothesis of discrete space-time. This hypothesis took various forms, but it usually included the notion of a smallest length and the corresponding notion of time atoms or minimum durations. Around 1930, several physicists suggested the idea of temporal atomicity and used it in attempts to explain phenomena in the cosmic radiation, the maximum number of chemical elements, the infinities in quantum electrodynamics, and some other problems. As mentioned in chapter 13, de Broglie, Schrödinger, Infeld, Pauli, Landau, and Heisenberg all seriously discussed the proposal that space and time might be subjected to absolute quantum uncertainties, typically taken to be $\Delta q = h/mc$ and $\Delta t = h/mc^2$.

Other physicists of less reputation published theories of discrete space-time which, however, made little contact with either experimental physics or current theoretical problems in quantum theory. In some cases, such theories were inspired by Eddington's attempt at establishing a unitary theory of the very small and the very large. The most elaborate attempt at formulating a new physics based on the notion of discrete space-time was perhaps that of the Polish-American physicist Ludwik Silberstein, who published a monograph on the subject in 1936. According to Silberstein, all the laws of physics, including the theory of relativity, broke down at time intervals of the order of 10^{-18} seconds. He did not find this problematic for, as he wrote, "All modern physicists are inclined to believe that our usual, molar physics, including our space and time concepts, are inapplicable in such circumstances" (Kragh and Carazza 1994, 460). Although the general idea of abandoning continuous space-time at extremely small distances was not foreign to the leading theoretical physicists of the 1930s, few of them took the idea very seriously. The hypothesis was never developed into an empirically fruitful theory and it remained unconnected with fundamental theory. Nonetheless, neither Eddingtonian fundamentalism nor theories of discrete space-time vanished entirely from the scene of physics. After World War II, they continued to be cultivated by a minority of physicists undeterred by the long record of failure.

Milne and Cosmophysics

Edward Milne, a brilliant astrophysicist who was appointed professor of applied mathematics at the University of Manchester at age 29 and from 1929 worked as a professor at Oxford, developed his own, original system of cosmophysics. In some respects, his system differed markedly from that of Eddington. For example, whereas the unification of cosmic and atomic physics was at the heart of Eddington's program, Milne did not find such attempts at unification interesting or useful. Yet, when it came to the fundamental issues of legitimate scientific reasoning, the two systems had much in common. Both were grand and ambitious projects of reconstruction, more like *Weltanschauungen* than ordinary physical theories. And both Eddington and Milne argued for deductivism, a sort of synoptic thinking based on a priori principles from which the laws of nature were deducible by rational reasoning. Milne began to develop his cosmological theory in 1933 and gradually extended it to cover also other aspects of physics. In 1948 he gave a comprehensive presentation of his theory, but by that time, interest in the theory was declining.

Milne found the general theory of relativity to be mathematically, as well as philosophically, monstrous. For him, space was not an object of observation but a system of reference, and thus could have no structure, curved or not. In his systematic 1935 presentation of his theory, *Relativity, Gravitation, and World-Structure*, he argued that all the fundamental laws of cosmic physics can be deduced from a very few principles, most of which may be obtained from analyzing the concepts used to order temporal experience and to communicate them by means of optical signals. The physics arising from such considerations would be a "kinematic relativity," as Milne called his system, because it involved no initial appeal to dynamic or gravitational assumptions.

The basic postulates of Milne's world system were the constancy of the velocity of light and the cosmological principle, according to which the world is homogeneous and isotropic on a large scale. Based on these two principles, and without making use of the field equations of general relativity, Milne derived a uniformly expanding world model and also found that Newton's constant of gravitation was not really a constant. According to Milne, it increased slowly with time, namely, as $G \sim t$. However, the physical meaning of this proportionality was not very clear, and Milne did not think of it as testable. The reason was that Milne's entire system was thoroughly conventionalistic and that he applied different time scales, each with its own physical implications but all being conventions that could, in principle, be used freely. The relationship would hold in "kinematic time" (t), but not in the usual Newtonian time frame (τ), which Milne called "dynamical time." The two time scales were connected logarithmically by the rule $\tau = \log(t/t_0) + t_0$, where t_0 signifies the present epoch. Thus, in the dynamical time scale, G reduced to a constant and the universe remained stationary and had an infinite past age. As to the question of which of the time scales gave the true representation of the universe, Milne considered it meaningless. According to him, the two descriptions were merely two different versions of the same physical reality, and he declared what "really happened" a scientifically illegitimate question.

Milne's system of kinematic relativity diverged distinctly from traditional physics, not only in its results but also in its methodology, which was a peculiar mixture of positivism and rationalism. He thought the universe contained an infinite number of particles, but realized that this was not a verifiable result and emphasized that his world-physics ultimately depended on philosophical reasoning. For example: "Whilst observation could conceivably verify the existence of a finite number of objects in the universe it could never conceivably verify the existence of an infinite number. The philosopher may take comfort from the fact that, in spite of the much vaunted sway and dominance of pure observation and experiment in ordinary physics, world-physics propounds questions of an objective, non-metaphysical character which cannot be answered by observation but must be answered, if at all, by pure reason; natural philosophy is something bigger than the totality of conceivable observations." And in 1937 he wrote about his program: "I have endeavoured to develop the consequences [of my theory] without any empirical appeal save to the existence of a temporal experience, an awareness of a before-and-after relation, for each individual observer" (Kragh 1982a, 78). Although Milne's system was concerned primarily with cosmology, he also believed it allowed him to deduce dynamics, electromagnetism, and even atomic theory from purely kinematic principles. He emphasized that this deductive approach contrasted with the ordinary empirical-inductive approach and resisted reductionist attempts to understand the laws of the universe in terms of atomic theory. Outside cosmology, however, Milne's attempt to reconstruct physics was singularly unsuccessful and was simply not taken seriously by his colleagues in physics.

In June 1938 a conference on "New Theories in Physics" was held in Warsaw under the aegis of the International Institute of Intellectual Cooperation (a commission under the League of Nations) and cosponsored by ICSU. Here, the cosmophysicists met with their more orthodox colleagues. Reports on various aspects of quantum theory were given by Bohr, von Neumann, de Broglie, Kramers, Klein, and Brillouin and were included in the 1939 conference proceedings, *New Theories in Physics*. Eddington's report on "Cosmological Applications of the Theory of Quanta" included derivations of the radius of the Einstein universe (400 megaparsecs [Mpc]), the Hubble parameter (432 km s^{-1} Mpc^{-1}), and the cosmical number (3.145×10^{79}), as well as numerical relationships such as $N/R^2 = 50 m_p m_e c/3\hbar^2$. The presentation was met with objections from Kramers, von Neumann, Rosenfeld, Bohr, and others, none of whom accepted Eddington's use of quantum mechanics. Milne's report on "A Possible Mode of Approach to Nuclear Dynamics," read in his absence by Charles Darwin, was no more acceptable. In agreement with Eddington, Milne argued that "the relation of any given [elementary] particle to the rest of the universe cannot be ignored in such discussions [of interactions between particles]" (p. 207). Contrary to Eddington, Milne simply ignored quantum mechanics in his discussion of nuclear dynamics: "Unable to present wave-mechanics, electron-spin, etc. rationally to myself, I have to attempt to understand matters in my own way; the result is a partial, but to me satisfying, reconstruction of certain portions of physics on a rational basis, a reconstruction to which each equation is a proposition with a content" (p. 219).

If Milne's heterodox system might appear to be the work of a crackpot when seen from the perspective of quantum physics, its reputation and impact were quite different when it came to cosmology. In fact, Milne was arguably the most influential cosmologist of the 1930s, and it was to a large extent his views that set the agenda for discussions in theoretical cosmology. From 1932 to 1940, there appeared about 70 papers related in one way or another to Milne's theory, which means that the theory had a predominant position in the period. Although Milne established no school of cosmophysics many of the most important theorists in England (and a few elsewhere) worked within the problem area defined by Milne's ideas. Among these were Gerald Whitrow, William McCrea, Arthur Walker, and Georges McVittie, who all did important work in cosmology and were deeply influenced by Milne's system of the world. This system was also of considerable importance to the later steady state theory, but after the war it quickly ran out of steam and with Milne's death in 1950, interest in the system died out. Milne's theory of world-physics thrived as part of a larger spiritual climate that was receptive to these kinds of ideas, but with the shift in intellectual climate after the war, the theory began to look like a curiosity of the past. To Milne, his system of the world was not just a physical theory, nor was it just a philosophical theory; it included both physics and philosophy, as well as religion. Like cosmologists of the older schools, he believed that his cosmological system, and that alone, was able to prove the existence of the Christian God and Creator.

The Modern Aristotelians

In a brief note of 1937, clearly inspired by Eddington and Milne, Dirac suggested a reconsideration of cosmology based on the large dimensionless numbers that can be constructed from the fundamental constants of nature. Dirac's fundamental postulate was the large number hypothesis—namely, that all very large dimensionless numbers occurring in nature are interconnected. He paid particular significance to numbers of the orders of magnitude 10^{39} and 10^{78}, for the following reason: With a unit of time given by e^2/mc^3 the age of the universe, which he took to be two billion years, will be about 10^{39}, which is almost the same as the ratio between the electrostatic and gravitational forces between an electron and a proton, e^2/GmM. That is, $T/(e^2/mc^3) \approx e^2/GmM$, where T is the Hubble time (the inverse of the Hubble parameter). If the numerical agreement is significant and charges and masses do not change with time, which is what Dirac assumed, it follows that the gravitational "constant" decreases with atomic time as $G \sim t^{-1}$. That is, while Milne suggested a gradual increase, Dirac held that gravitation would decrease. Dirac agreed with Eddington that the regularities exhibited by the large dimensionless numbers were not purely fortuitous, but while Eddington and most others believed the constants to be independent of the cosmic expansion (or cosmic time), Dirac regarded them to be contingent quantities depending on the history of the universe.

Another consequence of the large number hypothesis followed from the constant $\rho(cT^3)/M$, where ρ is the mean density of matter and $cT = c/H$ is the radius of the observable universe, the Hubble radius. Since the value of the constant—giving the number of particles in the universe—is of the order of Eddington's cosmic number 10^{78}, and this is the square of the period in atomic time, Dirac concluded that the number of particles will increase with time according to the law $N \sim t^2$. Both of these suggestions—the gravitational constant decreasing in time and the spontaneous creation of matter—were highly unorthodox and in conflict with the general theory of relativity. The following year, 1938, Dirac decided that cosmic matter was conserved after all. He claimed that conservation of matter could be reconciled with the large number hypothesis and maintained the idea of a decreasing constant of gravitation.

Dirac's adventure into cosmophysics had little immediate impact on contemporary cosmology. Most astronomers and physicists received Dirac's unorthodox theory with silence, if not embarrassment. Yet in the long run Dirac's speculations proved more influential than Milne's much-discussed theory. The lasting impact of the theory, it turned out much later, was the large number hypothesis itself, not the idea of a varying gravitational constant or the particular universe model suggested.

The general idea of a fundamental interconnection between the large combinations of natural constants proved to be a source of continual fascination in postwar cosmology and fundamental physics. One of the few physicists who took up Dirac's idea in the 1930s was Pascual Jordan, who immediately started to develop it into a more comprehensive theory and sought to harmonize it with the general theory of relativity. In Jordan's version, not only did the gravitational constant decrease in time, but matter was also created spontaneously, albeit in such a way that it did not violate the law of energy conservation.

The theories of Eddington, Milne, Dirac, Jordan, and others were not received favorably by those scientists and philosophers who had observed the rise of rationalistic cosmophysics with increasing dissatisfaction. Most mainstream physicists preferred to ignore the trend, but in England, its only stronghold, it met with open hostility and caused a sort of *Kulturkampf* in the late 1930s. Herbert Dingle, an astrophysicist and philosopher of science, thundered against the new "pseudoscience of invertebrate cosmythology" and what he considered the unbalanced apriorist methods of the "modern

Aristotelians." In Dingle's view, the acceptance of the methods of Eddington, Milne, and Dirac would mean the end of empirical physics as it had been known since the time of Galileo. According to Dingle, experimental knowledge, not a priori principles, should form the basis of physics. Milne, on the other hand, maintained that a completely rational explanation of nature was both desirable and possible. In 1937, in response to Dingle, he stated his view as follows: "The universe is rational. By this I mean that given the mere statement of *what is*, the laws obeyed can be deduced by a process of inference. There would then be not two creations [one of matter, the other of law] but one, and we should be left only with the supreme irrationality of creation, in Whitehead's phrase. We can only test this belief by the act of renunciation, by exploring the possibility of deducing from some assumed description of just what is the laws which what is obeys, avoiding so far as possible all appeal to empirically ascertained laws. Laws of nature would then be no more arbitrary than geometrical theorems. God's creation would be subject to laws not at God's further disposal" (Kragh 1982a, 100).

The British debate about cosmophysics in the late 1930s was one of those rare occasions in modern physical science where the very foundation of physics was discussed, including its bearing on social and ideological issues. Dingle was concerned that the views of the modern Aristotelians would create an intellectual atmosphere that was harmful not only to science, but also to the critical and scientific spirit in a wider sense. "What will be the state of mind of a public taught to measure the value of an idea in terms of its incomprehensibility and to scorn the old science because it could be understood?" he asked. "The times are not so auspicious that we can rest comfortably in a mental atmosphere in which the ideas fittest to survive are not those which stand in the most rational relation to experience, but those which can don the most impressive garb of pseudo-profundity. There is evidence enough on the Continent of the effects of doctrines derived 'rationally without recourse to experience' " (Kragh 1982a, 102). Dingle probably had in mind the situations in Continental dictatorships, such as Nazi Germany and the Soviet Union. Dingle's warning was echoed by John Bernal, the Marxist scientist. Without referring specifically to Eddington and Milne, Bernal commented in 1938, in his important *The Social Function of Science*, on the "mysticism and abandonment of rational thought" that had penetrated far into science itself: "Scientific theories, particularly those metaphysical and mystical theories which touch on the universe at large or the nature of life, which had been laughed out of court in the eighteenth and nineteenth centuries, are attempting to win their way back into scientific acceptance" (Bernal 1939, 3). Despite the validity of the warnings of Dingle and Bernal, the trend in cosmophysics was short-lived and did not, in fact, threaten ordinary physics based on experiments. At any rate, British physicists soon had other things to worry about. After September 1939, they concentrated, successfully, on militarily oriented physics, an area that was light years away from the rationalistic and idealistic cosmophysics. In his introduction to *New Theories in Physics*, the chairman of the Warsaw congress regretted "the absence of many German, Italian and Russian colleagues, whom circumstances have prevented from attending in spite of their feelings of brotherhood with scholars all over the world." We shall now turn to these circumstances.

Chapter 16

PHYSICS AND THE NEW DICTATORSHIPS

IN THE SHADOW OF THE SWASTIKA

WITH ADOLF HITLER'S rise to power as Reich chancellor on January 30, 1933, a new and sad chapter started in European history. Lise Meitner was among the many who listened to the radio broadcast of one of Hitler's speeches shortly before the Nazi party turned Germany into a dictatorship. In a letter to Otto Hahn, she described the speech as moderate, tactful, and personal and wrote that "hopefully things will continue in this vein. . . . Everything now depends on rational moderation" (Hentschel 1996, 18). But rational moderation was not what characterized the following months and years. Meitner herself had to flee Germany five years later. Within a short period of time, the new rulers made it clear that a revolution had taken place in Germany and that the policy and ideology of the Nazi party would have to be followed strictly. This implied, among other things, that other political parties were forbidden and that Jews and socialists could not hold positions as civil servants in the Reich. According to the "Law to Restore the Career Civil Service" of April 7, 1933, a non-Aryan was defined as a person who had one parent or grandparent who was not Aryan. One such person was Bethe, who on April 11, 1933 wrote to Sommerfeld, "You probably don't know that my mother is Jewish. Therefore, according to the new civil service law I am of 'non-Aryan descent' and hence not fit to be a civil service official of the Deutsches Reich . . . I have no choice; I must act accordingly and try to find a place that will accommodate me in some other country" (Eckert and Schubert 1990, 83).

The first wave of dismissals according to the law of April 1933 included more than one thousand university teachers, including 313 full professors. Such a drastic measure—and it was only a beginning—could not avoid damaging German science and scholarship, but this was of minor importance to the Nazi rulers, who had neither understanding of nor interest in science. On one occasion, when Max Planck cautiously suggested to Hitler that the dismissal of Jewish scientists would be harmful to Germany, the Führer (according to one version) flew into a rage and responded, "Our national policies will not be revoked or modified, even for scientists. If the dismissal of Jewish scientists means the annihilation of contemporary German science, then we shall do without science for a few years!" (Beyerchen 1977, 43). A large part of the German population may have agreed with the Führer. The dismissal of Jewish scientists was met with assent or indifference. An American scientist who visited Germany in June 1933 reported in a letter, "Most people don't give a darn; a large proportion is rather glad it all happened. Those extremely few who are upset by it are disinclined to say anything publicly or even privately" (Weiner 1969, 205).

German physicists, like most other groups, responded differently to the new situation. A few of them protested in public, but a far more common reaction was to express the worry quietly and try to reach some kind of understanding with the new rulers. Some Jewish physicists resigned in protest; most were dismissed. The majority of non-Jewish German physicists felt strongly bound to their country and had no wish to leave it, either in protest or for their own good. Planck, Laue, and Heisenberg were among the many who thought that it was their duty to stay, not only for the sake of the Fatherland but even more for the sake of German physics. When Einstein publicly resigned from the Prussian Academy of Sciences in March 1933, the academy, eager to appease the government, accused him of "agitatorial behavior." Laue suggested to Einstein that he might have acted less politically, with more self-restraint. But Einstein, now safely outside Germany, would have nothing of it: "From the situation in Germany you can see just where such self-restraint leads. It means surrendering leadership to the blind and irresponsible. Does not a lack of sense of responsibility lie behind this? Where would we be now if people like Giordano Bruno, Spinoza, Voltaire, and Humboldt had thought and acted in this way?" (Hentschel 1996, xliii). Of course, there were also physicists who welcomed the new regime, either because they believed in the Nazi cause, or at least sympathized with parts of it, or because they saw in the changed circumstances a way to further their own careers. Only a few scientists were members of the Nazi party (DNSAP) before 1933, but membership in any political party had been rare among scientists and scholars in the Weimar period. After 1933, Nazi party membership grew dramatically, especially among young scientists. (It is worth recalling that students were among Hitler's most eager supporters.) The *Uranverein*, the German nuclear power project, included 71 scientists, among them most of Germany's nuclear physicists. Of the physicists, chemists, and engineers involved in the work of the Uranverein, 56 percent were party members and of these, 8 percent had been members before 1933.

The scientific-political infrastructure and support system changed after 1933, but many of the institutions of the Weimar republic continued and, in general, they were not heavily politicized. The most important of the research funding agencies of the Weimar period, the Notgemeinschaft, continued to play a dominant role, though under the name of the Deutsche Forschungsgemeinschaft (German Research Society) from 1937 on. The Physikalisch-Technische Reichsanstalt was another institution that not only continued its existence, but also greatly expanded. After Paschen was forced to resign in 1933, the directorship was taken over by Johannes Stark, the Nazi physicist who was also president (1934–36) of the Notgemeinschaft. Stark reorganized the Reichsanstalt and in 1938 expanded its staff to 444, 138 of whom were scientists. In accordance with his low esteem for pure physics, Stark's Reichsanstalt concentrated on technical and military physics rather than the kind of mainstream physics that Paschen had sought to promote.

For the Kaiser-Wilhelm-Gesellschaft, with Planck as its president, the decade after 1933 was a period of growth. The Kaiser Wilhelm Institute for Physics, planned in 1917 but existing on paper only during the Weimar period, was finally completed in 1937, financed jointly by the government and by 2 million reichmarks from the Rockefeller Foundation. The large, well-equipped institute, the first director of which was the Dutch-born Peter Debye, quickly became Germany's premier institution for physics research. As far as money and institutions were concerned, German physicists had no reason to be dissatisfied with the new regime.

German physics suffered severely, both quantitatively and qualitatively, because of the regime's dismissal policy. A reliable estimate is that Germany lost 25 percent of its 1932 physics community. The total number of university-employed physicists fell from 175 in 1931 to 157 in 1938, but the decline was roughly countered by an increase in physicists working at the polytechnical institutes. Thus, in spite of the heavy drain of emigrating physicists, the physics population remained constant. On the other hand, compared with the situation in the United States, where the number of physics positions continued to increase, this stagnation was a sign of illness. More importantly, the physicists who left Germany represented a mass of talent, experience, and originality that simply could not be replaced. Take the example of Göttingen, the birthplace of quantum mechanics and a university with a large number of eminent Jewish physicists and mathematicians. Max Born, James Franck, Walter Heitler, Heinrich Kuhn, Lothar Nordheim, Eugene Rabinowich, Hertha Sponer, and Edward Teller were all dismissed or otherwise forced to leave the university, which meant that the physics institutes were left almost empty. Göttingen was exceptional, however; in fact, only in 15 of 36 German universities or polytechnics were teachers dismissed for political reasons. Yet it was the universities with the largest and most progressive physics institutes, especially Göttingen and Berlin, that suffered most. The polytechnical institutes were not hard-hit and at the university of Jena, a conservative university with no Jews among its physics staff, not a single physicist was dismissed. In general, the more theoretically oriented the institute was, the more dismissals: Of Germany's 60 university teachers with positions in theoretical physics, no fewer than 26 went into exile. The drain of scientific talent was remarkable. Among the physicists who were forced, or felt forced, to depart their positions in Germany between 1933 and 1940, six were Nobel laureates (Einstein, Franck, Hertz, Schrödinger, Hess, and Debye) and another eight would later receive the Nobel prize in either physics or chemistry (Stern, Bloch, Born, Wigner, Bethe, Gabor, Hevesy, and Herzberg). For a more comprehensive list of emigrés, see table 17.2.

In June 1933 John von Neumann reported to the Princeton mathematician Oswald Veblen about the situation in Germany, which he found "very depressing." He explained, "We have been three days in Göttingen and the rest in Berlin, and had time to see and appreciate the effects of the present German madness. It is simply horrible. In Göttingen, in the first place, it is quite obvious that if these boys continue for only two more years (which is unfortunately very probable), they will ruin German science for a generation—at least" (Weiner 1969, 205). As the realities of the Nazi regime's measures against Jewish scientists became known in other countries, many foreign physicists responded, first quietly by avoiding visiting Germany, canceling subscriptions to German periodicals, or resigning membership in German scientific associations. This, and the restrictions placed on Germans traveling abroad, contributed to an increasing international isolation of German physics. Starting in 1936, German scientists (along with Russians and Italians) were forbidden to participate in meetings associated with the League of Nations. In 1937 the Nazi regime forbade Germans to accept the Nobel prize. Samuel Goudsmit was more outspoken in his criticism of the Nazi policies than most physicists. In a bitter letter of resignation of 1936 to Walther Gerlach, he gave his reasons for discontinuing his membership of the German Physical Society: "I am disappointed in that the Society has never protested as a whole against the bitter attacks upon some of its outstanding members. Moreover, very few contributions to physics are coming from Germany nowadays. The main German export being propaganda of hatred" (Beyerchen 1977, 75). Several other foreign physicists followed Goudsmit, which contributed to the decline in membership that the society experienced through the 1930s. The German Physical Society had, in fact, succeeded in following a relatively independent course and avoiding the expulsion of its Jewish members. It was only in late 1938 that Debye, as president of the society, felt forced to request all Jewish members to withdraw membership. By 1940 Debye had had enough, and left Germany to take up a position at Cornell University.

The majority of foreign physicists were uncertain how to act and were careful not to break connections with their German colleagues. They were against Hitler and the Nazi policies, not against the Germans themselves. The American physicist and later Nobel laureate Percy Bridgman was an exception. In early 1939 he published a statement in *Science* in which he declared, "I have decided from now on not to show my apparatus or discuss my experiments with citizens of any totalitarian state." He realized, of course, that such a step went against the ideal of scientific internationalism, but argued that "the possibility of an idealistic conception of the present function of science has been already destroyed, and . . . perhaps the only hope in the present situation is to make the citizens of the totalitarian states realize as vividly and as speedily as possible how the philosophy of their states impresses and affects the rest of the world" (Hentschel 1996, 185).

German physicists and science administrators were well aware that Germany was no longer able to compete with the United States in many important areas of physics. They also knew, but could not say so openly, that the Nazi dismissal policy was, in large part, responsible for the decline. Carl Ramsauer was a recognized applied physicist and the director of the research division of AEG (Allgemeine Elektrizitätz-Gesellschaft), one of Germany's largest industrial corporations. In a 1942 memorandum to Bernhard Rust, secretary of science, education, and culture, Ramsauer warned that German physics had fallen clearly behind physics in the United States. Whereas American citations in the 1913 volume through the 1938 volume of the *Annalen der Physik* had risen from 3 percent to 15 percent, Ramsauer found that in the same period, German citations in the *Physical Review*—"the internationally acknowledged leading physics journal"—had decreased from 30 percent to 16 percent. In the area of nuclear physics, Ramsauer showed that Anglo-American research dominated: In 1927, the *Physikalische Berichte* listed 47 papers on nuclear physics written in German and 35 written in English; twelve years later, the figures were 166 in German and 471 in English. "The number of German papers on this most modern and promising field has thus risen 3.5-fold in this time, whereas the number of papers written in English have risen 13.5-fold. Yet, as every nuclear physicist will confirm, the quality of American papers is at the very least equivalent to that of German papers."

Later historians have analyzed data similar to Ramsauer's in order to provide a more detailed picture of the effect of the Nazi policy on physics. Within quantum mechanics, including quantum electrodynamics, 25 percent of all German-language publications between 1926 and 1933—and these made up 45 percent of all publications—were written by physicists who later left Germany (table 16.1). A similar pattern holds for nuclear physics, where the share was 18 percent. Moreover, the papers by the future emigrants were of more than average value, as indicated by the frequency with which they were cited in the physics literature. Citation studies show that papers by German emigrant nuclear physicists were cited more than three times as frequently as would be expected from the number of their articles. In other fields, the contributions of the future expatriates were less marked, generally with a larger share in the more theoretical and modern fields and a relatively low share in more experimental and traditional fields, such as acoustics and technical mechanics (table 16.2). The average contributions of the emigrés to all fields of German physics was 10.8 percent. Most of their

TABLE 16.1
Publications in Quantum Theory

	Publications		By later emigrants	
Year	*Total no.*	*Percent in German*	*No.*	*Percent in German publications*
1926	173	49.7	16	18.6
1927	319	50.8	42	25.9
1928	295	44.4	31	23.7
1929	286	43.4	28	22.6
1930	326	44.5	35	24.1
1931	208	50.5	33	31.4
1932	170	30.6	12	23.1
1933	153	38.6	20	33.9
total	1930	44.8	217	25.1

Note: Based on data in Fischer 1988.

publications were in the theoretically oriented, progressive *Zeitschrift für Physik* (where they made up 14.5 percent) and only few in the more conservative *Annalen der Physik* (5.9 percent).

In general, Nazi authorities were not opposed to science, and they channeled very large amounts of money to scientific research. German physics declined after 1933, but high-class research continued in both experimental and theoretical fields. Heisenberg and his group at the University of Leipzig were still among the world leaders in quantum field theory and cosmic ray research; von Weizsäcker did important work in nuclear theory and astrophysics; in solid state physics, Robert Pohl, Walter Schottky, and others con-

TABLE 16.2
German Publications in Physics, 1925–33, Select Fields

Field of physics	*Publications in German*	*By emigrants*	*Percent*
Quantum theory	864	217	25.1
Nuclei, radioactivity, particle rays	532	100	18.8
Spectra	958	123	12.7
Mechanics of fluids and gases	1740	163	9.4
Technical mechanics	1030	58	6.6
Acoustics	482	18	3.8
Total, all fields	23216	2505	10.8

Note: Based on data in Fischer 1988.

tributed significantly; and in nuclear physics and chemistry, the works of Meitner, Hahn, and Strassmann led to nuclear fission. Even in an environment increasingly poisoned by Nazi ideology, it was possible to conduct physics research of the highest value.

Aryan Physics

In 1936 Philipp Lenard, professor emeritus of physics, published a physics textbook in four volumes, entitled *Deutsche Physik* (German Physics). In the preface, he justified the unusual title: "'*German* physics?' you will ask.—I could also have said Aryan physics or physics of the Nordic type of peoples, physics of the probers of reality, of truth seekers, the physics of those who have founded scientific research.—'Science is international and will always remain so!', you will want to protest. But this is inevitably based upon a fallacy. In reality, as with everything that man creates, science is determined by race or by blood. . . . Nations of different racial mixes practice science differently" (Hentschel 1996, 100). The German or Aryan physics that Lenard advocated in his work had its origin back in the early 1920s, when Lenard and a group of other right-wing German physicists attacked Einstein's theory of relativity and, more generally, modern theoretical physics (see chapters 7 and 10). The view that there were different forms of physics, depending on race and nationality, agreed perfectly with the anti-internationalistic ideology of the National Socialists, as expounded in Hitler's *Mein Kampf* and Alfred Rosenberg's *Mythos des 20. Jahrhundert.* The loose group of right-wing antirelativists in the early 1920s included Lenard, Stark, and Gehrcke as its most notable members and of these, the two Nobel laureates turned to the Nazi party at an early date. Although Stark did not join the party until 1930, and Lenard in 1937, they were already devoted to the cause of Hitlerism many years earlier. In 1924 they coauthored a praise of Hitler and his allies, whom they described as "God's gifts from times of old when races were purer, people were greater, and minds were less deluded" (ibid., 9). The article contributed to their ostracism from the German physics community, but after 1933 they found themselves on the right side of the table of power and were ready to take their revenge. Whereas the aging Lenard, who had retired in 1931, acted mainly on the ideological front, the younger Stark worked politically and sought to take power over the German physics organizations.

What was the German physics that Lenard, Stark, and their allies wanted to install in place of the despised Jewish quantum-relativistic dogma? First of all, it was never formulated in a consistent or programmatic way and did not really make up a coherent worldview. Nor did it materialize in scientific practice. What characterized the Aryan physics view was rather what it was against, namely, modern physics with its complex mathematical apparatus, lack of visualizability, counterintuitive results, and dismissal of the classical worldview of Newton, Faraday, and Helmholtz. Basically, the Aryan physicists were antimoderns and romanticists, who longed for a return to a physics based on experiments and simple, understandable theory in agreement with intuition. Opposed to the specialization that had fragmented physics, they advocated a holistic view of nature. By then, Aryan physics was of course openly racist, claiming that all sound physics came from Aryans (although often stolen by Jews), whereas physicists of Jewish descent excelled in sterile theorizing, such as the theory of relativity. Lenard and his like believed that physics was truth-seeking, and that truth could be obtained only through experiments combined with mental images of reality. The most important of such images was the ether, which many of the Aryan physicists claimed was not merely an image, but a reality. In a fashion both simplistic and romantic, they maintained that the Nordic physicist, and only he, could establish a dialogue with nature through careful experiments and observation; nature would enter the dialogue and give him the answer to the questions posed. Advanced theory, on the other hand, would lead nowhere. As Lenard's pupil Alfons Bühl put it, "This exceedingly mathematical treatment of physical problems had undoubtedly arisen from the Jewish spirit. . . . Just as he [the Jew] otherwise—as in business—always has only the numerical, the credit and debit calculation before his eyes, so it must be designated as a typical racial characteristic even in physics that he places mathematical formulation in the foreground" (Beyerchen 1977, 132). Although all this may seem terribly naive, even ridiculous, the Aryan physicists were not fools without knowledge of science; they were supported by at least one philosopher of science of some prominence, Hugo Dingler. Since the Aryan physicists were kept out of the mainstream physics journals, such as the *Zeitschrift für Physik*, they had to publish their contributions and polemics elsewhere. A favorite journal was the *Zeitschrift für die gesamte Naturwissenschaft*, one of the more peculiar journals in the history of physics publication. Founded in 1935, the journal was the unofficial mouthpiece for proponents of Aryan physics and included a strange mixture of ordinary physics articles, anti-Semitic propaganda, polemics against established physics, and hagiographic history of science.

Aryan physics could also be called Nazi physics, and its ideas received considerable support from Nazi officials, such as Rosenberg. It was only because of this political support that Stark was able to secure a power base in German physics and, for a period, make Aryan physics look like a threat to conventional physics. The committed followers of Lenard and Stark were few, scarcely more than thirty in number, but because of the political circumstances and a good deal of sympathy among their students, their influence was relatively higher than their small number. They achieved a temporary success in 1939, when the question over Sommerfeld's chair in Munich was finally settled. Sommerfeld had wanted Heisenberg to follow him, but the Aryan physicists launched a vicious campaign against the two quantum theorists and accused Heisenberg of being a "white Jew." Although Heisenberg was rehabilitated (due to the personal intervention of Himmler), the campaign achieved its goal. Wilhelm Müller, a good Nazi and antirelativist with no knowledge of modern physics at all, was appointed professor of theoretical physics in Munich. It could have been a joke, had it not been real and had the background not been tragic. By the end of 1939, six proponents of Aryan physics had been appointed professors at universities or polytechnics (in Munich, Heidelberg, Karlsruhe, and Stuttgart).

In spite of occasional victories, the influence of the vociferous Aryan physicists was limited and short-lived. The large majority of the German physicists, even among those who sympathized with the Nazi cause, considered Lenard's and Stark's ideas to be unacceptable, ridiculous, and dangerous. Jordan, who was in favor of the regime and joined the Nazi party in 1933, would have nothing to do with the strange attempt to cancel the progress in theoretical physics attained during two decades. Aryan physics just was not able to replace the physics of Einstein, Bohr, and Heisenberg, and in practice, teaching of physics at many German universities was unaffected by the attempt to ban the theory of relativity. The theory was taught in textbooks and lectures, but Einstein's name was often left out. Aryan physics never obtained the same kind of power as Lysenkoism obtained in Stalin's Russia. After 1940, when Stark and his followers lost much of their political support, the movement began to disintegrate and three years later, it had practically ceased to exist. Perhaps the main effect of the Aryan physics movement was the impact it had on mainstream physicists like Heisenberg, von Weizsäcker, and Laue. They felt strongly that Stark and his comrades were on their way to destroying German physics and that it was imperative to fight the evil, which could be done only on German soil. It was a contributing reason for them to stay in Germany and to cooperate with the Nazi authorities. They did not fully realize that there were worse things going on in the Third Reich than the activities of a relatively harmless pack of Aryan physicists.

Physics in Mussolini's Italy

Mussolini came to power in 1922; three years, later Italy became a fascist dictatorship under Il Duce. At that time, Italian physics had little to boast of. There was no one to replace the old generation of experimentalists of high reputation, such as Augusto Righi, and in theory, the situation was even worse. Once an important nation in physics, Italy seemed on its way to becoming a provincial backwater. But then things changed. During the period 1925–38, the years of fascism, Italian physics flourished in a most remarkable way, and the country became one of Europe's most advanced nations in modern physics. The central figure in this process was undoubtedly Fermi, who became professor of theoretical physics in Rome in 1926, when he was only twenty-five years old. In the early 1930s, Fermi organized his group of young experimental and theoretical physicists, which included Edoardo Amaldi, Franco Rasetti, Bruno Pontecorvo, Emilio Segré, and others. Fermi and his group changed the landscape of Italian physics, and indeed of world physics, but they were not alone. At the same time as Fermi concentrated on nuclear physics in Rome, a strong group of young physicists in Florence made important contributions to the study of cosmic rays and related areas. The main figures in the Florence group were Bruno Rossi, Gilberto Bernardini, Giuseppe Occhialini, and Giulio Racah, all of whom enjoyed close relations with their colleagues in Rome. The Italian physicists were strongly international in their outlook and practice and often spent periods abroad. The modern physics they represented seems not to have caused any problems with the government or controversies with philosophers of a fascist inclination.

Fermi and most of his colleagues were absorbed in physics and not very interested in politics. Yet, although Fermi was basically apolitical, he knew very well how to maneuver in the political circles of Roman science, an area in which he had an important "godfather" in the physicist and politician Orso Maria Corbino. In 1929 Fermi was appointed as the only physicist member of the newly created Accademia d'Italia, an institution that Mussolini had established as an alternative to the old and distinguished Accademia dei Lincei, the members of which Mussolini suspected were hostile to fascism. As a member of the Accademia d'Italia, Fermi was entitled to wear the academy's fascist uniform and was given the title of "Excellency." He was also a member of the Italian National Research Council. Whether he liked it or not, Fermi was part of Italian politics and his research group in Rome, in reality, was protected by the fascist state. Fermi was never an active supporter of Mussolini's state, but neither was he an antifascist.

In spite of the ideological similarities between Italian fascism and National Socialism, Mussolini did not welcome Germany's change to the Third Reich. Many Italian physicists decided that collaboration with a nation that dismissed Jewish colleagues must be cooled down. This was the background for Fermi's decision to publish the important papers on neutron-induced nuclear reactions in English and not, as would have been natural under other circumstances, in German. In general, physicists in Rome and Florence turned to England and the United States and away from Germany. For a while, physics in Italy continued to flourish, undisturbed by the grave situation in Germany. From 1937, however, Mussolini's Italy and Hitler's Germany moved increasingly together, with Germany as the senior partner in the political, military, and economic partnership. As a result, in the summer of 1938, the fascist government introduced racial laws modeled after the notorious German Nuremberg laws. One of the consequences was that Jewish scholars and scientists were dismissed from Italian universities and, in general, life became difficult for Italian Jews. Fermi was affected primarily because his wife was Jewish, but also because he was publicly accused by fascist extremists of "having transformed the physics institute into a synagogue" (Segré 1970, 98). He decided to leave the country and soon found an opportunity, namely, his receipt of the Nobel prize on December 10, 1938. Instead of returning to Rome from Stockholm, he and his family went to the United States. Several other Italian physicists left the country, including Rasetti, Amaldi, and Segré. Rossi, who was Jewish, faced the same situation as Bethe in Germany five years earlier. He managed to obtain a passport and went to Copenhagen, where Bohr helped him financially and found him a temporary job. Blackett invited him to Manchester with a fellowship of the Society for the Protection of Science and Learning and in June 1939, Rossi, following in Fermi's footsteps, arrived in Chicago.

Physics, Dialectical Materialism, and Stalinism

Among the political weaknesses of Aryan physics (as well as related attempts in mathematics, chemistry, and biology) was that it could not appeal to an accepted philosophical foundation of National Socialism. There simply was no such foundation. National Socialism was built on action and emotion, not on a coherent system of ideas. This was contrary to the situation in the Soviet Union, where the regime was based ideologically on the socialist corpus of writings of Marx, Engels, and Lenin. From the earliest days of the new Soviet Union, philosophers found it important to analyze science from the point of view of the official Marxist philosophy, that is, the dialectical materialism that could be distilled from the works of Engels and Lenin. Marxist enthusiasts sought to promote a "proletarian science" that differed from the bourgeois science in terms of methods, aims, and approach. The movement attracted considerable philosophical and political interest, but failed to convince the physicists that they ought to change their science toward a more proletarian direction. In general, there were few political activists among Soviet scientists. According to a 1930 survey, the Soviet Union included about 25,000 "scientific workers," among whom were some 1,000 physicists. Only 44 of the physicists were members of the Communist Party.

During most of the 1920s there were no serious conflicts between party philosophers and the physicists, but in the 1930s the debate sharpened, especially in connection with the question of the interpretation of quantum mechanics. It was not so much an attempt from the side of the regime to impose its view on the physicists as it was a conflict between physicists and philosophers, among philosophers themselves, and in some cases, among physicists themselves. Because many philosophers regarded themselves as the intellectual guardians of true Marxism-Leninism, their voices were politically more important than they would have been under normal, democratic circumstances. If it could be argued that a certain physical theory, or a certain interpretation of it, was "idealistic," "subjective," or "Machist" (related to the ideas of Ernst Mach), it might lead to a potentially dangerous situation. For example, the philosopher Alexander Maksimov claimed in 1939 that Einstein, Schrödinger, Bohr, Dirac, and Heisenberg were all "idealists of the Machian variety" and that their views of quantum physics were ideologically unacceptable. "The struggle for Bolshevism in science is the struggle for a fundamental reconstitution of science," he declared (Vucinich 1980, 240). Party philosophers tried to engage the physicists in ideological discussions and to convince them about their errors, but they were not very successful. Many physicists simply ignored the philosophers and stayed away from questions concerning the relationship between physics and Marxism. To take up such questions might lead to difficulties and, after all, physics was much more interesting than philosophy. Others took up the challenge and argued, not unreasonably, that in order to criticize physics, one must understand physics; and the philosophers, they gleefully noted, did not. Or they argued that there were not, in fact, any contradictions between Marxist-Leninist dogma and current understandings of relativity and quantum physics. Abram Ioffe protested in 1934 that it was entirely unfounded to label Bohr and other proponents of the Copenhagen interpretation "idealists." Quite the contrary, he claimed that the astonishing insights reached by the Western physicists amounted to "a brilliant confirmation and enrichment of dialectical materialism" (ibid., 242).

The controversy was not simply between fanatic party philosophers and reasonable physicists. There were Soviet physicists whose scientific views were not more modern than those of, say, Lenard in Germany. They adhered to a Newtonian worldview, supported the existence of the ether, and rejected the theories of relativity and quantum mechanics. Among the conservative minority were Kliment Timiriazev, a professor of physics at the State University of Moscow, and Vladimir Mitkevich, a specialist in electrical technology. These and other "mechanists" accused progressive physicists such as Frenkel, Vavilov, Tamm, and Ioffe for promoting idealism, obscurantism, and clericalism. As in Germany, the attacks on the "new physics" included aspects of anti-Semitism. Another aspect of the dispute was regional, with Leningrad playing a similar role to that of Berlin in the German case. Leningrad was the stronghold of the quantum and relativity theorists, and the

Leningrad Physico-Technical Institute had as bad a reputation among communist hardliners as it had a good reputation in international physics. This was one of the reasons that caused a decline in the importance of Leningrad as a center of Soviet physics. In the mid-1930s, the Academy of Sciences was transferred from Leningrad to Moscow, and many physics institutions began to concentrate in the capital.

Although some philosophers and physicists attacked the physical theories themselves, most of the discussion in the 1930s concerned methodological and epistemological questions. There was broad (but not total) agreement that the theories of relativity and quantum mechanics were basically correct, and the attempts to create a specific communist or proletarian physics were feeble and not taken very seriously. Yet the questions of interpretation and method were politically delicate and could not be restricted to the purely academic level. Many leading Soviet physicists adhered to the views of the Copenhagen school and argued—indeed, had to argue—that these views could well be brought into harmony with dialectical materialism. This group of physicists included eminent theorists, such as Fock, Landau, Tamm, Frenkel, and Bronstein. A minority of physicists and a majority of philosophers disagreed violently. K. V. Nikol'skii, a physicist, attacked the Bohr-Heisenberg position, which he found was "totally incompatible with the progressive ideas in theoretical physics, for it was a consistent elaboration of idealistic or Machian principles" (Vucinich 1980, 245). In some of the aggressive arguments of the late 1930s, to brand a physicist as an "idealist" came unpleasantly close to Aryan physicists' labeling their enemies as "white Jews." Indeed, Ioffe noted sarcastically that Maksimov's stand, which included acrimonious attacks on Einstein and Bohr, had more than a superficial similarity to that of Lenard and Stark. The point was made explicit by Yakov Frenkel, the distinguished theorist, who had no respect for dialectical materialism and the supposedly deep insights in physics of Engels and Lenin. In a letter of 1937 to *Under the Banner of Marxism*, the party's theoretical journal, he noted that Maksimov's group "by virtue of its views, is amazingly similar to the group of reactionary physicists headed by Professor Stark. . . . In their treatment of modern physics, Mitkevich, Timiriazev, Kasterin, and together with them, Maksimov, differ from Stark, Leonard [*sic*], Herke [Gerhcke?], and other representatives of German obscurantism only in that they replace the term 'Jewish' by the term 'idealistic'" (Frenkel 1997, 215). Frenkel's letter was not published.

As in Germany, the well-established theory of relativity became a target of some philosophers' politically motivated criticism. They felt provoked by Einstein's attempt to geometricize physics and his emphasis on the role of pure thought in the establishment of physical theory. "The theory of relativity does not penetrate the depth of physical phenomena," Ernst Kol'man wrote in 1939. "All efforts to build physics on the geometry of continuous space are doomed to failure. . . . These efforts are steeped in groundless, metaphysical exaggeration, reinforced by the idealism of many theoretical physicists" (Vucinich 1980, 249). Most of the Soviet critics were both more open and more advanced than their counterparts in Germany, who totally denied the validity of relativity theory and rarely distinguished between the special and the general theories. On the whole, the dispute in the 1930s between philosophers and physicists was of a different nature from that taking place in Germany. Physicists and enlightened philosophers had no trouble advocating their views in opposition to the attacks of party philosophers, and there was no question of creating a specific Marxist physics as a counterpart to Aryan physics. The main reason why the debate occurred in a relatively free manner, without serious ideological restrictions, was that it was a debate between two groups of scholars, none of whose views were sanctioned by the political authorities. This situation changed after World War II, when Soviet intellectual life experienced a much harder ideological climate and when the question of the interpretation of quantum mechanics became part of the political-ideological game. Between 1948 and 1951, physics was heavily politicized, including a campaign against "reactionary Einsteinism," but even then it was mostly a question of what was the correct philosophy of physics. The dreadful case of Lysenkoism was not repeated in Soviet physics.

This does not mean that the Soviet physics community lived peacefully with Stalin's regime. After about 1933, the climate in the Soviet Union was marked by an unhealthy cocktail of xenophobia, suspicion, sycophancy, and fear of the secret police. Whereas Soviet physicists had earlier been active participants in the international physics life and main contributors to German physics journals, they were now increasingly forced into isolation. As an outlet for Soviet physicists, but also with occasional contributions from foreigners, the *Physikalische Zeitschrift der Sowietunion* was established in 1932. It included a strange mixture of technical and military physics, politically correct philosophy of science, and high-quality technical papers on quantum field theory and other fields of mainstream physics.

Stalin's victims came from everywhere, and they included physicists. The Russian despot was no more interested in physics than was Hitler, and nobody, scientist or not, could feel secure under the Great Terror. Between 1935 and 1941, millions of Soviet citizens were killed. As many as 18 million may have been arrested, and possibly half that number were executed or otherwise disappeared. Historians have estimated that more than one hundred physicists were arrested in the 1937–38 purges in the Leningrad area alone. This indicates that about 20 percent of all Soviet physicists may have been arrested. A few physicists, such as Gamow in 1933, escaped to the West. Others were arrested by the secret police, including Landau, who spent one year in prison until he was released after pressure from Peter Kapitza.

The Austrian physicist Fritz Houtermans worked at the Physico-Technical Institute in Kharkov. His communist conviction did not prevent him from being arrested, and he had to spend two years in prison before he was extradited to Germany in 1939. The same happened to Houtermans' colleague Alexander Weissberg, another Austrian physicist and dedicated communist. Viktor Bursian, a Leningrad theorist and specialist in quantum mechanics, was arrested in 1936 and died in prison ten years later. Thirty-two-year-old Matvei Bronstein, a Russian-Jewish theorist, was no more fortunate. He was arrested in 1937 and falsely charged not only with being a foreign spy, but also with "resolutely opposing materialist dialectics being applied to natural science" (Gorelik and Frenkel 1994, 145). His life ended before a firing squad. The same tragic fate befell Lev Rosenkevich and Lev Schubnikov, two of Landau's colleagues. Other Soviet physicists who were either executed or died in prison included Vsevelod Frederiks, a specialist in the theory of relativity, and B. Gerasimovich, an astrophysicist. The purges were of even greater consequence for Soviet astronomy and astrophysics than for physics proper. In the famous Pulkovo observatory outside Moscow, ten senior astrophysicists were arrested for "participation in a fascist, Trotskyist terrorist organization" (Josephson 1991, 316).

Whereas the Nazis expelled unwanted physicists, the Communists either shot them, jailed them, or did their best to keep them within the borders of the Soviet Union. They were not allowed to go abroad and attend scientific conferences. In the case of Peter Kapitza, the Russian specialist in magnetism and low-temperature physics who had spent many years with Rutherford in Cambridge, the Soviet authorities even went to the extreme of coming close to kidnapping him. On a visit to Russia in 1934 Kapitza, then living in Cambridge, was not allowed to leave the country. He was given ample funds to create a new physics institute and soon became one of the most influential physicists in the Soviet Union. Like Heisenberg in Germany, Kapitza quickly learned how to compromise with immoral political forces. The situation in the Soviet Union in 1937 may be glimpsed from a letter that Kapitza wrote to a high-ranking party official in protest of the arrest of Vladimir Fock: "I was greatly disturbed by the news that the physicist V. A. Fock was arrested yesterday. . . . It will distance our Soviet scientific circles still further from building socialism and may, moreover, undermine Fock's ability to work and so provoke a bad reaction from scientists here and in the West. It is said that, besides Fock, very many other theoreticians were arrested a few months ago in connection with the same affair. In fact, so many were arrested that in the university faculty of mathematics and physics no one could be found to lecture to students" (Boag et al. 1990, 337). Fock was released soon afterward.

Chapter 17

BRAIN DRAIN AND BRAIN GAIN

American Physics in the 1930s

FROM THE LATE 1920s, physics in the United States experienced a rapid growth, quantitatively as well as qualitatively. New institutes, departments, and graduate programs were established, often with the economic support of large private foundations. The most important of these was the General Education Board, a philanthropy founded by the Rockefeller Foundation, whose operations were restricted to the United States. Between 1925 and 1932, the General Education Board supported science departments at major American universities with the enormous sum $19 million.

Money was an important reason for the American progress, but not the only one. Another was the devotion of American leaders in physics (and the other sciences) to international cooperation and competition. This had traditionally taken place by sending American scientists to Europe, but in the 1920s the stream began to reverse, first with the invitation to several young European physicists to take up positions at American universities and with the creation of seminars and summer schools of the highest excellence. The annual University of Michigan summer school, starting in 1927, was the most important of these, but it was only one of several institutions that attracted visitors from Europe. Strong programs in physics were under development at many American universities, of which Princeton, Chicago, University of California at Berkeley, and the California Institute of Technology (Caltech) were among the most important.

The increasing attractiveness of American centers of physics is illustrated by the choice of institutions of recipients of postdoctoral fellowships from the Rockefeller Foundation or the International Education Board (IEB). IEB operated in Europe as a counterpart to the General Education Board; like that institution, it was funded by Rockefeller money. Of the 135 European physicists who received postdoctoral fellowships for studies abroad between 1924 and 1930, 44 chose to go to the United States. The most attractive European nations were Germany and England, with 26 and 25 physics fellows, respectively; 16 chose to go to Denmark, a choice that reflected the importance of Bohr's institute. By the early 1930s, American physics had progressed dramatically and was competitive with anything in Europe. When the American Physical Society held its annual meeting in Chicago in 1933, five months after Hitler had come to power in Germany, John Slater participated together with European notables such as Bohr, Cockcroft, and Fermi. Slater recalled that what impressed him most was "not so much the excellence of the invited speakers as the fact that the younger American workers on the program gave talks of such high quality on research of such importance that, for the first time, the European physicists present were here to learn as much as to instruct" (Weiner 1969, 201).

At the time of the Chicago meeting, American physics was, in fact, facing great troubles because of the effects of the Great Depression. The economic crisis hit American science with full force only after a delay of several years; the effects did not last long, but when the depression hit about 1933, the impact was serious. Financial support was severely reduced, the salary of faculty members was cut, and new Ph.D.s found it almost impossible to obtain positions. Yet, in spite of the troubles, the year of crisis did not result in a smaller number of Ph.D. students. In 1931, the last good year, American foundations donated a total of $5 million to research in the natural sciences; in 1934, the amount of foundation money had been reduced to $2 million. All physics institutions were affected, the National Bureau of Standards perhaps most brutally. Between 1932 and 1934, operating funds for this institution, the largest government employer of physicists, were cut by no less than 70 percent. Nor were industrial laboratories spared. By 1933, General Electric had fired 50 percent, and AT&T almost 40 percent, of their laboratory personnel. In addition to the financial troubles, many American leaders of physics were worried about the public image of science, which showed signs of changing from the traditional positive picture to a much more critical, or even negative, attitude. In the early 1930s, there was a kind of unorganized antiscience movement in American society that questioned the very rationale of the scientific endeavor. Like earlier and later movements of this kind, it was heterogeneous and drew on different, sometimes incompatible, kinds of dissatisfaction.

Some humanists deplored the lack of values in science and argued that science could not provide what modern man was most in need of: spiritual hope and moral guidance. They therefore suggested a moratorium on scientific research, or a strongly reduced science sector. The attitude was neither new nor restricted to the American arena. In 1927 the bishop of Ripon gave a sermon at the meeting of the British Association, in which he suggested, "At the risk of being lynched by some of my hearers, that the sum of human happiness outside scientific circles would not necessarily be reduced if for ten years every physical and chemical laboratory were closed and the patient and resourceful energy in them transferred to recovering the lost art of getting on together and finding the formula of making both ends meet in the scale of human life" (Bernal 1939, 2). Other critics believed that science and technology were to be blamed for the mass unemployment, for didn't the machines and automatized factories destroy more jobs than they created?

And were the labor-saving technologies not the products of science? Critics again argued that the fault of science was not that it produced technological innovations, but that it did not produce enough of them and not of the kind that benefited the ordinary citizen. Rather than using money on esoteric and useless subjects (such as the theory of relativity), physicists were advised to solve the basic needs of the people. As one critic put it: "With a small amount of such brains as are now focused on the speed with which the neutron penetrates the nucleus of the atom . . . the cost of the poor man's housing could be cut in half" (Kevles 1987, 247). This was a kind of criticism that shared elements with the one raised by the "proletarian science" movement in the Soviet Union.

The antiscience attitude was probably not very widespread, but it was taken seriously by leaders of the scientific community, who feared that it might harm the prospects of further progress when the effects of the depression ceased. Although the large philanthropies were far from being antiscience, they were influenced by the general appeal to a more humanistic and welfare-oriented science. This was a major reason for the Rockefeller Foundation's decision to change its areas of priority from the more basic physical and chemical sciences to the sciences dealing directly with man, such as biology, psychology, and the social sciences. "Uneasiness and even alarm are growing as the belief gains ground that the contributions of the physical sciences have outstripped man's capacity to absorb them," wrote the president of the Rockefeller Foundation in 1936. "There can be little doubt that a serious lag has developed between our rapid scientific advance and our stationary ethical development" (Kevles 1987, 249). The shift in Rockefeller policy was a serious matter for physicists, both in America and Europe. Millikan at Caltech and Bohr in Copenhagen, to mention just two examples, realized that they could get their physics projects funded only if they were relevant to biological problems. Consequently, they argued that this was the case, although in reality, their research in cosmic rays and nuclear reactions was pure physics.

Still, in 1935, Edwin Kemble of Harvard University described the job situation for fresh physics Ph.D.s as a "nightmare," but at that time, things began to improve. The effects of the Great Depression were no longer felt seriously in American science and a period of recovery quickly set in. The employment situation for physicists improved and the number of physics students increased dramatically. During 1931–40, more than 1,400 physics doctorates were awarded, double the number of the preceding decade. By 1940, nearly 200 graduate students received Ph.D. degrees in physics from American universities (see figure 2.1). Membership in the American Physical Society increased linearly, from about 1,300 in 1920 to about 3,700 twenty years later. By 1941 the number of physicists working in the United States reached about 4,600—almost five times the number of physicists in the Soviet Union. The number of industrial research laboratories rose from about 300 in 1920 to more than 2,200 in 1940. In a longer perspective, the years of depression were just a minor disruption of the general trend of growth in American physics. This growth and the general vigor of American physics were an essential factor in the country's ability to absorb the many European refugee physicists who arrived in the 1930s.

Many of the physicists employed in industry, estimated to be about 1,800 at the end of the 1930s, felt increasingly foreign to the culture of academic physics and decided to stay outside the American Physical Society. Quantum mechanics not only strengthened theory relative to experiment, but it also alienated a good part of the physics community. The threatened schism was avoided by the founding in 1931 of the American Institute of Physics, a kind of umbrella organization that took care of both applied and pure physics. *Physical Review* remained the flagship of American academic physics, and soon of world physics as well. Industrial laboratories continued to contribute substantially to the journal, but relatively less than in the 1920s. Many of their papers appeared instead in the *Journal of Applied Physics*, a journal established by the American Physical Society in 1931. Figures for the distribution of papers, according to sources, in the main British and American journals of pure physics are given in table 17.1. The figures are not directly comparable, but they do indicate the decrease in industrial physicists' share of *Physical Review* articles in the 1930s. This share was larger than in the case of Britain, but not markedly so in the mid-1930s. At that time, *Physical*

TABLE 17.1
Distribution of Papers, According to Sources, in the Leading Academic Physics Journals in the United States and Britain

		Source, percent		
	No. of papers	*Academic*	*Industry*	*Government*
USA (*Physical Review*)				
1930	311	88.7	7.7	3.5
1935	282	96.5	3.2	0.4
1940	290	93.4	5.5	1.0
Britain (*Philosophical Magazine* and *Proceedings of the Royal Society of London*, combined)				
1932	338	92.9	0.9	6.2
1936	301	93.7	2.3	4.0

Note: In the case of *Physical Review*, only papers originating from American institutions are included. These made up about 93 percent of all papers.
Source: Based on data in Weart 1979a and Bernal 1939.

Review was as heavily dominated by contributions from universities as its British counterparts.

Whether or not national styles in the sciences exist is a matter of debate among historians and philosophers. To many European visiting physicists in the 1920s and 1930s, it seemed clear that there was indeed an American style in physics that differed from that known in Europe. For example, the boundaries between physics and industry, between experimenters and theorists, and between physics and its neighboring sciences, such as chemistry and astronomy, were less rigid in the American system. Americans were less formal and often worked in groups in which there was no strict hierachy. Also, Americans seemed more willing to turn their scientific work into patents and, in general, to commercialize science. They were not afraid of publicity and tried actively to make the press interested in their work and use it for their advantage. In 1934, the National Association of Science Writers was formed in the United States and the American Physical Society started a propaganda campaign for physics. When the American Association of the Advancement of Science held its 1935 meeting, no fewer than sixteen reporters were present. American scientists and reporters cooperated in selling physics to the public, a phenomenon not yet known in Europe. But what most Europeans noted about American physicists was their pace of work—they worked at the laboratories even during weekends—and their passion for big machines. According to Franz Simon, a German low-temperature physicist who visited the United States in 1932, "Americans seem to work very well, only they obviously insist on making everything as big as possible." A Belgian physicist was impressed by the "richness of the laboratory" and inspired by the "constructive civilization of 'go ahead'" (Heilbron and Seidel 1989, 36). To many Europeans, visitors as well as emigrants, the Americans' predilection for big machines and complex technology was a sign of intellectual immaturity. Were the American experimentalists doing physics or engineering? Had they the time and capacity for thinking? According to Walter Elsasser, "Americans are mostly coarse types, very good workers but without many ideas in their heads. . . . Their number is impressive, it is true, but one should not worry too much about their technical facilities" (ibid., 350).

Intellectual Migrations

As a result of the political upheavals in Europe, a large number of physicists were dismissed, felt threatened, or decided for other reasons that they could no longer stay in the countries where they worked (see chapter 16). The majority of the emigrant physicists were German Jews, but there were also non-Jews and non-Germans among the European emigrants. Not all of them were refugees in the sense that they were dismissed or forced to leave their home countries, and several of the emigrants came to their new countries before 1933 or they left their homelands after 1933 without being dismissed or expelled. In practice, however, they were refugees nonetheless. They merely decided to leave before they were forced to, or what might be worse, and when they were first outside Germany they had in most cases no possibility of returning. For example, Wigner and von Neumann had already come to America in 1930 on a half-time basis, with the other half of their jobs being in Berlin; technically they were not refugees, but after 1933, when they were dismissed from their posts in Berlin, they could not return to Germany. George Hevesy, the Hungarian-Jewish chemist and physicist who was a professor at Freiburg University, was not dismissed and at first wanted to stay in Germany; but in May 1933, after having witnessed the first wave of dismissals of Jews, he decided that it was unsafe to stay and went to Copenhagen.

Many of the refugee physicists initially emigrated to nearby countries, such as Denmark, Switzerland, the Netherlands, or France, but in most cases, they stayed there only for a short time and then went on to either Britain or the United States; or, in many cases, first to Britain and then to America. Hevesy, who stayed in Denmark until 1943 and then fled to Sweden, was an exception. A smaller group of displaced scientists and scholars, including the physicist Richard von Mises and the astronomer Erwin Freundlich, went to the newly reorganized university in Istanbul, but in most cases only to move on to America or elsewhere. Conditions in Istanbul were unsatisfactory, among other reasons because all teaching had to take place in Turkish. Also, most of the physicists who had hoped to find a permanent refuge in the Soviet Union were disappointed. After 1937, they were dismissed, exiled, or put in jail. Yet another possibility for displaced Jewish scientists was the Hebrew University in Jerusalem, where Zionists (Einstein among them) sought to establish a strong science faculty. Although about thirty countries hosted displaced physicists between 1933 and 1945, by far the most important recipient nations of emigrant physicists were Britain and the United States. Table 17.2 lists a select number of physicists who emigrated to one of these two countries.

Many national and international academic organizations were quick to respond to the dismissal of German scholars and the suppression of academic freedom. For several years, however, there was a tendency to avoid direct criticism of a political kind and a wish to deal with questions of principle in a rather abstract way. Solemn declarations about intellectual freedom and the internationality and neutrality of science were common, direct action or criticism much less so. It was primarily from the scientists themselves that such criticism was openly voiced—rarely from their professional organizations. For example, in 1934 a group of distinguished European scholars, including

TABLE 17.2
Destinations of European Physics Emigrants

Britain	*USA*	*USA via Britain*
M. Born* (G)	V. Bargmann (G)	G. Beck (A)
P. P. Ewald (G)	F. Bloch* (S)	H. Bethe* (G)
H. Fröhlich (G)	L. Brillouin (F)	F. Ehrenhaft (A)
R. Fürth (G)	P. Debye† (N)	O. Frisch (A)
D. Gabor* (H)	M. Delbrück* (G)	K. Fuchs (G)
W. Heitler (G)	A. Einstein† (G)	G. Hertz† (G)
N. Kemmer (G)	W. Elsasser (G)	F. London (G)
N. Kurti (H)	E. Fermi† (I)	E. Rabinowitch (G)
K. Mendelssohn (G)	J. Franck† (G)	O. Stern* (G)
E. Schrödinger† (A)	G. Herzberg* (G)	L. Szilard (H)
F. Simon (G)	R. Ladenburg (G)	E. Teller (G)
	A. Landé (G)	
	E. Segré* (I)	
	L. Tisza (H)	
	V. Weisskopf (G)	
	E. Wigner* (H)	

Note: Persons denoted with a dagger were Nobel laureates when they emigrated; those with an asterisk received the Nobel prize after emigration. The letter following the name gives the nationality at the time of emigration: A = Austrian; F = French; G = German; H = Hungarian; I = Italian; N = Dutch; S = Swiss.

Ernest Rutherford, Paul Langevin, and Jean Perrin, condemned what they considered the misuse of science in Germany, namely that "the exact sciences have been openly degraded to jobbing for war industries" and that "only such investigations are favored which are likely to bring about a direct technical advance" (Weiner 1969, 209). Of course, within a few years, British and American physicists were themselves enthusiastically working to "degrade" the exact sciences in the service of the military. Another manifesto, signed by more than one thousand American scientists, condemned the Aryan physics movement as "an attack on all theoretical physics, and by obvious implication, on scientific theory in general" (ibid.). Even this was a rather lame criticism, but there was little scientists and their organizations could do to change the situation in Germany. What they could do was to help their unfortunate refugee colleagues with money and positions. In this respect, the physicists reacted swiftly and efficiently, proving in practice that "the international community of physicists" was more than just a phrase of celebration. In 1933–34 several aid organizations were formed by physicists outside Germany, most of them based on the initiatives of individual physicists and with no official support from government bodies. They were supported by individual gifts and donations from private foundations. Many scientists pledged to pay between 1 percent and 3 percent of their salaries to the noble cause. In 1933 German refugee scholars organized the *Notgemeinschaft deutscher Wissenschaftler im Ausland* (Emergency Society for German Scientists in Foreign Countries), first located in Zurich and later in London, where the organization received help from British learned societies.

In England, the Academic Assistance Council (AAC) was established in May 1933 as a coordinating body attempting to find temporary posts for refugee scientists. Its aim was to "defend the principle of academic freedom and to help those scholars and scientists of any nationality who, on grounds of religion, race, or political opinion, are prevented from continuing their work in their own country" (Weiner 1969, 211). The AAC was later renamed the Society for Protection of Science and Learning. Its first president was Rutherford and among the council's most active supporters was Leo Szilard, the Hungarian-Jewish physicist who had emigrated from Germany to England after Hitler came to power. According to one study, sixty-seven Central European physicists came to Britain and of these, almost half re-emigrated to other countries, in most cases the United States. Another study shows that 37 percent of the emigrant scientists and engineers at first sought exile in Britain and slightly fewer, 35 percent, in the United States. The larger and more dynamic American university system was better suited to absorb the emigrants, of whom 57 percent ended up in the United States and only 11 percent in Britain. This was in accordance with the policy of the Academic Assistance Council, which arranged for short-time support and openly encouraged refugee scientists to proceed across the Atlantic to seek permanent positions. The AAC described itself as a "clearinghouse" and stated in no uncertain terms that "the United States is the main terminal country" (Hoch 1983, 230).

The majority of the British immigrants lived on temporary research fellowships and only a few obtained permanent academic posts before the outbreak of the war. Integration into British physics was difficult, and there was only a very limited number of jobs. Perhaps the willingness to find jobs in Britain was limited too. When the German physicist George Jaffé tried to find a post in Britain in 1933, he was informed: "I gather there is a strong feeling that the [University] College has now absorbed its quota and that no further applications can be entertained. There is no doubt that this feeling is largely due to a fear that such admissions are eventually likely to react disastrously on the already scanty prospects of employment and promotion of our own graduates and teachers" (Rider 1984, 131). Even famous physicists had a hard time finding permanent positions. In 1933 Max Born accepted an offer to come to Cambridge University, without knowing for how long the appointment would be. He considered positions in faraway places like Bangalore and Moscow, and was greatly relieved when in 1936 he was offered the chair of natural philosophy (theoretical physics) at the University of Edinburgh.

The difficult situation in Britain was eased by the establishment of a program for displaced physicists and chemists set up by the large Imperial Chemical Industries (ICI) in 1933. The ICI provided a large number of grants for two or three years' duration, in part, but not only, to scientists with a expertise that might be of interest to the company. Among the ICI grantees were Schrödinger, Franz Simon, and Fritz London. One notable result of the ICI generosity was that low-temperature physics became a British speciality, mainly through the work of eminent refugee physicists like Kurt Mendelssohn and Simon. When war was declared, many of the German and Austrian refugees were interned as enemy aliens, often under harsh conditions. Some were interned in Britain, but others were deported to Canada, among them the young Austrian physicists Walter Kohn and Hermann Bondi. Kohn went on to the United States, where he became an important solid state theorist, whereas Bondi returned to England, where he started his distinguished career in cosmology and relativity.

Aid organizations in the United States followed the pattern of England, with the Emergency Committee in Aid of Displaced German Scholars corresponding to the AAC. The Emergency Committee typically provided grants to support emigré scientists at universities that lacked the funds for the position. Some of the money came from individual contributions from American scientists and a large part from grants provided by the Rockefeller Foundation or other philanthropies. The Rockefeller Foundation established a Special Research Aid Fund for Deposed Scholars which, between 1933 and 1939, provided $775,000 in grants. The Americans were keenly aware of the strained economic situations of their universities and also of the danger of conflicts between foreign scientists and young Americans seeking jobs. For this reason, the Emergency Committee's support was restricted to "mature scholars of distinction who had already made their reputations," whereas young scientists, who were likely to compete with American applicants, were given low priority. Most of the physicists from Central Europe who took up positions at American universities were in their thirties or forties.

One of the problems that the European refugee physicists had to face was the anti-Semitism that existed at many American universities. This was not a new phenomenon and, of course, it was not a problem only for the European emigrants. In 1937 Kemble recommended Eugene Feenberg, an American-Jewish physicist with excellent qualifications, for positions outside Harvard, where Feenberg had finished his Ph.D. under Kemble. Feenberg, wrote Kemble in an attempt to help, "is a tall rangy Texan and neither looks nor acts like a New York Hebrew." The letter of recommendation was of no help. "It is practically impossible for us to appoint a man of Hebrew birth . . . in a

Southern institution," Kemble was informed by the University of North Carolina (Kevles 1987, 279). James Franck, the refugee Nobel laureate, complained about a growing anti-Semitism in the United States and felt that the hostility toward Jews was no less than it had been in Germany before 1933.

In spite of the many problems that emigrant physicists faced when coming to America, assimilation went remarkably well for most of the about one hundred physicists who came from Europe to the United States between 1933 and 1941. A major reason for the success was the recovery that American physics experienced after the depression, which placed American physics departments in a better economic situation that most of their counterparts in Europe. Many of the emigrants were theorists and used not only to the high standards of German theoretical physics, but also to the stricter separation in Europe between theorists and experimentalists. When they arrived in America, the emigrants contributed to raising the interest for and standards of theoretical physics and they quickly learned to appreciate the lack of distinct boundaries between theorists and experimentalists that characterized many American universities. Hans Bethe, one of the most important of the refugee physicists, found the atmosphere at Cornell University much more stimulating than that at European universities. In Europe, he recalled, "It was customary . . . to let the professor address the class and talk and write formally on the blackboard and then leave. The students would listen and try to understand . . . here, whenever a student feels like it, he asks a question. I think it's much better" (Weiner 1969, 223). America gave much to Bethe, and he gave much to America. Stanley Livingston, who for a period worked with Bethe, recalled that "he [Bethe] gave me a feeling for the fundamentals of physics, and what was going on in nuclear physics. . . . I learned of many new kinds of concepts like magnetic moments and quantum aspects, that I had never heard of while with Lawrence [in Berkeley]. It was a different environment. I was now following a scholar and was really impressed" (Stuewer 1984, 34).

There can be no doubt that American physics became greatly strengthened as a result of the influx of the European emigrants. Especially in many theoretical fields, such as quantum electrodynamics, nuclear theory, relativity, and solid state theory, the emigrants were an invaluable asset. However, they could flourish in the American environment only because there already was a strong basis, both institutionally and intellectually, and both in experiment and theory. Contrary to what is often believed, the United States did not become the world's leading nation in physics research simply as a result of the brain gain. In 1936 *Newsweek* could proudly, and correctly, declare, "The United States leads the world in physics." The leadership was further enhanced by the wave of European emigrants, but it was created mostly by American physicists and the country's impressive achievements in higher education and scientific institutions. The emigrants were welcomed to American universities in part because of general humanitarian sentiments and in part because American physicists and science administrators realized that they would make a most valuable contribution to the research system. The motives were not political in the sense that America wanted to deprive Germany of its best brains, but when the war came, this was realized to be an extra bonus. In June 1941, at a time when the United States was still formally a neutral country, this strategic result of the intellectual migration was pointed out by a professor Gortner in a letter to the president of the Emergency Committee, Stephen Duggan. Gortner argued as follows: "I firmly believe that we could reduce the technological achievement of Central Europe to the basis of a technological achievement of Spain or Portugal if we could move out 1000 of their strategic men who are leaders in the field of natural sciences, and in the long run the battle for democracy would be won more cheaply by doing just this and the results would be much more permanent than can ever be accomplished by the billions of dollars which we are pouring into our own defense program" (Fischer 1988, 84). An interesting thought, but this was not the way things came to happen.

The migrating physicists were persons, not merely statistical figures. Consider as an example the fate of Fritz London, a Polish-born German physicist of Jewish descent born in 1900. London, who had started his academic career as a student of philosophy, did important work in quantum mechanics and worked in Zurich with Schrödinger, whom he followed to Berlin. During his Zurich period in 1927 he wrote, with Heitler, the pioneering paper in quantum chemistry, for the first time explaining the covalent bond in terms of quantum mechanics. During his years in Berlin, he was occupied mostly with problems of chemical physics. By 1933, London was known as an original and eminent physicist, although not quite of Nobel prize caliber (he was once nominated for the prize, but in chemistry). With the advent of the Nazi laws of 1933, he was forced to take a leave of absence from the University of Berlin, which in reality meant a dismissal. Like many of his colleagues, he received help through the informal physics network and in August 1933, was offered an ICI fellowship at Oxford University. When in England, he changed his focus to low-temperature physics and joined the group around Simon and Mendelssohn. Together with his younger brother Heinz, another refugee member of the Oxford group, London developed the first successful (macroscopic) theory of superconductivity. Although scientifically fruitful, London's stay in England was not happy and after three years, he was informed that the ICI fellowship had been terminated without possibility of extension. He then managed to obtain a research position at the Institut Henri Poincaré in Paris, for the first year of which he received a grant from the Comité Français d'Accueil aux Savants Etrangers (the French Committee for the Help of Foreign Scholars), the French counterpart of the AAC. In Paris he continued his studies of superconductivity and superfluidity. London liked Paris and declined an offer to come to the Hebrew University in Jerusalem, but in 1938 he accepted the post of visiting lecturer for 1938–39 at Duke University in North Carolina. Back in Europe, he received a new offer from Duke, now for a permanent position as professor of theoretical chemistry. He left for America on September 1, 1939, the day the German army invaded Poland. Many of the refugee physicists assimilated quickly into the American environment, but not all could do it as easily as Bethe, Fermi, or Weisskopf. London was deeply immersed in the European intellectual culture and strongly felt the difference between his world and that of the American south. To Frédéric Joliot, he wrote: "I am too European to be able to become enthusiastic about life here, which even for those child-like adults here is too calm. . . . It appears to me that people here are free of passions except bridge and football" (Gavroglu 1995, 169).

Chapter 18

FROM URANIUM PUZZLE TO HIROSHIMA

The Road to Fission

AS SOON AS THE neutron was discovered, physicists realized that the new particle, owing to its lack of electrical charge, might be used as an effective projectile in nuclear reactions. The earliest reported nuclear transmutations of 1932–34 made use of fast neutrons impinging on light target nuclei such as aluminium. The results were (n, α), (n, p) and (n, γ) processes, that is, the expulsion of either alpha particles, protons, or gamma radiation. At that time, Fermi and his group in Rome began a systematic study of neutron reactions with all the elements of the periodic system, from hydrogen onward. For a neutron source, they used a sealed glass tube containing beryllium powder and radon. In the course of this work, the Italian scientists discovered—purely accidentally—that neutrons that had passed through paraffin, wood, or water were much more effective in producing radioactive isotopes. They concluded that the neutrons had been slowed down by collisions with hydrogen nuclei. Further experiments confirmed that slow neutrons were more easily captured than fast neutrons. When the Italians bombarded uranium with slow neutrons, they were able to identify several beta-emitting products, one of them with a half-life of 13 minutes. Fermi, Franco Rasetti, and Oscar D'Agostino found that the activity could not be due to isotopes between uranium and lead, and that this negative evidence "suggests the possibility that the atomic number of the element may be greater than 92" (Wohlfarth 1979, 58) The announcement made headlines in the press, and in Italy it was celebrated as a great triumph of fascist culture. Although disturbed about the publicity, Fermi believed that he had manufactured the first transuranium elements. As late as December 1938, in his Nobel lecture in Stockholm, he spoke confidently about "ausonium" and "hesperium," the names used in Rome for elements 93 and 94.

The 1934 Rome announcement caused Otto Hahn and Lise Meitner at the Kaiser Wilhelm Institute of Chemistry in Berlin to engage in similar work. The institute, which had been founded in 1912, was funded at the time mainly by the chemical industry, directly or indirectly by the I. G. Farben Company, Germany's giant chemical corporation. Meitner and Hahn at first believed that they too had found transuranic elements and reported in 1935 that "it seems very probable that the 13 and 90 minute activities are elements beyond number 92" (Graetzer and Anderson 1971, 24). On the other hand,

Fermi's results were criticized by Ida Noddack (née Tacke), a German chemist who, together with her later husband Walter Noddack, had discovered the element rhenium in 1925. Ida Noddack found Fermi's conclusions completely unwarranted and denied that transuranic elements had been produced. After all, she argued, almost nothing was known about neutron-induced nuclear reactions, so why assume that the product belonged to the end of the periodic table? "It is conceivable," she wrote, "that in the bombardment of heavy nuclei with neutrons, these nuclei break up into several large fragments that are actually isotopes of known elements, but are not neighbors of the irradiated elements" (Wohlfarth 1979, 63). Noddack's anticipation of nuclear fission made no impact at all on the course of events. Although published in a chemistry journal (the *Zeitschrift für angewandte Chemie*), it was known to both Fermi and Hahn and Meitner, but they did not take the suggestion seriously. Not only was Noddack's paper highly critical and her suggestion speculative, but the author's scientific reputation was also somewhat undermined because of her controversial claim to have discovered element 43 (which she called masurium and is now known as technetium, first produced in 1937 by E. Segré and Carlo Perrier). Noddack was not "rehabilitated" as a precursor of the fission hypothesis until the 1990s.

From 1935, the centers of uranium research moved from Rome to Berlin and Paris, with the two groups entering what can be better described as a rivalry than a cooperation. Although the Paris and Berlin groups were by far the most important, they were not the only ones interested in neutron-irradiated uranium. For example, in Berkeley, Philip Abelson tried to identify the supposedly transuranic products by means of the tested and precise x-ray spectroscopic method. However, in looking for atomic numbers larger than 92, Abelson failed to interpret his x-ray lines correctly. When the fission hypothesis became known, Abelson quickly found evidence for tellurium and thus confirmed the hypothesis. In Berlin, Hahn and Meitner made numerous experiments, suggested elaborate decay schemes, and thought of a variety of hypotheses in order to clarify what happened when uranium was bombarded with neutrons. After two years of strenous work, their main conclusion was disappointing, namely, that irradiated uranium produced complex products of an unknown nature, probably including some transuranic isotopes. Yet not all their work was in vain. One of their hypotheses was that the uranium products were isomers of uranium, that is, isotopes with different half-lives but with the same number of protons and neutrons. At that time, nuclear isomerism was not generally accepted, the only known (and controversial) case being the "uranium Z" that Hahn had reported as a protactinium isomer in 1921.

The work of Hahn and Meitner proved the existence of isomers but did not solve the uranium puzzle. In Paris, Irène Joliot-Curie worked on the same problem but adopted a somewhat different approach. In 1937, together with Pavel Savitch, a Yugoslavian physicist working in Paris, she reported a substance with a half-life of 3.5 hours in irradiated uranium. At first they thought it was thorium, but after more work they concluded in October 1938 that it followed lanthanum in chemical separations and was therefore possibly actinium—although "on the whole the properties of R 3.5 hr are those of lanthanum." Then, in the third round, they suggested that the 3.5-hour substance could not be an actinium isotope, but was probably a new transuranic element. Had they suggested that the close chemical similarity with lanthanum was evidence of a lanthanum isotope with half-life 3.5 hours, they might have discovered fission. But they did not. The results of Curie and Savitch puzzled the Berlin team, which from 1935 had been extended with the inclusion of Friedrich Strassmann, an analytical chemist. While pondering how to understand the Paris experiments, Meitner decided in July 1938 to leave Germany; it was now left to Hahn and Strassmann to find a solution. However, they communicated by mail with Meitner, who unofficially still belonged to the Berlin group.

It was the attempt to explain the Curie-Savitch results that led Hahn and Strassmann to fission. Among the activities resulting from the bombardment of uranium with neutrons, they found one that was precipitated with barium and therefore concluded that it was probably a new radium isotope. It seemed to them that the lanthanum-like isotope might be actinium, created from the radium by beta decay. But could radium be produced from uranium by the emittance of two alpha particles? Bohr, Meitner, and other theorists said no, and Hahn and Strassmann returned to the laboratory. In early December 1938 they began to realize that what they had thought of as radium behaved very much like barium, much more than would be expected from the chemical similarity of the two elements. If so, the Curie-Savitch substance might be lanthanum, produced by beta-active barium. By December 18, 1938, they had experimental evidence that what behaved like barium was, in all likelihood, barium. But it seemed incredible that uranium could turn into a much lighter element, and Hahn did not easily draw the conclusion. "Perhaps you can propose some kind of fantastic explanation," he wrote to Meitner on December 19. "We ourselves know that [uranium] *cannot* really burst apart into barium" (Weart 1983, 112). Even in Hahn's and Strassmann's paper of January 6, 1939, the two authors avoided a definite statement that barium had been produced by neutron-irradiated uranium. "As chemists," they wrote, "we should replace the symbols Ra, Ac and Th . . . in [our] scheme . . . by Ba, La, and Ce. . . . [But] as nuclear chemists, closely associated with physics, we cannot decide to take this step in contradiction to all previous experience in nuclear physics" (Wohlfarth 1979, 58). But they now glimpsed a possible explanation. They had found among the supposed transuranium elements one that resembled rhenium. If "radium" was barium, then the "transuranic rhenium" might be a lower homologue of rhenium, that is, element 43 or masurium (Ma). As Hahn and Strassmann remarked: "The sum of the mass numbers Ba + Ma, thus e.g. 138 + 101, gives 239!"

The insight that the uranium nucleus may split when capturing a slow neutron was first reached by Meitner and her nephew Otto Frisch, both refugees from the Third Reich. Frisch worked with Bohr in Copenhagen and his aunt was in Stockholm, where she had a position at Manne Siegbahn's institute. When the two met in late December 1938 to spend the Christmas holidays at Kungälv near Gothenburg, they had not yet received a copy of Hahn and Strassmann's paper. But they knew about its results and tried to figure out what had happened in the uranium nucleus in the Berlin laboratory. Frisch recalled: "We walked up and down in the snow, I on skis and she on foot . . . and gradually the idea took shape that this was no chipping or cracking of the nucleus but rather a process to be explained by Bohr's idea that the nucleus was like a liquid drop; such a drop might elongate and divide itself" (Frisch and Wheeler 1967, 276). The liquid drop model of the nucleus went back to work performed by Gamow in 1929 and during the following decade, it was developed by Bohr, von Weizsäcker, and others. Bohr's 1936 version, known as the compound nucleus, was particularly important and well suited to illuminate the mechanism of neutron reactions. The theory of the compound nucleus was well known to Frisch, who realized that it might provide an explanation of the Hahn-Strassmann anomaly. The splitting process was termed "fission," a name suggested to Frisch by an American biologist working at Bohr's institute. Meitner and Frisch reported their fission hypothesis in a letter to *Nature* on January 16, 1939. The hypothesis was that the uranium nucleus "after neutron capture, divides itself into two nuclei of roughly equal size." Moreover, the fission would be a violent process: "These two nuclei will repel each other and should gain a total kinetic energy of about 200 MeV, as calculated from nuclear radius and charge. This amount of energy may actually be expected to be available from the difference in packing fraction between uranium and the elements in the middle of the periodic system" (Graetzer and Anderson 1971, 52). Meitner and Frisch also used the occasion to suggest that thorium underwent fission in a manner similar to uranium. They had privately suggested to Hahn and Strassmann to look for radioactive inert gases (krypton and xenon) as fission products, and when Strassmann found the gases, the fission hypothesis was substantiated. It is notable that the discovery of fission, one of the most important discoveries in twentieth-century physics, was made by two chemists working at a chemical laboratory, and not by nuclear physicists. Indeed, the discovery took the physics community by surprise. Not even the Berlin physicists were aware that something highly interesting was going on at the Kaiser Wilhelm Institute of Chemistry.

More than Moonshine

News about the splitting of the uranium nucleus spread rapidly in the international physics community. The route started in Copenhagen, where Frisch had discussed the matter with Bohr, who was preparing to leave for the United States. Bohr was greatly surprised, but immediately accepted the fission hypothesis. He was, Frisch wrote to Meitner on January 3, 1939, "only astonished that he had not thought earlier of this possibility, which follows so directly from the present conceptions of nuclear structure," that is, the model of the compound nucleus (Stuewer 1994, 78). Bohr and his collaborator, Léon Rosenfeld, arrived in New York on January 16, 1939, and Rosenfeld went straight to Princeton, where he discussed the conclusions obtained in Germany, Sweden, and Denmark. The announcement, made before Meitner and Frisch's paper had appeared, caused a sensation. Fermi, John Wheeler, and other American physicists immediately started working on the fission process. In late January 1939, Bohr attended the Fifth Washington Conference on Theoretical Physics, where he and Fermi discussed the new type of process and Bohr explained it qualitatively from the point of view of the liquid drop model. "The whole matter was quite unexpected news to all present," three American physicists reported in the February 15 issue of *Physical Review*. Fission was still a hypothesis, and the first phase of work, in both Europe and America, was concerned with verifying the suggestion of Meitner and Frisch. Using different methods, this was done within one or two months, first by Frisch in Copenhagen, who used an oscillograph to record the electrical pulses produced by the fission fragments in an ionization chamber. Shortly afterward, the Berkeley physicists Dale Corson and R. Thornton produced the first visual proof of fission by means of a cloud chamber photograph.

By the end of February, there was no longer any doubt about the reality of uranium fission, and a second phase, dealing with the possibility of a self-sustained chain reaction, had its beginning. The concept of a chain reaction had not occurred to either Frisch or Meitner. The possibility seems to have been first suggested to Frisch by Christian Møller in Copenhagen, but at first Frisch did not take it seriously. After all, there was as yet no indication of secondary neutrons. Yet, it was realized early on that if a fission reaction did not result merely in two nuclear fragments, but also in one or more neutrons, a chain reaction might be a possibility. John Dunning, a physicist at Columbia University, was among the very first to confirm the Meitner-Frisch fission hypothesis, which he did on January 25, 1939. As he wrote in his laboratory notebook from that date: "Believe we have observed new phenomenon of far-reaching consequences. . . . Here is real atomic energy! . . . *Secondary neutrons are highly important*! If emitted would give possibility of a self perpetuating neutron reaction which I have considered since 1932–35 as a main hope of 'burning' materials with slow neutrons and release atomic energy" (Badash, Hodes, and Tiddens 1986, 210; emphasis in original). The energy release per fission process, correctly estimated by Meitner and Frisch to be about 200 MeV, was measured by physicists at Columbia University and Princeton University in the spring of 1939. Both groups found that the two fission fragments had unequal masses and that the kinetic energy of the fragments was close to 175 MeV. That left about 25 MeV for other products, including extra neutrons. That such neutrons were produced was first shown in March 1939 by Frédéric Joliot and his collaborators Hans von Halban and Lev Kowarski, and slightly later by two American groups. The French physicists concluded that an average number of 3.5 neutrons were liberated per fission, a figure that was soon corrected to about 2.4. The important thing was that extra neutrons were produced in a number that might make a chain reaction possible. From theoretical considerations, however, Bohr suspected that the much more common uranium-238 isotope would not fission with slow neutrons, but only the rare (0.7 percent) uranium-235 isotope would. Bohr published his suggestion in a brief note on February 15. It was soon substantiated by further theoretical arguments, but lacked experimental confirmation. Only in March 1940 did experiments prove that Bohr was right. The new knowledge seemed to imply that any practical application of fission energy would be extremely difficult and costly.

In 1939, speculations about subatomic energy were far from new. Ever since the discovery of radioactivity, many people, scientists as well as nonscientists, had suggested that a new and powerful source of energy lay hidden in the interior of the atom. In 1903 Soddy described the earth dramatically as a "storehouse stuffed with explosives, inconceivably more powerful than any we know of, and possibly only awaiting a suitable detonator to cause the earth to revert to chaos." Eleven years later, in his book *The World Set Free*, the novelist H. G. Wells wrote about powerful atomic bombs. All the loose talk about atomic energy used for either peaceful or military purposes annoyed Rutherford. In 1933, in an address before the British Association, he said that "any one who says that with the means at present at our disposal and with our present knowledge we can utilize atomic energy is talking moonshine." Three years later, Bohr referred to "the much discussed problem of releasing the nuclear energy for practical purposes," concluding that "the more our knowledge of nuclear reactions advances the remoter this goal seems to become" (Rhodes 1986, 227). The discovery of fission did not make Bohr change his cautious attitude. In an address of December 6, 1939 to a Danish audience, he reviewed the latest developments in nuclear physics, including the great energy released in uranium fission. "One can understand what terrifying perspectives we would face if substantial amounts of uranium and thorium could be made to explode," he said. But there was no reason to worry: "A closer consideration shows that there is no cause for alarm in this respect, although one can hardly say with certainty that any large-scale release of atomic energy is entirely ruled out." He had in mind the difficulties of separating the two uranium isotopes. Yet, other physicists were quick to speculate about a possible uranium bomb. In February 1939, Oppenheimer wrote to Uhlenbeck: "I think it really not too improbable that a ten cm cube of uranium deuteride (one should have something to slow the neutrons without capturing them) might very well blow itself to hell" (Smith and Weiner 1980, 209).

By the end of 1939, more than 100 papers on fission had been published, and a large amount of knowledge accumulated by physicists in Europe and America. Review articles published in Germany, England, and the United States summarized the knowledge. An early review by Norman Feather of Cambridge University, completed in May 1939, concluded that "the possibility of a cumulative process of exothermic disintegration has to be considered" (Graetzer and Anderson 1971, 79). And a German review by Siegfried Flügge, titled "Can the Energy Content of Atomic Nuclei Be Made Technically Useful?" answered the question affirmatively, concluding that an "atomic machine" was indeed possible. Atomic energy was not yet a reality, but it was definitely not moonshine either. The theoretical understanding was still incomplete, but with Bohr's and Wheeler's detailed, semiempirical theory of September 1939, a foundation for further understanding was laid. The Bohr-Wheeler paper appeared in *Physical Review* on September 1, 1939, the same day as World War II began.

Although the possibility of a uranium bomb was not explicitly discussed during the first hectic weeks of 1939, physicists realized that fission research might some day lead to a bomb, and possibly first a German one. Many of the nuclear physicists in the United States who took up the study of fission were recent emigrants from Central Europe, and they were worried about the situation from an early date. Among them was Leo Szilard, the visionary Hungarian refugee who had worked in England and now lived in the United States, where he eagerly followed the work on fission. As early as 1934, Szilard had conceived the idea of a neutron chain reaction that might possibly lead to a violent explosion, but he had thought of using beryllium, not uranium. In order to prevent the Germans from producing a uranium bomb, Szilard suggested to his fellow physicists in February 1939 that they keep all uranium research secret. Szilard's unusual suggestion was met with skepticism, although several of the American emigrant physicists supported the idea. But there were also physicists who opposed it, because of priority reasons, because they found it unrealistic, or because they objected to the very idea of secrecy, so foreign to the ideals of science; and many found the possibility of a bomb so remote that they saw no point in even discussing it.

Nevertheless, Szilard was persistent and after some argument, most of the leading physicists were willing to support the secrecy plan. But not all physicists: Joliot and his group in Paris were unwilling to stop publishing and for this reason, among others, Szilard's idea could not be realized immediately. Physics journals continued to carry articles on fission throughout 1939, available to anyone who could understand them.

With the declaration of war the situation changed, however, and from 1940, physicists in Britain and the United States agreed to stop all publications of possible relevance to the use of atomic energy. In England, there already was a publication stop, and in April 1940 Gregory Breit became the chairman of an American censorship committee on uranium research. It is remarkable not only that the physicists agreed on such a drastic measure, but also that they did it purely voluntarily, without any pressure from their governments, and that they actually succeeded in keeping Western uranium research a secret to both German and, for a while, Soviet scientists. One of the last uranium papers that appeared in *Physical Review* was Edwin McMillan's and Abelson's announcement of the discovery of "Radioactive Element 93" (neptunium), which appeared in June 1940. The next, and much more important transuranic element, plutonium, was first produced by the Berkeley chemist Glenn Seaborg and his collaborators in 1941, but at that time the publication stop was effective. Seaborg's Nobel prize-rewarded discovery was first made public in 1946, when it appeared with the footnote "This letter was received for publication on the date indicated [January 28, 1941] but was voluntarily withheld from publication until the end of the war." Many of these footnotes were appended to papers in the 1946 issues of *Physical Review*.

The first serious attempt to explore the possibility of an atomic bomb took place in England, not in the United States. In March 1940, Frisch and Peierls made a quick study of how a uranium "superbomb" could be constructed in principle and how it would work. They estimated that a mass of one kilogram of metallic uranium-235 would be sufficient for a bomb and that "the energy liberated by a 5 kg bomb would be equivalent to that of several thousand tons of dynamite" and the radiation from it to "a hundred tons of radium." Apart from outlining the mechanism of the bomb, they also mentioned some of the political, ethical, and military aspects of the "practically irressistible" superbomb which, they feared, the Germans were already in the process of developing. Among these aspects were that "[o]wing to the spread of radioactive substances with the wind, the bomb could probably not be used without killing large numbers of civilians, and this may make it unsuitable as a weapon for use by this country" (Serber 1992, 81 and 86). As a result of the Frisch-Peierls memorandum, a British committee, named MAUD, was formed to work on the superbomb. The physicists associated with the committee considered various problems, in particular methods of isotope separation, the possible production of plutonium, and the neutron loss and multiplication in different volumes of uranium. Many of Britain's leading physicists were involved, including refugees like Frisch, Peierls, Kemmer, Simon, Kuhn, Kurti, and Klaus Fuchs. They were joined by Halban and Kowarski, who had fled Paris after the fall of France. The MAUD committee wrote its final report in the summer of 1941, concluding that an atomic bomb was feasible but also that a much larger organization was necessary for this work. The project would be huge and probably much too expensive for Britain alone. At that time, there was little cooperation between British and American physicists working with nuclear energy. Information about the British work in 1941 was passed on to Moscow by Fuchs, the German refugee communist physicist. Later during the war, Fuchs became a central figure in the Soviet network of agents who informed Moscow of the progress taking place in the American bomb project.

Toward the Bomb

The development of the American bomb program, generally known as the Manhattan Project, is well known and has often been described in detail. We shall merely recall the essential steps in the program that, following tradition, started with the letter that Einstein wrote to President Roosevelt in the summer of 1939. The famous letter was actually drafted by Szilard after consultations with Wigner and Teller. Einstein told the president of the United States that he had reasons to believe that "the element uranium may be turned into a new and important source of energy in the immediate future. . . . Now it appears almost certain that this [chain reaction] could be achieved in the immediate future." Furthermore, "[t]his new phenomenon would also lead to the construction of bombs, and it is conceivable—though much less certain—that extremely powerful bombs of a new type may thus be constructed. A single bomb of this type, carried by boat and exploded in a port, might very well destroy the whole port together with the surrounding territory. However, such bombs might very well prove to be too heavy for transportation by air" (Graetzer and Anderson 1971, 93). The Germans were possibly already working along this line—Einstein mentioned von Weizsäcker specifically—and for this reason, Einstein advised Roosevelt to take action. The letter had no immediate effect except that the president appointed an Advisory Committee on Uranium. About a year later, the United States began in earnest to prepare itself for war, which included the foundation of a National Defense Research Committee (NDRC) chaired by the engineer and physicist Vannevar Bush. The Uranium Committee was redefined as a subcommittee of the NDRC. In 1941, the NDRC was absorbed into a larger and more efficient organization, the Office of Scientific

Research and Development (OSRD), again headed by Bush. At that time, American physicists, chemists, and engineers had started working on uranium chain reactions, but were still only in an exploratory stage. The work was experimental as well as theoretical and involved, among other things, a general theory of controlled chain reactions developed by Fermi, Wigner, Wheeler, and others. A report prepared by Lawrence emphasized the possibility of using plutonium as a bomb material. "If large amounts of element 94 were available," wrote Lawrence, "it is likely that a chain reaction with fast neutrons could be produced. In such a reaction the energy would be released at an explosive rate which might be described as a 'super bomb'" (Smyth 1945, 65).

After the Japanese attack on Pearl Harbor, the nuclear program was expanded vastly and large sums of government money were allocated to research related to the future atomic bomb. A Metallurgical Laboratory, or Met Lab, was created at the University of Chicago, headed by Arthur Compton and with Fermi as leader of the experimental nuclear physics group. The goal was now clear: to construct an atomic bomb based on either uranium-235 or plutonium. Compton decided that a first step had to be a slow chain reaction, followed by a uranium reactor to produce plutonium, and then a bomb based on the plutonium produced from the reactor. According to his plan, the reactor should be ready by January 1943 and the bomb two years later. Work progressed satisfactorily, and theoretical and experimental studies indicated that a bomb with an energy corresponding to at least two kilotons of TNT could be ready at the scheduled time. But it would require an enormous investment and an organization of a scale and complexity that only the army could take care of. In early 1943, all the efforts were put together under a new military organization, code-named Manhattan Engineer District and headed by Brigadier General Leslie Groves.

The first step in the expanded program was to make a primitive reactor in order to determine if a chain reaction in uranium was possible. This was the work of Fermi and his collaborators at the University of Chicago, who used pure graphite bricks as moderators for the neutrons produced by naturally occurring, but high-quality, uranium. The Chicago "pile" known as CP-1 consumed 385 tons of graphite, 6 tons of pure uranium metal, and 34 tons of uranium oxide. The critical level, where the multiplication factor exceeds one, was obtained quickly and without major problems on December 2, 1942. Fermi was enthusiastic, not only because the work had succeeded but also because the pile was so easy to control with the neutron-absorbing cadmium rods. "To operate a pile is as easy as to keep a car running on a straight road by adjusting the steering wheel when the car tends to shift right or left," he wrote. That this first case of controllable nuclear energy produced only a minute amount of power was irrelevant, for its purpose was not to produce either heat or electricity. CP-1 was meant as a prototype of a plutonium generator. Its success made OSRD confident that an atomic bomb could be produced in time to be used in the war and implied the need for another upscaling of the project. In late December 1942, Roosevelt approved Bush's plan for using $250 million for factories producing uranium-235 and plutonium. It was uncertain how much plutonium could be produced and how quickly, and for this reason it was decided to go ahead with both types of fission materials. The most formidable problem with the uranium bomb was the separation of uranium-235 from natural uranium. Several methods were considered, and the gaseous diffusion method, where the gas uranium hexafluoride flows through a system of porous barriers, was found to be most practical. Another possibility, proposed by Lawrence, was electromagnetic separation by means of huge electromagnets or "calutrons," and this method also met approval. In 1944, plants using these two methods were supplemented with a thermal diffusion plant. The three methods were integrated in the huge factory system built at Oak Ridge, Tennessee, which started producing uranium-235 in April 1945. Oak Ridge included a pilot uranium reactor, but the three large water-cooled reactors that were to produce material for the plutonium bomb were built at Hanford, Washington.

The design of the atomic bomb was the job of the large group of physicists that began to be assembled at Los Alamos, New Mexico, in the spring of 1943. With J. Robert Oppenheimer as director of the laboratory, seven divisions were established, among them a theoretical division under Bethe, an experimental division under Robert Wilson, a bomb physics division under Robert Bacher, and an explosives division under George Kistiakowsky. Oppenheimer had not previously been engaged in uranium research and was considered a security risk by some military groups because of his occasional flirtations with communism some years earlier. But Groves had confidence in Oppenheimer's abilities as a leader, and his intuition proved right. Oppenheimer, respected by both the physicists and the army generals, was just the right man for an impossible job. The physicists arriving at Los Alamos were given the necessary technical background in a five-lecture course on "How to build an atomic bomb" by Robert Serber, a theorist and colleague of Oppenheimer. Just in case someone did not already know, Serber started the course with pointing out, "The object of the project is to produce a *practical military weapon* in the form of a bomb in which the energy is released by a fast neutron chain reaction in one or more of the materials known to show nuclear fission" (Serber 1992, 3). In an atomic bomb the neutrons have to be fast, not slow as in a reactor, and it was not known if a chain reaction could, in fact, occur with fast neutrons. Early experiments in Los Alamos proved that it could, and indicated the critical size of the bomb.

Another of the important problems studied by the physicists in the New Mexican desert was how to assemble a critical mass of fissionable material from two subcritical masses. One of the methods, a more sophisticated version of the one included in the Frisch-Peierls memorandum, was to shoot one of the subcritical masses into the other. Another method was proposed by Seth Neddermeyer, namely to surround a spherical subcritical mass with a chemical explosive and then "implode" it into a much smaller and denser mass; although of the same mass, the higher density would make it supercritical. The implosion method was untested and much more complex than the gun method, but it turned out that only the implosion method could be used in a plutonium bomb. The physicists discovered that plutonium-240, inevitably occurring with the ordinary plutonium-239, would undergo spontaneous fission and as a result produce too many neutrons for the gun method to work. (Spontaneous fission in uranium had been discovered by two Soviet physicists, Georgii Flerov and Konstantin Petrzhak, and reported in *Physical Review* in 1940.) Rather than relying on one type of bomb, a plutonium-implosion bomb or a uranium-gun bomb, it was decided to develop both kinds at the same time. The final stage of the bomb project took place in the summer of 1945, after the unconditional surrender of Germany and the end of the European war. The bomb had been planned to be used against the Third Reich, but the new situation changed nothing in the pace of the Manhattan Project. There were still the Japanese left, and the momentum of the giant project seemed beyond control.

Who were the physicists working in the Manhattan Project? It would perhaps be easier to list those who did not, for it included most of the Western world's most brilliant physicists, from legendary figures like Bohr to young up-and-coming physicists like Richard Feynman. Both extremes were equally valuable to the project. About Bohr, officially "Dr. Baker," Oppenheimer reported to Groves in early 1944 that "Dr. Baker concerned himself primarily with the correlation and interpretation of the many new data on nuclear fission and related topics . . . [but] very little with the engineering problems of our program although he is of course aware of their importance and their difficulty." About Feynman, Oppenheimer wrote later in 1944 that he "[is] not only an extremely brilliant theorist, but a man of the greatest robustness, responsibility and warmth, a brilliant and lucid teacher, and an untiring worker" (Smith and Weiner 1980, 270 and 276).

A large part of the most active of the Manhattan researchers were physicists who only a few years earlier had resided in Europe. Take a look at table 17.2 and one will find many of the bomb project's central figures. One of them was James Franck, the German Nobel laureate of 1925, who had left the country in 1935 and now worked in Chicago. He was one of the few scientists working in the bomb project who warned against its political and ethical consequences at an early stage. On June 11, 1945, before the explosion of the first atomic bomb, he and six of his Chicago colleagues wrote a report to the Secretary of War in which they explained their position. "Scientists have often before been accused of providing new weapons for the mutual destruction of nations, instead of improving their well-being," Franck wrote. "We feel compelled to take a more active stand now because the success which we have achieved in the development of nuclear power is fraught with infinitely greater dangers than were all the inventions of the past." Echoing many of Bohr's arguments, Franck looked to the future and warned against a nuclear armaments race, for "nuclear bombs cannot possibly remain a 'secret weapon' at the exclusive disposal of this country for more than a few years." More specifically, the Franck report advised that the American bomb be "first revealed to the world by a demonstration in an appropriately selected uninhabitated area" (Graetzer and Anderson 1971, 104). But this was not what happened. Franck and his small group had no sympathy among the military leaders and their worried attitude was far from shared by the majority of physicists working in the Manhattan Project. Most physicists had no moral objections against working with a weapon of mass destruction. There were some who refused, but the general attitude was that the work was justified in view of the war situation and the possibility that Hitler might get the bomb first. In addition, many of the physicists were simply "intrigued with a fascinating and difficult scientific and engineering problem," as David Anderson recalled some forty years later (Badash, Hodes and Tiddens 1986, 222).

The Death of Two Cities

"The most striking impression was that of an overwhelmingly bright light. . . . I was flabbergasted by the new spectacle. We saw the whole sky flash with unbelievable brightness in spite of the very dark glasses we wore. . . . I believe that for a moment I thought the explosion might set fire to the atmosphere and thus finish the earth, even though I knew that this was not possible" (Rhodes 1986, 673). This was the impression of Emilio Segré when he witnessed the first nuclear explosion in history, the so-called Trinity test in the Alamogordo desert on July 16, 1945 at 5:30 A.M. The bomb, placed atop a 100-feet steel tower, was of the plutonium-implosion type. The test was completely successful, with its energy corresponding to about 18 kilotons of TNT, which was more than most of the physicists expected. This, and not the worries of Franck, was what mattered to the physicists who observed the spectacular phenomenon. "Naturally, we were very jubilant over the outcome of the experiment," Victor Weisskopf recalled. "We turned to one another and offered congratulations, for the first few minutes. Then, there was a chill, which was not the morning cold. . . ." (ibid., 675). The plutonium bomb worked beautifully and there was enough material for the uranium bomb, which was better understood and therefore did not need a test. The time had come for real action, and that, it turned out, meant the destruction of Japanese cities. Up to that time, the development of the bomb had been left to the physicists, whose voices were also important in the political and military discussions. But it was the military and political leaders, of course, who decided what the bombs should be used for. There was much discussion among the physicists in Los Alamos, Berkeley, Chicago, and elsewhere, but they did not form a united front and on the whole, they were not inclined to disagree with the military leaders. The scientific advisory group, consisting of Compton, Fermi, Lawrence, and Oppenheimer, saw no acceptable alternative to letting the bombs explode over densely populated areas in Japan. At any rate, Truman, the new president, decided that the bombs should be dropped over Japan as soon as possible and according to the judgment of the generals. The wisdom of this decision has been the subject of endless discussion, but the important thing is that it was taken. At this stage, in early July, the physicists did not have much to say about the creatures they had constructed.

The first creature was "Little Boy," a ten-thousand-pound uranium bomb about ten feet long and thirty inches in diameter. Carried on the B-29 bomber "Enola Gay," it was brought to explosion over the city of Hiroshima on August 6, 1945 at 8:16 A.M. local time, about 2,000 feet aboveground. It did what it was designed to do, except that it did not force the Japanese to accede to the demanded unconditional surrender. Three days later, "Fat Man" took over. Dropped over Nagasaki from another B-29 (named "Bock's Car"), the plutonium bomb exploded at 11:02 A.M., at approximately the same height above the ground. It also did what it was designed to do (see table 18.1), and five days later the Japanese government and its emperor capitulated. World War II was over.

The $2 billion bomb project was the largest research project in history, involving more scientists and money than any previous or subsequent proj-

TABLE 18.1
Data on the Two Nuclear Bombs and Their Consequences

	Hiroshima	*Nagasaki*
Bomb type	uranium - gun	plutonium - implosion
Weight of bomb (tons)	4	4.5
Explosive power (kilotons TNT)	12.5	22
Population	285,000	270,000
Totally destroyed buildings	54,000	14,000
Dead in initial phase	105,000	65,000
Wounded	75,000	40,000
Area totally destroyed (square kilometers)	13	6.7
Deaths per square kilometers	8,100	9,700

Note: Based on figures given in *The Impact of the A-Bomb* (Tokyo: Iwanami Shoten, 1985).

ect. To most people, it proved dramatically that science, and physics in particular, was able to win wars and change the course of history. In reality, the Manhattan Project was completed too late to be of decisive importance with regard to the war. The real significance of the atomic bomb was political rather than military, and became clear only after the end of the war. From a military point of view, the less publicized research on radar, the other main area of allied military physics, was much more important.

Only the gargantuan American project succeeded in developing an atomic bomb, but militarily oriented nuclear research was pursued in other countries as well. In Japan, the navy established a uranium program with the purpose of developing a nuclear reactor for driving warships, but it withdrew from the project because it seemed too costly and too uncertain. At Tokyo University, a group under Nishina explored the possibilities of uranium-235 separation in order to make a bomb, but progress was slow. In the Soviet Union, a Commission on the Uranium Problem was created in August 1940 under the Academy of Sciences. The task of the committee was to study the possibility of using the energy from a chain reaction in uranium. The leading Soviet physicists working in the commission, Iulii Khariton, Igor Zel'dovich, Flerov, and Igor Kurchatov, duplicated independently the work of Frisch and Peierls in England. In early 1941, they calculated the critical mass of uranium-235 and found it to be about ten kilograms. They even included in their calculations a heavy neutron reflector. A little later, Kurchatov realized the importance of plutonium and emphasized that using the new element might be the best way to build a bomb. The Russian equivalent of Los Alamos, although much smaller, was "Laboratory No. 2," established for Kurchatov in the spring of 1943. A year later, it included seventy-four people, of whom twenty-five were scientists (about 2,000 people worked at Los Alamos). The Soviet bomb project was seriously hampered by lack of material, especially pure uranium and graphite. On the other hand, the Russians had the advantage of being informed about the secret American project by Fuchs, the German refugee physicist who worked in Los Alamos. Yet, by August 1945, the Russians were not even close to having an atomic bomb. They obtained their first micrograms of plutonium in August 1944, from the Leningrad cyclotron, and the first Soviet reactor (named F-1) went critical in the last days of 1946.

The German efforts, so much feared by the physicists in Britain and America, started at an early date. Already in the spring of 1939—right after the Paris announcement of secondary neutrons—German physicists pointed out the potential military applications of uranium physics and the *Uranverein* (Uranium Society) started a series of meetings. The group included Flügge, Paul Harteck, Fritz Bopp, Heisenberg, von Weizsäcker, and Walther Gehrlach, among others. Several nuclear research teams were established, at the University of Leipzig, the University of Hamburg, the Kaiser Wilhelm Institute of Physics in Berlin, and elsewhere. The general aim of the Uranverein was to study the possibility of using nuclear energy, primarily in the form of a reactor that could be used to drive submarines or even airplanes. Atomic bombs were initially on the agenda, but were not given high priority. However, the physicists were well aware of the possibility of a bomb. In a talk given in February 1942, Heisenberg mentioned that the isolation of uranium-235 would "lead to an explosive of unimaginable potency" and that a uranium machine "can also lead to the production of an incredibly powerful explosive" (Hentschel 1996, 300). During the first two years of the war, uranium research in Germany was on an equal level with that in Britain and America, but after 1942 progress declined and the military lost some of its interest in the project. Unaware of the progress taking place in the United States, Heisenberg and his fellow uranium researchers concentrated on producing a reactor. When the war ended and Heisenberg was taken prisoner, the primitive machine had still not operated at the critical level. Heisenberg, Hahn, and the other German physicists who were interned at Farm Hall in England were greatly surprised when they learned about the dropping of "Little Boy" over Hiroshima.

One can easily get the impression that the entire physics community was preoccupied with military science during the war years and that they had neither the time nor the desire to do pure science. That was far from the case, however. Although the amount of ordinary academic physics was greatly reduced between 1940 and 1945, there were still physicists working in areas of pure physics and producing a substantial output of papers in this category. Dirac refused to join the Manhattan Project and continued working on problems of quantum electrodynamics (but he also did work related to the British bomb project). Born stayed away from military physics. "I continue my work undisturbed," he wrote to Einstein in the spring of 1940. "Soon my department [in Edinburgh] will be the only spot in Great Britain where theoretical work is still done" (Kragh 1990, 159). At the new Dublin Institute for Advanced Studies in neutral Ireland, Schrödinger gave lectures and arranged colloquia on theoretical subjects, with participants including Heitler, Dirac, Eddington, and Born. Even Heisenberg, working hard with the uranium machine, found time to concentrate on pure theory. The last paper appearing in the *Zeitschrift für Physik* before the end of the war was the third part of Heisenberg's "The Observable Quantities in the Theory of Elementary Particles," a paper as remote from military applications as one can imagine. It was submitted on May 12, 1944. Japanese theorists, including Tomonaga and Sakata, likewise worked on foundational problems in the midst of the war. Tomonaga's important work, "On a Relativistic Reformulation of Quantum Field Theory," appeared (in Japanese) in 1943.

This is not to deny that the war had a very serious impact on academic physics, qualitatively as well as quantitatively. World physics survived in

TABLE 18.2
Number of Papers in Physics, All Journals, and Pages
in the Volumes of *Physical Review* and *Philosophical Magazine*

Year	*1938*	*1939*	*1940*	*1941*	*1942*	*1943*	*1944*	*1945*	*1946*	*1947*	*1948*	*1949*
Papers	5081	4705	3230	2737	3152	2968	2687	3148	3273	3765	4088	7500
Phys. Rev.	2965	2914	1677	1041	1008	428	417	945	1517	2240	2307	2275
Phil. Mag.	2237	1478	1130	1026	910	851	855	875	884	913	1008	1278

1944, but at a low level. The decrease in physics publications and the slow recovery after the peace is illustrated by the number of papers abstracted by *Physics Abstract* and the number of pages in two leading physics journals (table 18.2). The decline is visually displayed in figures 18.1 and 18.2, referring to the situation in Great Britain.

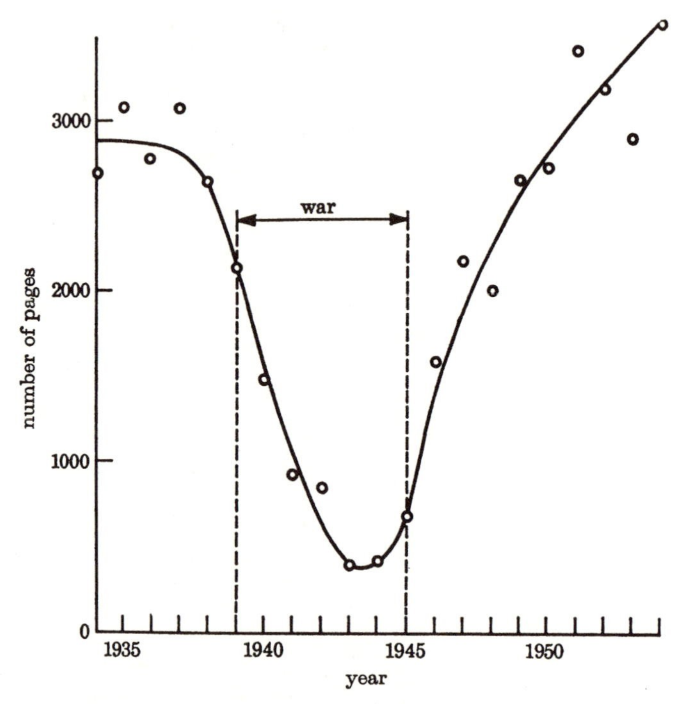

Figure 18.1. The number of pages published in each calendar year in the *Proceedings of the Royal Society of London*, section A. Allowance has been made for changes in the printed area of each page by normalizing to the format used in 1955. *Source:* E. Bullard, "The effect of World War II on the development of knowledge in the physical sciences," *Proceedings of the Royal Society A* 342 (1975): 519–36. Permission by The Royal Society.

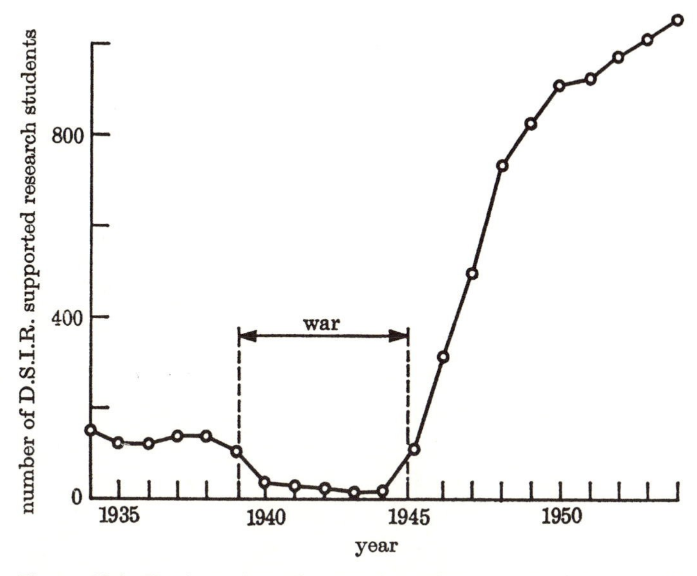

Figure 18.2. Total number of research students, most of them in the physical sciences, supported by DSIR (Department of Scientific and Industrial Research) each year. *Source:* E. Bullard, "The effect of World War II on the development of knowledge in the physical sciences," *Proceedings of the Royal Society A* 342 (1975): 519–36. Permission by The Royal Society.

PART THREE

PROGRESS AND PROBLEMS

Chapter 19

NUCLEAR THEMES

Physics of Atomic Nuclei

NUCLEAR PHYSICS had developed dramatically in the late 1930s and proved its value by providing the basis for the amazing atomic power based on uranium. After the war, as physicists returned to their universities and much of the research done from 1941 to 1945 was declassified, the young field of nuclear physics experienced a phenomenal growth. Lavishly supported by federal money, it became perhaps the most prestigious field of physics and one that attracted many of the brightest students.

The foundation of nuclear theory had been laid before the war, thanks to the work of physicists like Bethe, Bohr, Wheeler, Wigner, and von Weizsäcker. Yet the impressive progress was mostly phenomenological, and very little was known about the forces that held the protons and neutrons so strongly together in the nucleus. It was generally believed that Yukawa's theory of mesons exchanged between the nucleons was the best attempt at a theory of nuclear forces. During and shortly after the war, Pauli, Rosenfeld, Møller, and a few other theorists struggled to develop a meson theory of nuclear forces, but their complicated calculations failed to explain the strong forces satisfactorily. Two main reasons for the failure were that much too little was known about the Yukawa meson (pion) and its interaction with nucleons and also that the theories assumed the existence of only one kind of meson. Of course, it did not help that Yukawa's quantum was believed to be the light meson known from the cosmic radiation. Even with the new, strongly decaying meson of 1947, the problem of explaining the nuclear forces turned out to be exceedingly difficult. It was solved partially only in the early 1960s, with the discovery of new types of short-lived mesons. In his *Elementary Nuclear Theory* of 1947, Bethe wrote that meson theory "has so far not given any results in quantitative agreement with empirical facts on nuclear forces," and that "there are at present no trustworthy results of the meson theory of nuclear forces." Eight years later, in a volume celebrating Bohr's seventieth birthday, Landau concluded that "meson theories cannot be constructed without deep changes in the basic principles of modern physics." Yet the failure was not complete and it did not, at any rate, prevent physicists from exploring the structure of the nucleus, whatever kept it together.

The favored model of the late 1930s, and the one on which the understanding of the fission process was based, was the liquid drop model as developed by Bohr and Wheeler in particular. According to this model, nuclei were nearly incompressible droplets of extremely high density, with the nucleons strongly and collectively bound to other nucleons, somewhat like the molecules in a drop of water. The model was very successful, but there were nonetheless many experimental facts for which it was unable to account. For example, the liquid drop model could not explain the excited states found in the nuclear energy spectra. In 1949 it was challenged by the shell model, which, like the liquid drop model, had its roots back in the 1930s. As early as 1933, the German physicist Walter Elsasser, picking up a suggestion made a year earlier by the American T. Bartlett, had proposed that nucleons might occupy quantum levels in analogy with the atomic electrons. In this way, he could explain various regularities in the abundances of the chemical elements and the binding energies of their nuclei. Elsasser's idea of nuclear "shells" was well known, but lacked theoretical justification; also, it was unable to produce the closed shells at neutron numbers 82 and 126 that the empirical data suggested. For this reason, because it hinted at "numerology," and because it contradicted Bohr's ideas of nuclear structure, it was neglected by most physicists, who preferred to work with the empirically successful liquid drop model.

The reality of some kind of shell arrangement became clear only in 1948, when Maria Goeppert Mayer brought together evidence from diverse fields that showed the reality of what soon became known as "magical numbers." The name seems to have been coined by Wigner. The Polish-born German physicist Mayer (née Goeppert) had studied in Göttingen under Born, and in 1930 she emigrated to the United States. In 1946 Mayer moved to Chicago, where she worked with Teller and Fermi and obtained a half-time position at the new Argonne National Laboratory. It was only then that she became aware of the striking regularities associated with certain nucleon numbers and read Elsasser's 1933 paper. Mayer argued persuasively that nuclei with 2, 8, 20, 50, 82, and 126 protons or neutrons were particularly stable and hence that these magical numbers represented closed shells in the nucleus. Not only did these nuclei have large binding energies and the largest number of isotopes, but they were also markedly more abundant than nuclei in their neighborhood. The data on cosmic element abundances that the Swiss-Norwegian geochemist Victor Goldschmidt painstakingly collected during the 1930s were used to gain insight into both the micro- and the macroworlds; these data provided evidence for the structure of the atomic nucleus and were of crucial importance for Gamow's new "big bang" model of the universe. In fact, Mayer came to her shell theory of the nucleus via a failed attempt to understand how the elements were formed cosmologically. A revision of Elsasser's explanation of the magical numbers in terms of a shell model was proposed in 1949, independently by Mayer and the German physicists Hans Jensen and his collaborators Otto Haxel and Hans Suess. Jensen had been interested in the magical nuclear numbers for some years and during the war, he had discussed the subject with Goldschmidt in Oslo. In 1948 he discussed the matter with Bohr who, in spite of his vested interest in the liquid drop model, encouraged Jensen to develop his idea. However, as Jensen recalled in his Nobel lecture, when he sent his, Haxel's and Suess's paper to "a serious journal," it was rejected because "it is not really physics but rather playing with numbers." Only when he sent the paper to *Physical Review*, through Weisskopf, was it accepted.

By introducing a strong coupling between the orbital momentum and the spin momentum of each nucleon, Mayer and Jensen could explain a wide range of data related to the magical numbers. The particularly stable nuclei were those with closed shells with a total angular momentum of zero. Excited states were ascribed to a "valence nucleon" moving in a higher orbit. Contrary to the liquid drop model, the shell model assumed that each nucleon moved independently of the others in nearly undisturbed orbits in the nuclear field. For this reason, the simple shell model was also known as the independent-particle model. The theory's explanatory and predictive power indicated strongly that it was at least partially correct, and it was quickly accepted and developed by other nuclear physicists. It mattered less that the origin of the necessary spin-orbit coupling was something of a mystery. In 1963 the significance of the shell model was recognized by the Nobel committee, which awarded half a physics prize jointly to Mayer and Jensen. (The other half went to Eugene Wigner.) This was only the second time, and until now the last time, that the physics prize was awarded to a woman.

About 1950, then, there were two major theories of the nucleus: the liquid drop model and the shell model. Both were successful, but the successes were limited as they failed to account for certain experimental data. Even when various simplifying assumptions were removed and the models developed into more sophisticated versions, there were features none of the theories could explain. Moreover, the two models were in obvious mutual contradiction, one viewing the nucleus as a collective system, the other considering the nucleons as independent particles. This was a classical situation in the history of physics and the answer was, as in many similar situations, to consider the two models as extreme cases of a more general model. In this case, it would be a model that included liquid-drop as well as independent-particle aspects. James Rainwater of Columbia University realized in 1950 that evidence from nuclear electric quadrupole moments disagreed with the shell model in the rare earth group; this led him to suggest that the shape of the nucleus was not spherical, but distorted into a spheroid. At about the same time, Aage Bohr (the son of Niels Bohr) and the American-Danish physicist Ben Mottelson began to develop a model based on spheroidally shaped rotating nuclei in which surface oscillations (a liquid-drop feature) were combined with single-particle excitations (a shell-model feature). During 1950–53, Bohr and Mottelson developed their model of the nucleus, known as the collective or rotational model. The basic feature in this conception of the nucleus was the rotational motion of nuclei that were deformed and coupled to single-particle states. At that time, nuclear physics was strongly dominated by the United States, but most of the work of Bohr and Mottelson took place at Niels Bohr's institute in Copenhagen, where a group of nuclear physicists developed the collective model during the 1950s. One of the ways in which the Copenhagen physicists gained valuable information about the structure of the nucleus was by means of Coulomb excitation, that is, excitation of the atomic nucleus by bombarding it with low-energy charged particles that would not change the nucleus, but only act on it electromagnetically. In this and other ways, the Copenhagen group investigated rotational energy states systematically and used the data to confirm and refine the collective model. Rainwater, Bohr, and Mottelson shared the 1975 Nobel prize for "their work on the internal structure of the atomic nucleus." This was the last prize awarded so far for work in nuclear physics, reflecting that the classical era of this field of physics culminated around 1960.

Yet another Nobel prize rewarding a contribution to the understanding of the structure of the nucleus was for the American Robert Hofstadter's investigation of the details of the nucleus by means of electron scattering experiments. In principle, this method was analogous to Geiger's and Marsden's classical alpha scattering experiments that led in 1911 to Rutherford's nuclear atomic model. Bombarding the nucleus with fast electrons had the advantage that the effects of the badly understood nuclear forces could be ignored. Starting in 1953, Hofstadter used Stanford University's new linear accelerator to study the scattering of 116-MeV electrons from metals; three years later, he was able to use electrons with energy 550 MeV. He examined the scattered electrons with a huge, specially designed spectrometer and from the data, inferred how the charge density of the nuclei varied with the distance from the center. Encouraged by his results, in 1954 Hofstadter applied the method to hydrogen to find the charge distribution in a proton, and also to deuterium in order to examine the structure of the deuteron and the neutron. The Stanford experiments were thus of interest not only to nuclear physics, but also to elementary particle physics. The results showed what had long been assumed, namely, that the proton and neutron are different aspects of the same particle; and, moreover, that this particle, the nucleon, is an object of finite size, although not sharply delimited. Contrary to heavy nuclei, which have a well-defined radius, the charge density of nucleons was found to decrease smoothly from the center. The averaged, so-called root-mean square radius was found to be about 0.74×10^{-15} m. Hofstadter pictured the nucleons as being surrounded by clouds of mesons, with the clouds adding together in the proton and cancelling in the neutron. His experiments showed, as he said at the end of his 1961 Nobel lecture, that "the proton and neutron, which were once thought to be elementary particles, are now seen to be highly complex bodies."

Modern Alchemy

Fermi's belief in 1934 of having produced new chemical elements heavier than uranium turned out to be unwarranted, but seven years later, elements number 93 and 94 had become realities. This was only a beginning, although one of crucial importance. With the discovery of neptunium and plutonium, a new field of physics and chemistry was opened, the manufacture and science of transuranic elements. The recognized master of this field was Glenn Seaborg, who would later emerge as a strong advocate of nuclear power and in 1961 be appointed chairman of the Atomic Energy Commission. As early as 1944, elements number 95 and 96 were discovered by Seaborg's group at University of Chicago's Metallurgical Laboratory and named americium and curium, respectively. Whereas americium was identified as a decay product of plutonium-241, curium was produced by bombarding plutonium-239 with alpha particles in the Berkeley cyclotron. After the war followed elements with atomic numbers from 97 to 101, named berkelium, californium, einsteinium, fermium, and mendelevium (see table 19.1). Of these, the first two and the last one were manufactured "normally," with alpha particles accelerated in a cyclotron. Einsteinium and fermium, on the other hand, were first identified in the residues collected after the November 1952 thermonuclear test explosion at the Eniwetok Atoll. Scientists examining the residues at Argonne, Los Alamos, and Berkeley concluded that the isotopes had been produced as a result of neutron capture in the intense flux produced by the explosion. During their work in manufacturing and identifying new elements, the Berkeley scientists became experts in microanalytical techniques and learned to operate with what, at the time, was considered unbelievably small amounts of matter. For example, the identification of californium was accomplished with a total of about 5,000 atoms. Even this number was large compared with the production of mendelevium, which took place with—probably for the first time in the history of chemistry—an unweighable amount of the target element, einsteinium-253.

Up to and including element 101, the discovery game was completely dominated by the American—Californian, to be more precise—physicists and chemists, led by Seaborg. The manufacturing processes were based mostly on bombardment with alpha particles. But the Californians were not alone in the game. For example, the Swedish physicist Hugo Atterling and his collaborators used the cyclotron at the Nobel Institute of Physics to bom-

TABLE 19.1
Discovery of Transuranic Elements, 1941–61

Z	*Name*	*Symbol*	*Year*	*Location*	*Discoverers*
93	neptunium	Np	1940	Berkeley	E. McMillan, P. Abelson
94	plutonium	Pu	1941	Berkeley	G. Seaborg, E. McMillan, J. Kennedy, A. Wahl
95	americium	Am	1944	Chicago	G. Seaborg, R. James, L. Morgan, A. Ghiorso
96	curium	Cm	1944	Chicago	G. Seaborg, A. Ghiorso, R. James
97	berkelium	Bk	1949	Berkeley	G. Seaborg, A. Ghiorso, S. Thompson
98	californium	Cf	1950	Berkeley	G. Seaborg, A. Ghiorso, S. Thompson, K. Street
99	einsteinium	Es	1955	Berkeley etc.	G. Seaborg, A. Ghiorso, et al.
100	fermium	Fm	1955	Berkeley etc.	G. Seaborg, A. Ghiorso, et al.
101	mendelevium	Md	1955	Berkeley	G. Seaborg, A. Ghiorso, et al.
102	nobelium	No	1957	Stockholm	H. Atterling et al.
103	lawrencium	Lw	1961	Berkeley	A. Ghiorso et al.

Note: The data should not be taken too literally. Discoveries of chemical elements, even if these are manufactured, are complex events that cannot be easily reduced to definite dates and names.

bard uranium-238 with accelerated oxygen ions. In competition with the Americans, they reported the production of element 100 in early 1954. Three years later, in July 1957, a group of American, Swedish, and British physicists used the Stockholm cyclotron to produce a strong beam of carbon-13 ions ($^{13}C^{4+}$) and let it react with a sample of curium-244. They announced the discovery of element 102 and proposed to name it, for obvious reasons, nobelium. The California scientists, not used to competition in the discovery game, failed to confirm the Stockholm results and doubted whether element 102 had really been produced. They did find the element, but by a different method. As Seaborg wrote in a survey article in *Endeavour* in 1959, "In April 1958 a group consisting of Ghiorso, T. Sikkeland, J. R. Walton, and the author at the Radiation Laboratory identified the isotope 102^{254}. . . . Although the name nobelium for element 102 will undoubtedly have to be changed, the investigators have not, at the time of writing, made their suggestion for the new name." The controversy also involved a group of Soviet scientists, including Georgii Flerov, who questioned the American claim and produced element 102 by bombarding uranium-238 with neon ions. Contrary to Seaborg's expectation, the name nobelium was eventually accepted by the International Union of Pure and Applied Chemistry.

The alchemical experiments of the 1950s formed the framework for further work in this area. Synthesis of still heavier "transfermium" elements (with $Z > 101$) continued and by 1998, twenty artificial elements were recognized to "exist," the heaviest being $Z = 112$. The main players in the later phase of modern alchemy were the Lawrence Berkeley Laboratory, the Dubna nuclear research center in Russia, and the *Gesellschaft für Schwerionforschung* (Society for Heavy-Ion Research) in Darmstadt, Germany. Several more of the pioneers of nuclear physics were honored with elements named after them, including Rutherford (rutherfordium, Rf, 104), Bohr (bohrium, Bh, 107), and Meitner (meitnerium, Mt, 109). The old master of element synthesis, Glenn Seaborg, was honored with element number 106 (seaborgium, Sg). Normally, chemical elements are not named after living scientists, but in 1997 the International Union of Pure and Applied Chemistry recommended making an exception in Seaborg's case.

Hopes and Perils of Nuclear Energy

The basic theory of a nuclear reactor was developed during the war by Fermi, Wigner, and others and tested in the military reactors. After 1945, a large amount of work was spent in completing the understanding of fission processes and developing a detailed theory of the nuclear reactor. By 1958, with the publication of Alvin Weinberg's and Eugene Wigner's authoritative *The Physical Theory of Neutron Chain Reactors*, the work of the physicists was over in this area. A new species of engineer-scientists, nuclear engineers, were responsible for the second phase of the development of nuclear reactors. In the eyes of the public, though, they were physicists and the science of physics was closely related to the new technological wonders.

The earliest nuclear reactors, starting with the Chicago pile of 1942, were experimental (table 19.2). Power-producing reactors were first developed by British, American, and Russian engineers. At first, the new science-based nuclear industry developed slowly, but from the mid-1960s, the pace of development increased drastically. One might believe that America's political and industrial leaders embraced the new energy source and sought to use the nation's advanced position in the area to build up a civilian nuclear energy system. However, this was not the case. Whereas nuclear explosives were given high priority, American interest in nuclear power was limited initially. The country had large reserves of oil and coal, and the industrial corporations did not foresee a big market for nuclear technology. The world's first commercial nuclear power station was not American, but the British Calder Hall, which officially began operation on October 17, 1956. The reactor was

TABLE 19.2
Selection of Early Reactors, 1942–54

Country	*Location*	*Start*	*Moderator*	*Power (kW)*
USA	Chicago	1942	graphite	0.2
USA	Oak Ridge	1943	graphite	2,000
USA	Chicago	1943	heavy water	300
UK	Harwell	1947	graphite	100
Canada	Chalk River	1947	heavy water	10,000
UK	Harwell	1949	graphite	6,000
USSR	Moscow	1949	heavy water	500
USA	Brookhaven	1950	graphite	28,000
Norway	Oslo	1951	heavy water	300
France	Saclay	1952	heavy water	1500
USA	Idaho	1953	(none)	1400
USSR	Obninsk	1954	graphite	30,000

Note: All reactors were experimental, more or less, and all, except one, were of the thermal-heterogenous type. The exception was the 1953 reactor in Idaho which was a fast breeder, using 90 percent enriched uranium and no moderator. The power given in the last column is the thermal power.

based on experimental work done at Harwell by the Atomic Energy Research Establishment, Britain's innovative nuclear research center directed by John Cockcroft. More than a year later, the Calder Hall reactor was followed by an American nuclear station at Shippingport, Pennsylvania, designed to produce 100 MW of electric power. But it took some time until American nuclear industry had caught up. In 1966 Britain was still the world's largest producer of nuclear energy. Other countries followed more slowly, in many cases after having first built, or imported, experimental reactors (table 19.3). The first Soviet reactor was ready in 1949 and five years later, Soviet scientists and engineers had constructed the country's first nuclear power plant—at least of a kind, for it produced only 5 MW of electric power and must be described as semiexperimental.

Initially, American interest in nuclear power was limited mainly to military applications, which, as far as reactors were concerned, meant the Navy's program of nuclear-powered submarines. It was during this program that the first nuclear reactor ever to produce substantial amounts of electricity was constructed. The reactor, called Mark I, started up in the spring of 1953. Under the leadership of Admiral Hyman Rickover, the Navy's program was pushed ahead vigorously and in 1955, resulted in the launching of *Nautilus*, the world's first nuclear-powered vehicle. A couple of years later, Nautilus was followed by a semicivilian ship, the Soviet icebreaker *Lenin*. A few commercial carriers were later equipped with nuclear reactors, but the bright

TABLE 19.3
Growth in Nuclear Electrical Power, in GW (billion watts), for Major Users of Nuclear Energy between 1954 and 1978

	1954	*1960*	*1966*	*1972*	*1978*
USA	0.002	0.482	1.91	14.83	91.71
UK	—	0.414	2.97	4.50	10.95
France	—	0.08	1.17	2.71	5.58
USSR	0.005	0.305	1.02	2.62	10.01
West Germany	—	0.015	0.33	2.33	13.51
Canada	—	—	0.23	2.00	5.52
Japan	—	—	0.17	1.74	19.32
Italy	—	—	0.60	0.60	1.39
India	—	—	—	0.58	1.18
Sweden	—	—	0.01	0.45	5.58
World total	0.007	1.296	8.41	34.74	184.91

hopes of a world fleet of ships powered by uranium never even began to materialize. The use for nuclear reactors at sea was limited largely to the huge submarines. The reactor type favored by the Navy was the ordinary or "light" water type, in which water was used as both moderator and coolant. The type had certain technical advantages, and because it could be made relatively simple and compact, it was well suited for the submarines and carriers that Rickover planned. This was an important reason for the later dominance of the light-water reactor on the civilian market. The breakthrough in American nuclear industry came during the Kennedy administration, with the opening in 1963 of a 640 MW boiling-water reactor at Oyster Creek, ordered by Jersey Central Power & Light Co. and manufactured by General Electric. In the same year, the 570 MW Connecticut Yankee pressurized-water reactor, built by Westinghouse, began operation. Many of the reactors introduced in Europe and Asia in the 1970s came from these two companies.

Directly and indirectly, nuclear weapons continued to be an extremely important factor in the development of American physics. Shortly after the war, there was much discussion of how to organize nuclear research in peacetime, especially concerning the role of the military in the new organization and how to make international responsibilities agree with national defense purposes. Many physicists and other atomic scientists were active in these discussions and formed a "scientists' movement," whose aim was, among other things, to work for civilian control of nuclear energy. Their magazine, the *Bulletin of the Atomic Scientists*, became an important journal for science policy and discussions about the ethical consequences of the growing militarization of physics (see chapter 20). There was not, however, any consensus among the physicists and far from a united front against the influence of the armed forces. For example, the original bill of 1945 that emphasized the military interests in nuclear and atomic research was endorsed by leading physicists such as Lawrence, Fermi, Arthur Compton, and Oppenheimer. Yet the result of the political debate was the 1946 Atomic Energy Act, under which the Atomic Energy Commission (AEC) was formed.

The AEC was a powerful civilian organization, and it inherited from the military a large number of laboratories and installations from the war years. Although the establishment of the AEC was widely seen as a victory for civilian over military interests in atomic affairs, in fact much of the early AEC research was left to the armed services. The AEC quickly became one of the greatest benefactors of American physics, although not the greatest. In 1949 the AEC and the Department of Defense (DOD), in proportions of roughly two to three, together accounted for 96 percent of all federal money spent for university research in the physical sciences. The mix of military and civilian support to physics caused little concern among the majority of physicists, for whom money was money.

Already during the Manhattan Project, the possibility of a fusion or hydrogen bomb had been seriously considered and advocated, by Edward Teller in particular. The idea of using a uranium bomb to trigger a violent thermonuclear reaction seems to have occurred to Fermi as early as the fall of 1941. In the spring of 1946, a conference devoted to the possibility of a thermonuclear bomb concluded: "It is likely that a super-bomb can be constructed and will work" (Rhodes 1995, 255). However, the project was given low priority and later in 1946, fusion studies stopped at Los Alamos. It was only with the Soviet nuclear explosion of 1949 that Teller's idea of a thermonuclear "superbomb," supported also by Lawrence and Alvarez in Berkeley, was taken seriously and achieved political backing. In the fall of 1949, the influential scientific General Advisory Committee (GAC) of the AEC recommended not to proceed with the hydrogen bomb. The majority, consisting of Oppenheimer, Conant, and Lee DuBridge (a physicist and president of Caltech) argued in ethical and political terms that the bomb was unnecessary—indeed, unwanted—because "its use would involve a decision to slaughter a vast number of civilians." Fermi and Rabi, although issuing their own minority appendix, agreed in the rejection of what they considered to be "a danger to humanity as a whole" and "necessarily an evil thing considered in any light." The "Super," they wrote, "cannot be justified on any ethical ground which gives a human being a certain individuality and dignity even if he happens to be a resident of an enemy country" (Badash 1995, 83). As early as 1946, Arthur Compton had advised that "this development should *not* be undertaken, primarily because we should prefer defeat in war to a victory obtained at the expense of the enormous human disaster that would be caused by its determined use" (Rhodes 1995, 207).

There were other and louder voices, especially that of Teller, who argued passionately for a thermonuclear crash program. During the beginning of the Cold War, his voice reached the ears of many politicians and, not surprisingly, generals and admirals. The recommendation of the controversial GAC report was not accepted and in 1950, President Truman authorized the development of a superbomb based on fusion. Many of America's best physicists, including some of those (such as Oppenheimer, Bethe, and Fermi) who had argued against the superbomb, now engaged collectively in an effort to find a way of how to construct the bomb. The basic idea was to make a mass of deuterium and tritium fuse by the enormous heat produced by a fission bomb. For a time this seemed nearly impossible, but in the spring of 1951, Teller and the mathematician Stanislaus Ulam found a way to concentrate the radiation pressure from the fission bomb in such a way that it would compress and heat the fusion fuel. When he learned about the solution, known as the radiation coupling approach, Oppenheimer was said to have been impressed because it was "technically so sweet" (Kevles 1987, 378). With this theoretical breakthrough, work at Los Alamos proceeded as planned, and by 1952 a thermonuclear test device, called "Mike," was ready. It was successfully detonated at the Eniwetok Atoll in the Pacific on November 1, 1952. It took one and a half years of hard work to develop "Mike" into a real bomb that could be dropped from an airplane. The result was "Bravo," which exploded on March 1, 1954. The destructive yield was awesome, corresponding to about 15 megatons of TNT or more than 1,000 times as much as the 1945 Hiroshima bomb.

The hydrogen bombs, developed mainly for political reasons, were triumphs of American science and engineering, but less so for American world politics. In 1949, the Soviet Union exploded its first nuclear device, an equivalent of the American 1945 Trinity test and, like that one, working with plutonium. It was this bomb that, to a great extent, fueled the American efforts to develop an even more powerful bomb and thereby to achieve a decisive advantage in the Cold War that had recently been if not declared, then recognized as a reality. But the advantage was short-lived. The Soviet Union followed quickly in the new arms race, first in August 1953 with a thermonuclear test device and then, in 1955, with a hydrogen bomb dropped from an airplane. The yield, estimated to be a few megatons, was smaller than the American bomb, but a superbomb it was and, it turned out, was only a beginning. The largest manmade explosion ever was achieved in 1961, when the Soviet Union tested a new kind of hydrogen bomb. The explosive power was 58 megatons of TNT, corresponding to almost 5,000 Hiroshima bombs.

Controlled Fusion Energy

With the work done on the hydrogen bomb, it was natural to study the possibility of using fusion energy for peaceful purposes as well, in particular with the object of developing a fusion reactor as an analogue to the fission reactor. The first attempt in this direction, to confine a hot plasma in a magnetic field, was not inspired by the hydrogen bomb, but by Bethe's theory of thermonuclear reactions. As early as 1938, two American applied physicists, Arthur Kantrowitz and Eastman Jacobs, tried to obtain fusion in a rarefied hydrogen gas enclosed in a magnetized toroid. Not surprisingly, they failed. The possibility of controlled fusion was discussed during the late phase of the Manhattan Project, but it was only in 1951 that it was turned into a real research program under the AEC. The program, known as Project Sherwood, included a large number of America's leading physics laboratorics, and eight Sherwood conferences were held between 1952 and 1958. The most important Sherwood participants were Los Alamos (with 46 physicists participating in one or more Sherwood conferences,), Princeton University (with 41), University of California Radiation Laboratory (with 86), and Oak Ridge National Laboratory (with 22). Sherwood was big science, and it was expensive science. The chairman of the AEC between 1953 and 1956, Lewis Strauss, was a strong advocate of controlled fusion and was convinced that the program would succeed only if sufficient resources were provided. Money was no problem. From 1954 to 1958, Strauss increased the funding for the fusion program from $1.8 million to $29.2 million. "Surely there is no shortage of money for this work," a report on Project Sherwood noted in 1956. "In fact, one gets the feeling in visiting the various sites that the number of dollars available per good idea is rather uncomfortably large. There is certainly a feeling of some pressure to spend the money made available" (Bromberg 1982, 44). In the United States, the pioneer in controlled fusion was Lyman Spitzer, an astrophysicist who had worked part-time in the hydrogen bomb project and whose astrophysical studies had made him an expert in plasma physics.

The basic idea of controlled fusion energy is simple, namely, somehow to confine a hot heavy-hydrogen plasma by magnetic fields and heat the plasma to the many million degrees necessary for fusion processes to occur at a reasonable rate. At an early stage, the main problems were recognized to be: (1) to heat the plasma; (2) to sustain the fusion process, that is, to make the energy released in the plasma at least sufficient to maintain the temperature; and (3) to confine the plasma in a stable state. Spitzer suggested to use the deuteron-triton reaction $^{2}H + {}^{3}H \rightarrow {}^{4}He + n + 17.6$ MeV, which would be more easy to achieve than the deuteron-deuteron reaction leading to either $^{3}He + n + 3.2$ MeV or $^{3}H + {}^{1}H + 4.0$ MeV. Among the disadvantages of the first mentioned reaction was that it used tritium, which does not exist in nature, but this was more than outweighed by the larger energy output and the smaller ignition temperature necessary for a self-sustaining process. Whereas this temperature is about 45 million K for the deuteron-triton process, it is about 400 million K for the deuteron-deuteron processes.

The Americans attacked the confinement problem in different ways, of which we shall mention only the technical solution most favored in the early phase, known as the stellarator. Spitzer showed theoretically that a torus twisted into a figure-8 configuration (a stellarator tube) and supplied with external current-carrying coils would provide an axial magnetic field suitable for confining the plasma. The stellarator program was initiated in 1951, primarily at Princeton University. Experiments with the stellarator went through several modifications, and the design was improved gradually as the physicists gained more knowledge. By the end of 1954, Spitzer and his group were ready with a general design of a full-scale stellarator (called model-D), a more than 500-foot-long machine operating at a temperature of about 200 million K. The total investment was estimated to be close to $1 billion and the machine was judged capable of providing a net electrical power output of 5 GW.

However, it did not come that far. In 1954 Teller argued that instabilities would appear in the stellarator for large currents, and experimental evidence from 1955 seemed to confirm his prediction. Work with the model-D was stopped and it was decided to gain more experience with the smaller models in order to find ways to stabilize the plasma. By changing the geometry this proved possible, but only to some degree, and from 1956, the stellarator program was in a state of uncertainty, if not crisis. In spite of the many difficulties, the costly program continued in a somewhat artificial atmosphere of optimism. In 1957 work started on the design and construction of the model-C stellarator, designed to achieve a temperature about 100 million degrees K. At that time, the official attitude was still optimistic. Amasa Bishop, a high-energy physicist who, from 1953, was the AEC member responsible for Project Sherwood, wrote about the stellarator program: "Numerous experimental difficulties continue to plague the progress of the work; there is, however, reason to believe that these difficulties can and will be overcome in time." And about the Sherwood project as a whole: "With ingenuity, hard work, and a sprinkling of good luck, it even seems reasonable to hope that a full-scale power-producing thermonuclear device may be built within the next decade or two" (Bishop 1958, 167 and 170). After all, the stellarator concept was only one way of several to keep the plasma confined under the right conditions. Perhaps it was the sprinkling of good luck that was missing in the stellarator project. The project continued until the late 1960s, but never yielded what it promised. At that time a new concept, known as the tokamak and first developed by Russian researchers, began to look more promising. Consequently, the Princeton physicists changed their stellarator project into a tokamak project. The worldwide change about 1970 to tokamak machines has been described as a paradigm shift in fusion research.

During the first International Conference on the Peaceful Uses of Atomic Energy in Geneva in 1955, Homi Bhabha, the Indian physicist and chief of his country's Atomic Energy Commission, referred to the energy source of the future in tones no less optimistic than Bishop's: "The technical problems are formidable, but . . . I venture to predict that a method will be found for liberating fusion energy in a controlled manner within the next two decades. When that happens, the energy problems of the world will truly have been solved forever, for the fuel will be as plentiful as the heavy hydrogen of the oceans" (Bromberg 1982, 67). One of the reasons for the interest in controlled fusion was that it promised a practically inexhaustible supply of cheap energy. Even in the early 1950s, with no energy crisis threatening, it was realized that the resources of fossil fuels were limited and that nuclear fuel in the form of uranium could not be the final answer to the world's ever-increasing energy consumption. Bhabha's address was one of the first times that a physicist spoke openly of controlled fusion. Work in the American fusion program was classified, and the very existence of Project Sherwood was supposed to be a secret. This was also the case with similar projects in other countries. One of the reasons for the secrecy was the military significance of the intense sources of neutrons that controlled fusion would provide. However, in 1956 it was decided that there were neither military nor political reasons to keep the work secret. By 1958, at the time of the second Geneva conference, the United States declassified most of its work and released information about the stellarator and other concepts in controlled fusion.

The United States was not alone in the attempt to develop controlled fusion. In Britain, George P. Thomson thought about the possibility as early as 1945, and the next year, he drew up a patent application for a device in which deuterium gas was heated by a high-frequency alternating current. Thomson's idea was a "pinch" effect apparatus, in which the magnetic field confining the plasma was not external, but provided by the plasma current itself. British fusion experiments, using both Thomson's ideas and those of Peter Thonemann at Oxford University, started in 1949, before corresponding American work. They resulted in the ZETA (Zero-Energy Thermonuclear Apparatus) machine, a torus filled with diluted deuterium gas and ignited by a pulse of 100,000 amp or more. When the ZETA was first operated in 1957, a large number of neutrons were observed from the plasma, which was estimated to have a temperature of at least one million K. This was a promising result, but it was uncertain whether the neutrons were really of thermonuclear origin and fusion had thus been achieved. In 1958 it turned out that most of the neutrons were not thermonuclear. The ZETA shared the fate of the stellarator. They were both disappointments, although not wasted projects.

Even if the many resources used on controlled thermonuclear processes did not create a fusion reactor, they did create a new scientific community. Plasma physics is an older and much broader subject than fusion. Research in the area was done in the 1920s and 1930s, when Langmuir pioneered the experimental aspects and Landau the theoretical aspects. However, interest in the field was scattered and limited to plasmas of a low temperature. The combination of classical plasma studies with the new high-temperature magnetic plasma physics transformed the specialty into an important scientific subdiscipline. The declassification and free exchange of information among American, British, and Soviet researchers after the 1958 Geneva conference provided the conditions for the formation of an international community of plasma physicists. As the Soviet physicist Lev Artsimovich expressed it in his address to the 1958 Geneva conference, "This problem [controlled fusion] seems to have been created especially for the purpose of developing close cooperation between the scientists and engineers of various countries, working at this problem according to a common plan, and continuously exchanging their results of their calculations, experiments and engineering developments" (Post 1995, 1640). With declassification followed a flow of publications, many of them formerly classified. The fusion researchers now began to publish in ordinary physics journals, such as *Physical Review*, and also in the newly founded *Physics of Fluids* or, in the Soviet Union, *Atomic Energy*. In addition, they had the journal *Nuclear Fusion*, which was created in 1960 by the International Atomic Energy Agency (IAEA) and carried articles in English, Russian, French, and Spanish. The new situation resulted quickly in new university courses and graduate programs in high-temperature plasma physics and an increased output of Ph.D. physicists in this area. In the United States, the number rose from 14 in 1960 to 99 in 1963. Many of these came from Princeton University, where plasma physics had become the largest research program by 1958. Another indication of the strength of the new field was the creation of a Division of Plasma Physics within the American Physical Society. The change from a mission-oriented secret program to a normal scientific discipline implied a change in the methods and standards adopted by plasma physicists. The engineering and sometimes trial-and-error approach gave way to more systematic and rigorous methods, conforming with those of other academic physics disciplines. The process toward scientific respectability was not without problems, though, and there were physicists who looked on the new discipline with skepticism. For a period, plasma and fusion research had a low academic reputation, in part because of its heritage in a classified project relying heavily on engineering methods.

Forty years and several billion dollars after Thomson and Spitzer had first thought of how to obtain fusion power, physicists and engineers continued their attempts to develop a power-producing device. By 1990, the tokamak was still considered the most promising fusion power concept because of its excellent ability to confine plasma. At that time, the largest machines were Princeton's TFTR (Tokamak Fusion Test Reactor) and Europe's JET (Joint European Torus). In the late 1980s, both machines reached a confinement quality—the product of the plasma density and the time that energy remains in the plasma before leaking out—of 1.5×10^{20} sec m^{-3} and a temperature of about 20 keV (corresponding to about 150 million K). These are conditions close to the critical level, where more energy is produced than is used. As an alternative to the magnetic confinement methods, the old pinch effect technique was revived and developed greatly in the 1990s. The promising results led a leading American physicist and engineer to conclude that "if we can get started soon on design and construction of the next big step [the $400 million X-1 machine], we really think we can do the job in ten years" (*Scientific American*, August 1998, 27). Still, at the end of the millennium a power-producing reactor is not a reality and commercial fusion power remains a dream for the future, possibly even the distant future.

Chapter 20

MILITARIZATION AND MEGATRENDS

Physics—A Branch of the Military?

WORLD WAR II was in many ways a watershed for American science and scientists. It changed the nature of what it means to do science and radically altered the relationship between science and government . . . the military . . . and industry" (Forman 1987, 152). So wrote in 1984 Jerrold Zacharias, a leading physicist of the 1950s with great experience in military projects and science policy. The watershed caused by the war depended very much on a changed scale and structure of funding for science, in particular a spectacular rise in federal funding. Before the war, such funding was negligible, about $1 million for all of basic physics research in the United States. It has been estimated that the total amount of federal money allocated to basic physics in 1953 was about $42 million, or half that amount if counted in 1938 dollars. Real federal funds for basic physics in the period thus increased with a factor of no less than twenty. In addition, there was the support from the Rockefeller Foundation and other philanthropies, but this traditional source of support became relatively insignificant in an age dominated by massive funding from the federal government.

In the 1950s and 1960s, federal money was, to a large extent, synonymous with military money, or money from the Department of Defense and the civilian (but, in practice, militarily oriented) Atomic Energy Commission. Figure 20.1 gives a summary of how military expenditures for research and development (R&D) changed through the period 1935–85. It should be noted that R&D is a much broader category than science, which is again much broader than physics. The main point to note is the scale of military research, and the large part it constituted of the total research funded by the government. Of course, it was only a small part of the money that went to physics, but it was a very substantial part of the budget of American physics and, thereby, world physics. About 1950, some 70 percent of all research time of university physicists was spent on studies sponsored by the DOD or AEC. At the same time, 98 percent of the $22 million that the government used on academic physics came from either the DOD or AEC. With the rise of the National Science Foundation (NSF), the percentage decreased, but not drastically so. In 1960, it had dropped to 92 percent. Figures for federal support to fundamental solid state physics illustrate the general funding

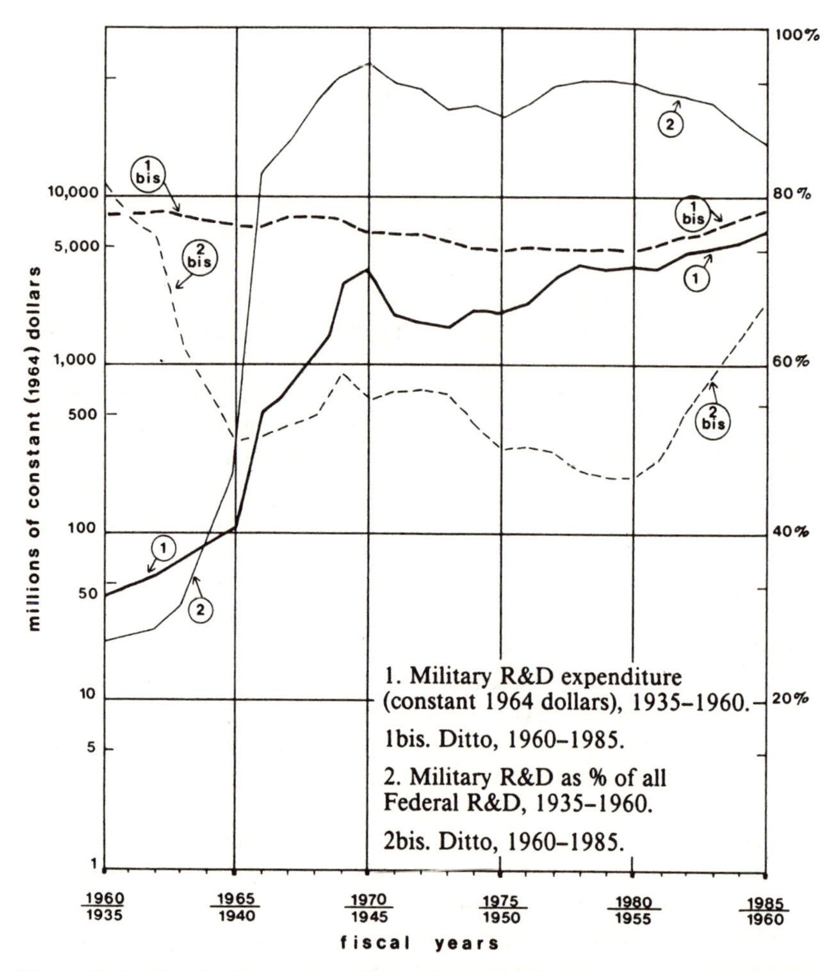

Figure 20.1. The development of U.S. military R&D expenditures, 1935–85. The figure is folded in order to facilitate a comparison between the two periods 1935–60 and 1960–85. *Source:* © 1987 by The Regents of the University of California. Reprinted from *Historical Studies in the Physical and Biological Sciences* vol. 18, figure p. 153, by permission.

structure. In 1953, this area received $3.1 million in federal support, of which only $10,000 came from the NSF. Two years later, support had increased to $4.5 million, with $3.8 million coming from the military, $0.6 million from the AEC, and $0.1 million from the NSF. The military support was divided among the Army, the Navy, and the Air Force in the proportion 7 : 6 : 5. Whatever the details, through the first two decades after the war, the dominant sources of funding for American physics were all related to the military system.

It was no wonder that some physicists thought that their science was dangerously close to becoming just a branch of the military. Philip Morrison, one of the young physicists of the Manhattan Project, voiced his concern in 1946, noting that "for every dollar the University of California spends on physics at Berkeley, the Army spends seven. . . . Some schools derive ninety percent of their research support from Navy funds." Morrison was worried by what looked like a militarization of physics, but realized the dilemma that he and his colleagues faced: "The physicist knows the situation is a wrong and dangerous one. He is impelled to go along, because he really needs the money. It is not only that the war has taught him how a well supported effort can greatly increase his effectiveness, but also that his field is no longer encompassed by what is possible for small groups of men. . . . He needs support beyond the capabilities of the university. If the ONR [Office of Naval Research], or the new Army equivalent, G-6, comes with a nice contract, he would be more than human to refuse" (Morrison 1946, 5). Morrison advocated that the civilian NSF, then under consideration, should replace the military as the principal benefactor of physics.

The philosophy of the NSF was to a large extent based on the ideas of Vannevar Bush in his 1945 report, *Science: The Endless Frontier.* In this influential work, Bush advocated a "civilian-controlled organization with close liaison with the Army and Navy, but with funds direct from Congress, and the clear power to initiate military research which will supplement and strengthen that carried on directly under the control of the Army and Navy" (Schweber 1989, 676). However, when the NSF came into being in 1950, it was with very limited sources and hence it was at first of relatively little importance. Only in 1972 did a major change occur with the passing in Congress of the Mansfield amendment, according to which the military should restrict its research efforts to projects of military relevance and abandon its engagement in basic research. Although there were always physicists who resented the military's role in civilian science, they were few and without much influence. With the advent of the Korean War, criticism was largely silenced. Many physicists who had been critical of the military dominance now accepted the situation and were ready to work for the military or receive its money.

The bonds between the military, the defense industry, and the physicists were further reinforced after the so-called Sputnik shock in 1957. In the same year, President Eisenhower established a new White House institution, the President's Science Advisory Committee, which was chaired by James Killian, president of MIT. Killian's committee was dominated by physicists and so was that of his successor, Jerome Wiesner, science advisor to President Kennedy. Both Killian and Wiesner were trained in physics and engineering and had close connections to the military. The establishment in 1958 of the National Aeronautics and Space Administration (NASA) produced new advantages for the physicists. The Cold War and the space and missile race created unique opportunities for the physicists who could now, in many cases, choose among military (DOD), civilian (NSF, NASA), and quasi-civilian (AEC) funding of their ever-more-expensive projects. In the 1960s American physicists were at the top of the world, economically, socially, and scientifically. The general atmosphere of science enthusiasm and boundless technological optimism has been vividly described by Daniel Kevles: "It was a time when Americans ranked nuclear physicists third in occupational status—they had been fifteenth in 1947—ahead of everyone except Supreme Court Justices and physicians; when physicists, among other scientists, were identified not only as the makers of bombs and rockets but as the progenitors of jet planes, computers, and direct dial telephoning, of transistor radios, stereophonic phonographs, and color television; when research and development in what President Clark Kerr of the University of California called this 'age of the knowledge industry' were believed to generate endless economic expansion. . . ." (Kevles 1987, 391).

Physicists were economically pampered indeed. In 1958, the population of U.S. physicists was 12,702, with the two largest fields being nuclear physics (2,622) and solid state physics (1,926). In terms of mere numbers, the physics community was not all that impressive. In the same year, the United States counted 35,805 chemists and 18,015 biologists; even the earth sciences, with 13,071 geologists, counted more scientists than physics. But although there were nearly three times as many chemists as physicists, chemistry received only half as much federal research support. Each physicist received an average of $11,000, while the corresponding figure for the chemist was $1,900; the average biologist received $4,900, and in geology and mathematics the amounts were $1,800 and $1,700, respectively.

The military's support was not limited to areas of direct—or even indirect—relevance to warfare, but covered all of physics, including areas that would seem to be utterly irrelevant to military interests. Numerous conferences, laboratories, summer schools, and research projects were wholly or partially supported by military agencies. The system soon mushroomed into a whole industry and became a natural part of the physicists' daily life. This is still the situation, more or less. As an example of military physics research in what would seem to be strictly a nonmilitary area, consider the Aeronautical Research Laboratories (ARL) at the Wright-Patterson Air Base in Ohio. Headed by Joshua Goldberg, in 1956 the ARL started a major research and support program in general relativity, at the time an even more esoteric and useless branch of mathematical physics than it is today. During its sixteen years of existence—it was killed by the Mansfield amendment—the program played a very important role in the revitalization of the theory of general relativity. For example, it supported leading theorists such as John Wheeler, Hermann Bondi, Pascual Jordan, Felix Pirani, Roy Kerr, and Alfred Schild. As one of its first activities, the ARL supported, together with the Office of Ordnance Research and the NSF, the 1957 Chapel Hill conference on "The Role of Gravitation in Physics," a support that included travel from Europe by military air transport for European participants. During later years, ARL organized more conferences, supported American and foreign theorists, and signed contracts with universities. The work that resulted from the program was scientifically important but hardly of military relevance. Typical titles of ARL reports were "Quantization of Covariant Field Theories" (by Peter Bergmann) and "Contributions to Actual Problems of General Relativity" (by Jordan).

The most important of the early military science-funding agencies was probably ONR, which in 1949 could boast that "the huge university research program of the Navy Department is the greatest peacetime cooperative undertaking in history between the academic world and the government." With a total expenditure of approximately $20 million, "[n]early 3,000 scientists and 2,500 college and university graduate students are actively engaged in basic research projects in the many fields of vital interest to the Navy" (Schweber 1988, 17). The expenditure of the ONR was less than 5 percent of the Navy's annual expenditure on research and development by 1950 (about $1.8 billion), but was still more than twice the total 1940 figure of $8.9 million. In other words, within the span of a decade, the Navy's expenditure on research and development had increased by the incredible factor of twenty. Among ONR's numerous activities was the funding of MIT's Laboratory of Nuclear Science and Engineering, with a 1950 budget of about $1.2 million. ONR also conducted conferences and summer schools, the "Project Hartwell" study of 1950 being a typical one. This was a joint MIT-ONR project, directed by Zacharias, the head of the Laboratory of Nuclear Science and Engineering. Its aim was to study undersea warfare, including its coordination with tactical atomic weapons. The brief and intensive study attracted some of the nation's most able physicists, including Luis Alvarez, Robert Dicke, Philip Morse, Charles Lauritsen, Edward Purcell, and Jerome Wiesner. It was just one of numerous projects of a similar kind that were conducted at the time and, indeed, through the remainder of the century. It was important for the early defense projects to involve outstanding physicists and, in this way, signal to the physics community at large that such projects were of great scientific worth and important as academic merit: If a Nobel laureate and leader of physics could join a military project, shouldn't you? The Navy and the other armed forces succeeded admirably in the strategy. One would be hard pressed to mention an important American physicist who consistently denied to work in military projects. Since 1950, more than thirty U.S. Nobel prize winners have drawn direct support from DOD. Participation in defense-oriented work soon became accepted and was even considered an academic qualification.

The United States was not the only country that mobilized physics in the service of the Cold War; so did the Soviet Union. For both scientific and military reasons, it was important for the Americans and their Western allies to know what went on within the Soviet laboratories. By far, most nonclassified Soviet research was published in Russian, and for that reason alone, was not easily accessible to Western scientists. In 1955 the American Institute of Physics, with the support of the NSF, started a program to translate Soviet journals so that Western physicists could keep abreast with the latest developments in the Soviet Union. By 1959, nine Soviet journals were being translated completely, making up some 13,000 pages altogether. The translated journals included the *JETP* (*Journal of Theoretical and Experimental Physics*), the review journal *Uspekhi*, and *Doklady*, the physics section of the Proceedings of the Academy of Sciences in the USSR.

What was the effect of the massive military patronage of American physics? First, of course, it led to real growth, not only in applied physics but also in fundamental physics, where the American position as a world leader was strengthened. Apart from the effect the military interests had on the scale, organization, and priorities of physics, it is also possible that it influenced the kind of physics under research and the physicists' attitude to their science—"the nature of what it means to do science," in Zacharias's words. Some historians of science have suggested that there was, in this period, a general orientation toward engineering aspects and applied physics, just the kind of physics that the military would be interested in. The military spirit is identifiable even in esoteric theory, they claim, namely in the pragmatic and instrumentalist attitude toward high-energy physics that characterized many American theorists. Whereas pragmatic theorizing was not an invention of the war, but was a characteristic feature even in the 1930s, it can be argued that the involvement of the physicists in military projects reinforced an already existing tendency and turned it into a worldwide norm. Whatever the validity of the claim (which is difficult to specify and even more difficult to test), it is a fact that American physicists were quite interested in those fields of science that were likely to be of military interest, including electronics, instrumentation, nuclear physics, and solid state physics. It is understandable that these fields received the lion's share of federal support. In 1953, for example, 64 percent of the federal funds for unclassified basic research in university physics departments was allocated to nuclear physics, as broadly defined. Solid state physics received 10 percent, cosmic-ray research 6 percent, atomic physics 5 percent, and low-temperature physics 1 percent (see also table 21.4, chapter 24).

The brief description given here has focused exclusively on the American scene, where the historical development of the military-physics relationship has been the subject of many detailed studies. Much less historical work has been devoted to the corresponding relationships in Europe and the Soviet Union. There is no doubt that the American case was of particular importance, not only because the relationship was closer and of a scale unknown in Europe, but also because the United States was the Western world's political, military, and scientific leader. The relationship between physics and the military was undoubtedly as close or even closer in the Soviet Union, but very little is known so far about the interaction between the two. The military also played an important role in some of the European countries, especially France and England, but its role was not nearly as important as in the United States. And in some of the minor European countries, the military played no role at all.

Whatever the different roles played by the military forces in the science systems of Europe of the United States, the two dominant regions of world science developed quite differently during the decade following the end of World War II. An important reason for the relative backwardness of European physics was simply that the European nations were much poorer than the United States. When an American nuclear physicist visited Italian physics facilities in 1950, he noted that these "are all greatly hampered by lack of money for research and low salaries for physicists." To the readers of *Physics Today*, he reported that the physics laboratory at the University of Rome, the pride of Italy, operated on an annual budget of $20,000, "a sum that would appear ridiculous to any comparable American university" (*PT*, January 1951). Yet the difference in productivity and quality between the two worlds of physics was not due to economic reasons alone. It was as much—or more—rooted in different mentalities, cultural traditions, and education systems. This was clearly recognized by a French physicist, André Guinier, after returning in 1947 from a two-month visit to American laboratories. In a revealing comparison between the two countries, he pointed out that American physicists specialized from an early age, whereas the French had a "taste for nonspecialization" that was unfavorable to the production of scientific results. "The real dilemma is to choose between having a profound but limited knowledge or having knowledge less deep but stretched over a variety of subjects," he wrote. "It is in this choice that Americans differ from us. We always consider the general culture superior. But we often forget to recognize that it is superficial." Guinier was impressed by the informal and cordial atmosphere he had seen in American laboratories and considered it a major reason for the success of American physics: "Between directors and workers, relations are very free. The young worker calls his boss by the first name. . . . On this particular point one can't but envy the American system: it is indispensable in scientific work that authority comes not from rank but from competence. This atmosphere makes possible work in collaboration. There are few articles in American scientific reviews which are not signed by two or more names. There are works compiled by the entire staff of large laboratories. . . . Few such attempts to work as a team in France have been crowned with succcss" (*PT*, May 1950).

Big Machines

Big money is not a sufficient, but it is a necessary condition for "big science;" therefore, it is not surprising that the big science era in physics coincided with the increase in government funding. Yet, it is important to point out that big science is not simply a matter of size or money, and also that big science is not purely a postwar phenomenon. Tycho Brahe's observatory (1580), Martinus van Marum's electrostatic machine (1784), and the Earl of Rosse's giant telescope (1845) were all very big and hugely expensive relative to the money available for research at the time. But they were exceptions, and it was only in the first half of the twentieth century that proto-big-science apparatus and laboratories began to influence the course of science. Earlier, we mentioned Kammerlingh Onnes's low-temperature laboratory in Leiden as an example, and with Lawrence's cyclotrons in the 1930s, we have the direct forerunners of postwar big science physics facilities. The big, expensive cyclotrons needed a new organization of researchers, a new spirit of work. The machines inspired group work and increased the demands of external funding, and they tended to lead to a new attitude of what constitutes successful science. These changes in internal and external contexts of science, rather than the scale itself, made the system of big science such an important phenomenon. Although postwar big science was far from limited to physics—it played a similarly important role in astronomy, for example—it developed most spectacularly in association with accelerators and other instruments used in high-energy physics. Rather than trying to map the entire development, we will exemplify it by mentioning a few of the more remarkable machines and laboratories. The growth in accelerator beam energy until the mid-1980s, and the many different families of accelerators, are displayed in figure 20.2.

The two most important accelerator research facilities in the early postwar period both emerged as direct consequences of the war efforts. The Brookhaven National Laboratory (BNL), mainly serving universities in the northeastern United States, was equipped with several accelerators, the most powerful of which was the Cosmotron. This 3-GeV proton synchrotron was based on the concept of phase stability that Edwin McMillan had found in 1945, and which was independently suggested by Vladimir Veksler in the Soviet Union. The other major accelerator center was on the West Coast, at the University of California's Radiation Laboratory, which was later known as the Lawrence Radiation Laboratory (LRL). Here, an even more powerful

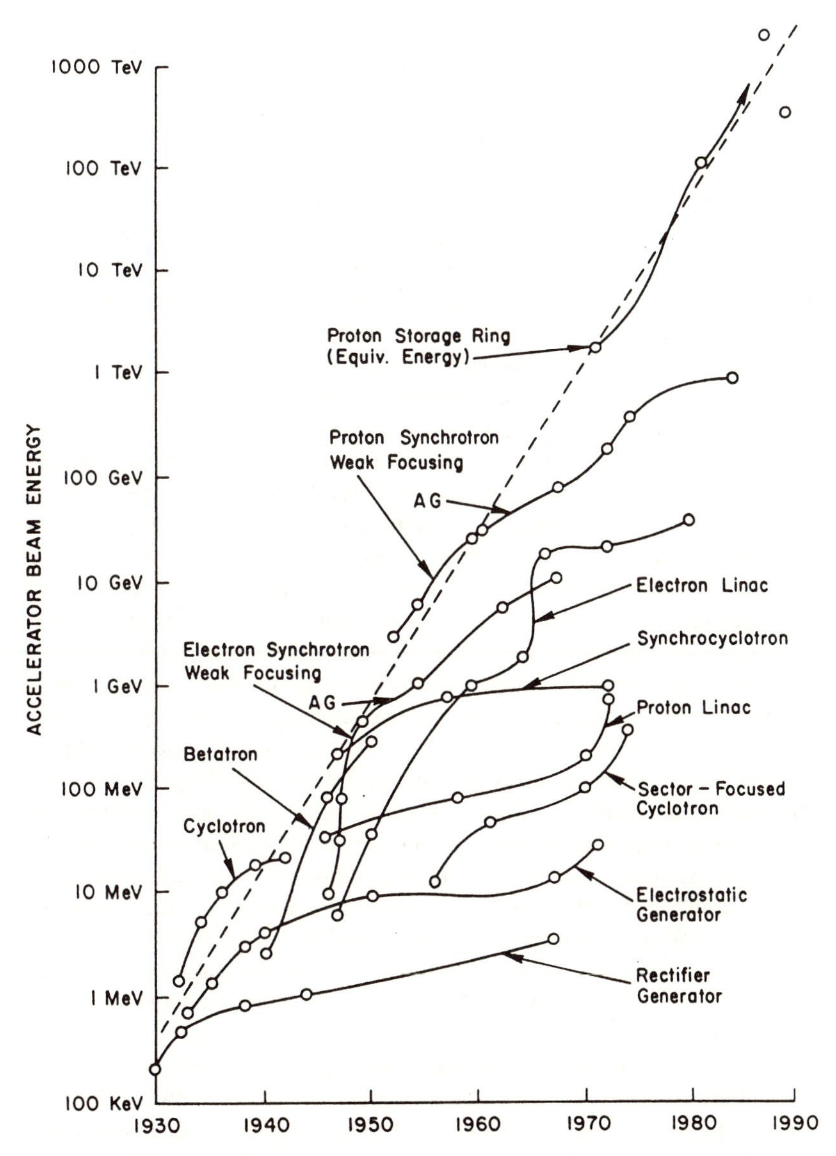

Figure 20.2. Growth of accelerator beam energy with time, 1930–84. *Source:* Reprinted with permission from *Physics Through the 1990s: Elementary-Particle Physics.* © 1986 by the National Academy of Sciences, Courtesy of the National Academy Press, Washington, D.C.

synchrotron was built, the $9 million Bevatron, which in 1954 accelerated protons to an energy of 6.2 GeV. The Cosmotron and the Bevatron dominated the early phase of accelerator-based high-energy physics, leading to important discoveries, such as pair production of strange particles in 1953 and the antiproton in 1955. Conventional synchrotrons more powerful than these machines would be prohibitively expensive, but a new type of accelerator, the alternating gradient machine, proved a worthy successor. The Brookhaven AGS (Alternating Gradient Synchrotron) of 1960 used a principle called strong focusing and became the next highly successful American accelerator. It was matched by CERN's independently designed proton synchrotron (PS), a $30 million machine completed in 1959 that could accelerate protons to 28 GeV. An even more powerful proton synchrotron was constructed at the Serpukhov center in Russia. With an energy of 76 GeV at its start in 1971, it was, if only for a brief period, the world's largest accelerator.

In 1957 Stanford physicists proposed to the AEC, DOD, and NSF to build a new linear electron accelerator, a further development of a type of linear accelerator (or "linac") already in operation at Stanford. The proposed machine was more than a continuation of an existing program; it was a step into a new range of big science. The accelerator, at the time the largest scientific instrument in history, was to be two miles long, and the construction cost was estimated at $78 million. Had it not been for the Sputnik shock, nothing might have come out of the proposal, which was justified only by its value to high-energy physics. Eisenhower wanted to demonstrate America's scientific strength, and in a 1959 talk characteristically titled "Science: Handmaiden of Freedom," he supported the project and decided to ask Congress for up to $100 million to realize it. The way through Congress was anything but smooth and it occurred to many, politicians as well as scientists, that the project was simply too expensive to be justified by its assumed scientific merits alone. "What is the practical result of this accelerator?" one Congressman asked. "What are the prospects of putting the knowledge that we will obtain from this accelerator into practical use? How will it raise the standard of living of our people?" (Wang 1995, 354). The answer was of course that SLAC—the Stanford Linear Accelerator Center—would not raise the standard of living. Yet the proposal passed Congress, which in 1961 appropriated $114 million and the following year, a contract was signed between Stanford and the AEC. Under the direction of Wolfgang Panofsky, SLAC was ready for operation in 1966. Even if the accelerator center did not raise the standard of living of most Americans, it turned out to be spectacularly successful from a scientific point of view. As we shall see in the following two chapters, SLAC was of crucial importance to the new high-energy physics of the 1970s.

Plans for building an American accelerator even more powerful than the Bevatron, the Brookhaven AGS, and the Serpukhov synchrotron started in the early 1960s. It was, like most of the other accelerators, to be financed by the AEC, an agency that in 1964 spent $135 million on high-energy physics. Because of the unprecedented expense of the new machine only one could be built; this led to a great deal of controversy between the groups in Berkeley and Brookhaven. After much discussion of where to locate America's accelerator center of the future, Batavia, Illinois won out over both Berkeley and Brookhaven. In 1967 President Johnson signed the bill for the National Accelerator Laboratory, or Fermilab (FNAL), a $250 million project built around a 200-GeV accelerator. Under the leadership of Robert Wilson, Fermilab's first director, the huge project was completed in 1972. Within months, Wilson forced the accelerator to produce a proton beam of 400 GeV, at that time the highest energy produced in a laboratory by far. But this was not the end. The method of storing beams in fixed orbits and then making them collide was first realized in 1959 with electron-electron beams (Stanford) and in 1963 with electron-positron beams (Orsay, outside Paris). Later and more powerful electron-positron colliders of great scientific value were constructed at SLAC in 1972 (called SPEAR) and at CERN in 1989 (called LEP). The acronyn LEP stood for Large Electron Positron collider, and with its 27-kilometer-long ring, it was indeed a large machine. Proton-antiproton colliders were developed in parallel with the electron-positron colliders and led in 1976 to CERN's SPS (Super Proton Synchrotron) collider, with which an energy of 500 GeV could be obtained. The American answer was Fermilab's Tevatron, which started operating in 1985 and which made advanced use of superconducting magnets to push the protons around in its 6.3-kilometer ring. The name of the machine referred to its maximum energy of one teva electron volt, 1 TeV = 1,000 GeV. In 1987 the Japanese joined the race in earnest toward higher energies with the TRISTAN collider, an electron-positron collider with beams up to 30 GeV, built at the National Laboratory for High Energy Physics (KEK) near Tokyo. Table 20.1 summarizes the development of accelerators between 1946 and 1985.

Many physicists felt, and presumably still feel, that a truly international accelerator facility would be the most rational choice in a field that is, after all, international. Why not skip the ridiculous and expensive competition between nations and work together? The idea of a "world accelerator for world peace" was first discussed in 1959 among American, Soviet, and European physicists. The idea—to build an international 1 TeV accelerator at a cost of about $1 billion—was supposed to lead not only to great scientific results but also, and no less important, to a climate of peace. According to Robert Wilson, one of the leading physicists behind the world accelerator idea, "The greatest force of such an international laboratory will be in developing our common culture in physical science . . . particles, accelerators, and society may interact again—this time to provide a force for international harmony" (Kolb and Hoddeson 1993, 106). The world laboratory continued

TABLE 20.1
The Rise of Big Science in High-Energy Physics

Name or type	*Location*	*Energy (GeV)*	*Year in operation*
synchrocyclotron	Berkeley	0.35	1946
Cosmotron	BNL	3	1952
Bevatron	Berkeley	6.2	1954
proton synchrotron	Dubna (USSR)	10	1957
PS	CERN	28	1959
AGS	BNL	33	1960
proton synchrotron	Serpukhov	76	1971
proton synchrotron	FNAL	400	1972
SPS	CERN	400	1976
Tevatron	FNAL	1000	1985

Note: Select proton accelerators 1946–85. Because of the different principles behind the accelerators, their energies cannot be compared directly.

to be discussed for several years and in the 1970s, the idea was upgraded to a 10 TeV machine. It was a pet idea among many physicists who wanted to combine worthy scientific goals with worthy political and ethical goals. But the romantic idea never got past the discussion level and was, in spite of all political rhetoric about internationalism in science, unrealistic.

Sensing that the initiative in high energy physics was on its way to being taken over by Europe, in 1983 American physicists suggested building the ultimate accelerator facility based on a Superconducting Supercollider (SSC). Appeals to nationalistic sentiments were among the arguments of the high-energy physicists who favored the SSC. Sheldon Glashow and Leon Lederman, two of America's leading particle physicists, argued as follows: "More and more, American accomplishments either recede into the past perfect or dangle in the future conditional while the Europeans pursue the present indicative. . . . Our concern is that if we forgo the opportunity that SSC offers for the 1990s, the loss will not only be to our science but also to the broader issue of national pride and technological self-confidence" (*PT*, March 1985, 34). The name of the SSC was no exaggeration. With a proton collision energy of about 40 TeV (or 40,000 GeV) and an underground ring of 83 kilometers in circumference, it was really to be a supermachine; the planned superconducting magnetic system included 41,500 tons of iron and 2 million liters of liquid helium. The expenses, estimated in 1990 to be about $10 billion, were to be super too. It would be the largest and most expensive physics facility ever built. The gigantic project was endorsed by President Reagan and in 1988, the site was announced to be Waxahachie, a small town in Texas. Also the next President, the Texan George Bush, liked the project, which he—in a rare fit of poetic mood—compared to "Louvre, the Pyramids, Niagara Falls all rolled into one" (Kevles 1997, 292).

There were good scientific reasons to build the SSC, but for an investment of this size, scientific reasons alone would not do. Among the other reasons that were suggested by SSC advocates was the misleading suggestion that work on the superconducting magnets might help in the development of medical techniques that might prove useful in the fight against cancer. As in some earlier cases of big accelerator projects, the opposition to the SSC came not only from politicians, but also from physicists and other scientists who argued that high-energy physics was economically overfunded and that a much greater payoff, both scientific and technological, would be obtained by allocating more money to other fields of science. One of the critics was Philip Anderson, the solid state theorist and later Nobel laureate, who as early as 1971 had suggested that "the pie is finite and what is 'pro' high-energy physics is 'con' to somebody else, so that it is now obvious that if we are to retain a healthy science we must look at all of its divisions critically." Anderson not only expressed doubts as to the technological spinoffs claimed to follow from high energy physics, but also objected to "the intellectual justification of elementary particle physics as almost the *only* direction in which 'really fundamental' or 'intensive' truths are to be found" (Anderson 1971). Twenty years later, Anderson reinforced his criticism in connection with the SSC project. Naturally, his arguments were countered by SSC advocates, including Leon Lederman, Steven Weinberg, and other leaders of the high-energy physics community. Work on the SSC continued for a while but in 1992, the project ran into serious troubles. It was finally rejected in the summer of 1993, after it had consumed nearly $2 billion. The reason for the rejection was both political and economic. Ten billion dollars was, after all, a staggering amount of money to spend to satisfy the curiosity of a small group of high-energy physicists. Under different political conditions, the project might conceivably have been accepted—namely, if it could be connected to questions of national security. But it was not thought to have any particular military value and, at any rate, the timing was bad: By 1992, the Cold War had ended, and the Soviet Union no longer existed.

One of the effects of the new kind of big science that flourished in the 1950s was a marked shift of the role of the physicist, from an individual researcher to a small wheel in a collective research effort. The physicist no longer did his own research, but participated in a research project. It was a shift that many physicists of the old school deplored. One of them was Percy Bridgman, the Nobel laureate and philosopher of science, who argued that the new style of physics was detrimental to creative ideas and intellectual freedom. Two decades later, this was a critique to be repeated and reinforced by a younger generation of physicists. Bridgman attributed this style to the research structure created during the war, when young scientists who "had never experienced independent work and did not know what it was like" joined the big research projects. "The result," wrote Bridgman, "is that a generation of physicists is growing up who have never exercised any particular degree of individual initiative, who have had no opportunity to experience its satisfactions or its possibilities, and who regard cooperative work in large teams as the normal thing. . . . The temper of the rising generation is recognizably different from that of the older" (Bridgman 1950, 299). Big science fostered—indeed, demanded—a cooperative as well as corporative spirit that was unheard of in the 1920s, when Bridgman did his important work.

Not only was there little room for individuality in large-scale experiments, but the expensive instruments also seemed to live their own lives and become more important than the physicists who worked with them. Were the machines monsters created by Frankenstein physicists? The changed relationship between physicists and machines was unintentionally spelled out by the director of the Brookhaven Cosmotron, Samuel Goudsmit (who, like Bridgman, was a Nobel laureate who had started his career in the 1920s). In an internal memorandum of 1956, Goudsmit emphasized, "In this new type of work experimental skill must be supplemented by personality traits which enhance and encourage the much needed cooperative loyalty." He went on, "Since it is a great privilege to work with the Cosmotron, I feel that we now must deny its use to anyone whose emotional build-up might be detrimental to the cooperative spirit, no matter how good a physicist he is. . . . I shall reserve the right to refuse experimental work in high energy to any member of my staff whom I deem unfit for group collaboration. I must remind you that it is, after all, not you but the machine that creates the particles and events which you are investigating with such great zeal" (Heilbron 1992, 44). The dangers—and perhaps the essence—of big science physics could not have been phrased more pointedly.

A European Big Science Adventure

About 1950, when nuclear and particle physics began to be increasingly dominated by accelerators, Europe lagged far behind the United States. At that time, the first initiatives to organize a joint European research project took place, with the French physicist Pierre Auger and the Italian Edoardo Amaldi among the prime movers. The result of many and difficult negotiations was the provisional founding of CERN (Conseil Européen pour la Recherche de Nucléaire) in 1952 and the establishment of a permanent organization in 1954. The initiative was mostly French and Italian, whereas the response of the leading European science nation, the United Kingdom, was cool. The British had their own accelerator, a 400-MeV synchrocyclotron in Liverpool, and they had no confidence in the grandiose continental plan, which Blackett described as "quite crazy." Moreover, there was no political interest in England for European cooperation, and even less for European unity. At first, the British government declined to participate and England joined fully only in 1954. Germany, or rather the new Federal Republic of Germany, presented another political problem, among other reasons because the country was not yet a member of UNESCO. The founding of CERN was widely seen as not just a scientific but also a political project, namely, a model for European cooperation. For Heisenberg and other German physicists, it was also a way to rehabilitate German physics after the war and forget the unpleasant Nazi past. By 1956, the CERN nations included Belgium, Denmark, West Germany, France, Greece, Italy, the Netherlands, Norway, Sweden, Switzerland, England, and Yugoslavia. The largest contributions came from France and England (each about 24 percent), followed by Germany and Italy (about 18 percent and 10 percent, respectively).

The aim of CERN was "to provide collaboration among European States in nuclear research of pure scientific and fundamental character, and in research essentially related thereto." The backbone of the collaboration was planned to be the world's largest accelerator, which in 1954 meant one more powerful than 6 GeV. It was specifically mentioned that "[t]he organization shall have no concern with work for military requirements and the results of its experimental and theoretical work shall be published or otherwise made generally available." Indeed, one of the major differences between American and European high-energy physics, and big science projects in general, was the very different roles played by the armed forces. Of course, one of the reasons for the indifference with which European military establishments regarded CERN was that there was no European military. The various national armed forces had no tradition of large-scale support of basic science, and no interest in a multinational organization like CERN. To underscore the nonmilitary, apolitical picture of CERN, it was decided not to include a nuclear reactor and to locate the laboratory near Geneva in neutral Switzerland. In the early phase of the project, Copenhagen had been considered, but Bohr and Kramers, the main advocates of the Copenhagen proposal, did not get it passed through.

CERN's first main project, the proton synchrotron, was originally thought of as an extrapolated version of the Brookhaven Cosmotron. In 1954, it was planned to have an energy of about 30 Gev and its cost was estimated at $16 million—about half the true cost. The machine was completed according to plan and during the 1960s, the laboratory expanded with new and larger accelerators and detector systems. It was a big science project from its very start and it grew bigger and bigger. For example, in 1955 the budget was 64 million Swiss francs (MSF), or about $16 million, and the staff consisted of 144 persons. Ten years later, the budget had increased to 275 MSF and the staff to 2,251 persons; and in 1975 the budget was 844 MSF, the number of staff members 3,788. In addition, the number of visiting scientists increased steadily, from 21 in 1955 to 315 in 1965 and 1,289 in 1975. The administrative and technical staff increased even more rapidly than the number of scientists (see table 20.2).

Although the joint European effort was successful in many ways, at first it did not succeed in seriously challenging the American leadership in high-energy physics. More often than not, the American accelerator laboratories came first with the important discoveries that were typically confirmed by the Europeans shortly later. The first ten or fifteen years were essentially a learning period for the CERN physicists, most of whom were unaccustomed to doing big science experiments the American way. In the United States, large-scale accelerator experiments, the entrepreneurial spirit, and the close cooperation among experimenters, theorists, administrators, and industry went back to the 1930s; this tradition was reinforced greatly by the military projects during the war. When European physicists first learned the new methods of doing and organizing physics, CERN became competitive with the finest of the American laboratories. Indeed, after some thirty years of massive American dominance in particle physics, in the late 1970s the balance of power shifted and European high energy physics laboratories began to take the lead (figure 20.3). The shift has been documented by means of bibliometric data. For example, whereas 53 percent of all publications in experimental high-energy physics originated in U.S. laboratories in 1970, and only 36 percent in Europe, in 1982 the figures were 35 percent from the United States and 56 percent from Europe. The corresponding data for the average number of citations per paper were 3.3 (USA) and 2.9 (Europe) in 1970, and 3.0 (USA) and 3.9 (Europe) in 1982. The main competitors during the 1970s were Fermilab and CERN. Whereas Fermilab earned nearly a quarter of all citations in experimental high-energy physics during 1973–76, the share fell to 9.3 percent during 1981–84. During the last five-year period, more than 15 percent of the citations were to papers from CERN's SPS program. In 1984, the average number of citations to papers originating from CERN's collider was 24, a very large number, which indicates high scientific value.

In late 1997, four years after the demise of the SSC, the United States

TABLE 20.2
Distribution of CERN Personnel

Category	*1955*	*1960*	*1965*
Scientists and engineers	83	170	349
Technicians	102	527	604
Ancillary staff	35	241	871
Administrative staff	49	127	316
Fellows	18	71	73

Note: Based on data from Hermann et al. 1990, 396–98.

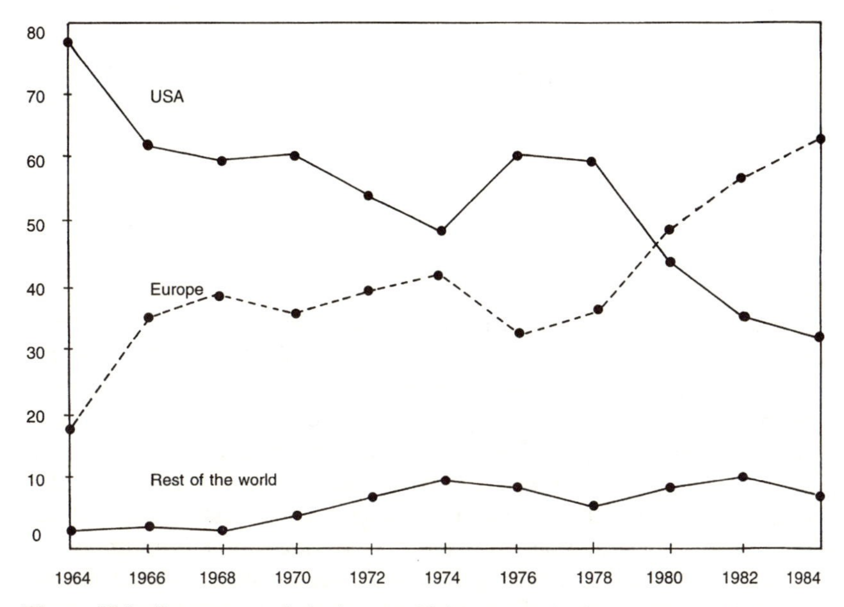

Figure 20.3. Percentage of citations to high-energy physics papers published in a given year and in the previous three years, by region and year. *Source:* Redrawn from Irvine et al. 1986.

reached an agreement with CERN to participate in the construction and use of the future Large Hadron Collider (LHC) and its associated particle detectors. The Americans agreed to contribute $531 million. The LHC, scheduled to operate in 2005, is estimated to cost about $6 billion and will be the world's most powerful accelerator. This latest phase of accelerator physics history demonstrates that the center of high-energy physics has now definitely shifted to Europe. It also demonstrates that very big, international particle physics was not killed with the SSC project. Although it is a European project, the LHC is not confined to the CERN member states, but involves a cooperation of forty-five different nations. In a sense, it is the realization of the old dream of a "world accelerator."

Chapter 21

PARTICLE DISCOVERIES

MAINLY MESONS

IN 1948 ELEMENTARY particle physics did not yet exist as a discipline, and the term high-energy physics had not yet entered the physicists' vocabulary. The study of nucleons, mesons, and other elementary particles was seen as a part of nuclear physics and activities in the field were often housed in new laboratories and institutes devoted to "nuclear studies." With the recognition in 1947 that the Yukawa π-meson was very different from the cosmic ray μ-meson, the stage was set for the investigation of still more new particles and eventually the emergence of a mature scientific subdiscipline of particle physics. In their 1949 monograph, *Theory of Atomic Nucleus and Nuclear Energy-Sources*, George Gamow and Charles Critchfield suggested "a convenient, even though not very sharply defined, boundary between *nuclear physics proper*, and the next, as yet rather unexplored, division of the science of matter which can be called tentatively *the physics of elementary particles*." (Emphasis in original.) While the first eight of the annual Rochester conferences, from 1950 to 1957, dealt with either "meson physics" or "high-energy nuclear physics," the continuation in Geneva in 1958 had dropped the "nuclear" and become a conference on "high-energy physics." At that time, a need was felt for greater international coordination of the many conferences on high-energy physics. At the 1957 General Assembly of the IUPAP, a High-Energy Physics Commission was established. The first commission consisted of Robert Marshak and Wolfgang Panofsky from the United States, Cornelis Bakker and Rudolf Peierls from Europe, and Igor Tamm and Vladimir Veksler from the Soviet Union.

In the early 1950s, an increasing reliance on complicated, powerful, sophisticated, and expensive instrumentation precipitated the transformation of the subfield into the big science of high-energy physics. As described in chapter 20, the subfield soon became characterized by heavy government funding and collaborations often involving dozens of researchers. In 1950, when the number of detected and predicted elementary particles and antiparticles stood at around twenty, the cosmic radiation was still the principal source of particles, but a few years later, the new generation of accelerators and detectors changed the experimental situation significantly. The first particle ever to be discovered artificially, manufactured in an accelerator rather than to be found in nature, was the neutral pion. The existence of a neutral

Yukawa meson had been proposed as early as 1938 by the German-British physicist Nicholas Kemmer, and in 1947 Oppenheimer suggested that the hypothetical particle would decay very quickly into two gamma rays. The particle was first detected at the Berkeley synchrocyclotron in 1950 and identified by the conversion of gamma rays into electron pairs ($\pi^o \rightarrow \gamma\gamma$ followed by $\gamma \rightarrow e^+e^-$). Shortly afterwards, British physicists found the particle in the cosmic radiation. (From a more philosophical point of view, it is not obvious when an object belongs to nature and when it is artificially produced. It can be argued that the electron appearing in J. J. Thomson's cathode ray tube was the first artificially produced elementary particle.)

The first evidence of a heavy meson was reported before the discovery of the pion. In 1944 the French physicists Louis Leprince-Ringuet and Michel l'Héritier published in *Comptes Rendus* a paper on "The Probable Existence of a Particle of Mass 990 m_o in the Cosmic Radiation," possibly what was later called a *K*-meson or kaon. However, little notice was taken of the single event, and the Frenchmen were not recognized as discoverers. After the war, heavy mesons were first reported by the two Manchester physicists Clifford Butler and George Rochester, who in October 1946—about a year before the pion was discovered—found a V-shaped event from the cosmic radiation in their cloud chamber. They interpreted it as the result of the decay of a heavy neutral particle. Another V-event arising from a charged particle was detected in 1947, but it took until 1950 before the discoveries were confirmed by more observations. These were made by Carl Anderson at Caltech, who obtained a large number of V-events. By 1952, the V-particles were firmly established and recognized to be of four kinds. The neutral particles were shown by Robert Thompson of Indiana University to decay into either $p\pi^-$ or $\pi^-\pi^+$, called Λ^o and θ^o, respectively (or V_1^o and V_2^o). Many more decays were found, some of them suspected of belonging to particles not hitherto known. In order to compare the many observations, determine the number of particles compared with the decays, and standardize the nomenclature, conferences were held in the United States and Europe. At a meeting in France in 1953, the International Congress on Cosmic Radiation suggested dividing the strange particles into two groups, the *K* or heavy mesons with mass smaller than the nucleon, and the hyperons with mass larger than the nucleon but smaller than the deuteron. The *L* or light mesons included pions and muons. Table 21.1 summarizes the new particles as they were known by 1957. At that time, the entire subject of the number and classification of particles was in a state of flux and there were several names and symbols in use for some of the particles.

Whereas most of the new "strange" particles, as they became known, were originally identified in the cosmic radiation, from 1953 on, the big accelerators increasingly dominated the field; and whereas the pioneering discoveries had been made by French and British physicists, from that time the field

TABLE 21.1
Strange Particles as Known around 1957

Class	*Symbol(s)*	*Mass* (m_e)	*Lifetime (seconds)*	*Strangeness*
K-mesons	K^+	966.5	1.2×10^{-8}	+1
	K^-	966.5	1.2×10^{-8}	−1
	K_1^o	965	$\sim 10^{-10}$	(+1)
	K_2^o	965	$\sim 10^{-7}$	(+1)
Hyperons	Λ^o (V_1^o)	2182	$3 - 10^{-10}$	−1
	Σ^+ (V_1^+)	2327	0.7×10^{-10}	−1
	Σ^- (V_1^-)	2343	1.5×10^{-10}	−1
	Σ^o	2325	$<10^{-10}$	−1
	Ξ^-	2585	$\sim 2 \times 10^{-10}$	−2

became dominated by American researchers. The 3-GeV Cosmotron at Brookhaven National Laboratory started producing pion beams in 1953 and in 1954, the Bevatron at Berkeley's Radiation Laboratory produced beams up to 6 GeV. The accelerators heralded a new era in the young science of high-energy physics.

No less important than the accelerators were the new methods that were introduced to detect particle events. In the first phase of high-energy physics, the tested cloud chamber was the favorite detector, which was developed in the early 1950s into various special versions suitable for accelerator experiments (high-pressure and diffusion cloud chambers). At the same time, new types of photographic emulsions came into widespread use. The nuclear emulsion method was developed in the 1930s, in particular by the Viennese physicist Marietta Blau, but it was only after the war that this method became of decisive importance in particle physics. This was a result of cooperation between British cosmic-ray physicists and Ilford, Ltd.

The traditional, low-cost methods, such as cloud chambers and emulsions, were supplemented with and eventually replaced by the bubble chamber technique, invented by Donald Glaser at the University of Michigan in 1952. Glaser's invention was soon developed into a versatile instrument, in particular, by Luis Alvarez and his Berkeley group. Bubble chambers with liquid hydrogen as the working fluid proved ideally suited for use with accelerators and were used extensively from the late 1950s. They differed from earlier detecting systems not only technically, but also by their size, cost, and complex operation. It was a kind of detector that small laboratories could neither afford nor make much use of—in short, an "undemocratic" detector. Alvarez's seventy-two-inch hydrogen bubble chamber of 1959 cost $2 million and required a $1 million computer to analyze the data. As early as 1957, the first reading machine that could automatically follow and measure tracks was constructed. It was developed by Jack Franck, a Berkeley engineer, and appropriately nicknamed "Franckenstein." Measurements of bubble chamber pictures developed into a minor industry. In 1967 it was estimated that some two million bubble chamber events were being measured annually. In the 1964 experiment leading to the discovery of the Ω^- hyperon (see below), about 100,000 pictures were taken and partially analyzed.

Very soon after the discovery of the first heavy particles, physicists began to wonder how to understand them theoretically, which initially meant classifying them into families. A much-discussed problem was the relationship between the production and decay of the new particles, indicating that they were strongly interacting, but had long lifetimes that did not agree with strong decays. Because of this puzzling behavior, the particles were often referred to as "strange." In 1951 Abraham Pais, and, independently, a group of Japanese physicists, suggested the idea of associated production, which implied that the *V*-particles could be produced only in pairs. Pais introduced a kind of quantum number, analogous to parity, in order to explain observed and nonobserved decays. Pais's quantum number was multiplicative and functioned as a selection rule, not unlike the quantum numbers introduced in early quantum theory in order to make sense of optical spectra. Its purpose was to act as an ordering principle by collecting the new particles in groups or families. In his 1951 paper, Pais mentioned that "the search for ordering principles at this moment may ultimatively have to be likened to a chemist's attempt to build up the periodic system if he were only given a dozen odd elements" (Pais 1986, 519). It may have been the first, but it was definitely not the last time that rules of particle classification were described as analogous to Mendeleev's chemical system. Yet, in spite of the popularity of the analogy, it was fundamentally flawed. Particle physicists were not merely engaged in a game of ordering particles; they also wanted to understand them as manifestations of a few, really fundamental particles, such as quarks. Their classification schemes were guided by theoretical notions and expressed in the language of mathematical physics. In this respect, their work differed fundamentally from that of Mendeleev. The Russian chemist devised his famous periodic system purely empirically, as a rational classification of the known physicochemical properties of the elements. Contrary to some other chemists of the period, he warned against conceiving the system as a reflection of some internal subatomic harmony in the elements.

An improved explanation scheme of the strongly interacting particles was soon developed by twenty-six-year-old Murray Gell-Mann at the University of Chicago, and also by Kazuhiko Nishijima and Tadao Nakano in Japan. Gell-Mann introduced an additive quantum number, which he called "strangeness" (S)—a name dating from the fall of 1953—and which was zero for pions, muons, and nucleons but nonzero for the new strange particles. For example, the K^+ was assigned $S = +1$ and the Λ was given S

$= -1$. In all strong interactions, the strangeness would remain unchanged, and thus the concept worked as a guide as to which reactions could occur and which not. A reaction such as $\pi^- + p \rightarrow K^- + \Sigma^+$ was not allowed, because it would violate strangeness conservation (S being changed from 0 to -2). Various predictions of the Gell-Mann—Nishijima theory were verified by experiments and by 1960, the idea of strangeness, although not well understood theoretically, was firmly established as a most useful guide in the jungle of elementary particles. The theory implied a controversial reclassification of the neutral K mesons which were now, according to Pais and Gell-Mann, to be understood as a mixture of K^o and its antiparticle, this being different from K^o and with opposite strangeness. K^o and $\bar{K}^o$ were connected through weak interactions and the decaying particles interpreted as a "mixture" or superposition of K^o and $\bar{K}^o$ with different decays. The short-lived K_1^o was known to decay into two pions and the long-lived K_2^o supposed to decay into three particles. The assumption was confirmed in 1956, when the K_2^o was first observed in an experiment at the Brookhaven Cosmotron.

Mesons, hyperons, and resonance particles were not the only new particles of the 1950s. The antiproton, predicted by Dirac in 1933, had long been expected to exist and even been reported a few times. For example, Soviet physicists claimed in 1946 to have detected the particle in the cosmic radiation, but neither this nor other claims made before 1955 were accepted. In order to produce antiprotons by proton-matter collisions, the energy of the protons had to be at least 5.6 GeV. The maximum energy obtainable with the 10,000-ton and almost $10 million Bevatron was 6.2 GeV, well above the threshold energy. Although the Bevatron was an ideal antiproton machine, it was not built with the antiproton in mind and it was only in 1955 that it was used to manufacture the particle. The detection of the antiproton was first achieved in the fall of 1955 by the Berkeley physicists Owen Chamberlain, Emilio Segré, Clyde Wiegand, and Thomas Ypsilantis. Their scintillators and Cerenkov counters showed about 60 antiproton candidates, but the ultimate proof of the particle, its annihilation with an ordinary proton, was not immediately confirmed. That came in 1956, when a group of Italian physicists, led by Amaldi and cooperating with the Berkeley group, reported the stopping of an antiproton in an emulsion stack. The next step, using antiprotons from the Bevatron to produce antineutrons, was taken in 1957 by another Berkeley team. None of the discoveries came as a big surprise.

There was more in the wake of the Bevatron experiments than scientific progress. The experimental success relied on the use of a system of quadrupole magnets first suggested by the Brookhaven physicist Oreste Piccioni, who felt that the Berkeley physicists had stolen the idea from him. In 1972, thirteen years after Segré and Chamberlain were awarded the Nobel prize for their discovery, Piccioni took the rather extreme step of bringing forward his charges in a lawsuit. Nothing except press coverage came out of the unusual initiative. Piccioni's lawsuit was not the only bizarre aftermath. In 1979, J. C. Cooper accused Segré and Chamberlain of deliberately having suppressed data in their 1955 report that showed evidence of tachyons or faster-than-light particles. "The Segré experiment is to the physics community what the Watergate tapes were to ex-President Nixon," he dramatized (Franklin 1986, 239). Unfortunately for the drama, but fortunately for physics, the accusation lacked evidence and few physicists took it seriously.

Weak Interactions

The neutrino survived detection four years longer than the antiproton. Contrary to the latter particle, which played almost no role at all in theoretical physics, the neutrino (or rather the antineutrino) became highly important after Fermi's 1933 theory of beta decay and was used routinely by nuclear theorists by 1940. Physicists believed in the existence of the neutrino irrespective of whether the particle had been detected or not. And what did it mean that a particle existed? In a 1952 paper in the *Bulletin of the Atomic Scientists* titled "Does the Neutrino Really Exist?" the theoretical physicist Sidney Dancoff argued from a positivistic perspective that it was a meaningless question and that concepts like "neutrino" and "electron" were simply convenient ways of organizing experimental data. As far as Dancoff was concerned, the neutrino, whether it had been detected or not, was as "real" as the electron. For more than a decade, it was widely assumed that the neutrino would remain undetected forever because of its exceedingly weak interaction with matter. There was no great interest in trying a detection experiment, but at Los Alamos, Frederick Reines and Clyde Cowan realized in 1951 that the intense beta decays from a nuclear explosion might provide enough (anti)neutrinos for detection through interaction with protons. Reines and Cowan seriously thought of using an atomic bomb for their experiment, but decided on the more peaceful radioactive environment of a nuclear reactor. After many difficulties, they designed an experiment in which the effect of neutrinos in water would produce a distinct signal of gamma ray coincidences. To measure the signal, they developed a complex detector system of 330 photomultiplier tubes immersed in an organic liquid. In 1956, using the Savannah River reactor as a neutron source, they found signals that were the unmistakable signs of neutrino-proton reactions. The discovery of the neutrino was even less surprising than the antiproton discovery made the previous year.

A few years after Reines' and Cowan's successful experiment, other experiments showed that reactions that occur with the neutrino do not occur with the antineutrino, and that the two particles were thus different. This was in agreement with the conservation of lepton number (L), an idea implicitly introduced by Emil Konopinski and Hormoz Mahmoud in 1953. Both the electron and the negative muon have $L = +1$, as have the associated neutrinos. But whereas the decay of a muon into an electron and a neutrino-antineutrino pair was well known, the decay into an electron and a gamma quantum was not observed; and yet it is allowed according to lepton conservation. Formally, this was accounted for by introducing yet another conservation law, that the muon number is conserved. This amounted to suggesting that the neutrino associated with the $\pi\mu$ decay would be different from the ordinary neutrino associated with the electron in neutron decay. If this is the case, $\nu_\mu + n \rightarrow \mu^- + p$ would be allowed, while $\nu_\mu + n \rightarrow e^- + p$ would be forbidden. The difference between these two reactions, and hence the existence of a muon-neutrino separate from the electron-neutrino, was shown experimentally in 1962 by a collaboration between Columbia University and Brookhaven National Laboratory led by Leon Lederman, Melvin Schwartz, and Jack Steinberger. The success of the experiment was secured by the detector system, a monster of a spark chamber weighing ten tons.

By far the most sensational development in weak-interaction physics in the 1950s was the proof that parity is not conserved in weak processes. The principle of parity invariance, or left-right symmetry, was carried over into quantum mechanics by Eugene Wigner in 1927–28, and immediately obtained a paradigmatic status. Parity, it should be noted, is in this context specifically a quantum-mechanical concept. When Hermann Weyl, in 1929, proposed a two-component Dirac wave equation for particles with zero mass and spin one-half, Pauli rejected the equation because it did not satisfy parity invariance. The Weyl equation was not rehabilitated until much later, after the discovery of parity nonconservation. Until the mid-1950s, physicists did not question the parity conservation dogma and the few experiments that, in retrospect, indicated violation of the conservation law were not interpreted as such.

With the discovery of the strange particles it appeared that one of them, named θ^+, decayed into two pions ($\pi^+\pi^o$); another, the τ^+ particle, into three pions ($\pi^-\pi^+\pi^+$). Moreover, the two particles had the same mass and the same lifetime, and so would appear to be just two different decay modes of the same particle. However, as shown by Richard Dalitz and others in 1954, the spin and parity state of θ was different from that of τ, hence they were presumably different. The puzzle was widely discussed in 1955–56 and various suggestions to solve it were made. In 1956 Tsung Dao Lee and Chen Ning Yang, two young Chinese physicists who had come to the University of Chicago shortly after the war, suggested parity nonconservation as a realistic possibility. They found to their surprise that conservation of parity was not well supported by earlier experiments: "[I]t became obvious to us that not only was there, at that time, not a single evidence for parity conservation in β-decay but that we must have been very stupid!" (Franklin 1986, 14). Having "stopped calculating and started to think," they suggested late in 1956 that parity conservation was violated in weak interactions, although they cautioned that "This argument is . . . not to be taken seriously because of the paucity of our present knowledge concerning strange particles." The thinking of Lee and Yang led them to suggest two kinds of experiments, one on beta decay and the other on $\pi\mu e$ decay, which should be able to reveal the suspected violation of parity conservation. The experiments were performed by three groups of physicists in early 1957: Chien-Shiung Wu at Columbia University investigated the beta decay; the meson decay was examined by Richard Garwin, Leon Lederman, and Marcel Weinrich, also at Columbia University, and, using a different technique, Jerome Friedman and Valentine Telegdi at the University of Chicago. The results proved that Lee and Yang were right: Parity is not conserved in weak interactions. The breakdown of the law of parity conservation was, in spite of its revolutionary nature, accepted quickly by the physics community. Pauli, who was a great believer in symmetry principles and who, as late as January 1957, had written to Victor Weisskopf that "I do not believe that the Lord is a weak left-hander," was no exception. In another letter to Weisskopf, he described his reaction: "Now, after the first schock is over, I begin to collect myself. Yes, it was very dramatic. . . . I am shocked not so much by the fact that the Lord prefers the left hand as by the fact that He still appears to be left-right symmetric when He expresses Himself strongly" (Franklin 1986, 25).

The discovery of parity nonconservation contributed to a general change in the intellectual climate of fundamental physics, leading to a tendency to question the absolute validity of other conservation laws as well. As Philip Morrison expressed it in 1958, "[it is] inescapable to regard all the fundamental symmetries as open now to reasonable doubt, and as candidates for experimental test" (Kragh 1997c, 204). Even the principles of energy and charge conservation could be questioned, as they were in connection with some of the cosmological models discussed in the period. The Lee-Yang explanation showed not only that parity (*P*) was violated in weak interactions, but also that the same was the case with charge conjugation (*C*), that is, the particle-antiparticle symmetry. This invariance property goes back to Dirac's introduction of antiparticles in 1931, was formulated as a theorem by Wendell Furry the following year, and was stated as a general invariance principle by Kramers in 1937. Landau, though accepting the result of Wu and others, was among those who were not happy about parity nonconservation because it seemed to be in conflict with the isotropy of space. "In my opinion a simple denial of parity-conservation would place theoretical physics in an unhappy situation," Landau wrote in 1957. He therefore suggested that "we still have invariance with respect to the product of the two operations [*CP*], which we call combined inversion . . . [and which] consists of space reflection with interchange of particles and antiparticles" (Franklin 1986, 79).

As had been shown by Pais and Piccioni in 1955, the particle-mixture theory of kaons proposed by Gell-Mann and Pais predicted that a beam of the long-lived K_2^0 particles after passage of matter would "regenerate" the short-lived K_1^0 particles which had originally been mixed with the K_2^0 particles. The phenomenon was confirmed in 1961 and a few years later, the Princeton physicists James Cronin, Jim Christenson, and Val Fitch, collaborating with the Frenchman René Turlay, set out to investigate it in greater detail. At the same time, they studied the decay of K_2^0 into two pions, a process that was forbidden according to *CP* conservation. The results of the elaborate experiment, carried out at Brookhaven's new Alternating Gradient Synchrotron, were published in 1964. They showed 45 two-pion decays out of almost 30,000 analyzed decays. As the Princeton group pointed out, this proved that in this particular decay, even *CP* conservation was violated. This important result, later rewarded with a Nobel prize, was mentioned only briefly in the paper of the Princeton physicists. "The presence of a two-pion decay mode implies that the K_2^0 meson is not a pure eigenstate of *CP*," they wrote. That was all. But it was enough.

What is perhaps the strongest of all the invariance principles, the *CPT* theorem, states that all processes are invariant under the combined operations of *C*, *P*, and time reversal (*T*). It was formulated in different versions in the 1950s, but cannot be ascribed to a single physicist or a definite year of birth. Paternity is usually credited to Pauli, Gerhard Lüders, or Julian Schwinger, but John Bell and Bruno Zumino should also be considered as candidates. Because it is founded solidly on the most fundamental concepts of relativity and quantum mechanics, the *CPT* theorem has a very high, almost sacred status in physics. With the proof of *CP* violation in K_2^0 decay, it was realized that either the *CPT* theorem cannot claim absolute validity or time reversal must fail. In spite of this dilemma, leading to a choice between equally unappealing consequences, most physicists accepted the Princeton results as proof of *CP* violation. Alternative explanations were suggested, but none of them won wide acceptance. The lesser of the two evils, it seemed to most physicists, was to accept that time reversal symmetry is broken in the two-pion decay of K_2^0. Although physicists have looked for direct evidence for *T*-violation since the mid-1960s, no definite confirmation has been found.

The significance of the 1957 parity violation and the 1964 *CP* violation experiments was not restricted to the cognitive implications. The events were very important in the formation of the subdiscipline known as weak interaction physics, a field that may be traced back to Fermi's theory of 1933–34, but which only obtained a social meaning from about 1957. Bibliometric data clearly show the significance of the two events for the rise of weak interaction theory. As a percentage of the total number of publications in physics, weak interaction publications rose from about 1 percent to 3 percent in 1958, and then fell back to its "normal" amount of 1 percent within two years. During 1950–70, the percentage of papers in weak interaction physics that were predominantly theoretical relative to those that were predominantly experimental rose from about 0.8 to 5.8 (Vlachy 1982, 1066). Sudden increases in the ratio in about 1958 and 1965 reflect the impact of the two pathbreaking symmetry violation discoveries.

QUARKS

According to physics folklore, Fermi is said to have answered a question concerning the names of the elementary particles with "Young man, if I could remember the names of these particles, I would have been a botanist." That was in 1954, at a time when a dozen particles were known. Ten years later, a review article noted that "Only five years ago it was possible to draw up a tidy list of 30 sub-atomic particles, . . . since then another 60 or 70 subatomic objects have been discovered" (Pickering 1984a, 50). Indeed, during these years the number of strongly decaying mesons and baryons exploded as a result of the innovations in accelerator and detector technology. With the increase in particles followed a need to understand them or, at least, classify them according to some theoretical framework. Ideas for organizing elementary particles, in the sense of reducing them to fewer objects, went back to the late 1940s and were based on the isospin formalism. In 1949 Fermi and Yang speculated that the pion might be conceived as a combination of a nucleon and an antinucleon, and in 1956 the Japanese physicist Shoichi Sakata suggested a theory according to which the two nucleons and the Λ particle—sometimes known as "sakatons"—were the building blocks of the heavy mesons and the hyperons. The interest of Sakata and other physicists at Nagoya University in the compositeness of elementary particles was guided by the Marxist methodology of dialectical materialism. Although Sakata's model appeared promising and was useful in understanding the meson states, by the early 1960s it no longer attracted much attention outside Japan.

The most successful of the early classification schemes was (like Sakata's) based on the SU(3) symmetry group, originally as part of a gauge field theory of strong interactions. The SU(3) group proposal was made in 1961, independently by Gell-Mann, then at Caltech, and Yuval Ne'eman, a young Israeli theorist who, at the time, was serving as a military attaché in London. The idea became known as the "Eightfold Way," a term introduced by Gell-Mann. According to this way, or classification system, particles were grouped or "represented" into multiplets characterized by definite spin and parity numbers and including particles with different isospin and strangeness. For example, the particles with spin one-half and positive parity formed an octet consisting of p, n, Λ, Σ^+, Σ^-, Σ^o, Ξ^-, and Ξ^o, with each of the eight particles being characterized by its isospin and strangeness. The octet was the basic entity of the Gell-Mann—Ne'eman theory, which also included decuplets and, for the mesons, nonets. At the time when the Eightfold Way was proposed, there were several "holes" in the patterns in the form of unknown particles, and there was uncertainty with regard to the proper place of some of the recently detected particles. In the octet consisting of pions and kaons, one particle was missing. When the η resonance was found in the fall of 1961 and turned out to fit nicely into the scheme, it gave valuable support to the Eightfold Way. The theory was discussed at the 1962 Rochester conference and also at a CERN conference the same year. At both conferences, the baryon multiplet of spin 3/2 and positive parity was among the topics. This multiplet contained nine particles, some of which, the Σ^* and Ξ^*, were first announced at the CERN conference. If it was a decuplet, it had to contain one more—and still unknown—particle. Gell-Mann, and independently Ne'eman, argued that the tenth member had to have strangeness -3 and a mass of about 1680 MeV; moreover, it would be an ordinary particle with a relatively long lifetime, rather than a resonance. Gell-Mann named the predicted particle the Ω^- hyperon and left the scene for the experimentalists. (The mass unit MeV, shorthand for MeV/c^2, is close to two electron masses.)

Physicists in Europe and the United States were equally eager to find, or rather make and then find, the predicted particle. The search quickly turned into a competition between CERN in Europe and the Brookhaven Laboratory in the United States, with the Americans as winners of the race. In early 1964 a group of thirty-three Brookhaven physicists, led by Nicholas Samios, were able to announce a completely successful experiment. They found, in their hydrogen bubble chamber exposed to kaons, a track that they concluded was the result of the decay of Ω^- produced as $K^-p \rightarrow \Omega^-K^+K^o$. Not only did the particle have the right strangeness (-3) and lifetime (0.7×10^{-10} sec), they also found the mass to be 1686 ± 12 MeV, in excellent agreement with the predicted value. The confirmation was of course a great triumph for SU(3) and the Eightfold Way classification scheme, the soundness of which could no longer be doubted. With the success of the scheme, physicists naturally wondered why it worked so well, a question that had been considered by Gell-Mann and others even before the Ω^- success. There was another dimension to the experiment, as characteristic of modern particle physics as the purely scientific dimension. Before the Brookhaven team submitted their paper to the fast-publishing *Physical Review Letters*, they called a news conference with instructions that the newspapers were to publish only after the paper had appeared. Somehow a leak occurred, and news about the discovery appeared in *New Scientist* before the publication date. One of the Brookhaven physicists recalled: "It was a big mess. It killed our publicity. We would have got first-page *New York Times* Sunday, which is a very good thing to get. After all, where do we get national money?—from Congress. It makes a great deal of difference if we get first-page in the *New York Times* since most congressmen read the front page. It's a factor, and we can't deny it, so it matters an awful lot to us" (Gaston 1973, 86).

In 1964 the interpretation of the Eightfold Way in terms of a composite model of hadrons (strongly interacting particles) was proposed by Gell-Mann, who, inspired by a passage in James Joyce's *Finnegan's Wake*, called the constituent particles "quarks." A very similar suggestion, but using the term "aces" rather than "quarks," was made independently by George Zweig, a twenty-seven-year-old Russian-born American physicist and student of Gell-Mann working at CERN. Physicists adopted Gell-Mann's more literary name. According to the quark idea, all hadrons consisted of combinations of two or three quarks and their antiquarks, meaning that the hundreds of hadronic particles and resonances were reduced to combinations of three fundamental entities. These were the "up" quark and "down" quark, both with zero strangeness but with opposite isospin, and the "strange" quark with $S = 1$. For example, the proton was composed as *uud*, and the negative pion as $d\bar{u}$. The most remarkable feature in the quark picture was that the new hypothetical objects had fractional electrical charges, namely, $+2/3$ for the u quark and $-1/3$ for the d and s quarks. Were the quarks more than a bookkeeping notation, a useful mnemonic aid? Did they exist as dynamic objects that could be detected? These were questions that were asked immediately by physicists who recognized the classificatory success of the simple quark picture but were suspicious with regard to the ontological status of the quarks. Gell-Mann, the inventor of the quark, emphasized that his invention did not exist. "It is fun to speculate about the way quarks would behave if they were physical particles of finite mass," he wrote in his 1964 paper. "A search for stable quarks of charge $-1/3$ or $-2/3$ and/or stable diquarks of charge $-2/3$ or $+1/3$ or $+4/3$ at the highest energy accelerators would help to reassure us of the nonexistence of real quarks" (Pickering 1984a, 88). Gell-Mann later claimed that he used the term "real," as opposed to "mathematical," not to deny that quarks existed but to avoid philosophical discussions concerning the meaning of existence of permanently confined objects.

The quark model was not particularly well received in its early years. One of the reasons was the experimentalists' persistent failure in detecting free quarks, another that the quark model was widely seen as theoretically ill-founded and in disagreement with more popular alternatives at the time. Quarks worked finely on the phenomenological level but to many physicists they were merely simplistic expressions of the dynamics of a not-yet-understood hadronic world. The less-than-enthusiastic response did not prevent experimentalists from attempting to disprove Gell-Mann, that is, to show that quarks existed, rather than to show they did not exist. A 1977 survey of quark search experiments listed about eighty such searches. Among the many searches following the 1964 theory, the most interesting was undertaken by William Fairbank and collaborators at Stanford University. In 1977, after several years of work, the Stanford group reported that it had found fractional charges in Millikan-like experiments, namely, three cases of tiny niobium balls carrying one-third of the electron's charge. The claim was controversial both experimentally and theoretically, for at that time, theorists had become convinced that free quarks do not exist but are forever confined within the hadrons of which they are parts. Suffice it to say that the claim was not confirmed by other experiments and that it was, after much discussion, rejected by the elementary particle physics community.

The quark was not the only elementary particle that escaped detection. Many other exotic particles have been predicted without ever having been found, either because they are too difficult to find or because they do not exist. The magnetic monopole is an example. Dirac showed in 1931 that isolated magnetic poles are possible, that is, consistent with the basic laws of physics. For more than three decades the possibility was ignored, but in the 1970s monopoles became interesting objects after Gerardus t'Hooft in the Netherlands and Alexander Polyakov in the Soviet Union showed that certain gauge field theories predicted massive magnetic monopoles. In 1975 Paul Buford Price and his coworkers announced that they had found in the cosmic radiation a monopole with the predicted strength. The discovery claim, announced at a press conference, caused a sensation. Price mentioned (ironically, perhaps) that "you might drive ships across the seas by putting a few monopoles in the ship and having the Earth's magnetic field tug it across the ocean." And this was not all, for according to the American Institute of Physics, the discovery might also lead to such useful applications as "new medical therapies in the fight against diseases such as cancer, and new sources of energy." When confronted with the remarkable statement, the AIP spokesman said that "people expected you to be able to say what use a discovery might and he [the spokesman] was simply doing the best he could" (Kragh 1981b, 160). Neither magnetic energy nor a cure for cancer resulted. Within a year, consensus decided that the discovery claim was a mistake.

The failure of Price's claim did not stop the monopole hunters, who continued to search for the elusive particle. In 1982 another sensational discovery claim was made, this time by Blas Cabrera at Stanford University. He recorded a change in the magnetic flux of a superconducting ring and interpreted it as the result of a monopole passing through the ring. Although the event was not explained in terms of other sources and was not otherwise proved to be a mistake, neither did Cabrera or others succeed in confirming it. A single event was not sufficient to change the status of the monopole from being a well-known missing particle to being a real particle.

The Growth of Particle Physics

From a social and quantitative point of view, the first quarter-century of high-energy physics was characterized by two features: a strong growth and a strong American dominance—not "Americanization," for that term would imply that the United States took over an already established field, and this was not the case; high-energy physics was, to a large extent, an American invention. (In this section we shall, in accordance with convention, refer to high-energy physics as HEP.) There are no sound data for the international distribution of HEP in the period 1950–70, but a study by the American Institute of Physics indicates that in the mid-1960s, 32.3 percent of the theoretical high-energy physicists were American or Canadian, 35 percent were from non-communist Europe, and the major part of the remainder was from the Soviet Union and their allies in Eastern Europe. The contribution from Japanese physicists was significant, that from third-world countries negligible. The American dominance is beyond discussion and, given the situation before the war and the course of events during and shortly after the war, not very surprising. Yet there are two other important factors: The United States invested heavily in HEP, but not really more heavily than some countries in Europe. For example, in 1967–68, the United States spent about $160 million on HEP, a large sum of money indeed. This was about four times that of Great Britain, but taking into account the differences in population and per capita income, the end result is that Britain invested more resources on HEP than did the United States. The other thing to recall is that many of the physicists working in America, and hence contributing to the physics share of this country, were Europeans or Asians. To the 30.4 percent American citizens that contributed to the above-mentioned 32.3 percent North Americans should be added about 12 percent who were not Americans but worked in the United States. During the first decades of the postwar period (and still today, if to a lesser extent), the United States acted as a magnet in many fields of physics and contributed to an internationalization—some would call it an Americanization—of physics. Yet the American dominance was never total, qualitatively or quantitatively. The 1981 ranking of countries, as measured by the proportion of the world's total publications in HEP, was headed by the United States as a clear number one, with almost 30 percent. There followed, in this order, the Soviet Union, Germany, Japan, the United Kingdom, France, and Italy, all with 4 percent or more. In the "second division," contributing between 1 percent and 4 percent, came India, China, Canada, Poland, Spain, and Israel (see table 21.2). A count from the same year, 1981, shows that, as measured by the number of particle physics papers per million

TABLE 21.2
Regional Distribution of Publications in High-Energy Physics, 1982

Rank no.	*Country or region*	*Percent of publications*
1	Western Europe	34
2	North America	31
3	Soviet Union	14
4	Japan	6
5	Eastern Europe	5
6	India	3
—	Other	7

Note: In that year, the total number of publications in the field was about 4,000. Data from Vlachy 1982.

inhabitants, Israel comes in as a clear first, followed by Switzerland, Germany, Denmark, and the United States.

Interesting as such figures may be, they should not be taken too seriously. Like most other branches of physics, HEP is international, and what really counts is not so much the national origins as the institutions where the research was carried out. In this respect, as we pointed out in chapter 20, CERN and other European laboratories took the lead from about 1980. The shift depended on different rates of growth in the production of high-energy physicists. In 1962 the European HEP community consisted of 685 scientists and the American community 798 scientists. By 1975 Europe had passed the United States as far as manpower was concerned. In that year, the American population of physicists with a Ph.D. in HEP had increased to 1,732, while Europe could boast of 1,806 Ph.D. physicists in HEP.

In spite of the internationalism, naturally there were trends characteristic of particular national and cultural settings. For instance, in Japan, particle physics and quantum field theory had a strong position before the war, a position that to a large extent must be ascribed to the works of Nishina and Yukawa. During the war years, Yukawa, Tomonaga, Sakata, and others continued their work under increasingly difficult conditions. With the American occupation followed a decree that "all research in Japan of either a fundamental or applied nature in the field of atomic energy should be prohibited" (Konuma 1989, 536). In spite of the many problems, theoretical HEP flourished in Japan in a most remarkable manner. The growth is illustrated by the *Progress of Theoretical Physics*, the international Japanese journal that was founded by Yukawa in 1946 and became a leading journal of HEP (see figure 21.1).

The general impression is that HEP "exploded" in the two first decades after 1945. This is not an unfounded impression. There are no reliable figures before 1964, but between 1964 and 1968—in many respects, the boom

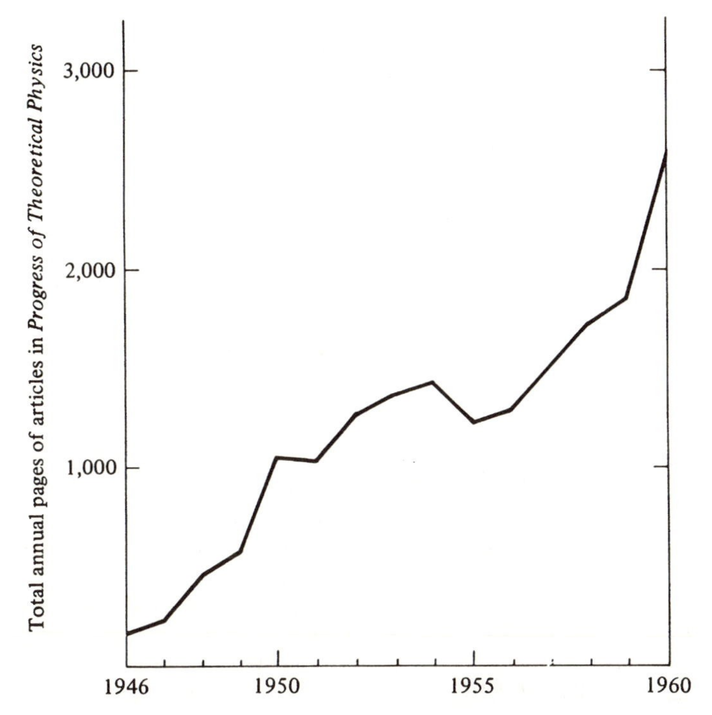

Figure 21.1. The expansion of *Progress of Theoretical Physics* 1946–60 as shown by the total annual pages. *Source:* Reproduced from Konuma 1989, with permission from Cambridge University Press.

years of HEP—publications in the category "nuclear physics and high-energy physics" rose from 5,486 papers to 8,242 papers. At first sight, this may seem to be a drastic increase, but in fact it did not even match the average increase in physics: The percentage of HEP papers of all physics papers fell from 19.1 to 16.3 (table 21.3). The main reason is that what until then was considered the big brother of HEP, nuclear physics, became less popular, not that interest in HEP declined. Quite the contrary, during this four-year period HEP experienced a dramatic boom, from 1,529 papers to 4,776 papers, a growth rate of an unprecedented 212 percent. Table 21.3 lists the growth in different fields during 1964–68, and table 21.4 the growth in some specialties in the same period. This is the period when HEP and solid state physics increasingly dominated the picture and when the fashion of the earlier decade, nuclear physics, lost its privileged position. From the mid-1930s to the mid-1950s, nuclear physics was the fashionable branch of physics par excellence, with HEP merely following in its footsteps. In 1939, 33 percent of all papers in *Physical Review* were in the category "nuclear

TABLE 21.3
Publication Distribution of Scientific Fields According to *Physics Abstracts* 1964 and 1968

Subject field	*1964* Number	*1964* Percent	*1968* Number	*1968* Percent	*Growth per year (percent)*
Solid state physics	9024	31.5	16992	33.7	22
Nuclear physics and HEP	5486	19.1	8242	16.3	13
Electricity and magnetism (inc. plasma physics)	3781	13.2	4572	9.1	5
Atomic and molecular	2156	7.5	4327	8.6	24
Fluids	1487	5.2	3612	7.1	36
Astrophysics	1124	3.9	2288	4.5	25
Geophysics	1110	3.9	2596	5.1	35
Optics	798	2.8	1407	2.8	19
Biophysics	206	0.7	83	0.16	−15
Total	28,656		50,477		19

Note: Not all fields are included. Based on Anthony, East and Slater 1969, 723.

physics" and 10 percent in "high-energy physics." In 1949 the proportions were 51 percent and 12 percent, respectively, but during the following decade, the propotion of HEP rose rapidly and that of nuclear physics decreased correspondingly.

There is more to the picture than simply an explosion. Consider first the funding of U.S. high-energy physics, primarily from the Department of Energy and the National Science Foundation (figure 21.2). In the decade from 1955 to 1965 there was an uninterrupted growth, reaching a peak in about

TABLE 21.4
Growth in Select Subdisciplines, 1964–68

Subject	*Number 1964*	*Number 1968*	*Annual growth (percent)*
High-energy physics	1529	4776	53
Magnetic properties of solids	910	2563	46
Electrical properties of solids	1232	2808	32
Optical properties of solids	985	2009	26
Molecular physics	1437	2329	16
Defects in solids	858	1229	11
Plasma physics	962	1373	11
Nuclear reactions	1043	1301	6

Note: Based on Anthony, East and Slater 1969, 724.

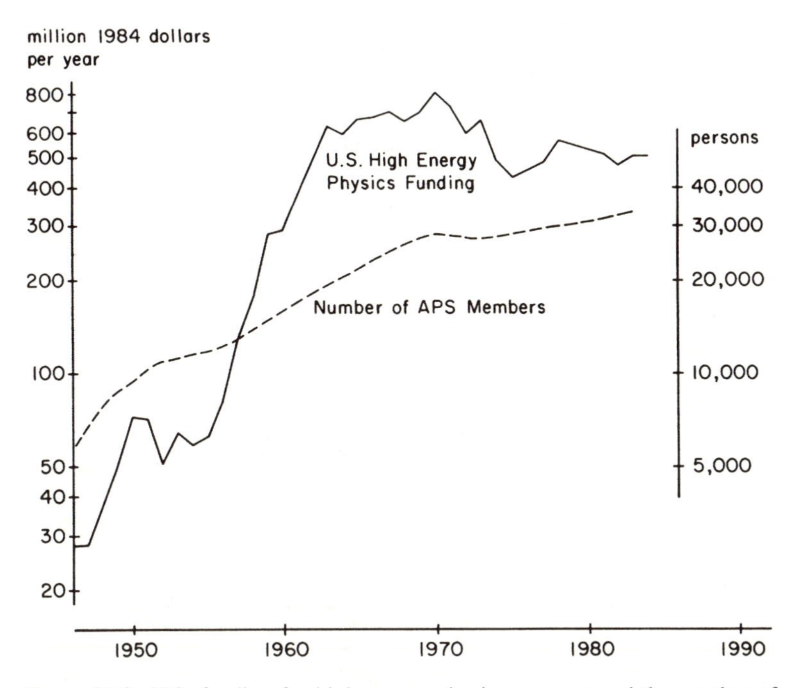

Figure 21.2. U.S. funding for high-energy physics per year and the number of members in the American Physical Society. *Source:* Reproduced from Yang 1989, with permission from Cambridge University Press.

1970, when the field received almost $800 million annually (1984 prices). But from that time on, HEP experienced a decrease in funding, although not a drastic one. Still, there were periods when the HEP community had reasons to feel worried (see also chapter 26). In 1975 American HEP received only half the money it had received five years earlier. During this brief period of crisis, even the funding in current dollars decreased. From about 1970 there was a real decrease in interest in American HEP, reflected, for example, in the number of doctorates granted in the field. HEP had been accustomed to growth and in 1971 the number of HEP doctorates peaked, with about 280 graduates receiving their Ph.D.s in HEP. Then the number rapidly dropped to about 130 in 1975, and it stayed in this neighborhood for the next fifteen years. The trend corresponded to a decrease from 19 percent of all physics doctorates in 1963 (the record year) to about 12 percent in the period following the bad year of 1975. Funding is one thing, Ph.D. students another; what about publications, widely considered an objective measure of the health of a scientific discipline? By and large, the number of publications in HEP rose steadily from 1960 to 1990, but the increase is to some extent an illusion, for measured as a proportion of all physics publications, the number decreased from almost 14 percent in 1960 to 6 percent in the years following 1975 (figure 21.3). These are figures for American HEP. Worldwide, the proportion of HEP articles of all physics papers was about 4 percent in 1981, with the United States slightly below the average and some

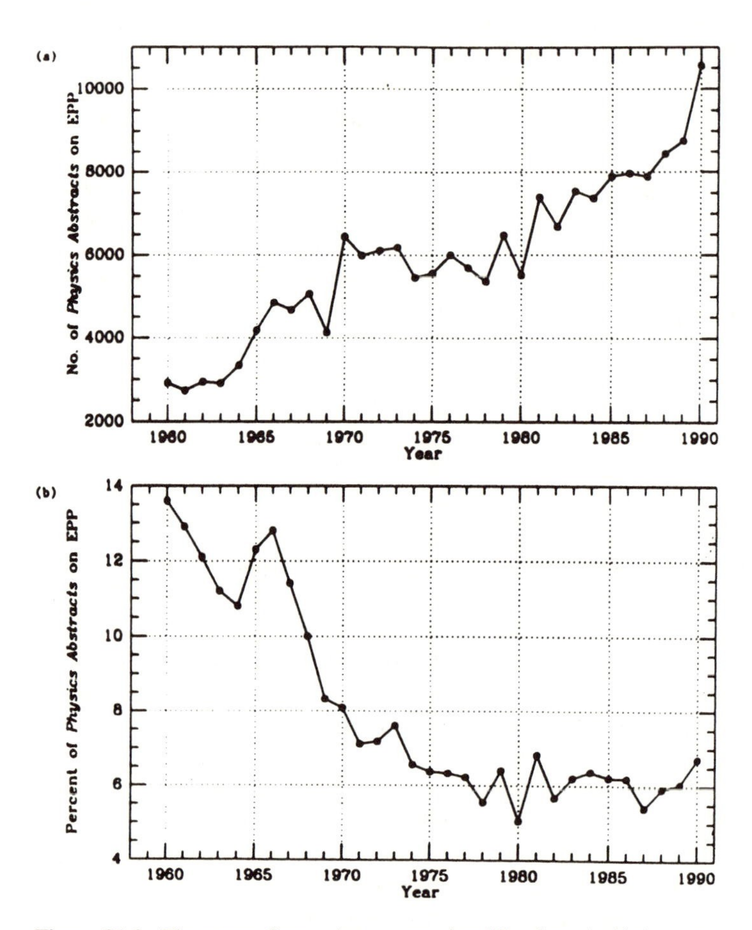

Figure 21.3. The upper figure shows annual publications in high-energy physics, as indicated by the number of entries in the sections of *Physics Abstracts* devoted to elementary particles, quantum field theory, cosmic rays, and particle accelerators and detectors and related instrumentation. In the lower figure, the same data are plotted as a percentage of all entries in *Physics Abstracts*. *Source:* Reprinted with permission from R. Corby Hovis and Helge Kragh, "Resource letter HEPP-1: history of elementary-particle physics," *AJP* (1991) 59: 779–807. © American Association of Physics Teachers.

smaller countries, such as Spain, Italy, Israel, and Pakistan, well above the average. There is, of course, a third-world aspect in HEP, as there is in science in general. In a nutshell, developing countries have scarcely been visible at all in HEP, and when they have been, their physicists have often worked and lived in Europe or North America. For example, whereas 6 percent of the world's high-energy physicists were Indian in about 1965, only 2.4 percent were employed in India.

Where did the high-energy physicists publish their papers? It turns out, not unexpectedly, that there was a fairly distinct hierarchy, with some journals being judged more prestigious (or otherwise desirable) than others.The rank of the relevant HEP journals in the 1960s was this: *Physical Review Letters*; *Physical Review* (both by the American Physical Society); *Physics Letters* (European, Netherlands-based); and *Nuovo Cimento* (Italian Physical Society).

Experimental HEP quickly developed into the model collaborative science, characterized by many-scientist collaborations, and was almost never the work of individual physicists of the Rutherford or Curie type. This was, of course, a result of the increasing dependence on big science facilities, such as accelerators and bubble chamber detectors. By the late 1960s, it was not unusual for a team of twenty or thirty physicists to appear as authors of papers, and this was only a beginning. Another feature to note is the changing relationship between experiment and theory in HEP. The number of theorists grew more rapidly than the number of experimenters, and the two groups became increasingly separate. By 1968, 316 of 682 advanced American graduate students in HEP were theorists. In only very few cases did young theorists migrate to the experimental culture, or vice versa. Of the young physicists who published two or more papers on weak interaction physics about 1970, the approach of 94 percent in their second paper was the same as in their first paper (that is, either experimental or theoretical).

Chapter 22

FUNDAMENTAL THEORIES

QED

As MENTIONED IN chapter 13, the quantum theory of electromagnetic interactions—quantum electrodynamics, or QED—was the subject of much discussion in the 1930s. At that time, many physicists felt that it would not be possible to develop the conventional theory into a theory that was both theoretically satisfactory and empirically fruitful. As long as the infinite quantities that turned up in the applications of the theory could not be avoided, there was little hope of fundamental progress. The problems did not receive much attention during the war, when most physicists were busy with other matters, but after 1945 it was time to launch a fresh attack on the infinities. Although the breakthrough was not caused directly by new experimental results, theory and experiment went hand in hand in the important developments of 1947–48 that led to a new theory of QED.

Dirac's 1928 theory of the hydrogen atom reproduced the experimentally confirmed Sommerfeld fine structure formula and was generally considered beyond criticism. The theory ignored effects associated with the interaction of the electron with its own field, however, and experiments in the 1930s indicated that the structure of the hydrogen H_α line did not fit the theory's predictions exactly. There were several conjectures that the discrepancy, if it was real, was caused by deviations from Coulomb's law. In 1938 Caltech theorist Simon Pasternack suggested what ten years later became known as the Lamb shift, but at that time, the experimental situation was unclear and Pasternack's idea remained uncultivated. It was only after the war that a definite experimental proof of the spectral shift was given. Using a sophisticated microwave technique with which he had become familiar during wartime work, Willis Lamb at the Columbia Radiation Laboratory clarified the question. Together with his student Robert Retherford, he showed in the spring of 1947 that the two hydrogen states $2S_{\frac{1}{2}}$ and $2P_{\frac{1}{2}}$, supposed to have the same energy according to Dirac's theory, were separated by a small energy corresponding to the wave number 0.033 cm^{-1}.

The significance of the Lamb shift was immediately recognized by the participants at the Shelter Island Conference on "The Foundations of Quantum Mechanics" near New York in early June 1947. Among the participants in this important conference were Bethe, Weisskopf, Kramers, Pais,

Schwinger, Feynman, and also Lamb, who reported on his results. The theoretical significance of the Lamb shift was that it might be due to the electron's interaction with the radiation field and that it could thus serve as a guide to an improved theory of QED. That the Lamb shift was indeed a QED effect was first shown by Bethe, immediately after the conference. Bethe showed in a provisional calculation that the major part of the Lamb shift could be explained in this way. His calculation was based on a simple form of mass renormalization theory, an idea that had already been proposed before the war, by Kramers in particular. Bethe's calculation was not fully relativistic, and the next step was to extend and refine it by making use of a relativistically invariant approach. Such calculations were made by several physicists in 1947–49, including Lamb and Norman Kroll, Weisskopf and Bruce French, Schwinger, Feynman, and Tomonaga and his group in Japan. The result was a complete agreement with experiments and a confirmation of the renormalization technique.

The seminal contributions of Julian Schwinger to the new QED were associated with another delicate experiment, the measurement of the electron's anomalous magnetic moment. In 1947 a group of Columbia University physicists showed that the value of the magnetic moment was slightly larger than that predicted by Dirac's theory. In the same year, the twenty-nine-year-old Schwinger began developing his version of a consistent and covariant QED and applying it to calculate both the Lamb shift and the anomalous magnetic moment of the electron. Schwinger's theory was exceedingly complex mathematically, but out of the mathematical jungle came numerical values that agreed beautifully with experiments. For example, the electron's magnetic moment is given by the factor g, which according to Dirac's theory was exactly 2, but experimentally was known to be slightly larger. Whereas experiments in 1948 gave the value 2.00236 for the electron's g-factor, Schwinger found theoretically that $g = 2.00232$. Schwinger presented his covariant reformulation of QED in the spring of 1948 at a meeting at Pocono Manor, a sequel to the Shelter Island meeting. Published versions of the theory appeared in a series of papers in *Physical Review* between 1948 and 1951. In the first of these papers, a preliminary note on the magnetic moment calculations, Schwinger noted that in his new QED formalism, "the interaction between matter and radiation produces a renormalization of the electron charge and mass, all divergences being contained in the renormalization factors." Sam Schweber has summarized the significance of Schwinger's work as follows: "His method was general in the sense that it yielded a consistent quantum electrodynamics to order $e^2/\hbar c$, and exhibited a divergence-free Hamiltonian to order $e^2/\hbar c$ that could be taken as the starting point to describe quantum mechanically any system consisting of electrons, positrons, and photons in the presence of an external Coulomb field. What previously had been pieces of a theory became welded and unified into a consistent and coherent quantum electrodynamics to order α [$= e^2/\hbar c$]" (Schweber 1994a, 309).

Unknown to Schwinger and the other participants in the Shelter Island conference, a theory rather similar to Schwinger's had already been developed in Japan by Sin-Itiro Tomonaga and his colleagues. Inspired by some of Dirac's theories, Tomonaga developed his covariant mass-renormalization approach to QED during and shortly after the war, in part together with Ziro Koba and under very difficult external circumstances. It is telling that the Japanese physicists obtained their first information about the Lamb shift through *Newsweek*. After having contacted Oppenheimer (who had organized the Shelter Island conference and at the time acted as a QED clearinghouse), in 1948 Tomonaga and his collaborators published in *Physical Review* a summary of their work, including a calculation of the Lamb shift. Tomonaga expressed his view of renormalization theory as follows: "The mass and charge of the electrons which we could actually observe [are] the corrected quantities $m + \delta m$ and $e + \delta e$. Consequently, [even] though the theoretical values of $m + \delta m$ and $e + \delta e$ may be infinitely large, their actual values are finite. And consequently again, the infinity-difficulties in . . . [QED] calculations can be evaded by substituting the finite experimental values instead of the theoretical ones $m + \delta m$ and $e + \delta e$. . . . In a word, we may say that we have lumped all the infinity-difficulties into the self-energy of a free electron, and the problem of vacuum polarization" (Schweber 1994a, 271). The important work of Tomonaga paralleled to some extent that of Schwinger, but it occurred isolated from the creative phase of American QED and did not influence either Schwinger or Feynman.

The third version of QED was that of Richard Feynman, another brilliant young theorist. Feynman earned his doctorate under Wheeler in 1942, and his first scientific work dealt with a new formulation of classical electrodynamics based on direct interactions between particles. Feynman's original idea was first to solve the divergence problems in classical theory by means of his field-free interpretation, and then to hope that the problems would also disappear when the theory was carried over into quantum mechanics. Although it did not happen this way, his early work helped familiarize him with a general space-time point of view and later led him to the idea of formulating quantum mechanics in terms of amplitudes for paths. In 1947, as a result of the Shelter Island conference and Bethe's Lamb shift calculation, Feynman began developing his alternative, highly original theory of QED. The theory was based on the path integral approach and calculations aided by a diagrammatic technique that soon became known as Feynman diagrams. He developed the theory in papers between 1948 and 1951 and included the rules for the famous diagrams in his classic paper, "Space-Time Approach to Quantum Electrodynamics," which appeared in *Physical Review* in 1949. In this paper, he noted with satisfaction that although there were still theoretical problems, "[n]evertheless, it does appear that we now have available a complete and definite method for the calculation of physical processes to any order in quantum electrodynamics."

By the summer of 1948, there existed two different versions of renormalization QED, the Schwinger-Tomonaga formulation and the Feynman formulation. Although the two formulations gave the same results, the relation between them was far from obvious. The situation was not unlike the one between matrix mechanics and wave mechanics shortly after Schrödinger's theory had appeared twenty-two years earlier. That the two theories of QED were in fact equivalent was shown in 1949 in an important work by Freeman Dyson, a twenty-five-year-old English mathematician who had come to Cornell University two years earlier to work with quantum field theory. In 1948 Dyson had completed the calculation of the Lamb shift and recognized that the theories of Schwinger and Tomonaga were just two different formulations of the same physical theory. The following year, he derived Feynman's theory in his own way, formulating it for the first time as a field theory, and proved that the theories of Schwinger and Feynman were equivalent. With Dyson's synthetic theory, the new renormalization QED was essentially complete.

The importance of the events of the late 1940s that resulted in the new QED is illustrated by the fact that five Nobel prizes were awarded to physicists contributing to the development: Lamb and Polykarp Kusch received it in 1955, the latter for his precise determination of the magnetic moment of the electron, and Schwinger, Feynman, and Tomonaga received the prize in 1965; Although the renormalization QED was clearly the work of four physicists—Schwinger, Tomonaga, Feynman, and Dyson—these were not the only ones who worked in the field and made contributions to it. But contrary to the situation in the mid-1920s, when quantum mechanics was born, European physicists and institutions played no significant role in the formative phase of renormalization QED. Pauli, who had returned to Zurich from the United States in 1946, followed the development of the new theory closely but did not contribute to its formation. The only European who made original contributions to QED (except Dyson, who in this context counts as an American) was the eccentric Swiss physicist Ernst Stueckelberg who, in works around 1946, reached many of the results obtained later by Schwinger and Feynman. Stueckelberg's writings, most of them in French, were considered very difficult and unclear, and their value was recognized only in the 1950s, after his death.

The early history of postwar QED invites several general observations. First, Schwinger and Tomonaga, living in two very different cultural environments, independently produced almost the same theories of renormalized QED. Together with other examples from the history of science, this supports the hypothesis that the cognitive content of physics does not depend significantly on the sociocultural environment. Second, rarely had an important theoretical breakthrough been so conservative in nature as the renormalization QED of the late 1940s. It was not, in fact, a new theory in the sense of a theory replacing or negating the prewar quantum electrodynamics, but rather the old theory in an improved form. The young key physicists were well aware of the continuity, and strove consciously to keep to the principles of the established theory. In sharp contrast to the revolutionary attitudes of the older generation (people like Dirac, Bohr, and Heisenberg), the outlook of Schwinger, Feynman, Tomonaga, and Dyson was essentially conservative and pragmatic. They took quantum mechanics and special relativity for granted and asked how the two theories could be used to form a consistent and useful theory of QED. Third, the conservative attitude went hand in hand with attitudes that were pragmatic and basically nonphilosophical. To Feynman, physics was a matter of calculation and comparison between calculated and experimental results—no more and no less. Feynman agreed with Dyson's positivistic credo, "You can't really understand anything unless you can calculate it." In 1953, Dyson wrote about QED: "It is the only field in which we can choose a hypothetical experiment and predict the result to five places of decimals confident that the theory takes into account all the factors that are involved. Quantum electrodynamics gives us a complete description of what an electron does; therefore in a certain sense it gives us an understanding of what an electron is. It is only in quantum electrodynamics that our knowledge is so exact that we can feel we have some grasp of the nature of an elementary particle" (Schweber 1994a, 568).

THE UPS AND DOWNS OF FIELD THEORY

After the successful developments in QED from 1947 to 1949, hope was high that similar methods of quantum field theory (QFT) might also be applied to the other fundamental interactions. However, these hopes were not immediately fulfilled. In the words of Steven Weinberg, the eminent theorist, what happened in the 1950s was instead this: "[I]t was not long before there was another collapse in confidence—shares in quantum field theory tumbled at the physics bourse, and there began a second depression, which was to last for almost twenty years" (Weinberg 1977, 30). There were several reasons for the crisis that in the mid-1950s manifested itself in a widespread disenchantment with QFT in the high-energy physics community, despite the theory's brilliant success in QED. For one thing, a few physicists simply disliked renormalization QED because they felt it was theoretically ill-founded and hence could not be a truly fundamental theory. It was widely believed that renormalization was a mathematical trick, but most of the young physicists didn't mind as long the trick worked. Premier among those who did mind was Dirac, whose work in the 1920s and 1930s had been essential for any version of QED. Dirac felt strongly that a theory that got rid of the infinities by means of renormalization procedures was "ugly" and hence wrong, notwithstanding its instrumental successes. Another veteran of quantum mechanics, Lev Landau, was no less critical. Together with a group of Russian theorists, Landau argued that QFT was suspect because the theory built on unobservable concepts, such as causality, local field operators, and continuous space-time on the microphysical level. At a conference in Kiev in 1959, Landau suggested giving up QFT as a fundamental theory and replacing it with a theory based on observable quantities, such as scattering amplitudes, and the concept of equally elementary compound particles.

Landau's radical suggestion was nourished by the failure of extending QED-like field theory to the weak and strong interactions. Fermi's theory of beta decay, as well as the "V − A" theory that Feynman and Gell-Mann published in 1958 as a parity-nonconserving extension of Fermi's theory, proved to be non-renormalizable. Also, when physicists sought to apply QFT to the study of strong interactions, they met with little success. There were no particular problems in formulating renormalizable QFTs of strong interactions, but these theories were of practically no use because they did not lead to reliable predictions and thus were effectively beyond testing. For these and other reasons, QFT was at a low point in about 1960, when many physicists were ready to follow Landau's advice and abandon the theory. According to Dyson, "Many people are now profoundly skeptical about the relevance of field-theory to strong-interaction physics. Field theory is on the defensive against the now fashionable *S*-matrix." Dyson believed that "[i]t is easy to imagine that in a few years the concepts of field theory will drop totally out of the vocabulary of day-to-day work in high energy physics" (*PT*, June 1965, 21). Weinberg, another leading field theorist, was no more optimistic. In 1964 he wrote, "It is not yet clear whether field theory will continue to play a role in particle physics, or whether it will ultimately be supplanted by a pure *S*-matrix theory" (Cushing 1990, 160).

The lack of faith in QFT, especially with regard to the physics of strong interactions, was aggravated by the presence of a strong alternative theory, the *S*-matrix theory. The basis of this kind of theory was formulated by Heisenberg as early as 1943 in an attempt to develop a relativistic quantum electrodynamics without infinite quantities. Heisenberg's 1943 program was consciously modeled on his quantum mechanics of 1925 and, like this theory, based wholly in terms of observable quantities. As such quantities, he chose the scattering or *S* matrix representing the transition of a physical system from an initial state ψ_i to a final state ψ_f. Formally, $\psi_f = S_{fi}\psi_i$ where the square of the scattering matrix S_{fi} gives the transition probability. After the war, Heisenberg's theory was developed by Christian Møller, Ernst Stueckelberg, Walter Heitler, and others, but by 1950 the original program had largely come to a halt. With the advent of renormalization QED, Heisenberg's mathematically complex theory no longer seemed necessary. All the same, the *S*-matrix idea lived on and, from the mid-1950s, became incorporated in the theory of strong interactions and known as "analytic *S*-matrix theory." Some physicists, including Gell-Mann, considered *S*-matrix theory to be complementary to QFT or an alternative way of formulating a quantum field theory of the strong interactions. But other theorists, starting around 1960, took a much more radical position and soon came to view *S*-matrix theory as antithetical to QFT. Foremost among this group of anti-field theorists was Geoffrey Chew of the University of California, Berkeley. In a 1961 lecture, Chew outlined his ideas of a clean break with QFT and heralded a coming revolution in fundamental physics. "Conventional field theory," said Chew, "is sterile with respect to strong interactions and, . . . like an old soldier, it is destined not to die but just to fade away" (Cushing 1990, 143). Two years later, Chew expressed his belief by using another metaphor: "The new mistress [*S*-matrix theory] is full of mystery but correspondingly full of promise. The old mistress [QFT] is clawing and scratching to maintain her status, but her day is past" (ibid., 175).

The essence of Chew's *S*-matrix theory, also known as bootstrap theory, was developed from 1961 to about 1966. It can be summarized as follows: First, Chew denied that there were elementary particles in the ordinary, reductionist sense; that is, he considered all particles to be equally composite or, if one likes, equally elementary. This idea of *nuclear democracy* included the notion of a "bootstrap" mechanism, namely, that all hadrons should be self-generated from the mathematical structure of the theory; for example, the mass of the proton must have the value it does as a consequence of the dynamics of the interactions. (This feature is reminiscent of other highly ambitious theories, such as Eddington's.) Second, the theory was based directly on particle momenta. In stark contrast to field theory, it ignored the alleged metaphysical concept of a microphysical space-time continuum. Although the theory was fully consistent with quantum mechanics, it did not operate with a wave function $\psi(x, y, z, t)$ defined at each point in space-time. Third, from a methodological point of view, Chew's *S*-matrix theory was as ambitious as it was radical. According to the bootstrap hypothesis, all physically meaningful quantities could be derived uniquely from the *S* matrix as requirements of self-consistency. The masses, spins, and charges of the strongly interacting elementary particles, as well as the number of such particles, should be fixed by the theory rather than assigned empirically: "The bootstrapper seeks to understand nature not in terms of fundamentals but through self-consistency, believing that all of physics flows uniquely from the requirement that components be consistent with one another and with themselves. No component should be arbitrary" (Chew 1970, 23). In principle, Chew suggested, a theory like *S*-matrix theory would explain the laws of physics as the only possible laws. He even spoke of *S*-matrix theory as "the precursor of a new science, so radically different in spirit from what we have known as to be indescribable with existing language" (Cushing 1990, 180). However, Chew advocated a limited bootstrap hypothesis, that is, one valid for hadrons only. He denied that the bootstrap hypothesis could be extended to physics as a whole.

S-matrix theory was very influential in strong-interaction physics in the 1960s. It appealed philosophically to many physicists and scored several scientific successes. Although very few physicists followed Chew all the way in his anti-QFT crusade, the general ideas of *S*-matrix theory, enjoyed great popularity. Gell-Mann and other physicists were inspired by, and contributed to, *S*-matrix theory but without subscribing to the idea of nuclear democracy so obviously in conflict with the quark concept. To them, there was no irreconcilable conflict between QFT and a moderate version of *S*-matrix theory. Chew's *S*-matrix program is an interesting example of a failed revolution in physics. By the late 1960s, it ran into trouble and with the successful developments in quark physics and gauge field theory in the early 1970s, the ambitious program more or less dissolved. The *S*-matrix theory was in many ways a grand and impressive theory, but it promised more than it could deliver. It soon became so hopelessly complex that more than one physicist in the late 1960s compared it with Ptolemy's astronomy shortly before the Copernican revolution. Another problem was that the theory was effectively limited to strong interactions and had nothing to say about the weak and electromagnetic forces. As one critic argued, *S*-matrix theory "is inconsistent with the existence of the electromagnetic field. Because the electromagnetic field is involved in our very means of measurement and observation, this is a serious shortcoming" (Stern 1964, 43). As interest in QFT faded in the 1950s, so interest in *S*-matrix theory faded in the 1970s. In neither case was the shift a result of the theories being proved wrong, but rather lay in their inability to generate new experimentally relevant results—an inability that, in the case of QFT, was temporary. It is ironic that Chew's 1961 comparison of QFT with an old soldier who "is destined not to die but just to fade away" should become valid, not with respect to QFT but with respect to *S*-matrix theory itself.

Gauge Fields and Electroweak Unification

In the early 1970s, high-energy physics experienced a stormy development that transformed the subject into a state sometimes referred to as "the new physics." The novelty of the particle physics of the 1970s was based mainly on two theoretical innovations, gauge theories of quantum fields and quantum chromodynamics. Whereas the first kind of theory was a general field theory with applications to weak interactions in particular, quantum chromodynamics was an extension of the quark theory of strongly interacting or hadronic particles.

The concept of gauge (or scale) invariance was originally introduced by Herman Weyl in his failed 1918 attempt to unify the theories of electrodynamics and general relativity. When quantum mechanics arrived, Weyl and others reformulated the idea to mean, briefly put, that the choice of phase of a wave function does not affect the wave equation. Weyl's theory of quantum gauge invariance appeared in 1929, but played a very small role in the subsequent development of quantum theory. It took twenty-five years until the concept was turned into a powerful dynamic principle. In 1954, Yang and Robert Mills at the Institute for Advanced Study in Princeton attempted to construct a locally gauge-invariant field theory of strong interactions closely modeled on QED. The Yang-Mills theory was governed by a gauge symmetry principle, which secured that the basic equations were invariant with respect to certain transformations depending on position and time. Yang and Mills found that a triplet of heavy vector bosons (spin 1, positive parity) was associated with the theory, in the same way that the photon was associated with QED. Unfortunately, none of these particles was known experimentally. Because of the close analogy between QED and the Yang-Mills theory, the vector bosons were assumed to be massless, corresponding to a long-range force. Since the weak and strong forces were both of short range, it was not clear to which area of nature the theory was applicable, if it was applicable at all. The theory of Yang and Mills was considered mathematically interesting, but of little or no physical use. It took a decade until it was realized that this "useless" theory was, in fact, of basic importance to a new gauge-field theory tradition destined to change high-energy physics.

As early as 1938, at the Warsaw conference, Oskar Klein had suggested that a spin-1 particle mediated beta decay and played a role in weak interactions similar to that of the photon in electromagnetism. Klein's hypothesis was part of an attempt to formulate a unified field theory that included strong, weak, and electromagnetic forces, to use a later terminology. The speculation was scarcely noticed, but almost twenty years later, it was taken up by Schwinger in an attempt to produce a unified theory of the weak and electromagnetic interactions. Schwinger's tentative theory of 1957 included the idea that the photon and the two charged Yang-Mills exchange particles (W^+, W^-) were members of the same family. Four years later, Sheldon Glashow, another Harvard physicist, suggested a gauge theory of weak interactions with three massive exchange particles. In addition to the W^+ and W^-, it included a neutral Z^0 particle. Glashow's theory, and a similar one produced by Abdus Salam and John C. Ward in England, described both electromagnetic and weak interactions, the first with parity conserved and the second with a violation of parity conservation. The masses of the intermediate vector bosons were not derivable from theory, but chosen to ensure agreement with experimental knowledge. In fact, the theory was unable to explain why the vector bosons, contrary to the photon, had mass at all.

An improved version of the Glashow-Salam-Ward model was developed in 1967, published first by Steven Weinberg and slightly later by Salam, a Pakistani theorist working at Imperial College, London. The Nobel prize-winning electroweak theory of Weinberg and Salam made use of a concept known as spontaneous symmetry breaking which originally, in 1961, had been introduced in the context of superconductivity theory by Yoichiro Nambu and others. Inspired by Nambu's work, in 1964 the British theorist Peter Higgs suggested a mechanism by means of which particle masses in Yang-Mills gauge theories could be generated. A similar suggestion was made independently by Robert Brout and François Englert in Belgium. Weinberg and Salam applied the Higgs mechanism to determine the particle masses in their electroweak, spontaneously broken gauge theory. Higgs's paper on mass-generating spontaneous symmetry breaking, now recognized as a landmark paper in theoretical physics, was first rejected by *Physics Letters*. It came at a time when particle physics was heavily dominated by *S*-matrix theory and QFT was out of fashion. "Realizing that my paper had been short on salestalk," as Higgs later put it, he reworked the paper and submitted it to *Physical Review Letters* (Hoddeson et al. 1997, 508). This time, it was accepted.

In the 1970s the Weinberg-Salam theory became recognized as a pioneering contribution to the nascent unification program, but during the first years, the theory was largely ignored. The early lack of impact is illustrated by the number of citations, here given in parentheses, to Weinberg's paper: 1967 (0), 1968 (0), 1969 (0), 1970 (1), 1971 (4), 1972 (64), 1973 (162), 1974 (242). The figures include self-references, and thus indicate that even Weinberg himself did not find his work to be particularly important at first. The change in reception that occurred in 1972 was related to the recognition that the Weinberg-Salam theory was not only a coherent electroweak gauge theory, but it was also renormalizable and thus shared with QED the desirable properties of predictability and calculability. This was first shown in 1971 by the twenty-five-year-old Dutch physicist Gerardus t'Hooft, who proved that a broad class of (non-Abelian) gauge symmetric theories, to which the electroweak theory belonged, were renormalizable. Another version of the proof was produced slightly later by Benjamin Lee, a Korean-American physicist. The proofs of t'Hooft and Lee were extremely important. Only then was the Weinberg-Salam theory turned into a workable theory and, more generally, gauge field theories recognized to be profound theories of interactions. As the physicist Sidney Coleman expressed it in a paper on the Nobel prizes awarded to Glashow, Weinberg, and Salam, t'Hooft's paper "revealed Weinberg and Salam's frog to be an enchanted prince" (*Science*, 14 December 1979). The "gauge field revolution" that started in 1971 gave rebirth to QFT.

The Weinberg-Salam theory was renormalizable and theoretically appealing. Was it also correct in the sense of agreeing with experiments? Like the earlier Glashow model, the theory predicted a massive, neutral partner of the photon that would mediate weak interactions without charge transitions between incoming and outgoing particles. Such "neutral current" processes, involving the hypothetical Z^0 as a carrier particle, distinguished the Weinberg-Salam unified model from earlier theories of weak interactions. Until neutral currents had been established experimentally, the Weinberg-Salam theory could not be considered an empirically satisfactory theory. The neutral currents were sought both in the United States (Fermilab) and Europe (CERN). In the summer of 1973, CERN experimenters announced their discovery of the weak neutral current based on analysis of processes between neutrons and neutrinos. The crucial instrument in CERN's success was "Gargamelle," a French-built twenty-ton bubble chamber. The CERN results were eventually confirmed by neutrino experiments made at Fermilab in 1974, although the confirmation did not come easily. There was considerable doubt about the results and, for a period, American physicists—and with them the majority of high-energy physicists—believed that neutral currents did not exist and that the CERN physicists had made a mistake. Yet, by the summer of 1974, the uncertainties had disappeared and from that time, neutral currents were firmly established. The discovery of neutral currents amounted to a confirmation of the 1967 Weinberg-Salam gauge theory.

The theory of Weinberg and Salam was a theory of electroweak interactions and, via the quark model, it had an obvious relation to the strong interactions. At first, from about 1967 to 1974, it seemed that the four leptons (e, ν_e, μ, ν_μ) nicely mirrored the four quarks (u, d, s, c), in agreement with what one would expect from a unification point of view. In 1974, Martin Perl and his collaborators analyzed data from electron-positron collisions collected at SPEAR, the recently completed electron-positron storage ring, or collider, at the Stanford Linear Accelerator Center (SLAC). The Stanford group found evidence of a new particle, possibly a superheavy lepton (tau, τ) produced in pairs and decaying into muons or electrons (for example, $e^+e^- \rightarrow \tau^+\tau^- \rightarrow \mu^-\mu^+\nu_\tau\bar{\nu}_\tau\ \nu_\mu\bar{\nu}_\mu$). The data analyzed by Perl and his group were not immediately accepted as proof of the existence of a τ lepton and as late as the end of 1975, Perl and his 35 collaborators wrote cautiously that they had "no conventional explanation" for the data; as an unconventional explanation, they suggested "the production and decay of a pair of new particles, each having a mass in the range of 1.6 to 2.0 Gev/c^2" (Cahn and Goldhaber 1989, 300). They knew they had discovered something new, but were not sure what it was. As an indication of their uncertainty, they sometimes referred to the new particle as *U*, for "unknown." (Incidentally, forty years earlier, Yukawa had introduced another U particle, later named the pion.) The confusion ended a year later with a series of experiments at Hamburg's PETRA (Positron-Electron Tandem Ring Accelerator), part of the DESY (Deutsches Elektronen Synchrotron) high-energy facility. The German experiments confirmed the Stanford results and provided final proof of the existence of a lepton as heavy as 1.8 GeV and with a lifetime of the order 10^{-13} seconds. The accompanying tau neutrino was assumed, rather than detected. The acceptance of the tau lepton (or tauon) implied that the symmetry between quarks and leptons—or between strong and weak interactions—was no longer satisfied. Two new quark flavors were needed and, as we shall see shortly, they were later found. During the late 1970s, quantum chromodynamics and QFT became closely associated with electroweak theory. Naturally, the challenging task was now to repeat the successes of QED and extend the electroweak Weinberg-Salam theory to the realm of strong interactions (see chapter 27).

In his previously mentioned review of 1979, Coleman described the development of electroweak theory through the 1970s in the following way:

> In 1973, experiments at CERN and Fermilab detected neutral current events . . . of a form and magnitude consistent with the theory. The next 5 years were a confusing period of exhilaration and disappointments, alarms and excursions. Experiment confirmed the theory; experiments denied the theory. Enormous theoretical effort was devoted to producing grotesque mutant versions of the theory consistent with the new experimental results; the new experiments were shown to be in error; the mutants were slain. In the last few years though, the experimental situation seems to have stabilised in agreement with the original 1971 version of the theory. The Weinberg-Salam model is now the standard theory of the weak interactions.

The "mutants" mentioned by Coleman referred to theories designed to explain anomalies that could not, at first, be brought in agreement with the Weinberg-Salam theory.

The success of the electroweak theory was incomplete insofar as the three "weakons"—the intermediate vector bosons—were still hypothetical particles. This situation changed in the early 1980s, when new proton-antiproton colliders at CERN and Fermilab produced the energies necessary to generate the massive particles. The CERN experiment was relatively low-cost, based on a technology that made use of existing facilities for storing antiprotons in a ring and making them collide with protons in the same ring. The CERN collider started producing collisions in the summer of 1981 and in the fall of 1982, the first indications of massive W particles appeared. On January 21, 1983, CERN physicists announced that they had found ten candidates for W particles decaying into an electron and a neutrino. The W mass was found to be about 80 GeV, in agreement with the theory. Half a year, later the Z particle was detected through its decay into e^+e^- and $\mu^+\mu^-$, and was assigned a mass of about 95 GeV. None of the discoveries were great surprises. They were quickly confirmed in further experiments at CERN and Fermilab, and CERN physicists Carlo Rubbia and Simon van der Meer were speedily awarded the 1984 Nobel prize for the discovery. The Italian Rubbia was the driving force behind the experimental program and the pioneer of the proton-antiproton collider; the essential technology of producing the projectiles in concentrated beams was developed by van der Meer, a Dutch technical physicist. There were, of course, many more physicists involved in the discovery than the two prize winners. The scale of modern high-energy physics is illustrated by the fact that the two discovery papers (one from each of the detector groups named UA1 and UA2) were authored by no fewer than 126 physicists from eleven different institutions.

Quantum Chromodynamics

As mentioned in chapter 21, in 1964 Gell-Mann and Zweig proposed that hadrons (strongly interacting particles) consist of fractionally charged particles called quarks. In the mid-1960s, the quark model did not arouse great interest, and it took most of a decade until it came to occupy a central position in strong-interaction physics. Experiments performed in 1967 by physicists from SLAC and MIT with inelastic scattering of electrons on protons gave results that confused the theorists, until Feynman interpreted the unusual scattering cross sections as an indication that the proton contained pointlike scattering centers. Feynman, who first published his theory in 1972, suggested that the proton, as well as other hadrons, consisted of an indefinite cloud of hard point-particles, which he called partons. In high-speed collisions, the partons would essentially act as independent particles. Feynman's parton model provided a convenient framework for understanding many experiments and was soon generally accepted. Many physicists tended to identify the partons with the quarks and disregard the considerable differences that existed between the two pictures of hadrons. For example, whereas the quarks were tightly bound within the hadrons, Feynman's partons were essentially free entities. The parton = quark hypothesis received support in 1971, when experiments at SLAC were interpreted in terms of partons with spin one-half. Still stronger support came from neutrino-proton scattering experiments at CERN, where the new Gargamelle bubble chamber produced convincing evidence for the parton model. The results published by the Gargamelle group in 1973–75 were generally accepted as proof that partons were fractionally charged quarks, a conclusion that was also supported by experiments with hadron reactions.

Further support for the parton model came from developments in theory. In 1973, American theorists David Politzer, Frank Wilczek, and David Gross discovered that gauge field theories of the Yang-Mills type were "asymptotically free," meaning that at very short distances (or very high energies), the strength would gradually decrease and tend asymptotically toward zero. This important result explained how Feynman's model with the free partons could be so successful: It was consistent with all known facts of hadron physics, and it made detailed calculations possible. In general, asymptotic freedom gave strong support to the validity of QFT and paved the way for the quark gauge field theory. In the original quark model, the quark was characterized by the "flavor" quantum number, which could be either *u*, *d*, or *s*. That a second quantum number was desirable was suggested later in 1964 by Oscar Greenberg at the University of Maryland. Greenberg pointed out that the quark composition of some elementary particles, such as the Ω^- (*sss*), did not agree with the Pauli principle if the three quarks were identical. His suggestion was developed the following year by Nambu and other gauge theorists. According to Nambu, quarks carried "color" in addition to flavor, so that each quark flavor related to three colors ("red," "green," and "blue"). The color was seen as an analogue to the electric charge, but was not expected to be of any relevance to the known hadrons, which were regarded as colorless. For this reason, the quark color was, for a time, considered to be of theoretical interest only. Experiments in the early 1970s showed, however, that this was not the case. The experiments indicated that color did exist, and that colored quarks were fractionally charged in the same way as flavored quarks.

The status of quantum chromodynamics—the gauge field theory of strong interactions—changed drastically in the fall of 1974, at a time when the term quantum chromodynamics and its acronym QCD (in analogy with QED) had not yet been coined. (The names seem to have been proposed by Gell-Mann and appeared first in 1978.) On November 11, 1974, two groups of American physicists announced that they had discovered a highly unusual elementary particle, which was considered to be a manifestation of the quark flavor "charm." That such a flavor should exist had been suggested by Glashow and James Bjorken shortly after the development of the original quark theory of Gell-Mann and Zweig. The charmed quark would have two-thirds of the positron's electrical charge, and whereas charm (like strangeness) would be conserved in strong and electromagnetic interactions, it would not be conserved in weak interactions. However, the Glashow-Bjorken hypothesis was ignored for several years because it lacked experimental support. The discovery of charmed quarks in the "November revolution" not only confirmed the quark theory, but also meant a great triumph for gauge theory in general and quantum chromodynamics in particular.

Data on electron-positron annihilation from 1972–73 disagreed with theoretical expectations and were widely seen as a threat against the parton-quark model. But the uneasy situation changed abruptly, from a threatening failure to a complete success, when the J/ψ particle was discovered in November 1974. The particle was detected in proton-nuclei collision experiments at Brookhaven by a group led by Samuel Ting, and in electron-positron collision experiments at Stanford's SPEAR by a group led by Burton Richter. Ting's group had evidence for the heavy meson several months earlier, but was not sure if the evidence was real or due to some artifact. Only when Ting heard about the SPEAR results did he realize that he had made a discovery. Ting and Richter shared the 1976 Nobel prize for the discovery of the particle that Ting called J and Richter ψ—hence it is generally known under the composite name J/ψ. The particle was unusually long-lived and, with a mass of 3.1 GeV (three times that of a proton), unusually massive. It was produced by the gamma quanta from electron-positron annihilation and decayed into hadrons. A few weeks later, another new vector meson (ψ') of mass 3.7 GeV was found and other massive, long-lived particles belonging to the same family followed during the next years. The reason that the J/ψ was considered so important was that it, and the other new particles, were seen as manifestations of the new quark flavor called charm. For example, the J/ψ was seen as a combination of a charmed quark and its antiquark. To put it briefly, charm was discovered in the November revolution and the discovery turned quantum chromodynamics and the quark model into realities. It amounted to a discovery of (confined) quarks as real constituents of hadrons. The existence of charm was further corroborated when D-mesons were discovered in 1976; these were interpreted as states of naked charm, that is, combinations of "ordinary" (d) quarks and charmed quarks. The large majority of physicists concluded that quarks had been discovered as real particles, not merely mathematical objects. Few listened to the remaining advocates of nuclear democracy and even fewer to the aging Heisenberg, who in the 1970s argued that a unified theory should be based on the underlying symmetries of the equations, rather than the notion of elementary particles.

By 1976, then, four types of quarks were recognized to exist: u, d, s, and c. With the discovery of the tau lepton and the recognition that it belonged to a "third generation" of fundamental particles, the quark-lepton symmetry was broken once again. In order to restore symmetry, two new quark flavors were needed. One of the desired flavors was promptly produced. In 1977 Leon Lederman of Columbia University announced that his group at Fermilab had found in $\mu^+\mu^-$ annihilations evidence for a very heavy particle of mass 9.5 GeV (almost ten times the proton mass). The particle, named upsilon (Y), was seen as a manifestation of a fifth quark flavor. As the 1974 experiments were interpreted as a discovery of charmed quarks, so the 1977 experiment was seen as the discovery of the b quark (b for "bottom" or "beauty") with electrical charge $-1/3$. The composition of the upsilon quark was interpreted as $b\bar{b}$. The sixth and last quark, the t quark (t for "top" or "truth"), did not turn up in experiments until much later, but in spite of the lack of experimental evidence, it was generally assumed to exist. By 1980 the number of accepted quarks had increased to six (u, d, s, c, b, and t), corresponding to the six leptons (table 22.1). To complete the story, a group of Fermilab physicists finally discovered the top quark in 1995. The extremely large mass of this quark, found in the 1995 experiment to be about 176 GeV, agreed with theory and the discovery was thus one more success of quantum chromodynamics and electroweak theory.

With the discovery of the new mesons and quarks, quantum chromodynamics came to dominate strong-interaction physics. According to this theory, hadrons consisted not only of quarks, but also of the massless "gluons" that kept the quarks together. As theoretical entities, the gluons were studied from about 1971, when some physicists interpreted evidence of neutral partons as the glue that held fractionally charged quarks together inside hadrons. The gluons acquired a measure of reality in 1979, when experiments at the new PETRA storage ring identified "three-jet events" that were interpreted as interactions involving quarks and gluons. The colliding electrons and positrons were thought to react according to the scheme $e^-e^+ \rightarrow q\bar{q}g$, where g denotes a gluon. Although European physicists emphasized that they had not really "discovered" the gluons, this was how the experiments were widely interpreted in the press, and not without the help of American physicists. The situation caused the British journal *New Scientist* to ask, "The evidence [for the existence of gluons] is therefore weak, so why

TABLE 22.1
Fundamental Particles as Known around 1995

Name(s)	*Symbol*	*Electrical charge (e)*	*Rest mass (MeV)*
Quarks			
up	u	+ 2/3	360
down	d	− 1/3	360
strange	s	− 1/3	540
charmed	c	+ 2/3	1500
bottom	b	− 1/3	5000
top	t	+ 2/3	176,000
Leptons			
electron	e	− 1	0.51
muon	μ	− 1	107
tauon	τ	− 1	1784
electron neutrino	ν_e	0	0
muon neutrino	ν_μ	0	0
tauon neutrino	ν_τ	0	0

Note: Antiparticles are not included.

have these results been hailed as so remarkable, particulary in the US?" The journal's answer was this: "The only conclusion seems to be that American particle physicists are trying hard to keep up the momentum for federal funding for their expensive form of research. The battle is already on for the next generation of accelerators, to go to higher energies, so physicists need to prove that future expenditure is well invested" (Pickering 1984a, 344). Yet, the discovery turned out to be real enough. By the end of 1979, further experiments at PETRA and other laboratories had confirmed the interpretation, and it was generally accepted that the gluon had been detected.

Chapter 23

COSMOLOGY AND THE RENAISSANCE OF RELATIVITY

Toward the Big Bang Universe

IF 1932 WAS THE *annus mirabilis* of nuclear physics and the beginning of particle physics, the year 1917 might be celebrated as the birth of rational cosmology. The scientific study of the universe at large was put on a new basis when Einstein suggested his cosmological field equations based on his recently proposed general theory of relativity. The title of Einstein's pioneering work, published in the *Proceedings of the Prussian Academy of Sciences* during the middle of World War I, was "Cosmological Considerations Concerning the General Theory of Relativity." Eighty years later, Einstein's work was still considered the foundation of scientific cosmology.

At the time that Einstein published his theory, observational knowledge of the statistics and motions of the galaxies was meager and the gap between theoretical and observational cosmology deep. Although the first measurements of galactic recession go back to 1912, and a relationship between the distance and the recessional velocity was suspected by some astronomers in the 1920s, in general observations played no great role in the first phase of relativistic cosmology. In this phase, between 1917 and 1930, it was generally accepted that the universe was static, and the main problem that occupied the few theoretical cosmologists was concerned with comparing the two static models satisfying the field equations. According to the Einstein model the universe was closed, whereas it was open and infinite (but devoid of matter) according to the model suggested in 1917 by Willem de Sitter in the Netherlands. It was only in 1930, after Edwin Hubble had established that the galaxies fly apart with a velocity proportional to their distance, that the static paradigm broke down and it was realized that the universe is expanding. This important insight had already been reached by the Belgian Georges Lemaître in 1927, and the possibility was theoretically argued by the Russian Alexander Friedmann as early as 1922. The works of Friedmann and Lemaître were ignored until they were rediscovered in 1930. During the next two decades, the majority of astronomers and physicists accepted that the universe was expanding in accordance with Hubble's data and the Friedmann-Lemaître solutions of the relativistic field equations. The most popular cosmological model was probably the Lemaître-Eddington model, according to which the universe started its expansion from a static Einstein state infinitely long ago. Another evolutionary model was suggested by Lemaître in 1931, namely, that the expansion started from a "primeval atom" and that the universe could therefore be ascribed a definite age. This first big bang model was initially received coolly, but in the late 1930s the general idea of an evolutionary, finite-age universe described by the laws of general relativity gained increasing respectability. The relativistic foundation was not accepted by all specialists, however, and there were numerous rival theories, of which Edward Milne's alternative was the most discussed (see chapter 15).

Cosmologists between the wars had no professional identity. They consisted of an uneven mix of mathematicians, theoretical physicists, astronomers, and physical chemists who worked part-time with cosmological problems. Although physical aspects of the universe played a subordinate role compared with the geometrical or space-time aspects, considerations of physical processes were not totally absent. For example, in a series of papers between 1928 and 1933, the American physical chemist and relativity specialist Richard Tolman investigated the thermodynamics of both static and expanding universes. As early as 1922, Tolman had studied the equilibrium between hydrogen and helium in an unsuccessful attempt to explain the relative abundances of the two elements. Quantum physics first entered cosmology, although in a vague and speculative manner, with Lemaître's brief note in 1931 on big bang theory, significantly titled "The Beginning of the World from the Point of View of Quantum Theory." The Belgian physicist pictured the original universe as "a unique quantum" in which "a kind of super-radioactive process" occurred with the production of super-heavy, radioactive elements as a result. Particles emitted by the hypothetical super-atoms would still be with us in the form of cosmic rays, he suggested. Lemaître's speculations were audacious, visionary, and poetic; unfortunately, they were also unconnected with the progress in nuclear physics that started accelerating at the time. The fruitful development that eventually established cosmology as a branch of physical science had its beginning in the late 1930s, when a few nuclear physicists turned to astrophysical problems, such as stellar energy production and the abundance distribution of elements.

Nuclear astrophysics was pioneered in the late 1920s by Atkinson, Houtermans, and Gamow; the subfield obtained its first striking success with Bethe's celebrated theory of 1938–39 (see chapter 12). Bethe's work was a theory of stellar energy production, not of element formation, and for this reason, it was not directly relevant to cosmology. The first nuclear-cosmological theory was included in von Weizsäcker's theory of 1938, a more primitive version of Bethe's slightly later theory. Von Weizsäcker greatly developed the nuclear-archaeological program—that is, the attempt to reconstruct the history of the universe by means of hypothetical nuclear processes and to test these by the resulting pattern of element abundances. Or, as he phrased it in the *Physikalische Zeitschrift*, "to draw from the frequency of distribution of the elements conclusions about an earlier state of the universe in which this distribution might have originated." According to von Weizsäcker: "It is quite possible that the formation of the elements took place before the origin of the stars, in a state of the universe significantly different from today's . . . [and with] a temperature of an order of magnitude of that which would come about through the complete transformation of the nuclear binding energy into heat [about 2×10^{11} K]. The accompanying density is likewise already in the neighbourhood of the density of the nucleus" (Kragh 1996b, 98). Von Weizsäcker's picture was thus that of a big bang universe. Although it was developed independently, his picture had much in common with Lemaître's primeval atom hypothesis.

The program initiated by von Weizsäcker was developed independently by Gamow into a nuclear-physically based model of the early universe. It is noteworthy that neither Gamow, Bethe, nor von Weizsäcker had any formal training in astronomy. Confident of the power of nuclear and quantum theory, they entered the field as physicists and learned whatever astronomy was necessary along the way. By 1939, Gamow had taken the step from the stars to the universe, entertaining the notion of a big bang universe but still with no idea of how the elements were formed from the hypothetical state of primeval hydrogen. The problem was to reproduce an element distribution corresponding to that known empirically, which at the time meant the data published by Victor Goldschmidt in 1937. At the eighth Washington Conference on Theoretical Physics, held in April 1942—four months after the United States had declared war on Japan, Germany, and Italy—Gamow and other American physicists reached the conclusion that a big-bang universe was necessary in order to account qualitatively for Goldschmidt's data for the heavier elements. According to the conference report, "the elements originated in a process of explosive character, which took place at the 'beginning of time' and resulted in the present expansion of the universe" (Kragh 1996b, 105).

The big bang picture was gaining momentum, and the driving force was nuclear physics. It was now realized among a small group of physicists and astronomers that the gross material of the present world was probably the result of what happened in a highly compressed and hot primeval state some two billion years ago (the currently accepted value of the Hubble time, roughly the age of the universe). This was a major conceptual change, but at that time it made no headlines. Not only was there a war going on, but the conclusion was also tentative and speculative. What it lacked in order to develop into a proper big bang cosmology was a connection between the nuclear physics of the early universe and relativistic models of cosmological evolution. During World War II, such esoteric problems had low priority, but shortly after the end of the war Gamow turned to the problem in earnest. On

October 24, 1945, Gamow congratulated Bohr on his sixtieth birthday. "[I am] studying the problem of the origin of the elements at the early stages of the expanding universe," Gamow told Bohr. "It means bringing together the relativistic formulae for expansion and the rates of thermonuclear and fission reactions. One interesting point is that the period of time during which the original fission took place (as estimated from the relativistic expansion formulae) must have been less than one millisecond, whereas only about one tenth of a second was available to establish the subsequent thermodynamical equilibrium (if any) between different lighter nuclei" (Kragh 1996b, 106).

Gamow's new approach, as presented in 1946, was based on a picture of the early universe as consisting of a relatively cold neutron gas expanding according to the Friedmann-Lemaître equations. A much improved version was developed in 1948, mainly in a collaboration with his Ph.D. student Ralph Alpher and by making use of new data on the reaction rates of neutron capture cross sections, which until then had been classified. The very early universe was now pictured as a hot, highly compressed neutron gas that somehow started expanding and decaying into protons and electrons. Some of the protons would combine with remaining neutrons to form deuterons and from these nuclei, heavier elements were assumed to be synthesized by successive neutron capture and beta decay. Neither in this nor in later versions did Gamow attempt to answer the question of what caused the expansion or the initial decay of the neutrons two billion years ago. Gamow and Alpher wanted to avoid questions about the origin of the universe, and simply took the starting conditions as given. They considered the beginning at $t = 0$ to be outside the realm of physics and, for this reason, they never used the term "big bang" for their theory.

Later in 1948, Gamow and Alpher realized that the primordial universe, of temperature about 10^9 K, must be dominated by radiation rather than matter. This affected the details of the calculations and, more importantly, led Alpher and his collaborator Robert Herman to conclude that fossils of the cooled primordial radiation should still be with us. In a brief paper in 1948, they calculated the present background temperature of the universe to be 5 K. Gamow, Alpher, and Herman reported the prediction of a cosmic background radiation seven times between 1948 and 1956, but in spite of its being well known, their result attracted no attention at all. Remarkably, it was ignored and eventually forgotten until it was revived in the mid-1960s, at a time when Gamow and his two associates were no longer active in cosmological research. All the same, the Gamow model of the universe continued to be developed and refined, primarily by Alpher and Herman but with occasional input also from other nuclear physicists, including such notables as Fermi and Wigner. The original assumption of an initial universe consisting only of neutrons turned out to be untenable, as first argued by the Japanese physicist Chushiro Hayashi in 1950. According to Hayashi, nuclear processes other than neutron decay had to be taken into account, which led him to suggest that the earliest universe consisted of a mixture of protons, neutrons, and photons. On this basis, Hayashi made a crude calculation of the present distribution between hydrogen and helium, with results in fair agreement with the rather uncertain observational data existing at the time. A still more refined model was developed by Alpher, Herman, and James Follin in 1953, in a paper that marked the zenith of the classical big bang theory. Making use of the most recent advances in nuclear and particle theory, the three physicists provided a detailed and comprehensive analysis of the early universe, starting at a time of 10^{-4} seconds after the initial explosion, when the temperature was about 10^{12} K. Among the results obtained by Alpher, Herman, and Follin was a present weight percentage of helium of about 32 percent, a figure that matched reasonably well with the one estimated from spectroscopic data.

In spite of the impressive advances made in big bang theory between 1948 and 1953, the theory was unsuccessful in attracting wide interest and was de facto abandoned for more than a decade. Why was this essentially correct theory, as we now think of it, disregarded until the mid-1960s? A theory as grand and ambitious as the Gamow-Alpher cosmology naturally faced problems, among which were its apparent inability to account for galaxy formation and, more seriously, its failure to explain the formation of the heavier elements. In order to build up elements heavier than helium, some way of bridging the gaps at mass numbers 5 and 8 (for which no nuclei exist) had to be found. The problem was that in spite of many attempts, no satisfactory solution—that is, one corresponding to the physical conditions of the early universe—was found. The failure seemed to imply that the original rationale of Gamow's theory, the cosmological formation of elements, had to be abandoned. On the other hand, the failure did not amount to a refutation of the theory, for it was quite possible to assume that although helium was produced cosmologically, the other elements were the later results of nuclear reactions in the interior of the stars. It is difficult to avoid the conclusion that the lack of interest in Gamow's big bang theory after 1953 was in part the result of sociological factors unrelated to the qualities of the theory itself. One of these reasons was undoubtedly that the theory had no clear disciplinary affiliation, but involved two fields of physics which, at the time, were seen as widely different. It was a physical rather than astronomical theory, but by combining nuclear physics with general relativity, it went against the trend of specialization that characterized American physics in the 1950s. The barrier separating the trend-setting nuclear and particle physicists from (what were considered to be) the dusty cosmologists implied that cross-disciplinary research programs such as Gamow's had difficulty recruiting new people and attracting interest from existing specialties. So when Gamow drifted from cosmology to molecular biology and other matters in the mid-1950s, and when Alpher and Herman at the same time moved to scientific careers in private industry, there was no one to take over where they left off.

Cosmology has traditionally been thought of as a study midway between science and philosophy and, for this reason, one that required a methodology different from that governing ordinary physical research. Gamow and his small group of nuclear cosmologists disagreed. Being physicists by training and spirit, they saw the early universe as something that could be dealt with by ordinary methods of physics. They considered it a difficult problem, but nonetheless a problem that did not differ qualitatively from other problems of nuclear physics. Confident that advanced computer-assisted calculations with input of nuclear-physical laboratory data would give the right answer, they saw no need to introduce new principles or discuss the conceptual state of cosmology at any length. In short, their attitude was pragmatic and empirical. Gamow conceived of himself as a cosmo-engineer, and on one occasion compared the cosmologist with the engineer designing a new car: As the engineer had to rely on known laws and materials, so should the cosmologist look for models of the universe satisfying known laws and agreeing with experimental data. This was an attitude strikingly different from that predominating in cosmology, and it possibly contributed to the alienation of Gamow's program from the kind of theoretical cosmology cultivated in Europe. On one hand, Gamow's cosmology-as-engineering approach agreed nicely with the pragmatic spirit that permeated American physics at the time (see chapter 22). On the other hand, his research program missed the one essential ingredient that might have attracted the physicists' interest, namely, experimental data and the possibility of testing.

The Steady State Challenge

1948 was not only the year of Gamow's big bang theory, but it was also the year when a radical alternative to relativistic evolution cosmology was proposed in Cambridge, England. This theory, the "new cosmology" or steady state theory of the universe, was suggested in two different versions, one by Fred Hoyle and the other by Hermann Bondi and Thomas Gold. Although the two versions differed in their philosophical outlooks, they shared the same foundation and led to the same observational results. Hoyle, Bondi, and Gold had in common with Gamow, Alpher, and Herman that they were physicists with no formal training in astronomy. This was about the only similarity between the two groups of cosmologists.

The steady state theory was motivated by a methodological dissatisfaction with the evolutionary cosmologies based on the theory of relativity, and especially with those assuming a beginning or creation of the universe. Hoyle pointed out that the creation could not be causally explained and so, as he wrote in his 1948 paper, these theories went "against the spirit of scientific enquiry." Apart from the philosophically based objections, the three Cambridge physicists stressed the so-called time-scale difficulty, a problem common to most relativistic theories of the big bang type. According to these theories, the age of the universe is related to the Hubble parameter in the sense that the age is smaller than the inverse constant of recession (the Hubble time), which in about 1950 was thought to be 1.8 billion years—considerably smaller than the 3 billion years that reliable radioactive dating methods indicated for the earth. This embarrassing discrepancy disappeared later in the 1950s with improved measurements of the Hubble parameter, but in 1948 it was real enough and a problem that relativistic cosmology could avoid only by pretending that it did not exist.

Hoyle's, Bondi's, and Gold's solution was to base their alternative on the postulate that the universe is not only spatially but also temporally homogeneous, that is, it looks the same at any location and at any time. In order to make a stationary and infinitely old universe agree with the recession of the galaxies, they assumed that elementary matter (such as hydrogen atoms or neutrons) is created continually throughout the universe. The creation of matter had to take place at such an exceedingly slow pace that the process was impossible to observe directly, but it was nonetheless a drastic hypothesis because it violated both the time-honored principle of energy conservation and the respected general theory of relativity. Building on these assumptions, the steady state cosmologists deduced that the universe was an exponentially expanding Euclidean space with a constant mean density of matter given by $\rho = 3H^2/8\pi G$, where H is the Hubble constant and G Newton's constant of gravitation. Among the other deductions following from the theory was that the average age of galaxies in any large region of the universe was one-third of the Hubble time, or about 600 million years. The theory remained essentially unchanged during the 1950s, except that the British physicist William McCrea in 1952 reinterpreted it in close analogy with the general theory of relativity and argued that the continual creation of matter did not necessarily violate energy conservation. McCrea's ingeneous interpretation included important insights that would later be rediscovered in relativistic cosmology, but at the time it attracted little interest.

From its beginning in 1948, the steady state theory attracted fierce opposition, not only because of its unconventional nature, but also because Hoyle used it ideologically in attacks on the big bang theory and what he claimed was a religious foundation of this view. That there was an unholy alliance between Christian belief and cosmological ideas of the creation of the universe seemed confirmed in 1952, when Pope Pius XII argued that modern big bang cosmology was in deep harmony with Christian dogmas and provided strong support for the existence of a transcendental Creator. Although religious and philosophical arguments were hotly debated, however, the course and outcome of the controversy between the two cosmological theories did not depend on these themes. What really mattered were arguments of a more convential scientific nature, which first and foremost meant observational tests. The steady state theory scored minor victories in providing plausible mechanisms of galaxy formation and, in particular, the basis for a successful nuclear theory of the stellar origin of elements heavier than helium. This important theory, a milestone in nuclear astrophysics, was developed in 1956–57 by Hoyle in collaboration with Margaret and Geoffrey Burbidge from England and the Caltech nuclear physicist William Fowler. The Burbidge-Burbidge-Hoyle-Fowler theory built on a process of bridging the mass-8 gap that would work only in the interior of stars, and thus contradicted the assumptions of Gamow's big bang theory. However, although the B^2HF theory (as it was called) was widely seen as an argument for the steady state theory, it did not rule out the rival big bang theory. The victory that Hoyle scored was more psychological than real.

Whereas evolutionary cosmologies predicted that the rate of galactic recession was proportionally larger for distant galaxies, according to the steady state model the velocity would increase in direct proportion to the distance. The Hubble diagram for very distant galaxies should thus be able to discriminate between the two kinds of models. Allan Sandage at the Mount Wilson Observatory collected data that indicated a slowed-down expansion in agreement with the evolutionary view and concluded that the data contradicted the steady state theory. However, the data were not certain enough to constitute a crucial test that would be accepted by both sides.

The general situation in the late 1950s was characterized by the inability of the different observational tests to discriminate clearly between the two theories. Some of the tests seemed to favor one kind of theory, others to favor the other, but none of them were decisive. When radio astronomical methods were applied to the same problem, first in 1955 by Martin Ryle of Cambridge University, the story at first seemed to repeat itself. Ryle found that the distribution of radio sources with intensity disagreed with the steady state theory, but his conclusion was premature and contested by radio astronomers in Australia. It was only in 1960 that new measurements resulted in a consensus among radio astronomers that the distribution of radio galaxies clearly contradicted the steady state theory. This did not immediately lead to the theory's fall, however, and for a couple of years, Hoyle and his collaborators attempted to avoid the conclusion, either by reinterpreting the data or by inventing new versions of the steady state theory that agreed with the radio source counts. From that time onward, however, the steady state theory was no longer considered a serious alternative by the majority of astronomers.

The steady state theory was not only different from the big bang theory in the conventional sense of offering a different picture of the universe and leading to different predictions. The two theories also differed markedly in a philosophical and sociological sense. They represented contrasting "styles" in cosmology and, indeed, in science as such. Bondi, Gold, Hoyle, and their followers denied that cosmology was simply a special case of physics and they stressed the difference between terrestrial physics, with its repeatable and law-governed phenomena, and the science of the universe. Because of the uniqueness of the concept of the universe, cosmology was held not to be an explainable, but only a describable, science. Moreover, an understanding of the universe needed to be based on principles not derived from local physics, and if these led to consequences in contrast to accepted knowledge, the Cambridge physicists were willing to sacrifice the absolute validity of laws of nature. The steady state theory was an attempt to revolutionize cosmic physics, but the attempt failed. It is no accident that steady state cosmology emerged in England and that it was only in this country that the theory found widespread support and aroused serious discussion. In some respects, the spirit of the steady state theory was a continuation of the a priori theories of cosmophysics that were popular in British physics and astronomy in the 1930s and that we described in chapter 15.

Cosmology after 1960

Although the radical revolution envisaged by the steady state theorists failed, a kind of revolution in cosmology did occur in the 1960s. But it was a conservative revolution that built on established physics and had strong roots in the past. To speak of a "renaissance" rather than a "revolution" may be more appropriate. The change was associated with, and depended on, a number of spectacular discoveries that demonstrated the existence of new objects and phenomena of cosmological relevance. These discoveries depended again on the rapid development of instruments and technologies such as radio- and microwave methods, rockets, and artificial satellites. Even the heavenly science of cosmology was deeply influenced by instruments and methods originating in a military physics context. As the unexpected discoveries in the 1890s led to a new picture of the microcosmos, so did unexpected discoveries in the 1960s lead to a new picture of the macrocosmos—although, in the latter case, it was more the completion of an already known picture than the founding of an entirely new one.

Among the new revelations were that the sky is filled with x-rays, both in the form of discrete sources and as diffuse background radiation. This discovery, made in 1962 by the American Riccardo Giacconi and his collaborators, became the starting point of an important new branch of astronomy. More important for cosmology was the discovery of quasistellar objects that the Caltech astronomers Maarten Schmidt and Jesse Greenstein made in early 1963. They found a number of starlike objects that differed from normal stars and galaxies by their unusual spectra, varying intensity, enormous output of radio energy, and very high redshifts. It is interesting to note that quasars, as the objects were soon named, had been observed on photographic plates years before it was realized that they were an entirely new kind of radio objects. In this respect, the case may remind us of other discovery episodes in the history of physics, such as the discovery of x-rays and the discovery of the positron (chapters 3 and 13). These cases illustrate the obvious point that recording something is not identical to discovering something. Quasars became immediately fashionable, as well as controversial, objects of study among astrophysicists. What physical processes could account for their gigantic energy output? Did their redshifts indicate that they were cosmological objects, or could they otherwise be explained? It was soon agreed that quasars were indeed at great cosmological distances, which implied that they were probably at variance with the steady state theory, according to which objects that existed only very far away and long ago were prohibited. In order to settle the question, astronomers devised a test by mapping the redshifts of quasars against their flux densities. Results published in 1966 contradicted the distribution predicted by the steady state theory. Although it was possible to explain away the contradiction, as Hoyle preferred, most astronomers and physicists accepted it as a genuine refutation of the steady state theory.

At that time, the theory had already been shaken by the even more important discovery of the cosmic microwave background radiation. As mentioned previously, as early as 1948 Alpher had predicted such a blackbody radiation of temperature 5 K, but the prediction was effectively forgotten. Sixteen years later, the Princeton physicist Robert Dicke reached a similar conclusion independently and suggested to his colleague James Peebles that he examine the question. Peebles estimated the present temperature of the hypothetical background radiation to be 10 K and in the spring of 1965, Dicke and Peebles started a collaboration with experimenters to detect the radiation. Before they had obtained any results, they learned about experiments performed by two physicists at Bell Laboratories, Arno Penzias and Robert Wilson. The two AT&T physicists had used a radiometer to measure signals from the Milky Way and in the course of their measurements, they realized that there was a systematic excess temperature in their antenna of 3.3 K, independent of the direction in which the antenna pointed. Penzias and Wilson could not explain the excess temperature, but Dicke and Peebles immediately realized that what had been detected was, in fact, the microwave background from the big bang. The discovery, soon confirmed at other wavelengths, was announced in the July 1965 issue of the *Astrophysical Journal*. Here was a discovery that effectively undermined the steady state theory and gave solid support to the big bang theory, which predicted this kind of radiation exactly. To most physicists and astronomers, the discovery of the microwave background was an *experimentum crucis*, indeed, a proof that the world had started in a big bang. The 3 K radiation—its temperature is now measured as 2.735 K—is still considered the most impressive argument in favor of the big bang theory. After the discovery in 1965, it was realized that the cosmic background radiation was important in calculations of the helium abundance, a problem that could not be solved by theories of stellar element formation. Based on big bang assumptions and a microwave background of 3 K, in 1966 Peebles found a helium abundance of 27 percent, in excellent agreement with what had been estimated from observations. This was yet another triumph of the new big bang cosmology.

Penzias and Wilson received the 1978 Nobel prize for their discovery, in spite of the fact that they had not actually discovered the radiation—that is, identified it as of cosmic origin—but had only detected something for which they could not account. At the presentation speech, the head of the physics Nobel committee said that the discovery had turned "cosmology [into] a science, open to verification and observation." The feeling that a truly scientific or physical (as opposed to mathematical) cosmology emerged only with the events of 1963–66 was widespread in the late 1960s, and it is still part of the working history of cosmologists. According to this history, events prior to about 1963 belong to a prescientific, semimythical stage, governed by either sterile mathematics or superstitious ideas, such as continual creation of matter. Apart from the obvious falsity of this version of history, it is important to note that the big bang picture was not a result of the mid-1960s, but can be found fully developed in the earlier works of Gamow and his collaborators.

This is not to deny that cosmology experienced a kind of extra takeoff in the 1960s, both scientifically and socially. For example, the annual number of scientific articles on cosmology increased between 1962 and 1972 from about 50 to 250 (figure 23.1). Even an annual output of 250 papers is a small number for a scientific subdiscipline, however. The modest scale of cosmology is further illustrated by the fact that by 1972, fewer than 0.3 percent of all publications abstracted in *Physics Abstracts* were within the entries "cosmology" or "cosmogony."

More important than the number of publications, with the elimination of the steady state rival there appeared in the mid-1960s a consensus among cosmologists with respect to the main problems to be solved and the criteria to be used. The "hot" big bang relativistic theory obtained a paradigmatic status, and alternative interpretations were marginalized. At the same time that cosmology became cognitively institutionalized, it achieved a social institutionalization that made the subject a full-time professional occupation with an increased scientific respectability. There was a growing integration of the subject into university departments, and not only in the mathematics

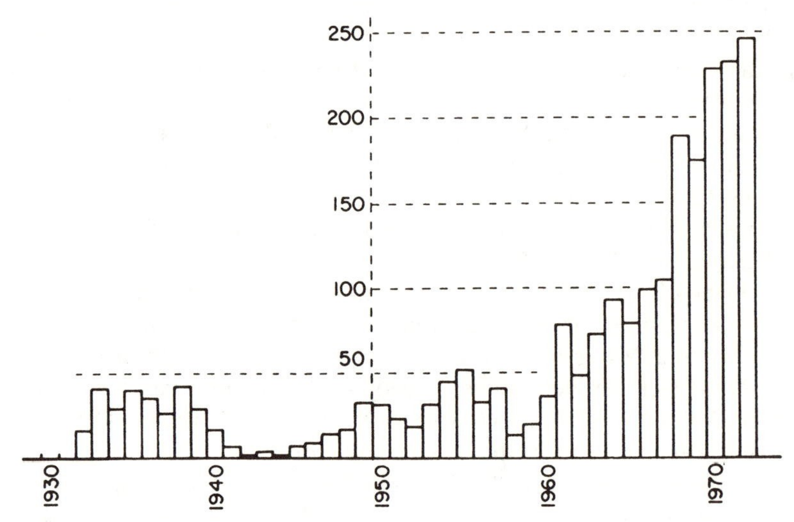

Figure 23.1. The growth of cosmology, as indicated by the annual number of publications listed under "cosmology" and "cosmogony" in *Physics Abstracts*. *Source:* Reprinted with permission from M. P. Ryan and L. C. Shepley, "Resource letter RC-1: Cosmology," *AJP* (1976) 44: 223–230. © 1976 American Association of Physics Teachers.

departments, where the few courses in the 1950s had typically taken place. From the 1970s, cosmology became increasingly taught in departments of astronomy, physics, or space sciences, and students were brought up in a research tradition with a shared heritage. The national differences that had characterized earlier cosmology also disappeared. Originally, big bang theory had been an American theory, steady state theory belonged to the British, and the Russians had hesitated to do cosmology at all. Now, the field became truly international. It was no longer possible to tell an author's nationality from the cosmological theory he or she advocated.

The collaboration between nuclear physics and cosmology that started with Gamow in the 1940s accelerated in the 1970s, when elementary particle physics became an important ingredient in the new cosmology. For example, detailed calculations made in 1977 by Gary Steigman, David Schramm, and James Gunn at the University of Chicago showed that the number of different neutrinos could not be larger than three, or possibly four, if the hot big bang theory was correct. The prediction was later confirmed by high-energy accelerator experiments and served to increase confidence in the basic correctness of the big bang model. The most remarkable contribution of particle physics to modern cosmology was the inflationary theory that the 31-year-old American physicist Alan Guth suggested in 1981. (A form of inflation had been proposed by the Russian physicist Alexei Starobinsky in 1979, but had failed to attract attention.) Based on the concept of "false vacuum," Guth devised a model according to which the very early universe underwent extreme supercooling and expanded suddenly by a gigantic factor—an increase by a factor of 10^{30} within a period of 10^{-30} seconds. After the initial explosion, the expansion slowed down in agreement with the standard big bang theory. In 1982 Guth's theory was improved by Andrei Linde in the Soviet Union and, independently, by Andreas Albrecht and Paul Steinhardt in the United States. The new inflationary universe model explained, among other things, the large-scale homogeneity of the universe, the absence of magnetic monopoles, and the near-flatness of space, neither of which could be explained by the standard theory. Although the inflationary model is not unproblematic, it has been highly successful and caused a major change in cosmological thinking. By 1997, more than 3,000 papers had been published on the inflation theory. We shall follow up on a few more aspects of the particle physics-cosmology interface in chapter 27.

The Renaissance of General Relativity

By the mid-1920s, Einstein's general theory of relativity was reasonably well confirmed and accepted by most physicists. However, only a handful of physicists and mathematicians worked with the theory as a research topic; the large majority found general relativity to be as experimentally empty as it was mathematically abstruse. Whereas the special theory of relativity was used routinely in atomic and particle physics, general relativity seemed to have almost no connection with experiment and appeared irrelevant to most branches of physics. In the wake of Dirac's special-relativistic quantum mechanics of 1928, many physicists sought to unify quantum mechanics and general relativity, but none of the attempts proved viable. To the extent that the general theory of relativity was cultivated at all, it was by a small group of mathematicians, theoretical physicists, and astronomers, who investigated the theory's mathematical structure or derived results from it that could not, however, be tested experimentally. In the 1930s and 1940s, it was a decidedly unfashionable theory, not least when compared with quantum theory and nuclear physics. The only area of physical research in which the theory played a major role was cosmology, but most physicists found cosmological studies to be on the periphery of science, if scientific at all.

What has been called the low-water mark of general relativity lasted until the early 1950s. From that time onward, and especially from about 1960, the field began to attract new interest and soon experienced a remarkable renaissance. Small groups of young relativists began to form around theorists such as John Wheeler and Peter Bergmann in the United States, Bondi in England, Leopold Infeld in Poland, and Vladimir Fock in the Soviet Union. One sign of the renewed interest was a series of conferences on problems in general relativity and associated areas, the first one being the 1955 Bern conference celebrating the fiftieth anniversary of Einstein's theory. It was followed by the Chapel Hill conference in the United States (1957), a conference in Royamont, France (1959), and one in Warsaw (1962). The subjects dealt with in the early conferences were of a theoretical and mathematical nature or referred, in a few cases, to astronomical observations. In 1970, publication of a journal called *General Relativity and Gravitation* started under the auspices of the International Committee on General Relativity and Gravitation, which was founded in 1960 as a result of the Royamont meeting. In the first issue of the journal the editor, André Mercier, called attention to "the extraordinary and very satisfactory combination of astrophysics and GRG [general relativity and gravitation] which has arisen throughout the years." Concerning this combination, he wrote: "GRG has definitely saved cosmology from the 'too-hypothetical,' and astrophysics combined with particle physics has made GRG theories very concrete, whereas, during one or two decades, they had remained despised by so many physicists on the pretext that they were unphysical." At physics departments, courses in general relativity multiplied and new textbooks were written. The publication in 1973 of *Gravitation*, a massive (1,280-page) textbook written by Charles Misner, Kip Thorne, and Wheeler, was another sign that general relativity had changed and moved into mainstream physics.

There were basically four reasons why the status of and interest in general relativity changed so drastically in the 1960s. First, the rejection of the steady state theory was widely seen as a triumph of general relativity. Second, new discoveries in astronomy stimulated the application of relativity to astrophysical problems. Third, Einstein's theory was challenged by a new and much discussed theory of gravitation. Fourth, and most important, new methods of experimental physics turned the theory of general relativity into a laboratory science. "Einstein's theory of gravitation, his general theory of relativity of 1915, is moving from the realm of mathematics to that of physics," concluded the American theoretical physicist Alfred Schild in 1960. "After 40 years of sparse meager astronomical checks, new terrestrial experiments are possible and are being planned" (Kragh 1996b, 318).

One of the pioneering experiments to which Schild referred was made in early 1960 by Robert Pound and Glen Rebka of Harvard University, who measured the "Apparent Weight of Photons," as their paper was titled, or put differently, the gravitational redshift. This prediction of general relativity had been confirmed by astronomical measurements in the early 1920s, but rather inaccurately and not very decisively. In order to measure the shift in frequency caused by the variation of the gravitational field of the earth over merely twenty meters (the height of the Harvard laboratory building), Pound and Rebka made sophisticated use of a new method of narrowing the shape of spectral lines, thereby producing a gamma ray with an extremely well-defined frequency. The method was based on the Mössbauer effect, named after the German physicist Rudolf Mössbauer, who had discovered it in 1958. Mössbauer's effect quickly found use in a variety of fields, from nuclear chemistry to general relativity, and the young discoverer was awarded the 1961 Nobel prize for his work.

The Pound-Rebka experiment confirmed the relativistic prediction within 10 percent and, in an improved experiment of 1965, the agreement was narrowed down to 1 percent. The importance of the Pound-Rebka experiment was not only that it provided support for Einstein's theory, but also that it ushered in a new era of experimental relativity. The combined result of experiments in the 1970s, making use of atomic clocks, rockets, satellites, computers, and other advanced electronics, was a complete confirmation of the theory of general relativity. By the end of the twentieth century, experimental general relativity had become big science.

The theoretical aspects of general relativity were studied by John Wheeler, Bryce DeWitt, and Roger Penrose, to mention a few. In 1960 Penrose, a British mathematician, introduced new and powerful topological tools in the theory, and in 1965, he proved that a gravitationally collapsing star will inevitably end in a space-time singularity. Further work by Penrose, Stephen Hawking, Robert Geroch, and others resulted in 1970 in a comprehensive singularity theorem, according to which a universe governed by the general theory of relativity must necessarily have started in a space-time singularity.

Astronomical discoveries in the 1960s and 1970s were another important source for the revival of general relativity (see table 23.1). The discovery of quasars in 1963, and the chance discovery of pulsars four years later by Jocelyn Bell and Anthony Hewish at Cambridge University, resulted immediately in attempts to understand the two remarkable phenomena theoretically. It turned out that explanations were possible only by making use of the theory of general relativity. In agreement with a suggestion first made by Thomas Gold at Cornell University, pulsars were explained as neutron stars rotating with extreme speed and regularity. Neutron stars, arising from the gravitational collapse of massive stars, had been discussed theoretically by Oppenheimer and Hartland Snyder in 1939, but it took two decades until the subject attracted widespread attention. In 1967, Wheeler coined the name "black hole" for a spherical mass collapsed into a singularity, and in 1974 the possible discovery of a black hole in the x-ray source Cygnus X-1 was reported. Although black holes are generally assumed to exist in abundance, there has still been no unequivocal confirmation of these remarkable cosmic objects.

Yet another unexpected discovery in astrophysics was made in 1974, when Joseph Taylor and Russell Hulse detected radioastronomical signals

TABLE 23.1
Important Astrophysical Discoveries, 1962–79

Phenomenon	*Year*	*Discoverers*
X-ray stars	1962	Bruno Rossi, Riccardo Giacconi (US)
Quasars	1963	Maarten Schmidt, Jesse Greenstein (US)
Cosmic masers	1965	Harold Weaver, S. Weinreb, A. Barrett (US)
Microwave background	1965	Arno Penzias, Robert Wilson (US)
Infrared stars	1965	Gerry Neugebauer, Robert Leighton (US)
Pulsars	1967	Jocelyn Bell, Anthony Hewish (UK)
Superluminal sources	1971	Irvin Shapiro (US)
Binary pulsars	1974	Russell Hulse, Joseph Taylor (US)
Gravitational lensing	1979	Dennis Walsh (UK)

that they interpreted as emitted by a binary pulsar—that is, a pulsar orbiting around an invisible companion, perhaps another pulsar. The finding of Taylor and Hulse was more than just one more cosmic discovery: It caused great excitement among relativists, who realized that binary pulsars were ideally suited as testing objects for the theory of general relativity. The theory agreed excellently with observations. Even more important, in 1978 Taylor showed that data from the binary pulsar system strongly suggested the emission of gravitational radiation at a rate agreeing with the prediction of general relativity. That accelerated massive bodies emit gravitational radiation was first argued by Einstein in 1916, but the suggestion remained a matter of theoretical speculation until Taylor's demonstration. In 1993, Taylor and Hulse received the Nobel prize "for the discovery of a new type of pulsar, a discovery that has opened up for new possibilities for the study of gravitation." Incidentally, this was the first time that a Nobel prize was motivated in part by reference to gravitational physics. Even before the Taylor-Hulse discovery, gravitational waves were looked for experimentally, primarily in a series of experiments conducted by Joseph Weber at the University of Maryland. Weber believed that signals in his massive resonant-bar detector indicated the reception of gravitational waves and announced the discovery of the waves in 1969. His claim was contested, however, and the majority of physicists concluded that Weber had not discovered the gravitational radiation. Weber disagreed, and continued his experiments. Still, by 1998, gravitational waves have not been directly observed, and by implication, neither has the graviton, the quantized version of gravity waves.

Einstein's general theory has never been without rivals in the form of alternative theories of gravitation. To mention just a few, Dirac and Jordan sought unsuccessfully to develop theories with a gravitational constant varying in time, and in 1964 Hoyle and Jayant Narlikar proposed a non-Einsteinian theory based on direct interactions between particles. The most serious challenge to general relativity in the 1960s was perhaps a theory developed by Carl Brans and Robert Dicke at Princeton University in 1961. The Brans-Dicke theory was much discussed around 1970, when it was realized that it led to a number of geo- and astrophysical predictions different from those of general relativity. For example, the theory of Brans and Dicke predicted a perihelion shift of mercury smaller than the 43″ general-relativity prediction. Dicke argued from solar observations that part of the observed value, and thereby the agreement with general relativity, was due to the sun being more oblate than hitherto believed. The extent of the sun's oblateness continued to be a matter of discussion for a decade, but by the mid-1980s, it had become clear that the Brans-Dicke theory disagreed with experiments, whereas general relativity did not.

The combination of laboratory experiments, astronomical observations, and advances in the mathematical foundation of general relativity resulted in a new and exciting subfield, relativistic astrophysics. The field was introduced at the first Texas Symposium on Relativistic Astrophysics in 1963, the first of an important series of conferences that were the relativists' counterpart to the particle physicists' series of Rochester conferences. The topic of the first Texas Symposium was gravitational collapse and the new quasars. It was a truly interdisciplinary meeting, including among its participants nuclear physicists, relativity theorists, cosmologists, and astronomers. Fred Hoyle, William Fowler, Kip Thorne, Allan Sandage, Edwin Salpeter, Roy Kerr, and Maarten Schmidt all presented papers. Whereas the 1963 symposium included about three hundred participants, more than eight hundred physicists and astronomers took part in the ninth symposium in 1978. The many symposia and summer schools in relativistic astrophysics were followed by textbooks and proceedings volumes. Among the first and most comprehensive books in the new field was *Relativistic Astrophysics* (1971) by Yakov Zel'dovich and Igor Novikov, two eminent Soviet physicists. The revival of interest in gravitation physics and relativistic astrophysics was further reflected in the venerable Solvay congresses. The eleventh congress in 1958 dealt with "Astrophysics, Gravitation, and the Structure of the Universe," and included addresses by Hoyle, Lemaître, O. Klein, Wheeler, and others. Six years later, the thirteenth congress was devoted to "The Structure and Evolution of Galaxies," and the theme of the sixteenth congress in 1973 was "Astrophysics and Gravitation." Topics discussed during the 1973 conference included x-ray sources, neutron stars, quasars, pulsars, and black holes.

Chapter 24

ELEMENTS OF SOLID STATE PHYSICS

THE SOLID STATE BEFORE 1940

AROUND 1930, MANY physicists were occupied with investigating the properties of solid bodies. Yet, in spite of the considerable activity within these areas of research—many of them with roots back in the nineteenth century—there was no discipline of solid state physics in either a social, institutional, or cognitive sense. From a sociological and historical point of view, solid state physics did not exist. It was only after World War II that the new science of the solid bodies, later to be renamed condensed-matter physics, took off and absorbed several specialties which, until then, were not thought to belong naturally to the same area of science. The 1930 volume of *Physics Abstracts* (then still named *Science Abstracts*, section A) included no entry among its subjects for solid state physics or related terms. The main subject groups were General Physics, Light (including radioactivity), Heat, Sound, Electricity and Magnetism, and Chemical Physics. Each of these groups included papers that a later generation would recognize as belonging to the field of solid state research. Ten years later, in 1940, the subject index of *Physics Abstracts* included an entry on Solids that was divided into the main categories Structure and Theory. The entry on Solid State Theory was further subdivided into Solids, Theory; Crystals, Lattice Dynamics; and Quantum Theory. From that time onward, one can begin speaking of solid state physics as a separate discipline, the practitioners of which were in the process of forming a scientific community.

The scientific core of the new discipline, the theoretical framework that gave it the necessary cognitive coherence, was the application of quantum mechanics to the solid state of matter. Before quantum mechanics, physicists (as well as chemists, crystallographers, and metallurgists) had studied the mechanical, optical, magnetic, electrical, and crystalline properties of solids, but there was no common theoretical denominator to these studies, which consequently appeared to be only loosely connected, if connected at all. As late as 1930, it was common to include large parts of chemical physics, including discussions of molecular spectra and chemical reactions, in books and review articles devoted to the physics of condensed matter. One of the first review articles including "solids" in its title was written in 1937 and appeared, characteristically, in the *Journal of Applied Physics*. The authors, Frederick Seitz and Ralph Johnson, wrote, "Until recently, the various theories of different solids were conspicuously lacking in unity. To interpret the distinctive properties of the three solids copper, diamond, and rocksalt, for example, one had to begin with three widely differing pictures of their internal constitution." Now the situation had changed, and a unified theory of the solid state based on quantum mechanics was a realistic possibility. As the authors wrote, "the quantum theory has succeeded in interpreting many of the observed properties of solids which the classical pictures left unexplained" (Weart 1992, 628). Three years later, Seitz wrote one of the first textbooks ever in the new solid state physics, appropriately entitled *Modern Theory of Solids*.

Among the many specialties making up the proto-solid-state physics of the 1930s, the theory of metals was one of the most important. The early theories of Drude, Riecke, Lorentz, and Bohr, which were all based on the free-electron gas model, gave reasonably good results in some areas, such as the relationship between thermal and electrical conductivity. However, they did not give the right temperature dependence of the resistance of pure metals, nor did the old theories succeed in explaining the magnetic properties of metals satisfactorily. According to the classical theory, one would expect the specific heats of metals to be much greater than those of insulators, yet experiments proved that this was not the case. It was evident that some new idea was needed, and that the new idea had to come from quantum mechanics.

The first physicist to apply the new quantum mechanics to the study of metals was Pauli, who in 1926–27 realized that free electrons in a metal must obey Fermi-Dirac statistics. When first Fermi—and then, independently, Dirac—introduced the new statistics in 1926, it was far from clear whether it was this form of statistics or the alternative Bose-Einstein statistics that applied to matter. For example, both Fermi and Dirac believed originally that gas molecules obeyed the same statistics as electrons—that is, they were fermions. (The names "fermion" and "boson" were introduced by Dirac in 1945.) The question was clarified only with Pauli's work. On the basis of the exclusion principle and the related Fermi-Dirac statistics, he developed an important theory of paramagnetism, which in many ways became the starting point for all later quantum theories of metals. It is ironic that Pauli's work attained this status, for Pauli was not particularly interested in solid state theory and considered the field to be much less "pure" than the exciting fundamental physics of quantum field theory. In the early 1930s, he sometimes referred to solid state physics as "dirty physics." In a letter of July 1, 1931 he wrote to Peierls, who had just calculated the residual resistance in metals, "The residual resistance is a dirt effect and one shouldn't wallow in dirt." All the same, Pauli played an important role in the early phase of solid state theory and seems to have had no difficulty in doing "fundamental" and "dirty" physics at the same time. Neither did Heisenberg,

Bethe, Peierls, Bloch, Mott, Frenkel, and Landau. They were not solid state physicists, but physicists with an interest in the solid state as one of several areas of application of quantum mechanics.

Pauli's work was followed by the work of his former professor, Arnold Sommerfeld who, in the same year, made use of Fermi-Dirac statistics to give a much improved theory of electrical conduction in metals. Sommerfeld showed that the electron gas must be completely degenerate and that only a small fraction of the electrons can therefore contribute to the specific heat of the metal. In this way, the relatively low specific heat was explained qualitatively. The works of Pauli and Sommerfeld opened up a new chapter in the history of solid state physics, but they were only a beginning and although they built on quantum statistics and spinning electrons, they did not employ the full apparatus of quantum mechanics. Although Sommerfeld's theory was a notable step forward, its agreement with observations was far from perfect. Yet the imperfect theory stimulated much criticism and refinement and formed the nucleus of an entire research program. For example, in 1928 the *Zeitschrift für Physik* contained fourteen papers directly related to Sommerfeld's electron theory. It soon turned out that the main problem of the Sommerfeld theory was not its disagreement with some experimental data, but rather that it worked so well after all.

A truly quantum-mechanical theory of metals was developed during the next four years, first of all by young physicists associated with Pauli, Sommerfeld, and Heisenberg. The pioneering phase of this development took place in Munich, Zurich, and Leipzig, but soon afterward important contributions were coming from England, France, the Soviet Union, and the United States as well. By the late 1920s, the mathematical and conceptual structure of quantum mechanics was largely completed and physics students were increasingly advised to apply the theory to new areas, of which solids were seen as particularly promising. One of the students was Felix Bloch, who worked under Heisenberg in Leipzig. As part of his doctoral dissertation, in 1928 Bloch began a quantum-mechanical study of the wave functions of electrons in a lattice. Important concepts such as "Bloch's theorem" and "Bloch states" date from this work. Drawing on ideas from the new quantum chemistry, due to Hund, London, and Heitler, Bloch examined the behavior of electrons, not in individual atoms or molecules but in the overlapping electric fields of the many atoms making up a crystal. He assumed a potential in which the electrons were tightly bound to the lattice. It then turned out that the energy states were not discrete, but continuous bands of allowed energies. Bloch's work, and the almost simultaneous works of Peierls and Bethe, laid the foundations of band theory, possibly the most important part of solid state theory. It followed from the early band theory that high conductivity is not merely a matter of a high degree of electron mobility. The electrons need to find free states in the energy bands; if such states do not exist, the crystal will be an insulator. Along this line of reasoning, Bloch succeeded in explaining, for the first time, the difference between metals and insulators.

In 1929, on Heisenberg's suggestion, Peierls investigated the anomalous or "positive" Hall effect, in which a current in some metals is influenced by the magnetic field as if the conduction carriers were positively charged. In connection with this work, he noted that an electron near the band edge would behave in a peculiar way, namely, move as if it were positively charged. This was the basis of the idea of the "hole"—or defect electron, as Peierls called it—as a vacancy near the top of an otherwise filled band. Such a hole, Heisenberg showed explicitly in 1931, would behave like an electron of positive charge and with a positive effective mass. (The solid-state hole of Peierls and Heisenberg had some similarities with Dirac's anti-electron hole introduced at the same time, but the two ideas were independent.) Another important solid state concept was introduced by the Paris physicist Léon Brillouin, who in 1930 developed Peierls's ideas into the geometrical technique that since then has been known as Brillouin zones. He later recalled, "At first I did not realize that I was doing something that might become really important. I did it for the fun of it, following my own line of investigation by sheer curiosity" (Hoddeson, Baym, and Eckert 1992, 119). Band theory was further developed by Alan Wilson, a Cambridge physicist who went to Leipzig to work with Heisenberg and Bloch, in part because he found physics in Cambridge to be too concerned with the atomic nucleus. In 1931 Wilson published his pioneering papers on semiconductors, in which for the first time he explained a semiconductor as an insulator with a band gap across which electrons can pass as a result of excitation. Wilson's model was a stage on the way toward more realistic solid state calculations, and hence toward a greater unity between theory and experiment. The trend continued throughout the 1930s and included as an important element the application of band theory to sodium and other real metals. The first such application was made in 1933 by Wigner and his student Frederick Seitz at Princeton's new Institute for Advanced Study (founded in 1932). The Wigner-Seitz calculations were further developed by John Slater and his students at MIT.

The early phase of the quantum-mechanical electron theory of metals was completed in 1933, when the entire field was reviewed by Bethe in the *Handbuch der Physik*. The nearly 300-page-long article appeared under the names of Sommerfeld and Bethe but was almost entirely written by Bethe, the junior author. Other articles in the volume covered "Dynamic Lattice Theory of Crystals" and "Structure-Sensitive Properties of Crystals" and indicated that solid state physics was on its way to forming a distinct scientific subdiscipline. The new subdiscipline was German by birth, but spread rapidly to other countries. In the United States, the first centers of solid state physics were Slater's department of physics at MIT and the one at Princeton University, where Wigner gave lectures and supervised Seitz and John Bardeen. The most successful of the new solid-state-oriented institutions was perhaps the University of Bristol, which until the 1930s had not been one of England's important centers of physics. In 1930 the Department for Scientific and Industrial Research (DSIR) approved a program in theoretical solid state physics at the university, an initiative that characteristically was more strongly supported by industry and government circles than by the university. In 1932 Nevill Mott, a nuclear physicist with no previous contributions to solid state physics, became a professor at Bristol. Although his self-professed ignorance of metal theory was "profound," he quickly transformed the physics department into a world center of solid state research. Mott recalled that he "was fascinated to learn that quantum mechanics could be applied to problems of such practical importance as metallic alloys, and it was this as much as anything else that turned my interest to the problems of electrons in solids" (Eckert and Schubert 1990, 91). In Bristol, as elsewhere, the practical importance of solid state physics was given high priority and was a main reason for economic support. Yet it was academic physics, not engineering physics, that interested Mott and his colleagues. *Theory of the Properties of Metals and Alloys*, written by Mott and Harry Jones in 1936, used quantitative quantum-mechanical calculations adapted to real metals and freely mixed empirical considerations with approximate methods. Although it was "dirty physics" according to Pauli's scale of values, the book became a standard work for a new generation of metal physicists. The Bristol group had close contacts with colleagues in the United States, from both the academic world and the commercial world, and in England the group collaborated with the National Physical Laboratory and private industries. The close links between industry and solid state physics became highly important after World War II, but even before the war, such links were well established in countries like England, Germany, the Netherlands, and, not least, the United States.

Semiconductors and the Rise of the Solid State Community

There are substances that belong neither to the typical metallic state nor the typical nonmetallic state. Such semiconductors, either as chemical elements or compounds, had been known since the mid-nineteenth century. As early as 1833, Michael Faraday had observed that the resistivity of silver sulfide decreases as the temperature increases, contrary to the behavior of metals. Later in the century, photocurrents were discovered in selenium and in 1874, the German physicist Ferdinand Braun observed that contacts between some materials had the effect of rectifying the current. The term "semiconductor" (the German *Halbleiter*) seems to have been coined in 1911. In the 1930s, most of these and other semiconducting properties were reasonably well explained by Wilson's band theory as it was developed by a growing number of solid state physicists. Semiconductors were understood as being insulators with a small forbidden band between the valence band and the conduction band. At the same time, semiconductors were studied experimentally, not least in Göttingen, where Robert Pohl had established a strong school in experimental solid state physics. By the late 1930s, it was known that the rectifying property was a junction effect, taking place at the interface of the metal and the semiconductor, and that semiconductors came in two types, the *p*-type and the *n*-type. These now-familiar names were introduced in 1941 by Jack Scaff of Bell Laboratories to replace the previously used names defect and excess types. In *n*- or excess-type semiconductors, the majority of charge carriers are negative, whereas they are positive in the case of *p*- or defect-type semiconductors. Although there were various theories for the rectification in both types of junctions (metal to either *p*- or *n*-type semiconductor), a satisfactory explanation proved difficult. A largely successful theory of rectifying contacts was developed around 1940, primarily by Walter Schottky in Germany, Mott in England, and B. I. Davydov in Russia. Semiconductor physics was a very small part of the period's physics. In 1933, about 60 papers on the subject were published and during the following years, the number of papers decreased steadily, to reach a low point of 20 in 1940.

The war meant a great increase in semiconductor research, in particular because pure silicon and germanium were of crucial importance as radar detectors. It was during the course of war-related work that the first *p-n* junction was investigated in the United States and proved to be an excellent rectifier. This kind of rectifier had already been suggested by Davydov in 1938, but his suggestion attracted little attention at the time. The single most important event in the history of semiconductor research was undoubtedly the invention of the transistor in late 1947. The idea of extending the analogy between a semiconductor and a vacuum diode to one including a triode was well known, and a working solid-state triode had even been constructed before the war by Pohl and Rudolf Hilsch in Göttingen. As a result of inquiries from the large electrotechnical company AEG (Allgemeine Elektrizitäts-Gesellschaft), in 1938 the two physicists performed experiments with electric currents controlled by crystals. Their device used a potassium bromide crystal and amplified the current more than one hundred times. It was anything but practical, however, and was neither patented nor further developed. Although Pohl had good connections to industry, he was not interested in turning his work into technological innovations or taking out patents.

Ernest Braun, a solid state physicist and historian of science, has summarized the development of semiconductor physics in the following apt way:

"The history of semiconductor physics is not one of grand heroic theories, but one of painstaking intelligent labor. Not strokes of genius producing lofty edifices, but great ingenuity and endless undulation of hope and despair. Not sweeping generalizations, but careful judgment of the border between perseverance and obstinancy. Thus the history of solid-state physics in general, and of semiconductors in particular, is not so much about great men and women and their glorious deeds, as about the unsung heroes of thousands of clever ideas and skillful experiments—progress of a purposeful centipede rather than a sleek thoroughbred, and thus a reflection of an age of organization rather than of individuality" (Braun 1992, 474). Yet, in spite of the generally unromantic development of semiconductor physics, the field did not lack either great men or glorious deeds. The invention of the transistor may well be called a glorious deed.

The transistor was the baby of Bell Laboratories, which in 1945 established a research program in solid state physics that included a subgroup in semiconductor physics led by William Shockley, an MIT physicist who had worked with AT&T since 1936. The long-range goal of Shockley and Walter Brattain, a Bell experimentalist, was to make a solid state amplifier that could be used, among other things, as a switch in telephone systems. It was not a development program, however, and the primary goal of Shockley, Brattain, and their colleagues was simply to extend the scientific knowledge of the properties of silicon and germanium. At first their attempts failed, but in 1947 John Bardeen, who had come to the Bell Laboratories two years earlier, analyzed theoretically the problems involved in the experiments. Bardeen concluded that energy states on the semiconductor surface had to be taken into account, because the electrons might be trapped at the surface. The theory of surface energy states had previously been suggested for the free surface of a solid, but not for semiconductors. According to Bardeen, "The novel feature was not the idea of surface states . . ., but to apply the idea to understand the real surface of a semiconductor" (Braun 1992, 468). After further exploration, an experimental breakthrough followed the theoretical breakthrough. On December 23, 1947 Bardeen and Brattain observed the transistor effect in the first point-contact transistor, which consisted of two wire electrodes on a germanium crystal. The name "transistor" was used as an abbreviation of "transfer resistor." After a secrecy period of half a year, the solid state amplifier or crystal triode, as the invention was often called, was made public at a press conference and it was reported, without fanfare, in the *New York Times* of July 1, 1948. However, although it was a groundbreaking invention, the point-contact transistor had a short life and was in fact obsolescent as soon as it was produced. It was quickly replaced by the *p-n* junction transistor based on a theory that Shockley developed in 1948, but which the *Physical Review* refused to publish because the theoretical foundation was found wanting. It was this work that made Shockley share the 1956 Nobel prize with Brattain and Bardeen for "their investigations on semiconductors and their discovery of the transistor effect." And it was this work, as included in Shockley's 1950 book *Electrons and Holes in Semiconductors*, that formed the basis of modern solid state electronics.

The celebrated invention of the transistor was one of the events that led semiconductor physics, and solid state physics in general, to flourish in the 1950s. The explosive growth is illustrated by the publications on semiconductor physics: Between 1948 and 1955, the number of papers per year increased from 20 to almost 400. The 1950s was the decade in which solid state physics truly became both an independent scientific discipline and a social community of a new breed of physicists. Indications that a scientific specialty has developed into a community will normally include textbooks, graduate courses, designated chairs, conferences, professional societies, and specialized journals. Some of these indicators existed before the war, but most of them came about only in the 1950s. During the war years, there was much discussion in the American Physical Society about the formation of new divisions, which was a wish especially among applied physicists, who felt that their areas of research were not sufficiently well represented or respected within the existing organization. A proposal for forming a division of the solid state, or a division of metal physics as it was originally named, was made in 1944. The division was finally approved in 1947, and from that time one can speak of "solid state physicists" as scientists not only dealing with a particular area of research, but also belonging to a scientific community distinct from the other communities of physicists. It took several years until the new discipline gained momentum, one of the reasons being that it had to compete with the nuclear physics discipline, which at the time was developing with extreme speed and vigor. Indeed, nuclear and particle physics seemed, for a short period, to monopolize American physics. Not only was that field generously supported by federal and private money, but it also had an intellectual and cognitive reputation as a "heroic" science, with which no other field could compete. Solid state physics, on the other hand, had traditionally been considered a somewhat less worthy subject because of its connections to industry and a less fundamental domain of nature. It still had, among some physicists, an aura of "dirty physics." Slater recalled that at MIT, "our department was often looked down on, to some extent, by those who felt that no physicist of any imagination would be in any field except nuclear and high-energy physics" (Weart 1992, 656).

Yet the new solid state community turned out to be highly competitive, even with the more glamorous discipline of nuclear physics. For a long time, solid state physics did not have its own journal. Nor did it really need one, for the ever-expanding *Physical Review* served as an outlet for a large part of the most valuable solid state literature. By 1964, the journal was published in two sections, one of which dealt primarily with the solid state and the other with nuclear and particle physics in particular. When it split into separate journals in 1970, *Physical Review B* was devoted to solids and was larger than *Physical Review A*. As far as scientific personnel were concerned, there were about 350 American Ph.D. physicists in solid state in 1951, or about one-tenth of the American population of Ph.D. physicists and less than half the number of Ph.D. nuclear physicists. Ten years later, the picture had changed substantially, not because nuclear physics had become less popular, but because solid state physics had exploded. Now there were about 2,000 Ph.D. solid state physicists in the United States, almost as many as the number of nuclear physicists. It has been estimated that the total budget for American solid state physics at that time was on the order of $100 million.

The phenomenal growth of solid state physics in the period from 1950 to 1968 is illustrated in figure 24.1. Note that the growth rate was somewhat

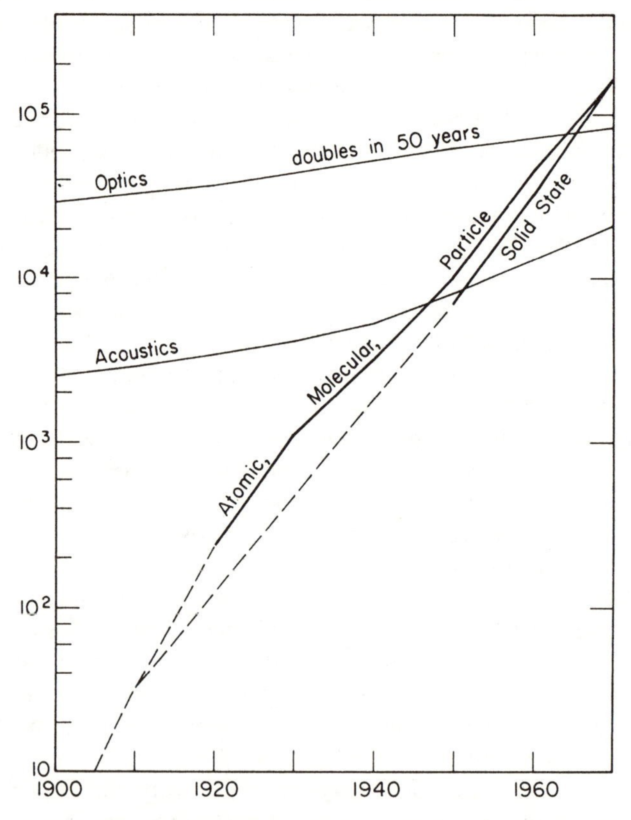

Figure 24.1. Cumulative output of papers in solid state physics and several other subfields of physics. *Source:* Menard 1971. Reprinted by permission of the publisher from *Science, Growth and Change* by Henry W. Menard, Cambridge, Mass.: Harvard University Press, Copyright © 1971 by the President and Fellows of Harvard College.

higher than in the category "Atomic, Molecular, Particle" (including nuclear physics) and that by 1965, the number of papers in solid state had exceeded those in the particle category. An important factor in the growth was the rise of the semiconductor industry and other industries where solid state physics was found useful. Whereas in American physics as a whole, the ratio between physicists employed in industry and in higher education was about 1:2, the proportion among solid state physicists was the reverse, about 2:1. The affiliation with industry is further illustrated by the structure of financial support. Although high-energy physics attracted the largest amount of federal support, solid state physics was far the largest recipient of industrial money (table 24.1). The figures given here are from American physics, which was the largest national community and the only one from which data are available. The situation in other countries was not substantially different, but it was on a lesser scale and the trends in development came somewhat later. According to a study of solid state publications in 1961, the United States and the Soviet Union each accounted for about one-quarter of the papers. Great Britain and Japan accounted together for another quarter, and France and Germany together for about one-tenth. Thereafter followed the minor scientific nations, including Italy, the Netherlands, Switzerland, and Denmark.

Breakthroughs in Superconductivity

As noted in chapter 6, superconductivity was a complete puzzle in the 1920s. With Bloch's 1928 theory of metals, new hopes of understanding the phenomenon were raised and many theoretical physicists started applying the new ideas of the solid state to the case of superconductivity. The names of those engaged in the effort make up an impressive list of élite physicists. Bohr, Heisenberg, Bloch, Landau, Brillouin, Frenkel, Kronig, Bethe, Casimir, and Pauli were among those who optimistically tried to solve the problem. However, all their clever ideas failed. Bloch, who worked hard on su-

Table 24.1
The Four Largest U.S. Basic Physics Subfields in 1970

Field	*Federal*	*Industrial*	*Total*	*Percent*
High energy	150	—	150	27
Solid state	56	80	136	24
Plasmas and fluids	77	10	87	16
Nuclear	73	2	75	13

Note: The amounts are in millions of dollars.
Source: PT, July 1972.

perconductivity around 1931 (but never published his work), later recalled, "I was so discouraged by my negative result that I saw no further way to progress and for a considerable time there was for me only the dubious satisfaction to see that others, without noticing it, kept on falling in the same trap. This brought me to the facetious statement that all theories of superconductivity can be disproved, later quoted in the more radical form of 'Bloch's theorem': Superconductivity is impossible" (Gavroglu and Goudaroulis 1989, 77).

The failure of the theorists was, to some extent, compensated by the progress made by the experimentalists, the most important being the discovery of the Meissner effect in 1933. Walther Meissner and Robert Ochsenfeld, two physicists at Berlin's Physikalisch-Technische Reichsanstalt, discovered that when a solid cylinder of tin or lead was cooled below its transition point in a constant magnetic field, the lines of force were suddenly expelled from the metal. The diamagnetic character of superconductors was completely unexpected, and presented the theorists with yet another phenomenon to explain and incorporate in their theoretical models. The Meissner effect provided an important stimulus for the first satisfactory theory of superconductivity, a macroscopic theory developed in 1935 by the brothers Fritz and Heinz London, who at the time were refugee physicists in England. The theory provided electromagnetic equations that described the vanishing of resistance in a superconductor, as well as the diamagnetic character, which the London brothers considered to be an intrinsic property of superconducting metals. In spite of this success, it was realized that the London-London theory was not the final answer to the problem of superconductivity, for the theory was phenomenological and did not explain the phenomena on a microphysical basis.

After 1945, theoretical work on superconductivity continued with renewed vigor, now assisted by more refined theoretical tools not only from solid state theory, but also from quantum field theory. The contributors to the new phase included celebrities like Landau, Born, Heisenberg, and Feynman, but the problem appeared to be as difficult to solve as it had been before the war. Feynman shared with Bloch his fascination with superconductivity, as well as his frustrations with not being able to solve the problem. Shortly before his death, Feynman recalled that he "spent an awful lot of time in trying to understand it and doing everything by means of which I could approach it. . . . I developed an emotional block against the problem of superconductivity, so that when I learned about the BCS [Bardeen-Cooper-Schrieffer] paper I could not bring myself to read it for a long time" (Mehra 1994, 430).

It was a less famous physicist who made the first significant step toward a microscopic theory. Herbert Fröhlich, a German physicist who had emigrated to England during the Nazi reign, suggested in 1950 that superconductivity might be the result of an interaction between electrons produced by the quantized lattice vibrations (so-called phonons). Based on this idea and making use of methods from quantum field theory, he developed a theory that implicitly contained the result that the critical temperature would decrease with the atomic mass of the superconductor. Earlier attempts to measure a dependence of the temperature on the mass had given negative results and until 1950, there were no theoretical reasons to believe in an isotope effect. The effect was nonetheless discovered in 1950, and Fröhlich now realized that his theory included the prediction that the critical temperature would vary as the inverse square root of the atomic mass. When experiments performed later in 1950 confirmed the $M^{-\frac{1}{2}}$ variation, they were naturally taken as strong support of Fröhlich's theory. Yet, although it was a notable advance, the theory was not acceptable because it failed to explain the Meissner effect, among other reasons. Whereas Fröhlich had considered only electron-phonon interaction, in 1955 Bardeen and David Pines also took into account the direct Coulomb repulsion between two electrons and showed that even under these more realistic conditions, there would be an attraction at low energies.

The real breakthrough came in 1957 and combined electron-phonon interaction with the insight that electrons with opposite spin may form bound boson pairs in metals as a result of the attractive interaction between electrons mediated by the lattice. Such "Cooper pairs" had been suggested as early as 1946, in a chemical context, but it was only in 1956 that Leon Cooper gave a theoretical justification of the idea. Moreover, he suggested that the pairs should be treated not as discrete entities, but collectively. The following year Cooper, together with his University of Illinois colleagues John Bardeen and John Schrieffer, developed the electron-pair hypothesis into a detailed microphysical theory of superconductivity. Bardeen had left Bell Laboratories for Illinois in 1951 and Schrieffer was his Ph.D. student, working on superconductivity. The three men's theory, subsequently known as the "BCS theory," ascribed the superconducting state essentially to a condensation of electrons into Cooper pairs having a common momentum and being represented by a single coherent wave function. The theory of Bardeen, Cooper, and Schrieffer explained all experimentally known facts of superconductivity and made a number of novel quantitative predictions, which were quickly confirmed. As a result, the BCS theory was accepted as the correct, if not the final, answer to the long-sought question of why some metals are superconductors. As Bardeen recalled: "If there was a discrepancy, it was usually found, on rechecking, that an error had been made in the calculation. All of the hitherto puzzling features of superconductors fitted neatly together like the pieces of a jigsaw puzzle" (Bardeen 1973, 35). The three American physicists received the 1972 Nobel prize in recognition of their theoretical breakthrough—Bardeen for the second time, as the only double physics laureate ever. Another Nobel prize followed in the wake of the BCS theory when the twenty-two-year-old British physicist Brian Josephson predicted that the superconducting quantum state should be able to leak through a barrier between two superconducting materials. The Josephson effect, predicted in 1962, was quickly confirmed experimentally and used in advanced measurement technology.

The BCS theory was not only a triumph of quantum solid state theory, but it also substantiated the universal validity of nonrelativistic quantum mechanics. By the mid-1950s, a microscopic theory of superconductivity had already been sought for a quarter of a century, and the lack of success could reasonably be suspected to indicate some flaw in the fabric of quantum mechanics itself. In a letter to Landau in 1954, Feynman noted that, apart from gravitation, superconductivity was the only phenomenon that still defied explanation in terms of quantum mechanics. Three years later, the list of problematic phenomena was reduced from two to one. Of course, the success of the BCS theory did not mean an end to theoretical work in superconductivity; on the contrary, it greatly stimulated new work, both in superconductivity and in other areas of theoretical physics. Throughout most of its history, the theory of superconductivity made use of theoretical advances in other fields, especially in quantum field theory. The theory transfer was eased by the fact that several of the scientists interested in superconductivity came from, or had a solid training in, quantum field theory and elementary particle physics. Fröhlich and Cooper, for example, were experts in these areas, which were apparently so different from condensed matter at very low temperatures.

Superconductivity, especially after the BCS theory, attracted theoretical physicists who found it worthwhile to investigate the connections between the theory of superconductors and the much more general theory of quantum fields. For example, Yoichiro Nambu explored the analogy between the two theories systematically and applied the methods of quantum electrodynamics to superconductivity. The BCS theory was not gauge invariant and several theorists, finding this to be a defect, therefore tried to formulate a gauge invariant version without losing the theory's explanatory power. Nambu proved in 1960 that the lack of gauge invariance is not a defect, but reflects the nature of superconductivity, namely, the gauge-dependent energy gap. Nambu was not only interested in exploring superconductivity by means of quantum field theory, but also realized that there is a striking mathematical analogy between the two theories and that methods from the theory of superconductivity could therefore be used to solve problems in quantum field theory (see chapter 22).

The status of the BCS theory was not affected significantly by the refinements and generalizations that took place after 1957, and by the 1970s, the field of superconductor research might have seemed to be largely completed—at least as far as theory was concerned, for the experimentalists were busy extending the empirical knowledge and searching for new superconductors with higher critical temperatures (T_c). This kind of work had little to do with theory and much to do with systematic experiments and knowhow. John Hulm and Bernd Matthias, an Englishman and a Swiss who both worked in the United States, investigated low-temperature properties of transition metal compounds and similar materials. The new class of superconductors ("type II") had very large critical magnetic fields compared to ordinary superconductors. In 1953 Hulm and Matthias found materials (V_3Si and Nb_3Sn) with a T_c about 18 K, a modest improvement over the previous record of 15 K recorded by German physicists in 1941. Subsequent experiments with other materials raised the record to 23 K, a temperature obtained in 1973 for a niobium-germanium compound. Prospects for high-T_c superconductors, and then possibly a breakthrough in the practical use of superconductor technology, looked dim until the mid-1980s, when J. Georg Bednorz and K. Alex Müller at the IBM Research Laboratory in Zurich concentrated on so-called perovskites, a class of metal-oxide (ceramic) minerals. The German (Bednorz) and Swiss (Müller) physicists found in 1986 that a Ba-La-Cu-O system had a T_c of 35 K, a discovery that stimulated a flurry of work in low-temperature laboratories all over the world. It was also a challenge to the theorists, for the BCS theory failed to explain the new kind of superconductivity. Teams led by Chung-Wu Chu in Houston and his former student, Maw-Kuen Wu in Alabama, investigated a Y-Ba-Cu-O material and reported in early 1987 that it was superconducting even in ordinary liquid nitrogen. One of the first things Chu did was to draft a patent application. The reported $T_c = 90$ K was soon confirmed and started an explosive development in high-T_c superconductivity research, with the March 1987 meeting of the American Physical Society as a first climax. More than 4,000 physicists attended the tumultous session, known as the "Woodstock of Physics," on the new superconducting ceramics. By August 1987, *Physical Review Letters* had received more than 300 manuscripts on high-T_c superconductivity. The next year a Japanese group announced superconductivity at 110 K and later that year, an American group obtained $T_c = 125$ K in a Tl-Ca-Ba-Cu-O ceramic. By 1995, the record was about 164 K, held by a Hg-Ba-Ca-Cu-O material under pressure (figure 24.2). The importance of Bednorz's and Müller's discovery was recognized by the Nobel committee in Stockholm, which awarded the prize to the two physicists in November 1987, making it one of the fastest recognitions in the history of the prize.

A main reason for the enthusiasm with which high-T_c superconductivity was received was the possibility that the discovery might be developed into technologies for the commercial market. As early as July 1987, when the Federal Conference on Commercial Applications on Superconductivity convened in Washington, D.C., the nascent field had given birth to a booming business in science publications, including seven weekly newsletters cover-

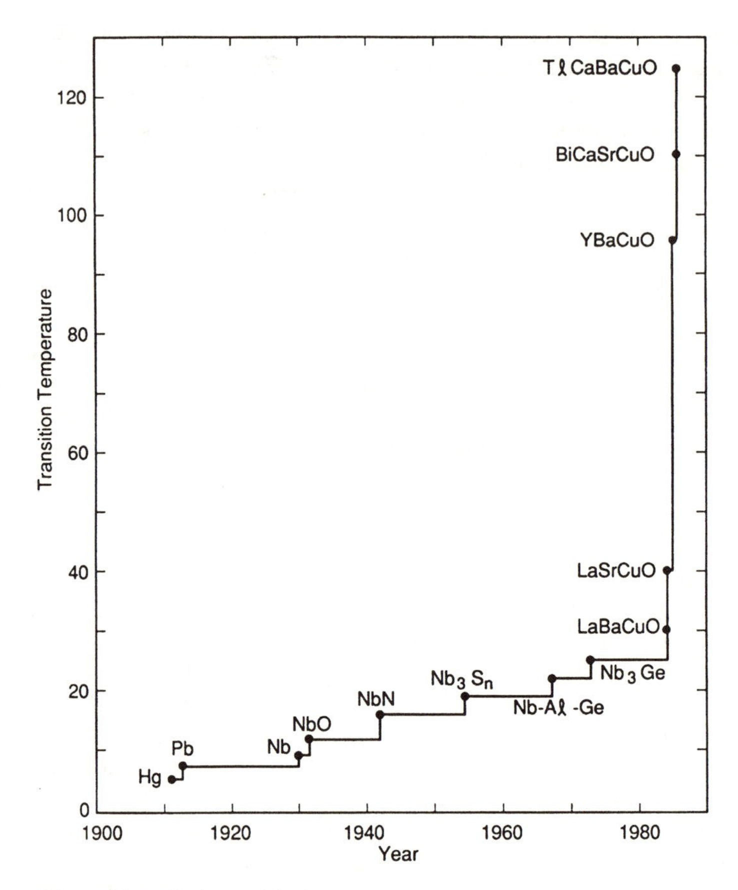

Figure 24.2. Highest critical temperatures and the years they were obtained. *Source:* P. F. Dahl, *Superconductivity: Its Historical Roots and Development from Mercury to the Ceramic Oxides* (1992). Copyright by Springer-Verlag. Used with permission.

ing the latest developments. Japanese scientists and business interests were particularly active in the race to commercialize superconductivity and invest the resources thought necessary to win the race. Government and private industry founded the International Superconductivity Center (ISTEC) with the purpose of developing commercially useful technologies based on high-T_c superconductivity. ISTEC was a consortium open to companies worldwide, but in fact, almost all the member companies were Japanese. To participate fully in the program, a company would have to pay about $800,000 initially and $110,000 per year subsequently. In this way, ISTEC was guaranteed an initial funding of $35 million in 1988 and $5 million annually thereafter. The

Japanese companies found such investments to be justified; they predicted that superconductivity would become a $30 billion annual market by the year 2000. The prediction was much too optimistic. As in the case of fusion research, the road from scientific discovery to technological innovation turned out to be much more difficult and uncertain than expected.

Chapter 25

ENGINEERING PHYSICS AND QUANTUM ELECTRONICS

IT STARTED WITH THE TRANSISTOR

ACCORDING TO A survey made in 1951 of members of the American Physical Society, 42 percent worked in so-called modern physics, defined as quantum theory, nuclear physics, electronics, and atomic and molecular physics; only 27 percent worked in the "old physics" of classical theory, acoustics, and optics. The most popular field turned out to be electronics, which 18 percent of the respondents identified as their "leading field of specialization." At number two was nuclear physics, with 15 percent. The fascination with electronics, at that time still meaning vacuum tube circuits, accelerated and took a new turn as solid state devices began to replace the vacuum tubes. Bell Laboratories' germanium point-contact transistor of 1948 was a triumph of solid state physics, but it was not a great commercial success (see chapter 24). Not only was it difficult to develop into an industrial product—Western Electric started to manufacture it in 1951—but the transistors were also expensive, and operated rather poorly compared with the miniaturized electron tubes with which they competed. For example, they were noisy, not very reliable, and sensitive to humidity and mechanical disturbances, and one transistor often varied in characteristics from another in the same series. Three years after Brattain's and Bardeen's invention, it was not obvious that the transistor had any practical application.

A possible solution to the problems was to replace the point-contact with a junction transistor, but for a period, Shockley's idea remained theoretical. As two historians have put it, "Shockley could design any number of semiconductor amplifiers employing these junctions in every configuration imaginable, but his sketches would remain mostly a form of mental masturbation until the art of fabricating junctions finally caught up with his explosive inventiveness" (Riordan and Hoddeson 1997, 177). The practical problems were eventually solved and the junction transistor became an invention in 1950. Shockley quickly produced a theoretical understanding of the *p-n-*junction transistor, first published in the *Bell System Technical Journal* and later, with his collaborators Gordon Teal and Morgan Sparks, he gave an account of the new transistor type in *Physical Review.* At first, the invention attracted little attention. It was only after further development that Shockley and his collaborators at Bell Labs were able to present the first practical microwatt junction transistor in 1951. General Electric followed in the same year with its own type of junction transistor. The new low-power and low-noise device proved superior to the point-contact transistor and became the real starting point of the transistor industry.

Still, whether of one type or another, the first generations of transistors were so expensive that they had little chance on the commercial market. Fortunately for the pioneer transistor manufacturers, there was one customer for which the prize did not matter and that was more than willing to support the development of the transistors. During the war, electronics had become increasingly important to the U.S. armed forces, especially the Navy and the Air Force, and military officers recognized the need for miniaturized electronic circuits in radar, navigation aids, and missile systems. For this reason, they were eager to promote the development of transistorized equipment, no matter what the costs. The military did so in two ways, by purchasing a large number of transistors and by funding research and development in transistor electronics. Of the 90,000 transistors produced in 1952, nearly all were purchased by the armed services. It is estimated that the U.S. military spent some $50 million for these purposes between 1952 and 1964. The early transistor industry depended heavily on the military, without which it would have developed much more slowly. The military money made it possible to develop new production methods that lowered the price and eventually made the transistor competitive on the civilian market, including as a component in scientific instruments. As late as 1967, the United States spent more on electron tubes than on transistors and other semiconductor devices, but from that time onward, the transistor industry exploded and dwarfed the tube industry (figure 25.1).

The early transistors used germanium as a semiconductor. The next important step was the shift to silicon, a step first taken by Texas Instruments, a small firm that originally manufactured geophysical instruments for oil prospecting and had entered the nascent transistor industry in 1952. The development of the first successful *npn*-junction silicon transistor was the work of Gordon Teal, the Bell Labs physicist who had cooperated with Shockley before leaving for Texas Instruments in 1953. Shockley, more than anyone the father of the age of microelectronics, left Bell Labs in 1954 and founded the Shockley Semiconductor Laboratory in Palo Alto, California, thereby initiating the famous Silicon Valley era.

Although Shockley's company was not commercially successful, indirectly it was of great importance by being the source of several other small, innovative firms established by former employees. One of the "traitors" (as Shockley called them) was Robert Noyce, who established a semiconductor company associated with Fairchild, an instrument firm. Scientists at Fairchild Semiconductor introduced new manufacturing processes for silicon

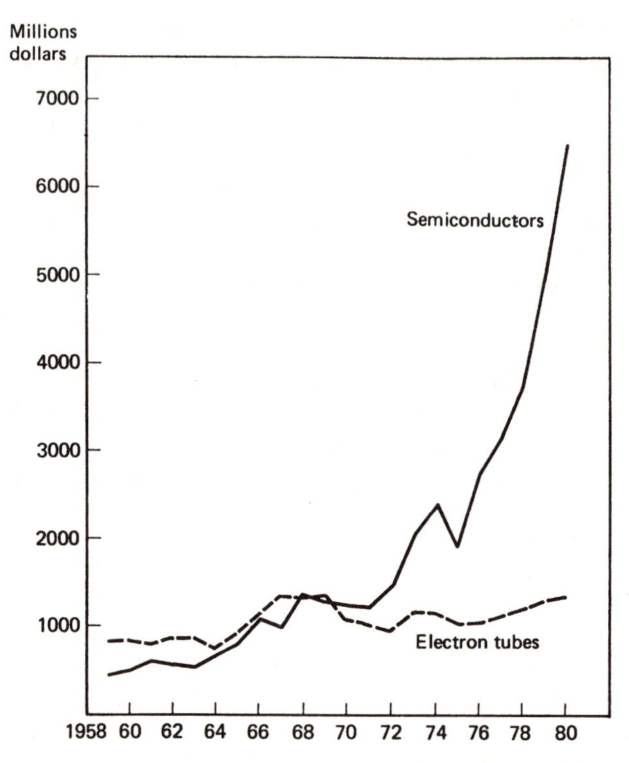

Figure 25.1. U.S. spending on electron tubes and semiconductor devices, 1959–80. *Source:* W. A. Atherton, *From Compass to Computer: A History of Electrical and Electronics Engineering* (1984). With the permission of Macmillan Press Ltd.

transistors, which led to the innovative, high-performing "mesa" transistor in 1957. In this device, the transistor's functioning was restricted to a thin oxidized surface layer by a method known as planar technology. The method was suitable for making integrated circuits of many silicon or germanium transistors in a single block of semiconductor—that is, without the use of connecting wires. The idea of an integrated semiconductor circuit was in the air and Noyce filed a patent for the invention in the summer of 1959. Unknown to him, earlier the same year another similar patent had been filed by Jack Kilby, a scientist recently hired by Texas Instruments. Kilby's first integrated "monolithic" circuit included three resistors, one capacitor, and a transistor constructed in one piece of germanium. The invention of the integrated circuit, made independently by Noyce and Kilby, initiated a new chapter in the history of microelectronics. This new chapter developed in a manner strikingly similar to the earlier transistor chapter: The first integrated circuits from Texas Instruments were prohibitively expensive and were bought almost exclusively by the military, the first one by the Air Force. In 1962 the average price of an integrated circuit was $50, and all circuits were purchased by the military. Total federal spending in electronics that year was about $10 billion, of which $9.2 billion were spent by the Department of Defense. When prices began decreasing, the military's share also decreased. By 1968, the average price had dropped to $2.30 and less than half of the integrated circuits went to the military.

During the 1960s, integrated circuit technology was developed to a great extent by Texas Instruments, Fairchild, and a number of other specialized firms. The idea of a field-effect transistor (or FET), that is, one in which the current through a thin semiconducting layer is modulated by applying an external electric field, was conceived by Shockley as early as 1945. However, it took fifteen years until scientists at Bell Labs succeeded in developing a practical field-effect transistor in the shape of the metal-oxide-silicon (or MOS) transistor. This type of transistor turned out to be well suited for integrated circuits because MOS transistors could be packed more densely and required less power. By the mid-1960s, integrated circuits were entering the civilian market and soon developed into microprocessors, which were used in computers and elsewhere. In 1963 the American electronics industry produced $252 million worth of transistors. Of these, 37 percent were used by industry, mostly in computers, and 16 percent by consumers.

This entire development, of such obvious importance for the modern world, started with the transistor, a product of solid state physics. In his 1950 textbook on semiconductors, Shockley speculated briefly about the future of transistor electronics. His speculation was closer to reality than he could have known: "Those who have worked intensively in the field share the author's feeling of great optimism regarding the ultimate potentialities. It appears to most of the workers that an area has been opened up comparable to the entire area of vacuum and gas-discharge electronics." The change that occurred with the shift to integrated circuits in the 1960s was less fundamental than the earlier shift from electron tubes to discrete transistors. Integrated circuits involved no new physics and were, contrary to the transistor, developed by engineers and industrial scientists rather than research physicists. Noyce and Kilby would hardly have been serious candidates for a Nobel prize (although they may have been nominated). Speaking about the integrated circuit, Kilby noted: "In contrast to the invention of the transistor, this was an invention with relatively few scientific implications. . . . Certainly in those years, by and large, you could say that it contributed very little to scientific thought." Another participant in the early development of integrated circuits, Douglas Warschauer, suggested that about 1960, "the engineers and technologists were by and large no longer listening to the basic scientists and really weren't even very much interested in what the basic scientists were doing" (Braun and MacDonald 1978, 103 and 138). Yet, although the semiconductor industry was only remotely related to developments in pure physics, Kilby's claim that the integrated circuit "contributed very little to scientific thought" is hardly correct. Indirectly, innovations in microelectronics were enormously important for scientific thought, namely, through the electronic instruments that revolutionized many branches of the experimental sciences. Any modern experiment in high-energy physics, any modern observation in astrophysics, and any important work in modern experimental condensed-matter physics depend crucially (if usually without acknowledgment) on the technique of integrated circuits invented by Kilby and others.

Microwaves, the Laser, and Quantum Optics

The laser, one of the most important and versatile scientific instruments of postwar physics, was the direct descendant of the maser, or microwave amplifier, and the maser itself was the child of wartime radar science and technology. Although microwave spectroscopy of gases was invented before 1940, radar work and abundant radar surplus equipment totally changed the speciality and had turned it into a sophisticated and rapidly developing field by 1950. According to a historian's evaluation, "Molecular spectroscopy through the absorption of microwaves in gases is, unquestionably, the premier example of a flourishing field of physical research created—in every sense—by radar" (Forman 1995, 422). Microwaves were not only of great interest to physicists, but they were of no less interest to the U.S. military, which wanted to know how to build a source of millimeter waves. This was not an easy task, for existing electronic devices, such as magnetrons and klystrons, were not suited to reach this range of wavelengths.

After having graduated from Caltech in 1939, Charles Townes spent eight years at Bell Labs, where he worked with radar and navigation devices. In 1948 he joined Columbia University's Radiation Laboratory. This laboratory was sponsored mainly by the military services (the Army Signal Corps and the Office of Naval Research, ONR), which were interested in magnetrons yielding very short wavelengths. By the end of 1949, researchers at the Radiation Laboratory had developed magnetrons that could generate 3–4 mm microwaves; in 1951, Townes reached a record-low wavelength of a little more than 1 mm. However, the millimeter magnetrons turned out to be largely useless for practical purposes. They were inefficient, very expensive, and had lifetimes that rarely exceeded two hours. In the spring of 1951, as an advisor to the ONR and chairman of its Advisory Committee on Millimeter-Wave Generation, Townes came up with the idea of how to generate the waves in an entirely different manner. Townes' idea was to use nature's own oscillators, molecules and atoms, and use stimulated emission in a beam of molecules as a source for strongly amplified millimeter waves. A similar scheme was suggested independently by Joseph Weber, who did not, however, develop his scheme into a practical device. (This was the same Weber we met in chapter 23, who claimed to have detected gravitational waves.) The scientific basis of Townes's innovation, the concept of stimulated emission, went back to Einstein's 1917 radiation theory, but it was a concept that did not receive much attention until the 1950s.

In an important experiment in 1950, Edward Purcell and Robert Pound demonstrated stimulated emission and also population inversion—that is, the existence at nonequilibrium conditions of a higher energy state more populated than the lower energy state. In order to bring molecules from the lower into the higher state, they can be optically "pumped," that is, excited by absorbing incoming photons. The method of optical pumping was originally proposed by the French physicist Alfred Kastler in 1950 and verified experimentally by his group in Paris two years later. Stimulated emission, population inversion, and optical pumping were key ingredients in the maser, but until 1951 they lived separate lives and played no great role in physics. The last crucial ingredient was the idea of using a resonant cavity to produce positive feedback and thereby obtain a net gain in energy, a process that Townes originally described as "a self-sustaining [molecular] chain reaction." In 1951 all the main ingredients of the maser method were well known, but before Townes, nobody had combined them with the purpose of amplifying microwaves. One of the reasons for the late development of the maser concept may have been that it conjoined knowledge of the physicist with that of the engineer. Townes wrote, retrospectively, that "the necessary quantum mechanical ideas were generally not known or appreciated by electrical engineers, while physicists . . . were often not acquainted with pertinent ideas of electrical engineering. It is understandable that the real growth of this field came shortly after World War II since this brought many physicists into the borderland between quantum mechanics and electrical engineering" (Bromberg 1991, 222).

Townes's new idea of May 1951 was the maser concept or "Apparatus for obtaining short microwaves from excited atomic or molecular systems," as the entry in his laboratory notebook was titled. Instead of having the molecules responding passively to the stimulus of an electron beam, Townes conceived the idea of resonating molecules that would both generate the microwaves and carry the energy and frequency with them. Other physicists at the Columbia Radiation Laboratory had no confidence in Townes's scheme and argued that an alternative method of generating microwaves, based on Cerenkov radiation, looked more promising. However, Townes stuck to his idea of molecular oscillators and in the spring of 1954, he achieved the first success in the form of the detection of oscillations from ammonia molecules.

With the 1.25-cm amplifier that he built along with James Gordon and Herbert Zeiger, Townes could demonstrate the first "maser," an acronym for *m*icrowave *a*mplification by *s*timulated *e*mission of *r*adiation. Or, as it was introduced in the first report to *Physical Review*, "[a]n experimental device, which can be used as a very high resolution microwave spectrometer, a microwave amplifier, or a very stable oscillator." The name maser was introduced in a more comprehensive article of 1955, titled "The Maser—New Type of Microwave Amplifier, Frequency Standard and Spectrometer." At that time, an improved ammonia beam laser was ready and so were Townes's patent application, Columbia University's press release of the invention of the maser (or "atomic clock"), and Townes's report to the Signal Corps. In this first practical maser, built according to Townes's ideas of 1951, ammonia molecules were sent through an electrical filter system (a "focuser"), in which molecules in the excited state were separated from those in the lower state. The beam of excited molecules then entered a cavity where they created an oscillating electromagnetic field emitted as output microwaves. It was only after experiments with their first maser that Townes and his collaborators realized one of its most valuable properties, namely, that it was exceedingly free of noise. Moreover, the amplification obtained by introducing radiation into the cavity was not originally considered to be an essential part of the maser concept. At first, Townes paid no attention to what is perhaps the most characteristic feature of masers and lasers, the coherence of the waves. Of course, it was not the first time that a scientist made a brilliant innovation without a clear understanding of what he had done.

Townes's ammonia maser, based on population inversion between two energy states, was quickly followed by new maser types and methods of obtaining population inversion. In 1955, Nikolai Basov and Alexander Prokhorov at the Lebedev Physical Institute in Moscow suggested a pumping scheme that allowed continuous inversion of population, and hence continuous amplification. The scheme made use of quantum transmissions in atoms with three energy levels, of which the intermediate level was metastable (i.e., it had a long lifetime). The following year, a similar scheme was analyzed in greater detail and applied to solid state masers by Nicolaas Bloembergen, a Dutch-American physicist working at Harvard University. Bloembergen's work had a lasting influence on maser and laser physics, and quantum electronics in general, because it brought into these areas concepts from magnetic relaxation and nuclear magnetic resonance (NMR). The first efficient paramagnetic masers based on Bloembergen's three-level theory appeared in 1958. The successful family of ruby masers was pioneered by Chihiro Kikuchi and his coworkers at the University of Michigan. Because of their low noise temperature, solid state masers found early use in radio telescopes and microwave antennas. For example, Penzias and Wilson made use of a ruby maser amplifier in their celebrated 1965 discovery of the cosmic background radiation. Such applications were, however, exceptions. Contrary to its offspring, the laser, the maser never became commercially significant.

After 1955, the idea of applying the maser principle to smaller wavelengths in the visible or infrared range was a natural one. It was originally thought that this would be very difficult and that a very, perhaps prohibitively, large power input would be needed. It was again on the instigation of the military—this time the Air Force Office of Scientific Research—that Townes took up the problem. In 1957 he investigated it theoretically and reached the conclusion that there probably would be no serious problems in applying maser techniques to the visible region. He joined forces with Arthur Schawlow of Bell Labs and in 1958, the two physicists published a detailed analysis in *Physical Review* on "Infrared and Optical Masers," concluding that "the prospect is favorable for masers which produce oscillators in the infrared and optical regions." At the same time, AT&T filed a patent application. The Townes-Schawlow paper was the theoretical basis for the laser, but at the time it was still only a theory.

During the next two years, several physicists and engineers, both at universities and corporate research laboratories, competed to develop a operable laser. One of them was Gordon Gould at the private company TRG (Technical Research Group), who in the fall of 1957 had figured out independently how a laser might be constructed. It was also Gould who first used the term "laser" as an acronym for *l*ight *a*mplification by *s*timulated *e*mission of *r*adiation. Characteristically for the period, shortly after the Sputnik shock, most of the research was lavishly supported by military agencies. One of these agencies was the Advanced Research Projects Agency (ARPA), which was interested in promoting work on the laser as a possible weapon. In 1959, for nearly $1 million, ARPA engaged TRG in a secret program with the purpose of developing a functional laser. Using different methods and approaches, scientists at Bell Labs and IBM's Watson Center also worked hard to construct a practical laser. And so did Theodore Maiman, a physicist at Hughes Research Laboratories, who focused on solid laser media rather than the gas lasers investigated by the other groups. Maiman worked with the chromium ions making up a small part of a ruby crystal and found a way of obtaining population inversion by exciting the ions optically by means of light from a pulsating xenon flash tube. Experiments confirmed his calculations and on May 16, 1960 he obtained his first laser emission. Shortly afterwards, Maiman had a working ruby laser ready and quickly submitted an article to *Physical Review* on "Optical Maser Action in Ruby." The editor turned down the article, however, possibly because the paper did not live up to the journal's policy of admitting only papers that "contain significant contributions to basic physics." For this reason, the first announcement of the laser—a brief paper entitled "Stimulated Optical Radiation in Ruby"—appeared in the British journal *Nature*. In 1961 Maiman and his collaborators submitted a more detailed account of their experiments to *Physical Review*. This time, the paper was accepted. Initially, there was some uncertainty as to whether Maiman had really obtained a ruby laser action, and it took some time before he was recognized as the inventor of the laser.

In 1960–61, several groups of scientists developed other types of lasers. In late 1960, Ali Javan and his group at Bell Labs succeeded in constructing a gas discharge laser operating with neon atoms excited by collisions with helium atoms. This was an important innovation, because it was the first continously working laser. Javan's first helium-neon laser operated at the infrared wavelength of 1.15×10^{-6} m. Another important result in the early history of the laser was the construction in 1962 of a laser using a gallium arsenide crystal, the first example of the quickly growing family of semiconductor lasers. It took several years until laser research became a hot field of physics, but by the mid-1960s, the new field had entered a state of rapid development. A bibliography compiled in 1964 included about 600 references to lasers, including surveys and other non-research literature. Two years later, the updated bibliography contained 3,390 references, with an index of 3,335 authors. Most of the early papers reporting on laser research were published in *Physical Review* or the new *Applied Physics Letters*. As the field burgeoned, publications appeared increasingly in more specialized journals, such as *Optics Communications*, *Applied Optics*, and *Laser Focus*.

The laser was welcomed and studied as much by industrial as by academic physicists. About 70 percent of American laser papers listed in *Physics Abstract* in the early 1960s were submitted by industrial laboratories. Not surprisingly, a large part of laser research in the United States, whether at universities or in industry, was funded by the military. It has been estimated that DOD expenditures for laser research in 1963 were about $20 million, and that as many as 80 percent of laser research articles acknowledged at least partial support from DOD. The laser was of great importance not only because of its use as an instrument in science and its many applications in military and civilian technologies, but also because it was a major factor in the revival of optics that occurred in the 1960s. Optics had stagnated for two decades and about 1950, it was widely considered a somewhat dull discipline with a great past, but without prospects of a great future (compare with figure 24.1). Together with holography, optical uses of semiconductors, and the rise of nonlinear optics, the laser created a new and vigorous development in the science of optics.

Unlike the maser, the laser also became a smashing success on the commercial market. In this respect, its trajectory was similar to that of the transistor. In 1993 it was estimated that the annual laser market totaled a value of about $20 billion, and that more than three-fourths of it was civilian. The commercial applications of lasers are widespread and well known, including not only surgery and medicine, printing, and optical communications, but also compact discs, welding of metal parts, smart-card readers, and entertainment; in addition, lasers serve an important role in a variety of scientific instruments. The most produced and most versatile of the many laser types has been the semiconductor laser. In 1988, about two hundred million lasers of this type alone were produced.

The development of masers and lasers occurred primarily in the United States, but there were also important contributions from Russian physicists and, after 1960, Japanese and European physicists. At the Lebedev Institute in Moscow, Basov and Prokhorov realized the possibility of an ammonia maser independently of Townes, suggested the three-level method, and in 1958, proposed the use of semiconductors for lasers. Townes, Basov, and Prokhorov received the Nobel prize in 1964 for their contributions. Two years later, Kastler was awarded the prize and in 1981, two of the other pioneers of laser techniques, Bloembergen and Schawlow, were similarly honored. Maiman, the inventor of the laser, was not found worthy of the award.

Optical Fibers

The first extensive communictions system, Claude Chappe's optical telegraph of the 1790s, was based on light signals, but subsequent innovations in telecommunications—from electrical telegraphy to radio—made use of widely different parts of the electromagnetic spectrum and transmitted the signals through either metallic wires or the atmosphere. From about 1860, optical telegraphy largely disappeared. Although the idea of using light sent through a glass tube as a means of storing and transmitting information was considered in the 1930s, it was only after the war that experiments with optical fibers began. One of the first studies was made in the early 1950s by a group of Dutch scientists associated with the Delft University of Technology. Their project was sponsored by the Netherlands' Defense Research Council and aimed at providing the country's new submarines with improved periscopes—it was not only in the United States that the military supported research in physics. In order to ensure total reflection, the Delft scientists developed glass and plastic fibers coated with a layer with a lower refractive index. There was neither military nor commercial use for the fibers, however, and the Philips company declined to purchase and develop the technology.

The first practical use of clad fibers was not in communictions but in medicine, namely, in the flexible gastroscope invented in 1957. At that time, it was realized that in order to make fiber optics work over longer distances, two critical problems had to be overcome: First, a coherent and strongly focused light source was needed; second, the signal loss was prohibitively large in existing glass fibers to make them practical waveguides. With the invention of the laser, it seemed that the first problem was on its way to being solved. Whereas gas lasers were not practical, the potential for semiconductor lasers was recognized at an early date, especially because this type of laser was compact and easy to modulate by means of an input current. The first laser sources used in fiber-optical experiments were gallium-arsenide lasers, and as receivers various kinds of solid state photodiodes found applications. The semiconductor diode lasers pioneered by Bell Labs in 1970 were found to be even more useful as sources of optical communications. Alternatively, another AT&T invention could be used, the light-emitting diode or LED. Although transistor technology did not play a direct role in the development of optical fiber systems, it did play an indirect role, as a source of the semiconductor lasers and photodiodes.

The laser sparked much interest in optical communications, a field that until then had been given low priority by engineers and industrial scientists. With the source problem solved in principle, interest focused on the medium of transmission. The crucial question was whether it was possible to design thin fibers, made of glass or other transparent materials, with an optical attenuation vastly smaller than the best types of glass known at the time. The optical glass industry was unprepared for the kind of scientific and developmental work that was needed in order to produce glass of the extremely fine quality required for optical transmission. The whole question was thoroughly investigated by two physicists working for Standard Telecommunications Laboratories in London, Charles Kao and George Hockham. In a paper published in the *Proceedings of the Institution of Electrical Engineers*, the two physicists calculated and discussed various modes of loss in a fiber waveguide of circular cross section. The seminal work was based on classical electromagnetic theory and consisted essentially of solving the Maxwell equations under the boundary conditions given by the structure of a cylindrical glass fiber. It was a work that would have been understood and appreciated by Lord Rayleigh and other specialists in electromagnetic wave theory around the turn of the century—in fact, the first complete theoretical analysis of electromagnetic propagation in dielectric cylinders was performed as early as 1910, by Peter Debye and his student D. Hondros.

Kao and Hockham predicted that the critical value for signal loss in an optical fiber was 20 decibels per kilometer (dB/km) or less, corresponding to about 1 percent transmission through one kilometer of fiber. Available glass fibers exhibited a loss of about 1,000 dB/km; thus, the Kao-Hockham calculation showed that improvements of two orders of magnitude were needed (and recall that the dB is a logarithmic unit). This was bad news, but Kao and Hockham nonetheless concluded that it was "difficult but not impossible" to develop such low-loss fibers. "Theoretical and experimental studies indicate that a fibre of glassy material constructed in a cladded structure with

. . . an overall diameter of about 100 λ_o [100 times the wavelength] represents a possible practical waveguide with important potential as a new form of communication medium," they wrote. "The realisation of a successful fibre waveguide depends, at present, on the availability of suitable low-loss dielectric material" (Bray 1995, 271).

The work of Kao and Hockham was a challenge to the glass industry, and it prompted private and government laboratories throughout the world to take up the challenge. Bell Labs, national telecommunications laboratories, technical universities, and private industries engaged seriously in the attempt to find suitable materials for fiber optics. Corning Glass Works in the United States, the largest and most research-oriented glass company in the world, established a research group to find a solution. The group was led by Robert Maurer, a physicist with a Ph.D. from MIT who had joined Corning in 1952 as an applied physicist. The Corning team focused on fused silica, to which it added oxides in order to increase the refractive index. Not only did Corning's new fibers have excellent electromagnetic properties, but they also were mechanically strong and chemically stable. Interestingly, the innovation was inspired by the transistor industry's work with pure and doped silicon. The first doped-silica fibers were ready in 1970 after four years of often-frustrating work. With an attenuation factor of 16 dB/km, they were highly promising and indicated that the development of practical optical fibers was just a matter of time. During the following years, Corning developed fabrication methods for the new kind of glass fiber and produced still more efficient versions. In 1975 Maurer and his team had reached 4 dB/km and by 1980, the loss had been reduced to 0.3 dB/km. At the same time, Japanese researchers reported a record low attenuation of 0.2 dB/km, very close to the theoretical limit for doped silica fibers. In the late 1970s, fiber-optic communications had been transformed from an invention to an innovation, and the first commercial fiber cables came into operation. Among the most important uses of the new technology were submarine cables, the first one being the 1986 Channel link between England and Belgium. Two years later, the first transatlantic optical fiber cable, named TAT 8, was opened for service between England and North America. The innovative project, a collaboration among AT&T, British Telecom, and France Telecom, had a capacity corresponding to 40,000 telephone conversations, or almost double the existing cable capacity across the Atlantic. Fiber optics continued its rapid development through the 1980s and 1990s. One of the most important innovations was the invention in 1987 of a way to amplify signals in an optical fiber by inserting, directly into the line, a fiber doped with ions of the rare-earth element erbium. Together with the development of fiber lasers, the erbium-doped amplifiers initiated a new phase in optical fiber communications.

Chapter 26

SCIENCE UNDER ATTACK—PHYSICS IN CRISIS?

SIGNS OF CRISIS

FROM ABOUT 1965, U.S. physics, and to some extent also European physics, seemed to steer into more troubled water. The years since the Manhattan Project had been one uninterrupted and unprecedented period of growth, optimism, political support, and public goodwill. In the late 1960s, the war in Vietnam became a highly controversial issue in the United States, and in academic circles antimilitary sentiments became more acceptable and common. Because such a large part of the American physics community worked on military projects or was funded by military money, physicists easily became identified with the military-industrial complex that at the time was under fire. At about the same time, nuclear energy began to lose some of its innocence and fascination. For the first time, people began to ask whether nuclear power stations were really the unproblematic road to cheap energy that until then had been taken for granted. Could one be sure that reactors were safe? What about the radioactive waste? And how was it possible to prevent dictatorships from using nuclear power technology to produce atomic bombs? The early unease with nuclear energy was easily extrapolated to an unease with the scientists most closely associated with the technology, the physicists. A more direct consequence was President Ford's 1974 decision to abolish the Atomic Energy Commission and replace it, in 1977, with a new Department of Energy (DOE). It was of more than symbolic importance that the word "atomic" disappeared. The Vietnam war and the increased interest in environmental problems were only two of the more important factors in the complex web of political, cultural, and emotional reorientations that shook the Western world in that period.

At the same time as the traditional cultural values came under attack, the relationship between government and science transformed. Politicians and administrators began questioning the maxim that what was good for science was good for society. A more disillusioned attitude to science resonated with changes in the American funding system that caused concern among many physicists. The Mansfield amendment of 1969 prohibited the Department of Defense from spending money on research not directly related to military purposes. Many physicists were in favor of the amendment, but they had reasons to fear its consequences. Although the National Science Foundation (NSF) received an extra sum of $85 million, it was unable to compensate fully for the loss of support from the military agencies. In addition, an increased fraction of the NSF money was allocated to socially useful research projects, some of them in the social sciences, rather than to basic science. In 1970 William Koch, director of the American Institute of Physics, commented that physics, "from being regarded as a science desperately needed for national survival and prestige," had become placed in "a more conventional social context, with new priorities" (Schweber 1994b, 143).

For two decades, it had been accepted wisdom that investment in basic science would pay off greatly in the industrial sector and thus secure economic growth. Now, some leading economists denied the rationality in public support for science as an innovation factor. According to Harry Johnson, a professor of economics, there was no clear correlation between the amount of money a country spent on science and the country's economic growth. Basic science, he argued in 1968, was "purely and simply a luxury good, accessible only to the very few privileged people in the society who have the education to appreciate its esoteric mysteries and who have been able to persuade a sufficient number of their fellow citizens to support their activities" (Kragh 1980b, 39). It was a characteristic that fitted a large part of the physics community well and an argument that, if accepted, would lead to a sharp decline in government support of basic physics research. The new issue was that physicists felt an increased pressure to justify their requests for money. It was no longer enough to appeal to purely internal criteria of scientific curiosity or to postulate, as Schwinger did in 1965, that "[t]he scientific level of any period is epitomized by the current attitude toward the fundamental properties of matter. The world view of the physicist sets the style of the technology and the culture of the society, and gives direction to future progress" (Schweber 1997, 664).

In constant 1983 dollars, federal obligations for basic research in physics dropped from $1.045 billion in 1967 to $690 million in 1976—a 34 percent decrease. Support then slowly increased, but in 1983 it still amounted to less than 85 percent of the 1967 level. During the same period, DOD and NASA expenditures for basic research decreased much more drastically than NSF expenditures increased (see table 26.1). As support declined, so did the pro-

TABLE 26.1
U.S. Obligations for Basic Research in Physics, in Millions of 1983 Dollars

Source	*1967*	*1976*	*1984*
NSF	94.8	126.4	151.6
AEC/DOE	519.8	379.5	563.9
NASA	216.8	107.0	101.6
DOD	179.6	60.4	126.1
Total federal	1045.2	690.2	960.4

Note: Based on data in Brinkman et al. 1986.

duction of new physicists. At the beginning of the 1970s, more than 1,600 physics Ph.D.s were awarded at American universities. Ten years later, the number had decreased to little more than 900 (figure 26.1). The number of Ph.D.s in the largest subfield, solid state physics, decreased from 442 in 1971 to 202 in 1980; in elementary particle physics, the second largest subfield, the decline was from 278 to 117. Only optics experienced a growth, from 25 to 43, no doubt in part a result of the "laser effect" and the increased importance of optics in high-tech industries. The job market was generally unfavorable, and an appreciable fraction of the new Ph.D. physicists were unemployed or had to take jobs not related to their education. As in the early 1930s, the job crisis was limited and of a relatively short duration. By 1980, the "good old days" had largely returned, or so it seemed.

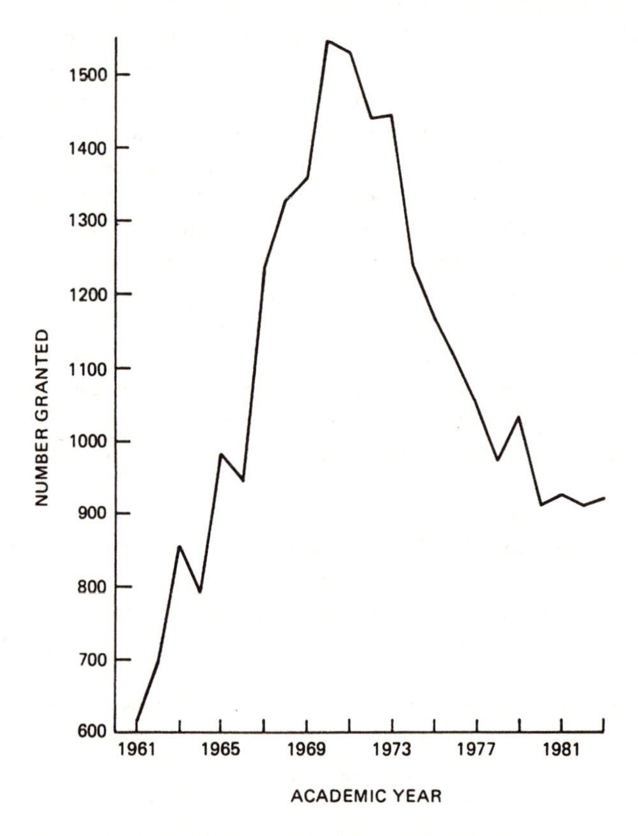

Figure 26.1. U.S. physics degrees, 1961–83. *Source:* Reprinted with permission from *Physics Through the 1990s: Elementary-Particle Physics*. © 1986 by the National Academy of Sciences. Courtesy of the National Academy Press, Washington, D.C.

Yet, the number of physics graduates continued to stay at a low level and, moreover, a large percentage of the graduates were foreign citizens, Europeans or Asians. It should be noted that American physics experienced a "crisis" only in a relative sense, namely, relative to the extreme support and goodwill it had enjoyed during the two previous decades. Moreover, the crisis concerned external factors and did not seriously affect the progress in scientific knowledge. Indeed, the years during which the crisis was most discussed, especially within the high-energy community, were just the years when high-energy physics made so many remarkable advances that people began to talk about "the new physics."

The crisis in American physics in the 1970s did not only manifest itself in terms of economic support and recruitment of new physicists. These problems were also connected with, and to some extent the result of, a more general "moral" crisis, which threatened to be even more serious than the crisis in material resources. Whereas physicists had been accustomed to being the heroes of the nation and physics the king of the sciences, from the late 1960s public confidence in the sciences fell rapidly. A 1971 poll found that only 37 percent of the public rated scientists "very favorable." And it was not only the public who began losing confidence in what was now referred to with increasing frequency as the "scientific priesthood." *The New Priesthood* was the title of a 1965 book written by Ralph Lapp, a nuclear physicist with a strong interest in the social and ethical consequences of physics. Dissident voices were heard, often loudly, from the physicists themselves, many of whom began to question the existing order of affairs. One of the main topics was the involvement of physicists, as a group, in social affairs. Were societal questions a legitimate field, if not of physics then of the physicists? More specifically, where did the American physics community stand with regard to military research, the arms race, and the war in Vietnam? Such questions were not new, but around 1970 they were asked more frequently and with more determination.

Charles Schwartz, a Berkeley physicist, suggested in 1968 that the American Physical Society (APS) should involve itself in public issues, and that concerns only with physics as physics were wrong. In Schwartz's view, the Society should speak out on the Vietnam war in particular, "not because we have any unique competence in this matter, but because we share an equal concern and responsibility with all other segments of American social structure" (*PT* January 1968, 9). The proposal aroused much discussion among APS members, many of whom supported Schwartz and demanded that APS take a stand. Others, of course, were horrified that APS should turn into a political instrument. As one physicist, employed at a military laboratory, put it, "We do not want the society to be degraded by falling into the same category as the 'hippies,' 'extremists' and other characters under outside influence" (*PT* February 1968, 15). That Schwartz's proposal was voted down by a solid majority did not imply that American physicists were socially indifferent, generally in favor of the government's policy in Vietnam, or happy to work for the Pentagon. There was a real and widespread concern with how to use and organize science in a more humane and socially just way. For example, in 1969 physicists established Scientists for Social and Political Action, an organization whose aim was to seek "a radical redirection of modern science and technology." The secretary of the organization was Martin Perl, a forty-two-year-old Stanford physicist who would later win a Nobel prize for his work in neutrino physics. The percentage of physicists who protested against the involvement in Vietnam was greater than that of any other academic discipline, social scientists and humanist scholars included. Surveys indicated that two-thirds of the leading academic physicists were sympathetic to radical student activism and that four out of five disapproved of classified research at the university laboratories. The opposition was not limited to the Vietnam war, but included the arms race and, in particular, the newly proposed antiballistic missile (ABM) system. At the 1969 APS spring meeting, more than one thousand physicists signed a petition to President Nixon, urging him to stop the ABM system and deploring "the beginning of a particularly dangerous, yet ultimately futile round of nuclear arms escalation when our expanding domestic crisis demands a reallocation of the national resources" (Easlea 1973, 311).

The decline in funding for physics in the 1970s and the new social awareness of many physicists were most marked in, but not limited to, the United States. In almost all European countries, government budgets for science as a whole leveled off, and within the budgets the physical sciences were relatively hard hit, although not as hard as in the United States. There was in this decade a general and universal switch of funding and public interest from physics to biology and the social sciences. It would hardly be correct to speak of a "crisis" in European physics, which, in spite of funding cutbacks and other problems, evolved steadily and fairly healthily. Of course, there really was no such thing as "European physics," only physics in the different European countries. The fragmentation among many nations was perhaps the greatest problem in what we shall nonetheless call European physics. The difficult economic situation around 1970 made Britain and France reduce their expenditures for research and development (R&D), and in Britain several large physics facilities were closed. Only in West Germany did physics experience a modest increase in support, and the country was also the only one that supported both CERN and its own national high-energy physics laboratory (the DESY facility).

An important step toward a more unified European physics system was taken in 1968 with the founding of the European Physical Society (EPS), with the Italian Gilberto Bernardini as its first president. The idea to establish such an organization came from a 1965 meeting of the Italian Physical

Society, where the participants agreed that physics' increasing involvement in social, political, and ethical problems necessitated greater unity among physicists in Europe. The EPS was basically a union, or umbrella organization, of existing national physics societies and scientific academies. In 1972 it included organizations from twenty-four countries, eight from communist Europe and fifteen from noncommunist Europe; as the only nonEuropean country, Israel was a member as well. The member organizations of the EPS represented about 36,000 members, making the EPS the world's largest professional organization of physicists.

Traditionally, each of the larger European countries had its own more or less international journals of physics. These often had a great past, but in the American-dominated postwar era, it was more than uncertain whether they would also have a great future. The first journal to be published entirely by the EPS, and in this sense the first truly European physics journal, was *Europhysics Letters*, which started in 1986. It incorporated the letter sections originally belonging to the Italian *Nuovo Cimento* and the French *Journal de Physique*. Although the new journal accepted manuscripts in English, French, German, and Russian, from the very beginning all papers were written in English. It had long been recognized that English was the international language of physics and that national journals in other languages had no real role to play in frontier physics research. By 1980, a leading physicist who could not speak and write English was a rarity, almost a contradiction in terms. After 1986, other European physics journals merged, closed down, or were reorganized under the EPS publication system. Rationalization was required and there simply were too many, and too unimportant, physics journals in Europe. The culmination came in early 1998, when two of Europe's most prestigious and historically important journals, the German *Zeitschrift für Physik* and the French *Journal de Physique*, merged and metamorphosed into the new *European Physical Journal*. Modeled after the *Physical Review*, the new journal appeared in five different sections. Of course, it appeared not only in paper, but also in electronic form.

By 1983 the crisis—if it really was a crisis—in U.S. physics had passed and Federal support began to show a substantial increase. The most drastic change was the amount of money allocated to defense research. President Reagan's 1984 budget requested a $6.9 billion increase for federal R&D, with almost the entire increase related to the Department of Defense. Whereas the defense part of federal investment in R&D was 48 percent in 1980, in the 1984 budget it was 67 percent. What happened during the 1980s was a remilitarization of science, with a role assigned to physics reminiscent of that of the 1950s—that is, as a supplier of new weapons technology and other militarily useful knowledge (see figure 20.1). The main reason for the renewed role played by military research was Reagan's Strategic Defense Initiative (SDI, or "Star Wars"), a technologically ambitious $26 billion research program aimed at developing a "nuclear shield" based on advanced laser and particle beam weapons designed to destroy attacking Soviet ballistic missiles. The initiative caused heated debates in the physics community, in many ways a replay of earlier debates related to the ABM system and the Vietnam war. Although many physics laboratories would be funded generously as a result of the SDI project, the majority of American physicists were less than enthusiastic about it. Within a year after Reagan's announcement, the Union of Concerned Scientists had enlisted support against the SDI from about 1,000 scientists, including 54 Nobel laureates. They argued that SDI was not only technologically naive and economically ruinous, but also "morally repugnant." As in earlier discussions about the role of physics in military projects, there was no agreement among the physicists. One of the Nobel laureates protesting against SDI was Robert Wilson, who in 1965 had discovered the cosmic background radiation. His codiscoverer and co-laureate, Arno Penzias, declined to sign the protest, arguing that it was "more polemic than science."

The upheaval in the world of physics included a new critical attitude toward physics research itself, both on the cognitive level and with regard to priorities and organization. Articles on "the crisis of physics" multiplied, both in the United States and in Europe. As a typical example of the new criticism, consider that of a young British physicist who, in 1974, attacked the big science system for leading to "elitism, competitiveness, 'grantmanship,' bureaucratisation, business management mentality, and concentration of economic and decision-making power"; and this was not all, for it also stifled "spontaneity, independence, originality, creativity, and even objectivity." In high-energy physics, he argued, the system had led to "a succession of fads that have become as popular as hula-hoops. . . . Younger physicists have come to realise that it is to their advantage to get in on one of these fads as early as possible, whether or not it really proves fruitful and whether or not they really believe in it" (Yaes 1974, 463). Some of the same themes were taken up by Michael Moravcsik, an American theorist, in a paper characteristically titled "The Crisis in Particle Physics." Moravcsik believed that elementary particle physics had gone all wrong, and that part of the reason was to be found in a mix of sociological and methodological factors. Conformism ruled, he claimed, the worst symptom being "fadism, that is, the concentration of efforts around a small number of fashionable topics and approaches, which are determined more by the personal qualities of some influential physicists rather than by the usual criteria of scientific merit" (Moravcsik 1977, 92). His criticism was answered by a representative of the particle physics establishment, the British theorist John Polkinghorne. Symptomatically, perhaps, in the 1980s Moravcsik went into studies of third-world science and Polkinghorne chose theology as a vocation instead of physics. Socially as well as cognitively, high-energy physics was in a state of flux in the 1970s.

The objections against the science system raised by American physicists were almost wholly concerned with the misuse of physics, the organization of physics, and the lack of social responsibility. It was a radical opposition, but only relatively so. In spite of all rhetoric, it did not question fundamental assumptions, such as the value of science or the accepted methods of scientific research. Moreover, it was a moral rather than an ideological criticism, based on explicit political analysis. The strong commitment of many European scientists to Marxism was not a feature found in the American debate, where "Marxology" was largely absent. Left-wing European physicists tended to be more directly inspired by Marxism and, in some cases, they demanded a more radical transformation of science. To mention but one example, the French physicist Jean-Marc Lévy-Leblond argued that ideology was inseparable from science and that part of the ideology of modern physics was the epistemological hierarchization of science that ascribed a privileged position to "fundamental" particle physics. (A similar point was raised by Philip Anderson and others in the United States, but without the ideological overtones.) Another part of the science ideology was the "traditionally elitist and meritocratic conception of science, . . . [and] the ideology of expertise and competence." How is it, asked Lévy-Leblond and his kindred spirits, that the more a science is involved in daily production and everyday life, the more it loses its "scientific" nature? Science, he went on in a classical Marxist argument, "has progressively moved away from its origins and ceased to draw its fruitfulness and inspiration from the mass of popular knowledge"; as an example, he mentioned water divining (an example of traditional folk knowledge) and suggested that this phenomenon should be investigated with the same seriousness as "a particular detail of the effective [cross-] section of proton-proton collision at high energy." And yet, even the radical Lévy-Leblond hesitated in advocating a specific Marxist physics. He admitted that high-energy physics was done essentially in the same way in the United States as in the Soviet Union, and probably even in Maoist China, and that "it is not easy to radically change a point of view on the internal content of any science" (Lévy-Leblond 1976, 166–69). Other prophets of the 1960s and 1970s were less hesitant in their claims for modes of science that differed fundamentally from the despised "capitalist science" epitomized by high-energy physics.

A Revolt against Science

Science criticism on a fundamental level, sometimes including the rejection of science as such, was not an invention of the 1960s. Throughout its history, science has been attacked continually and much of the criticism that became so visible in the 1960s and 1970s had its roots in the nineteenth century or even earlier. According to one strand of criticism, typically to be found in the writings of the British Marxist crystallographer John D. Bernal in the 1930s and 1940s, science is socially organized in the wrong way and used for the wrong purposes; it therefore needs to be changed. Yet Bernal and his comrades in the "red scientists" movement of the 1930s did not reject the values of science as such; they merely argued for deep changes in the social practice of science. Change the society—from a capitalist structure into one based on socialism—and science will be cured of its sickness. But there were other, older versions of science criticism that took a more radical position and the consequence of which was a negation of science rather than a redefinition of science. The romantic or utopian, and often anti-intellectual, tradition emphasized subjectively and intuitively gained insight in nature and opposed the canons of objectivity characteristic of ordinary science. Vitalism, antirationalism, and versions of *Lebensphilosophie* go back to about 1800, if not earlier. They experienced a revival in the 1970s.

Feeding on a variety of sources, some of them hardly compatible, there emerged from the late 1960s a new and vigorous form of science criticism which, in its more extreme form, argued for an abolition of science as ordinarily understood. Some of the criticism was inspired by the views of a group of mainly European sociologists and philosophers, including Herbert Marcuse in the United States, Jürgen Habermas in Germany, and Louis Althusser and André Gorz in France. In his influential *One Dimensional Man*, a classic of the student revolt, the German-American Marcuse argued that Western science was directed inherently toward domination of nature, as well as people. According to Marcuse and other philosophical gurus of the period, the essence of science was exploitation. Nature in her original or "anarchistic" state had been mutilated—raped—by the enforcement of scientific abstractions from outside; the scientific knowledge of nature had thereby become identical with dominance and exploitation. Moreover, the repressive technological society was founded on the physical sciences, and these sciences were therefore responsible for the repression and dehumanization that were characteristic of modern society. Marcuse claimed that "the mathematical character of modern science determines the range and size of its creativity, and leaves the nonquantifiable qualities of *humanitas* outside the domain of exact science" (Schweber 1994b, 144). That Marcuse and the other gurus of the period knew little about science, and even less about physics, is true but historically irrelevant. Their views were taken seriously by a large part of the younger generation, who rejected the scientific project and accepted the picture of physicists as soulless machines in the service of the military and industrial rulers. It is worth noting that there were distinct, although at the time unnoticed, parallels between the antiscience and radical-science movements of the 1970s and the attitude toward science that flourished in Germany in the early 1920s under the spiritual leadership of Oswald Spengler (see chapter 10). Likewise, there were parallels to the situation in the Soviet Union, the Third Reich, and elsewhere in the late 1930s. In 1942 the sociologist Robert K. Merton wrote about "a frontal assault on the autonomy of science" and warned against what he saw as a threatening anti-intellectualism. "The revolt from science," he wrote, "which [until recently] appeared so improbable as to concern only the timid academician who would ponder all contingencies, however remote, has now been forced upon the attention of scientist and layman alike" (Merton 1973, 267).

Belonging to a quite different tradition than the European social critics, Thomas Kuhn's 1962 analysis of the historical development of science unintentionally became another source for the discontent with science. With a Ph.D. in physics under John Slater, Kuhn had no wish to join the antiscience trend. In his best-selling *The Structure of Scientific Revolutions*, Kuhn argued against the positivistic view of science, suggested that there is no such thing as scientific progress across periods of revolutionary change, and hinted that science develops in a nonrational manner. The message, it seemed to many of his young readers, was that physics was no more scientific than psychology, art history, or literary criticism. Nor was modern astronomy to be believed any more than astrology. Developing some of Kuhn's themes, the Austrian-American philosopher Paul Feyerabend went further and attacked science and the scientific method as purely ideological notions in line with religion, myth, and propaganda; only science, contrary to religion and myth, had come to be the dominating dogma of modern time, executing a mental dictatorship on a par with that of the Roman Catholic Church in the Middle Ages. Feyerabend argued for an abolition of obligatory science instruction in schools and for a cessation of any kind of government support of science activities. "What's so great about science?" he asked. "What makes modern science preferable to the science of the Aristotelians, or to the cosmology of the Hopi?" (Feyerabend 1978, 73). According to Feyerabend, nothing at all.

The works of Kuhn and Feyerabend formed the background of other historical, philosophical, and sociological studies of science, the latest fashion being known as the program of sociology of scientific knowledge or social constructivism. Contructivist sociologists of the 1980s and 1990s denied that the scientific world view is grounded in nature and should therefore be given higher priority than any other worldview. Science, they said, is basically a social and cultural construction fabricated by negotiations, political decisions, rhetorical tricks, and social power. Since truth and falsehood are always relative to a given local framework, scientists' beliefs about nature are not inherently superior to those of any other group. In a book on the history of modern high-energy physics, the author, a young physicist turned sociologist, concluded, "The world of HEP [high-energy physics] was socially produced . . . there is no obligation upon anyone framing a view of the world to take account of what twentieth-century science has to say" (Pickering 1984a,

406). Such views, popular in large parts of the academic world, are not what most physicists like to hear. Social constructivists have been accused of contributing to an atmosphere of "higher superstition" and a revival of antiscience sentiments.

What has all this to do with physics? In a direct sense, very little. The views of Feyerabend and Marcuse were hardly what were most discussed in the physics laboratories. But indirectly, science criticism and the general discontent with science that resulted after about 1970 were also important to physics. As a hard and "masculine" science with a (deserved) reputation for close connections with military applications, physics came under attack more than most other sciences. Many bright students decided that studies in social sciences, or perhaps in the biological and environmental sciences, were more to their taste. In general, the popularity of physics among students fell markedly.

There were other ways to dissent from established physics than riding on the relativist bandwagon or claiming physics to be an expression of masculine and capitalist modes of thought. Fritjof Capra, an American particle theorist, believed that he had discovered a deep connection between modern quantum theory and forms of Oriental mysticism, such as Zen Buddhism. His *The Tao of Physics*, first published in 1975, became immensely popular because it resonated so well with the spirit of the decade. It was soon followed by a stream of other works in the same dubious genre. Capra's project was in a sense the very opposite of that of the science critics who accused physics of being a soulless, materialistic enterprise. According to Capra, the insights of quantum physics were the very same as those reached much earlier by Eastern mystics through meditation and intuition. The common insight included that physics was basically subjective, a reflection of the human mind rather than describing an independent nature made up of particles and fields. Moreover, Capra found in the "democratic" *S*-matrix theory the most convincing parallel with Eastern mysticism. That the *S*-matrix theory had been abandoned by the large majority of physicists was a fact that seemed to be of no concern to the author and was probably unknown by most of his readers. *The Tao of Physics* was representative of a trend connected with the counterculture that did not reject physics as such, but suggested alternative interpretations and extrapolations that, to many readers, were more appealing than the authoritative versions. David Bohm's alternative to the Copenhagen interpretation of quantum mechanics had its start many years before the 1970s, but it was only with the new cultural climate that he developed his ideas in a direction consonant with the mysticism and holism that characterized the "new age" movement. Bohm's ideas of an "implicate order" made him a cultlike figure in wide circles, but they made no impact at all on mainstream physics.

The End of Physics?

"Is the end in sight for theoretical physics?" was the title of Stephen Hawking's inaugural lecture when, in 1980, he assumed the prestigious Lucasian Chair of Mathematics at Cambridge University. Hawking considered it a realistic possibility that theoretical physics might indeed end in a not-too-distant future, possibly as early as the turn of the century. But what does it mean that physics, or science, comes to an end? According to Hawking, it meant "that we might have a complete, consistent, and unified theory of the physical interactions which would describe all possible observations." Because of recent progress made in fundamental physics, he felt that "there are some grounds for cautious optimism that we may see a complete theory within the lifetime of some of those present here" (Crease and Mann 1986, 410). Two points are worth noting. First, Hawking understood the phrase "end of theoretical physics" as the establishment of a unified and complete theory that encompassed, in principle, all special theories. Second, Hawking was optimistic that such a theory would be found; that is, he believed it would be a good thing. In the 1980s and 1990s, the end-of-physics theme became widely discussed and was the basis of a minor publication industry. Far from being a symptom of crisis, the popularity of the theme reflected the sense of progress that occurred in attempts to unify the basic laws of physics. After all, many physicists would say, the ultimate goal of fundamental physics is to produce a complete or final theory and then, in a sense, commit suicide. Having digested this final theory of the future, there would be nothing left to do, at least nothing of fundamental interest.

As Hawking was well aware, the end-of-physics theme has a long history. We recounted in chapter 1 how, in the 1890s, Michelson prophesied the end of physics, although what he had in mind was not a grand unified theory. Following the early triumphs of quantum mechanics, and especially Dirac's special-relativistic formulation of quantum mechanics in 1928, several physicists suggested that physics was approaching a state of completeness. The optimism was great and a general feeling prevailed that "physics was almost finished," as Peierls recalled. The feeling was famously expressed by Dirac in a paper of 1929. "The general theory of quantum mechanics is now almost complete," he wrote. "The underlying physical laws necessary for the mathematical theory of a large part of physics and the whole of chemistry are thus completely known, and the difficulty is only that the exact application of these laws leads to equations much too complicated to be soluble" (Kragh 1990, 267).

It soon turned out that Dirac's optimism was unfounded and that relativistic quantum mechanics instead gave rise to all kinds of serious problems.

Dirac's view of the final theory was, like Hawking's, reductionistic. A truly complete theory, in the shape of a set of equations, was expected to explain all phenomena in a deductive manner; in this sense, the phenomena could be reduced to special instances of the final theory. The ideal of theory reductionism can be found many years before Dirac and was, for example, the methodological basis of Gustav Mie's unified field theory from about 1912. Einstein, too, was a subscriber to the view that the ultimate aim of physics is "to arrive at those universal elementary laws from which the cosmos can be built up by pure deduction," as he wrote in 1918 in an address given in honor of Planck's sixtieth birthday.

Following the premature optimism of 1928–30, it took a long time until the quantum physicists resumed the discussion of the end of physics. Meanwhile, other physicists pursued the theme in a different manner. The deductive theories of Milne and Eddington, developed in the 1930s and 1940s, promised an a priori knowledge of the entire universe, including "explanations" of the natural constants, such as the fine structure constant. Eddington's unfinished "fundamental theory" was an ambitious attempt to provide a final theory and differed from earlier attempts only in being more unorthodox and obscure. In a 1949 article titled "Any Physics Tomorrow?" George Gamow asked if the future of physics would "present us with ever broadening horizons offering limitless possibilities for further explorations"; or, alternatively, would physics converge toward "a complete and self-consistent system of fundamental physical knowledge with all the ground thoroughly explored and with no new striking discoveries to be expected?" (*PT* January 1949, 17). Gamow argued that fundamental physics could, in principle, be based on four constants of nature (he chose c, h, k, and an elementary length), and suggested that if this happened physics would have met its end:

> If and when all the laws governing physical phenomena are finally discovered, and all the empirical constants occurring in these laws are finally expressed through the four independent basic constants, we will be able to say that physical science has reached its end, that no excitement is left in further explorations, and that all that remains to a physicist is either tedious work on minor details or the self-educational study and adoration of the magnificence of the completed system. At that stage physical science will enter from the epoch of Columbus and Magellan into the epoch of the National Geographic Magazine!

Gamow did not require a single unified theory from which all other theories could be deduced, only that the fundamental laws, whether connected or not, were known. In the same year as Gamow published his article, Einstein expressed in his *Autobiographical Notes* his deep belief that "Nature is so constituted that it is possible logically to lay down such strongly determined laws that within these laws only rationally, completely determined constants occur (not constants, therefore, whose numerical values could be changed without destroying the theory)" (Schilpp 1949, 63).

Richard Feynman was one more physicist who believed that scientific knowledge at the most fundamental level is finite, and that there will come a day when physics at this level ends. "This thing cannot keep on going so that we are always going to discover more and more new laws," he wrote in 1965. Like Gamow, he used the metaphor of geographical explorations to make his point clear: "We are very lucky to live in an age in which we are still making discoveries. It is like the discovery of America—you only discover it once. The age in which we live is the age in which we are discovering the fundamental laws of nature, and that day will never come again. . . . There will [in the future] be a degeneration of ideas, just like the degeneration that great explorers feel is occurring when tourists begin moving in on a territory" (Feynman 1992, 172). Although the ideas of Gamow, Einstein, and Feynman did not refer to a unified theory, their conceptions of the end of physics were quite similar to the views that became so popular near the end of the century. Steven Weinberg was one of many theorists in the 1990s who believed in a final theory from which flow all arrows of explanation. The theory would not be logically inevitable, but it would be logically isolated, that is, "every constant of nature could be calculated from first principles. . . . We would know on the basis of pure mathematics and logic why the truth is not slightly different" (Weinberg 1993, 237). Whatever the precise shape of the final theory, Weinberg was confident that it would be quantum-mechanical, a view shared by almost all unificationists.

Most particle theorists' visions of a final theory are reductionistic and based on a hierarchical view of the sciences that has been challenged by many philosophers, as well as a few physicists. Philip Anderson's criticism of the 1970s against the claimed epistemologic superiority of high-energy physics included the observation: "The ability to reduce everything to simple fundamental laws does not imply the ability to start from those laws and reconstruct the universe. In fact, the more the elementary particle physicists tell us about the nature of fundamental laws, the less relevance they seem to have to the very real problems of the rest of science, much less to those of society" (Anderson 1972, 393; see also chapter 21). Anderson pointed out that it is the solutions to the equations, not the equations themselves, that relate to physical phenomena, and the solutions are specific to the different levels of nature. Although atoms and molecules are built up of quarks and leptons, and these obey the laws of elementary particle physics, atomic physics, chemistry, and solid state physics cannot be reduced to elementary particle physics. If this view of "emergence" is accepted, there is no reason to expect that a final theory will be a theory "of everything." Hence the final theory will not herald the end of physics, only the end of fundamental high-energy physics.

"The end of physics" can be understood in a sense that has nothing to do at all with final, unified theories. Physics as we know it can end simply because people lose interest in it or because governments and funding agencies decide that there is no need to support basic physics research on a substantial scale. Such gloomy warnings are not new, but as the century came to a close, they were raised more frequently and more loudly than earlier. And they were raised by physicists who were worried about yet another period of crisis, enhanced by the lack of really exciting discoveries for more than two decades. To some, it seemed that the crisis of the 1970s had returned and that internal and external factors conspired to make life hard for the physicists. With the dissolution of the Soviet Union in 1990, the situation became particularly difficult for physics in Eastern Europe and Russia. It was little consolation to physicists in the Western world that they fared much better than their Eastern colleagues. Leo Kadanoff, a leading American condensed-matter theorist, noted in 1992 that "[s]cience is in low regard" and that the antiscientific attitudes of the 1970s were still politically important. "We are fast approaching a situation in which nobody will believe anything we say in any matter that touches upon our self-interest. Nothing we do is likely to arrest our decline in numbers, support or social value. . . . [T]oday when the public thinks of the products of science it is likely to think about environmental problems, careless or dishonest 'scientific' reports, Livermore cheers for 'nukes forever' and a huge amount of self-serving noise on every subject from global warming to 'the face of God.'" Serious as the situation was, in Kadanoff's view, it was not a prelude to the end of science; but it might well be a prelude to a much reduced role of science in the twenty-first century, and of physics in particular: "In recent decades, science has had high rewards and has been at the center of social interest and concern. We should not be surprised if this anomaly disappears. We will all be disappointed and hurt by this likely development. But if we can back away and look at the situation with some perspective, all of us in science can say that we have been lucky to be part of a worthwhile enterprise" (*PT* October 92, 10).

Chapter 27

UNIFICATIONS AND SPECULATIONS

The Problem of Unity

BY 1918 IT WAS known that Einstein's theory of general relativity described gravitation and the Maxwell-Lorentz theory electromagnetic phenomena. These were the two fundamental forces recognized at the time, and so it was natural to try unifying the two theories. The field theories of Weyl (1918) and Eddington (1921) were among the earliest unification schemes belonging to the new program of geometrization of physics. The approach of Theodor Kaluza, a mathematician at the University of Königsberg, was different from that followed by Weyl, but Kaluza's aim was the same. In a work of 1921, Kaluza attacked the problem of providing "a completely *unified picture of the world* . . . [which is] one of the great ambitions of the human spirit" (Vizgin 1994, 151; emphasis in original). Kaluza postulated a five-dimensional Riemannian space by adding to the four dimensions of the usual space-time an extra, hypothetical world dimension. The physical meaning of the fifth dimension was unclear, but Kaluza found that in this dimension, the trajectory of a particle was always a closed curve. Moreover, his five-dimensional field theory, based on fourteen potentials (ten gravitational and four electrodynamic) included both the general-relativistic theory of gravitation and the fundamental equations of electromagnetism. Einstein found the theory appealing at first, but soon he and most of the few other specialists in the field reached the conclusion that it had no connection to the physical world. Kaluza and the early unificationists did not incorporate the quantum world into their schemes and only vaguely realized that somehow, the quantum theory would change the game of unification. As early as 1916, in his paper on gravitational waves, Einstein commented that "the quantum theory must modify not only Maxwell's electrodynamics but also the new theory of gravitation." Kaluza likewise recognized that "any hypothesis that lays claim to universal significance is threatened by the sphinx of modern physics—quantum theory" (Vizgin 1994, 158). However, neither Einstein nor Kaluza made use of quantum theory in their unitary theories.

An extension of Kaluza's five-dimensional theory to cover quantum phenomena as well was suggested by Oskar Klein in 1926, shortly after the appearance of Schrödinger's wave mechanics. In Klein's theory, the fifth coordinate was nonobservable but nonetheless physically meaningful—namely, a quantity conjugate to the electrical charge. In this way, he hoped to be able to explain the atomicity of electricity as a quantum law and also account for the basic building blocks of matter, the electron and the proton. By assuming the five-dimensional space to be closed in the direction of the fifth coordinate with a certain period λ, Klein argued that "the origin of Planck's quantum may be sought in this periodicity in the fifth dimension." That is, he wanted to explain the quantum of action rather than accepting it as an irreducible constant of nature. As to the period, he suggested $\lambda = hc(2\kappa)^{\frac{1}{2}}/e$, with e as the numerical charge of the electron and κ Einstein's constant of gravitation. "The small value of this length," namely, $\lambda = 10^{-30}$ m, "may explain the non-appearance of the fifth dimension in ordinary experiments as the result of averaging over the fifth dimension," Klein wrote in *Nature* in the fall of 1926. The five-dimensional Kaluza-Klein theory attracted considerable interest in the late 1920s and was investigated by many theorists, including Vladimir Fock in the Soviet Union, Léon Rosenfeld in Belgium, Louis de Broglie in France, and Dirk Struik in the United States. From about 1930, however, most physicists lost interest in the theory, which seemed remote from physical testing and application. Like so many candidates for a unified theory, it remained on the periphery of physics for many years, considered to be speculative and of mathematical interest only.

Much later, however, the Kaluza-Klein theory experienced a revival. In the mid-1970s, Joël Scherk, Edward Witten, and others reconsidered the theory in connection with the new theories of supergravity that extended Einstein's general theory of relativity. Supergravity, a supersymmetric version of general relativity, was invented in 1976 by Daniel Freedman, Peter van Nieuwenhuizen, and others. Around 1980, the many-dimensional Kaluza-Klein theory was a hot area in theoretical physics, interacting and competing with new schemes of quantum gravity, such as superstring theory. Whereas Kaluza and Klein could do with five dimensions, the most popular of the reborn Kaluza-Klein theories sought to obtain unification by means of an eleven-dimensional space. However, although the supergravity version of the Kaluza-Klein theory was found to be mathematically fascinating, it predicted particles that had no similarity with those known in nature. After much work and considerable enthusiasm, it turned out that the eleven-dimensional Kaluza-Klein theory did not work any better than the original model from the 1920s.

The difficult problem of harmonizing quantum mechanics and general relativity was attacked from many different angles. One way was the attempts to quantize the gravitational field, typically in a manner similar to that of quantizing the electromagnetic field in quantum field theory. With the Heisenberg-Pauli quantum electrodynamic theory of 1929, it seemed for a while that weak gravitational fields might be treated in the same manner as electromagnetic fields. Heisenberg and Pauli claimed, overoptimistically, that a quantization of the gravitational field "may be carried out without any new difficulties by means of a formalism wholly analogous to that applied here [their 1929 paper]." It soon turned out that the problem was much more difficult than anticipated. The first one to apply the new field quantization procedures to gravitation was Rosenfeld, who in 1930 studied what he called "gravitational quanta." A fuller and more ambitious attempt to integrate quantum mechanics and general relativity was made by the Russian physicist Matvei Bronstein in works between 1933 and 1936. Bronstein discussed unified "*cGh* physics" and examined the quantum limits of the general theory of relativity at what later would be called the Planck length, $l_P = (hG/c^3)^{\frac{1}{2}} = \sim 10^{-35}$ m. However, Bronstein's works attracted very little attention at a time when the focus of physics was on the atomic nucleus, cosmic radiation, and the problems of quantum electrodynamics. It was only in the 1950s, with the renewed interest in general relativity, that the problem of quantizing the gravitational field was taken up by more than a few physicists. And it was also only then that the Planck scale, including the Planck mass—$(ch/G)^{\frac{1}{2}}$ or about 10^{-5} gram—were explicitly discussed in the physics literature, first by Klein and John Wheeler. Dirac, Peter Bergmann, and others worked on new, so-called canonical methods of quantizing gravitation, an approach that culminated in 1958, when Dirac succeeded in putting the general theory of relativity into Hamiltonian form. With the theory in this form, it was relatively easy to apply the rules of quantization to gravitational fields. Dirac suggested in 1959 that the resulting quanta of gravitation be called "gravitons," a name that immediately became part of the physicists' vocabulary. He was, apparently, unaware that the name graviton for a quantum of gravitation energy had already been introduced by Dmitri Blokhintsev and F. Galperin in 1934, in a paper published in Russian only.

Grand Unified Theories

By 1973, the electroweak theory and the new quantum chromodynamics (QCD) were in place and widely accepted as good theories for the weak, electromagnetic, and strong interactions (chapter 22). The symmetry associated with the Weinberg-Salam electroweak theory was known as a spontaneously broken SU(2) × U(1) symmetry, that of QCD a SU(3) symmetry. The SU(2) × U(1) gauge theory was the first successful unification in modern particle physics, but it was not a truly unified theory in the strong sense of the term. For example, it operated with two different interactions and associated coupling constants. The extension of the electroweak theory to cover strong interactions as well was most directly achieved by extending SU(2) × U(1) into SU(3) × SU(2) × U(1). In this case, the theory would deal with both leptons and quarks, but it would still be based on different groups, different coupling constants, and different elementary particles. In early 1974, Howard Georgi and Sheldon Glashow at Harvard University suggested in a programmatic paper that a proper unification of the three forces of nature should be based on a simple group known as SU(5), which was different from just a combination of the SU(3) and SU(2) × U(1) groups. "We present a series of hypotheses and speculations," the two theorists wrote, "leading inescapably to the conclusion that SU(5) is the gauge group of the world—that all elementary particle forces (strong, weak, and electromagnetic) are different manifestations of the same fundamental interaction involving a single coupling strength, the fine-structure constant" (Zee 1982, 46).

In this first example of a grand unified theory, Georgi and Glashow argued that within the SU(5) picture, the three coupling constants would merge into a single one at very large energies or, what amounts to the same thing, extremely small distances. Contrary to so many other unified theories, the SU(5) theory related directly to empirical physics. According to Georgi and Glashow, a certain experimentally determinable quantity in electroweak theory (the parameter $\sin^2\theta_w$, where θ_w is called the mixing or Weinberg angle) could be calculated to be 3/8. In the Weinberg-Salam electroweak theory the quantity could be determined only by experiment. Another appealing feature of the SU(5) theory was that it explained the numerical charge equality of protons and electrons, that is, why hydrogen atoms are neutral. Until the advent of SU(5), the charge equality had been considered just a contingent fact of nature, without any foundation in fundamental theory. In the early 1960s, in response to a suggestion by Hermann Bondi and Raymond Lyttleton that there might be a slight but cosmologically important charge difference, experiments proved that the charge excess, if it existed, was smaller than one part in 10^{20}. With the Georgi-Glashow theory, there were now theoretical reasons to believe that the charges were exactly the same. More dramatically, the theory predicted that the proton was unstable. This would also be the case with the neutron, even when bound in nonradioactive atomic nuclei. The SU(5) unified theory incorporated not only quarks, leptons, and the electroweak exchange particles (the photon and the heavy *W* and *Z* bosons), but also a group of superheavy, colored vector bosons known as *X* particles. These were required to have fractional electrical charges, 1/3 or 4/3 of both signs. Interestingly, the new particles would cause transitions between quarks and leptons, a kind of reaction not allowed by earlier theories. As an example, Georgi and Glashow mentioned the process in which a proton decays into a positron and a neutral pion by a mechanism where one of the proton's up-quarks transmits an *X* particle to its down-quark. This was indeed a remarkable prediction, and one of the reasons for Georgi's and Glashow's remark: "Our hypotheses may be wrong and our speculations idle, but the uniqueness and simplicity of our scheme are reasons enough that it be taken seriously."

The Georgi-Glashow approach was quickly developed by Georgi, Helen Quinn, and Steven Weinberg, who, a few months later, applied renormalization group techniques to calculate the effective coupling constants as functions of the energy. They realized that the original value for the mixing angle parameter, 0.375, was valid only at the extreme energy characteristic of unification (about 10^{15} GeV) and found a value of about 0.20 at low energies. In 1974 this predicted value was markedly smaller than the best experimental value (about 0.35), but later experiments gave results in much better agreement with the theoretical value. By 1980, the experimental value had moved down to 0.23 ± 0.02, which naturally encouraged physicists to take the theory seriously. The calculations of Georgi, Quinn, and Weinberg also led them to estimate the mass of the superheavy bosons to be as as large as 10^{17} GeV, not far from the Planck mass and therefore suggesting a possible connection to gravitation theory. As to the proton's lifetime, the three physicists supplied the first theoretical estimate, namely, between 10^{31} and 10^{32} years. In accordance with the suggestion of Georgi and Glashow, the proton decay was predicted to be mediated by an *X* boson that would cause the proton to change into, for example, a neutral pion and a positron. Because of the pion's decay into electron pairs generated by gamma photons, the decay would leave a clearly detectable track of electron trajectories ($p^+ \rightarrow \pi^o e^+$; $\pi^o \rightarrow \gamma\gamma$; $\gamma \rightarrow e^- e^+$). With the works of Georgi, Glashow, Quinn, and Weinberg, as well as independent work done by Abdus Salam and Jogesh Pati, the SU(5) unified theory was successfully launched as the first and prototypical example of a grand unified theory, or GUT for short. It was quickly taken up by many theorists who realized that this might be an important step toward the long-sought unification of the forces of nature. According to one physicist, "The idea of grand unification liberates us from our dreary existence amidst the debris of symmetry breaking, freeing us to dream about the 'true' physics at 10^{16} GeV in a magnificent leap of imagination probably unprecedented in the history of physics." The prediction of the proton's instability, and thereby the instability of all matter, was called "the most stunning prediction of our times" (Zee 1982, 260 and 242). By 1981, a small but expanding grand unification industry was in existence. A review of articles published in this year included around 800 references.

The possibility that protons might be unstable had been briefly mentioned, only to be rejected, by Weyl in 1929. In 1938 Ernst Stueckelberg stated explicitly, "No transmutations of heavy particles (neutron and proton) into light particles (electron and neutrino) have yet been observed in any transformation of matter. We shall therefore demand a conservation law of heavy charge" (Pais 1986, 488). Stueckelberg's empirical law of nucleon conservation led in the postwar period to the law of baryon conservation. (The name "baryon" was proposed by Abraham Pais in 1954.) Although this law was generally taken for granted, a few experimentalists looked for indications of proton decay many years before the topic became fashionable as a result of the SU(5) predictions. In 1954 Frederick Reines and Clyde Cowan were preparing the experiments that, two years later, would result in the detection of the neutrino. Reines and Cowan, collaborating with Maurice Goldhaber, realized that their system of detectors could also be used to set a lower limit on the proton lifetime, and they concluded in 1954 that the lifetime was greater than 10^{22} years. Later and more elaborate experiments performed two miles below the earth's surface in a South African gold mine in 1974 resulted in a lower limit of 2×10^{30} years. This was in the neighborhood of the theoretically predicted value, but the experiment, primarily aimed at neutrino detection, had no direct connection to the prediction of proton decay. In fact, by 1974 it had been running for almost three years in order to achieve the necessary sensitivity. All the same, when Reines and M. F. Crouch published the result in early 1974, they did relate their experiment to "recent interest in the possibility that baryon conservation may not be an absolute principle."

When the GUT prediction became known, and especially after new calculations of 1979 showed a lifetime of 10^{31} years to be probable (rather than 10^{37} years, as some theorists had calculated), interest in proton decay experiments grew dramatically. The result was a series of big science grand unification experiments in the United States, India, Japan, the Soviet Union, and Europe. One of the first, by a collaboration of American laboratories, was performed in a deep salt mine in Ohio, filled with 10,000 tons of very pure water. The detector system weighed about 8,000 tons and employed more than 2000 photomultiplier tubes. No decaying protons were found in the experiment; although other experiments in the 1980s reported a few possible candidates for proton decay, these could not be substantiated. The situation did not change materially during the following decade. The experiments grew bigger and more expensive, but no sure signs of proton decay were found.

That grand unified theory is not only mathematical calculations, but may involve some very big science as well, is illustrated by the Japanese-American Super-Kamiokande project. This is the world's largest neutrino observatory, a water Cerenkov detector located in the Kamioka mine north of Tokyo. The Super-Kamiokande water tank has a volume of about 50,000 cubic meters and is surrounded by about 13,000 photomultiplier tubes sensitive enough to detect single photons. The detector has looked for nucleon decay since it became operational in 1996, but so far the search has been in vain. Did the failure to observe proton decay mean that the grand unified theory was disproved? No; the experimental results showed that the original SU(5) theory was inadequate, but they were not interpreted as a failure of the GUT scheme as such. Many other forms of GUTs were on the market, leading to different lifetimes and decay modes of the proton, and so the GUT scheme itself was not easily falsifiable.

The predicted decay of the proton was not the only reason why GUTs were considered extremely interesting areas of research around 1980. The *X* bosons and their effects were undetectable, or nearly so, at ordinary temperatures, but they were expected to play important roles at the much higher temperatures governing the early universe. In order to study the physical effects of the *X* particles, the very early big bang universe was the ideal laboratory (and, since energies of the order of 10^{15} GeV were far beyond those produced by any conceivable accelerator, the only realistic one). As early as 1967, the Soviet physicist Andrei Sakharov—perhaps better known as a political dissident—had suggested that baryon number might not be conserved exactly and that the nonconservation might be cosmologically important. The suggestion did not attract much interest, and it was only in 1978, when the Japanese Motohiko Yoshimura used the new GUTs to predict a baryon-antibaryon assymmetry caused by primordial *X* bosons, that cosmology became an important part of GUT, and vice versa.

Although the Super-Kamiokande has not found any decaying protons, it has not operated in vain. In 1998 data from cosmic ray neutrinos indicated the existence of neutrino oscillations, that is, that muon neutrinos change into some other kind of neutrinos that are not ordinary electron neutrinos. The reports galvanized the particle physics community, for oscillations mean that at least one of the involved neutrinos must possess mass and that the standard model must therefore be modified.

Superstring Theory

Impressive and comprehensive as the grand unified theories were, according to some physicists they were not impressive and comprehensive enough. From a methodological point of view, it was seen as a blemish that they were not based on a fundamental principle and that they included several free parameters, such as the coupling constants and the quark and lepton masses. Also, GUTs did not include gravity and were therefore incomplete. The most popular candidate of the late twentieth century for a completely unified theory, the superstring theory, included a quantum theory of gravity and avoided free parameters. However, the theory was not originally developed as an attempt to solve the old puzzle of unifying quantum mechanics and general relativity.

Modern string theory had its beginning in the context of the *S*-matrix–inspired program for understanding the strong interactions. In 1968 the Italian theorist Gabriele Veneziano suggested a model of hadrons in order to explain certain collision processes. About two years later, Yoshiro Nambu and others interpreted Veneziano's theory as a string, rather than a particle, model. However, it quickly turned out that string models of the Veneziano-Nambu type were unphysical: They described only bosons, required a 26-dimensional space-time, and predicted a massless spin-2 particle as well as a tachyon as a ground state of the mass spectrum. Tachyons, hypothetical particles that move faster than light, have imaginary masses; although they are consistent with the theory of relativity, they are not acceptable quantum particles.

In the early 1970s, a somewhat more realistic string theory, a ten-dimensional "dual pion model," was found by John Schwarz and André Neveu. The new theory accomodated fermions as well as bosons. In order for the early string model to be a consistent theory, space-time had to be endowed with ten dimensions, one temporal and nine spatial. The inevitable occurrence of a massless spin-2 particle was originally seen as a flaw, for no such particle could be of any relevance to hadron physics. However, in 1974 Schwarz, in collaboration with the French theorist Joël Scherk, turned the difficulty into a virtue by realizing that the particle might be a graviton. Until then, string theory had been a theory of the strong interactions, with no reference to gravitation. The theory was now reinterpreted as a candidate for a unified theory of all the fundamental forces. It turned out that if the constant of gravitation was to have its correct value, the length scale of the strings needed to be close to the Planck length—that is, immensely smaller than the original hadronic strings. The objects of the theory, strings, were very different from the point particles (leptons, quarks) appearing in the standard model. Strings are one-dimensional curves of extension 10^{-35} meters that can exist in two versions, open and closed. The tensions of the strings are enormous, corresponding to energies of 10^{19} GeV. The excitations or vibrations of a string are interpreted as giving the spectrum of elementary particles, hopefully including those known empirically and definitely (and somewhat embarrassingly) including many others. In the superstring theories, there are many particles corresponding to a given type of vibration and, all of the known particles are supposed to be described by the string's ground state. The necessary ten dimensions are six too many, but the six extra dimensions are curled up or "compactified" in a manner similar to the fifth dimension in the Kaluza-Klein theories. They will thus not be observable.

Interest in string theory was limited in the 1970s, in part because of the successful development of QCD following the "November revolution" in 1974. Still, a small number of theorists continued working on string theory, which by 1980 had become superstring theory—that is, incorporating space-time supersymmetry. In supersymmetric theories (SUSYs), all particles have hypothetical partner particles, such as "squarks," "selectrons," "photinos" and "gravitinos." For example, a selectron is a spin-zero supersymmetrically transformed electron. Neither the selectron nor other exotic supersymmetrical partners have been found. One of the advantages of incorporating supersymmetry was that the tachyonic ground state would then disappear. Superstring theories, as they existed by the early 1980s, were mathematically interesting but not likely to be of much physical relevance. Even on the theoretical level there were severe problems, namely, that the theories were plagued by infinities and what are technically known as anomalies. Anomalies are terms that violate the symmetries or conservation laws when the theory is quantized, and therefore make the theory inconsistent. There were other consistency problems, such that the ten dimensions seemed incompatible with the correct form of superstring theory.

In the summer of 1984, Schwarz and his British collaborator Michael Green, drawing on work by Edward Witten and others, brought new life to the theory by showing that all of the anomalies would cancel each other out if the theory were governed by one of the internal symmetry groups known as SO(32) and $E_8 \times E_8$. In the SO(32) superstring theory, charge conservation arose as a result of including gravity, and it was this theory that Schwarz and Green first developed. Shortly after this breakthrough, Witten suggested how the SO(32) superstrings could be compactified to get a four-dimensional theory. The Schwarz-Green paper initiated what has been called the "superstring revolution," occurring almost exactly ten years after the "revolution" in conventional gauge field theory. Another important component of the 1984 revolution was the development of a new version of superstring theory, known as the heterotic theory, by David Gross and his collaborators at Princeton University. The heterotic string theory built on the $E_8 \times E_8$ group and permitted only closed strings. Being a mixture of the older 26-dimensional boson string theory and the new ten-dimensional supersymmetric theory, the heterotic theory was considered promising with respect to connections to the low-energy world. The remarkable thing about the three consistent superstring theories known by the late 1980s was that they were unique and completely free of adjustable parameters. The mathematical structure of the theory of superstrings was so tightly knit that it could not be changed without falling apart. This was a feature that appealed strongly to many theoretical physicists. In stark contrast to earlier unified theories within the gauge field tradition, the entire development of superstring unification was mathematical. There was no experimental input from the laboratories, and there were no suggestions of new experiments based on the superstring theory.

With the developments in the fall of 1984, superstring theory became very popular among mathematically minded theorists. A kind of bandwagon effect occurred. Whereas the annual number of papers dealing with string theory had been fewer than one hundred during the years 1975–83, in 1987 about 1,200 papers were published on the subject (see figure 27.1). Super-

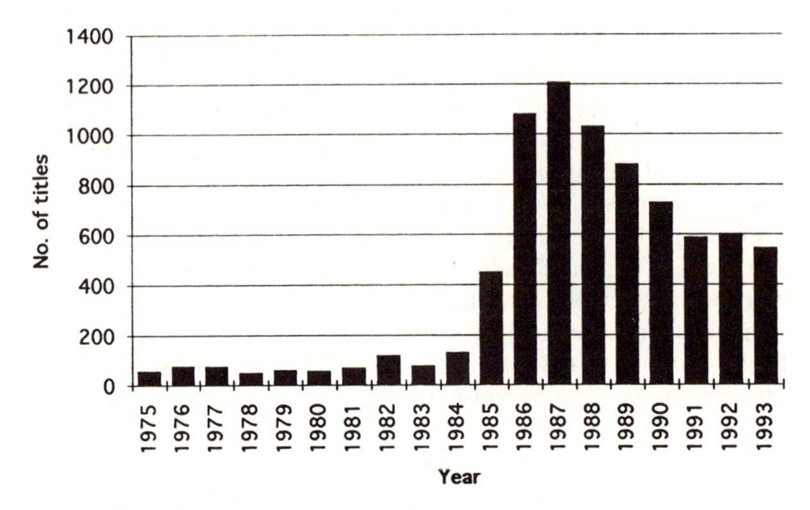

Figure 27.1. Number of papers on strings and superstrings. *Source:* Galison 1995. Reprinted by permission of Walter de Gruyter GmbH & Co.

string theory was hailed as a great accomplishment, the near-fulfillment of a century-old dream, the stepping stone to a new physics, the sought-after holy grail of a quantum theory of gravity. At least, this was how the string theorists themselves looked at the situation. Witten expected that the soon-to-be-expected proper understanding of string theory would "involve a revolution in our concepts of the basic laws of physics—similar in scope to any that occurred in the past" (Davies and Brown 1988, 97). He proclaimed that the theory "will dominate the next half century, just as quantum field theory has dominated the previous half century" (*PT* July 1985, 20). And Schwarz suggested that the heterotic theory could well be the mythic "theory of everything" (TOE) that, in principle, would imply the end of fundamental physics.

Not all physicists were happy about the self-proclaimed superstring revolution. The main problem with the superstring theory, it seemed to many, was its glaring lack of connection with experiment. The theory did not predict anything nontrivial that could be tested; although it was believed to include physics at lower energy, the particles and results of ordinary physics could not be actually deduced from the theory. It was no wonder that many high-energy experimentalists received the superstring revelation with less than complete enthusiasm. Several years after the superstring breakthrough in 1984, an often-heated debate began. Experimentalists were not alone in opposing a theory that could be "tested" only by means of mathematics; they were joined by theorists who disliked the theory's lack of contact with experiment. Among the opponents or skeptics were Feynman, Glashow, Schwinger, and Georgi, whereas Weinberg and Salam were positive and eagerly took up superstring theory. Glashow was one of the most outspoken opponents of the superstring theorists, whom he criticized for arrogantly replacing the empirical notion of truth with purely mathematical considerations of uniqueness and beauty. "Until the string people can interpret perceived properties of the real world, they simply are not doing physics," Glashow wrote in 1988. He asked, rhetorically, "Should they be paid by universities and be permitted to pervert impressionable students? . . . Are string thoughts more appropriate to departments of mathematics or even to schools of divinity than to physics departments?" (Galison 1995, 399). To Glashow, the string theorists were "kookie fanatics following strange visions" and the superstring fashion a disease "far more contagious than AIDS" (Davies and Brown 1988, 191). Weinberg recognized the difficulty in associating superstring theory with a physical picture that could be tested experimentally, but did not consider it a flaw:

> The final theory is going to be what it is because it's mathematically consistent. Then the physical interpretation will come only when you solve the theory and see what it predicts for physics at accessible energies. This is physics in a realm which is not directly accessible to experiment, and the guiding principle can't be physical intuition because we don't have any intuition for dealing with that scale. The theory has to be conditioned by mathematical consistency. We hope this will lead to a theory with solutions that look like the real world at accessible energies. (Davies and Brown 1988, 221)

Interest in string theory declined in the early 1990s, but by 1998, optimism and excitement had returned to the string community. The main reason for this was new developments in what became known as M-theory, P-brane theory, and D-brane theory, which are kinds of generalized string theories. The new work promised ways to connect the results of string theories and those of more conventional gauge field theories.

Quantum Cosmology

The idea of using particle physics to inform and constrain cosmological theories goes back to the very beginning of elementary particle physics in the early 1930s, and played an important role in the pioneering works of Gamow, Alpher, and Herman in the late 1940s. With the rapid development of high-energy physics in the 1960s and 1970s, the close connection between cosmology (and astrophysics) and particle physics strengthened. According to some physicists, the connection changed the entire status of cosmology: "It appears . . . that cosmology has become a true science in the sense that ideas not only are developed but also are being tested in the laboratory. . . . This is a far cry from earlier eras in which cosmological theories proliferated and there was little way to confirm or refute any of them other than on their aesthetic appeal" (Schramm and Steigman 1988, 66). Moreover, the confidence in the hot big bang model led people to use this model to gain insight into particle physics at the extreme energies of the very early universe, energies far beyond what any accelerator on earth could produce. In this way, the early universe became "the poor man's accelerator," as it was often called, and high-energy physics and cosmology entered an increasingly symbiotic relationship. Whereas nuclear astrophysics was a well-established field in the 1950s, particle astrophysics emerged as a vigorous new subfield of physics in the 1970s. For example, the symbiosis was manifest in the journal *Astroparticle Physics*, founded in 1992.

According to the big bang scenario, as it was accepted by many specialists around 1980, the initial phase of the ultradense and ultrahot Planck universe (the first 10^{-43} seconds) was governed by quantum gravity, and therefore outside the reach of current scientific knowledge. With expansion followed cooling and lower—if still gigantic—energies of 10^{16} GeV or more. In this phase the strong, weak, and electromagnetic forces were united, in accordance with the theories of grand unification. Further expansion, now at a time between 10^{-33} seconds and 10^{-2} seconds after time zero, implied a freezing out or symmetry breaking of first the strong and next the weak and electromagnetic interactions. At the end of this phase, production of both the X bosons and the W bosons were believed to have stopped and color forces come into existence in the form of nucleons. The subsequent cooler periods were of less interest to the high-energy physicist. They corresponded to energies obtainable in the laboratory.

As mentioned in chapter 23, one of the first and most important particle results arising from big bang calculations was that the number of neutrino species could at most be four, in order to agree with the observed abundance of helium, a result obtained in 1977 by Gary Steigman, David Schramm, and James Gunn. Further refinements led to a lower limit of three neutrino families. The prediction, based solely on cosmological arguments, was confirmed in 1993 when results from CERNs Large Electron Positron (LEP) machine indicated that there were indeed three species of neutrinos. As noted by Schramm: "In some sense this was the first time that a particle collider had been able to test a cosmological argument, and it also showed that the marriage between particle physics and cosmology had indeed been consummated" (Schramm 1996, xvii). The boundary on the number of neutrino species presumed that the neutrinos were massless, or almost so. The possibility of heavy tau neutrinos could not be excluded, but calculations from 1991 showed that the primordial nucleosynthesis required the mass of the tau neutrino to be less than 0.5 MeV. This constraint agreed with, but was finer than, the one obtained experimentally, and thus afforded another test of the early big bang scenario. The restriction of the number of neutrino families to three was generally taken to imply that the number of families or "generations" of all fundamental particles (leptons and quarks) was limited to three as well.

The mentioned cosmological calculations relied on and supported the standard electroweak theory, but not the GUTs including strong interactions. The new grand unified theories of the SU(5) type were immediately applied to address questions in cosmology that had so far remained beyond explanation. One of the questions concerned the almost complete predominance of matter over antimatter. Why is there not, as far as one can tell, any antimatter in the universe? Following up on Dirac's speculations of 1933, Maurice Goldhaber suggested in 1956 that the particle-antiparticle symmetry be extended to the entire universe. He proposed that the universe might consist of a cosmos and an "anticosmos" separated from the very beginning of their existence. A few physicists agreed with Goldhaber that the universe is symmetric between matter and antimatter on a very large scale, but this was a minority view. In the 1960s, the Swedish physicist and later Nobel laureate Hannes Alfvén suggested a cosmological theory based on equal amounts of the two forms of matter. His theory was ignored or rejected, not only because it was launched as an alternative to the big bang model, but also because there was no trace of the vast amounts of antimatter it required. By 1975, the question of antimatter had been discussed for two decades, but no explanation had been found. The accepted view was that matter asymmetry, or baryon number inequality, was for some reason part of the initial condition of the universe. That is, rather than explaining the asymmetry, it was taken as a contingent fact.

With the nonconservation of baryon number included in grand unified theories, an explanation of the asymmetry was readily obtained. In his work of 1978, Yoshimura assumed an initially symmetric universe and showed that the violation of baryon number conservation and *CP* invariance characteristic of GUT led to an excess of particles over antiparticles shortly after the "creation" at $t = 0$. The decay rate of X particles in the very early universe would differ from that of anti-X particles, and the result would be more quarks and leptons than antiquarks and antileptons. Annihilation would do the rest of the job, and the present composition of the world in terms of particles might thus be seen as a result of early grand unification. During the years 1978–80, Yoshimura, Weinberg, Frank Wilczek, and others further used GUT to solve another cosmological question, namely, to put it crudely, why the universe is so empty. The number of baryons in the observable part of the universe is around 10^{79} (Eddington's number) and that of photons around 10^{88}. The ratio is considered a fundamental quantity, because theory prescribes it to be constant in time. Why are there one billion times as many photons than baryons? According to GUT reasoning, this had not always been the case, but was the result of the slight asymmetry in quark-antiquark annihilation processes in the early universe. The application of GUT to cosmology led to brilliant successes, but also to severe problems. For example, a large number of primordial massive monopoles, about 10^{16} times as heavy as protons, were expected to be formed. Yet not a single monopole has ever been observed.

The most important cosmological implication of the GUTs, a kind of spin-off, was perhaps the inflation model introduced by Alan Guth in 1981 and improved by Andrei Linde and, independently, Andreas Albrecht and Paul Steinhardt in 1982. The inflationary universe model relied crucially on the phase transition connected with the spontaneous symmetry breaking predicted by GUT at a temperature of about 10^{27} K. Although this model was successful in the eyes of many cosmologists, there were also those who found that inflation was a chimera. How seriously should one take a cosmological theory that completely depended on a grand unified theory whose principal prediction, the decay of protons, had failed to be verified? It was not only in connection with superstrings that some physicists and astronomers were worried about the flight of theoretical fancies. "[The inflationary model] has no evidence to support it," wrote two cosmologists, "and yet, because the theory is pretty in a mathematical sense, many theorists have embraced it and chosen to disregard these issues. . . . cosmology is approaching the frontier where science is no longer based on experimental evidence and makes no testable predictions. Once this border is crossed, we have left the world of physics behind and have entered the realm of metaphysics" (Rothman and Ellis 1987, 22).

There were physicists who dealt with the quantum aspects of cosmology in ways very different from, and rather more speculative than, those of the GUT physicists. The idea to apply quantum mechanics directly to the entire universe, without having a quantum theory of gravity, goes back to Eddington and Schrödinger in the late 1930s. In 1939 Schrödinger obtained the remarkable result that particles might be created in an expanding universe solely as a result of the expansion, but his work was effectively forgotten. In the 1950s, Bryce DeWitt in the United States took up a related research program, building on some of Wheeler's ideas. The extravagant extrapolation of quantum mechanics to the entire universe led DeWitt in 1967 to a quantum equation, a sort of cosmic Schrödinger equation, which became known as the Wheeler-DeWitt equation. However, it was unclear how to interpret the equation, how to find the unique solution corresponding to the one and only universe, and indeed how to do anything useful with the equation at all. Alexander Vilenkin, developing an earlier suggestion made by Edward Tryon, argued in 1982 that the existence of the universe might be understood as a quantum mechanical tunneling process from a "nothingness" of quantum vacuum fluctuations. The title of Vilenkin's paper, "Creation of Universes from Nothing," must have surprised many readers of *Physics Let-*

ters. Many other physicists took up the idea of explaining the creation of the universe as a result of vacuum fluctuations; this idea caught the imagination of theologians and philosophers alike. However, the majority of physicists considered the fascinating idea wildly speculative, as of course it was.

Undaunted by the absence of an acceptable quantum theory of gravity, from about 1980, quantum gravity models were applied by several physicists to account for the creation of the universe from a "state" of nothingness. One of the best known of such models was proposed by James Hartle and Stephen Hawking in 1983. Taking their point of departure from the Wheeler-DeWitt equation, the two physicists developed a wave function of the universe, a concept they found legitimate, and argued that it represented the amplitude of the universe coming into existence from a finite quantum fuzz. According to Hawking, there would be no problem of creation because near $t = 0$, in the quantum fuzz, the very notion of space and time would lose its meaning. Thus, there would be no singularity and no edge of space-time. By being self-contained, the universe would neither be created nor destroyed. According to Hawking, in his best-selling *A Brief History of Time*, this view has profound implications beyond the realms of science:

> With the success of scientific theories in describing events, most people have come to believe that God allows the universe to evolve according to a set of laws and does not intervene in the universe to break these laws. However, . . . it would still be up to God to wind up the clockwork and choose how to start it off. So long as the universe had a beginning, we could suppose it had a creator. But if the universe is really completely self-contained, having no boundary or edge [in time], it would have neither beginning nor end: it would simply be. What place, then, for a creator? (Hawking 1988, 149)

—which may be an appropriate way of ending this section.

PART FOUR

A LOOK BACK

Chapter 28

NOBEL PHYSICS

THE WILL OF Alfred Nobel, the Swedish industrialist who made a fortune on the invention and manufacture of dynamite, is the foundation of the entire Nobel Institution. In 1900, following a complex legal process, the will was implemented into a code of statutes of the Nobel Foundation, and the first prizes were awarded the following year. Alfred Nobel wanted to establish annual prizes in physics, chemistry, physiology (or medicine), literature, and peace work, but his will was quite vague, so it was left to the executors and the later Nobel Institution to work out the details and create an organizational machinery. According to the will of November 27, 1895, the money should "be annually awarded in prizes to those persons who shall have contributed most materially to benefit mankind during the year immediately preceding." The share in physics was to be awarded to "the person who shall have made the most important discovery or invention in the domain of physics." Moreover, the will emphasized that "no consideration whatever [shall] be paid to the nationality of the candidates, that is to say, that the most deserving be awarded the prize, whether of Scandinavian origin or not" (Crawford 1984, 221).

The Nobel prizes are officially awarded at a ceremony taking place each year in Stockholm on December 10, the day of Nobel's death. Before this happens, an elaborate process takes place. The Swedish Royal Academy of Science, a corporation of mainly Swedish scientists organized in different sections, elects five members to each of the scientific committees, including the physics Nobel committee. One of the memberships is reserved for the president of the physics section of the Nobel Institute. The real power in the election process lies with these committee members. The physics committee collects nominations from a potentially large number of nominators, some permanent and some appointed on an ad hoc basis. Those permanently entitled to nominate include all members of the committee and the Academy, previous Nobel physics laureates, and professors of physics at the old Scandinavian universities and similar institutions. The ad hoc nominators include professors at select foreign universities and a number of individually chosen scientists.

Some nominators make use of their right to propose candidates to the committee, which then considers all proposals and prepare detailed reports on a few of them. The committee deliberates on which candidate(s) to recommend, but has no obligation to follow the proposal of the majority of the nominators. In fact, in most cases before 1940, the most-nominated scientists had not been recommended by the committee. For example, in 1910 Poincaré was suggested by thirty-four nominators and Planck by ten, but the committee decided to recommend van der Waals, whom only one nominator had proposed.

Although the committee's decision is the crucial part of the nominating system, it does not end there. The nomination of the committee needs to be confirmed by the Academy physics section, and the proposal is finally discussed by the Academy plenary, which has the final word. Although the Academy will almost always follow the recommendation of the committee, it is not obliged to. In 1908, for example, the committee recommended Planck and the Academy physics section endorsed the proposal. When the Academy met in plenary session, however, the physicists' candidate was rejected by a substantial majority. In the end, it was decided to award to prize to the French physicist Gabriel Lippmann for his invention of a method for color photography. A similar situation occurred four years later, when the plenary meeting turned down Kammerlingh Onnes, the candidate recommended by the physics committee and the Academy's physics section. Kammerlingh Onnes had to wait only one more year until he received the prize, however.

The result of the nomination procedure may well be that it is decided that none of the proposed candidates is worthy. In that case, the prize for the year is reserved for the next year and is usually awarded then. In the first half of the century this happened not infrequently, as it did in 1921, when the prize was reserved for 1922 and then awarded to Einstein. A prize can be reserved for only one year. If still not awarded, it will be "permanently reserved," that is, annulled. This has happened five times with the physics prize: in 1916, 1934, 1940, 1941, and 1942.

From its very start, the Nobel prize was a success and came to be considered the most prestigious of all scientific prizes. One reason for the status of the prize and the publicity surrounding it was the sheer amount of money involved. In 1901 it consisted of about $40,000, a purchasing power almost double its present value and, at the time, corresponding to thirty times the annual salary for a university professor. Adding to the continuing significance of the prize has been that the choice of a Nobel laureate has usually been acknowledged as reasonable by the physics community at large. Only in very few cases has the choice been considered "wrong" or "strange," that is, clearly at odds with the views of the majority of élite physicists. Also, the Nobel science prizes have maintained their reputations as truly international awards, not clearly influenced by political and ideological factors. In this respect, there is a great difference between the science prizes and the prizes in peace and literature. Of course, politics enters the decision process, but the prize is scientifically, not politically, motivated. Right from the beginning of the Nobel Institution, it was realized that the prize was a reward not only to an individual scientist, but also to the nation to which the scientist belonged. The prize competition was seen as a contest among nations, a view that has prevailed throughout the century. In periods of tension and war, the prizes could not avoid being associated with nationalistic sentiments, very much like the medals won during Olympic games. The Nobel physics committee has always, and for understandable reasons, feared to award a prize for an experiment that turns out to be wrong or a theory that either cannot be verified or may be proved wrong the next year. For example, in 1903 LeBon was nominated for his N-rays, and in 1905 Blondlot for his "black rays," in both cases with a nomination from only one scientist. One can easily imagine what it would have meant to the reputation of the Nobel prize if it had been awarded to one or both of the French scientists.

The ambiguities of the statutes of the Nobel Foundation, reflecting the ambiguities of the will, created opportunity for arbitrariness in the interpretation of what kind of scientific work merited a prize. For example, how seriously should the proviso of recency be taken? In the statutes of the Nobel Foundation, the proviso was softened and specified to mean that "the year immediately preceding" was to be understood in the sense "that a work or an invention for which a reward under the terms of the Will is contemplated, shall set forth the most modern results of work; . . . works or inventions of older standing to be taken into consideration only in case their importance have not previously been demonstrated." Although the criterion included in the will was softened already with the first award—Röntgen's discovery was then five years old—in general, recency was considered important during the early decades. It was a major reason why Boltzmann and Kelvin were not considered. The Nobel Institution was eager that the prize not be awarded for long and meritorius service in physics, but for a specific work of recent origin. This remained the official policy, but practice did not always follow the policy. Noteworthy examples are Reines, whose discovery of the neutrino antedated his prize by thirty-nine years, and Kapitza, who had to wait about forty years before he received the prize; Chandrasekhar's prize followed some forty-five years after his main work. Van Vleck's much-belated prize is yet another example.

Another, and more serious, problem was how narrowly the concept of physics was to be understood. The terms "invention" and "benefit of mankind" were originally taken to mean that advances in applied physics and technology could be rewarded alongside work in pure physics. Marconi's invention of wireless telegraphy was rewarded in 1909, and Gustav Dalén, a Swedish engineer, received the 1912 prize for his invention of automatic illumination of lighthouses. Although the early physics committee looked with sympathy to inventions, however, its sympathy did not extend to patents. In 1901 it decided that, as a rule, patented inventions should not be considered. Naturally, this policy excluded most successful inventors, from Edison to Maiman. Since 1912, no prize has been awarded for purely technological work. Among the unsuccessful nominees from the first two decades, one will find famous inventors such as Edouard Branly (wireless telegraphy), Valdemar Poulsen (arc generator), Carl von Linde (refrigeration machines), Ferdinand von Zeppelin (airships), Orville and Wilbur Wright (airplane), Michael Pupin (loading induction coils), and Thomas Edison (electric bulb, phonograph, and much else). The passus of "benefit of mankind" was no big problem. Following a long tradition in academic science, the committee members soon resolved that fundamental progress in science was, by definition, to the benefit of mankind. That is, utilitarian considerations did not have to enter the decision process. In this way, they were spared justifying how Lorentz's explanation of the Zeeman effect or, later, Schwinger's theory of quantum electrodynamics, benefited mankind in any practical way.

The policy of not granting Nobel prize status to technological innovations did not imply that scientists employed by private companies were prevented from receiving the coveted prize. As we have seen, since the 1920s, laboratories affiliated with private industries have increasingly engaged in pure science and contributed to the progress of physics also at the fundamental level. Early examples are Langmuir's chemistry prize of 1932 and Davisson's physics prize of 1937 (see chapter 9). Table 28.1 lists Nobel physics laureates who did their rewarded work while working at, or being principally associated with, private laboratories. The 1986 prize was awarded for pioneering work in electron optics and for the design of the first electron microscope. The invention went back to the early 1930s, when Max Knoll and Ernst Ruska at Berlin's Technical University constructed an operable electron microscope. A large part of Ruska's later work was done at a laboratory that Siemens & Halske set up in 1937, and at which Ruska and his colleagues produced the first commercial electron microscope in 1939. Half the 1986 prize went to Heinrich Rohrer and Gerd Binnig for their design of the scanning tunneling microscope. Rohrer joined the IBM Research Laboratory in Switzerland in 1963, and Binnig followed in 1978.

Originally, the committee defined physics broadly, more as physical sciences than pure physics. Astrophysics, meteorology, geophysics, and physical chemistry were included, but not astronomy. In the 1920s there was a change in opinion, however, resulting in an inofficial exclusion of astrophysics and the earth sciences. Whether some interdisciplinary subfield or sister science should be accepted as physics is of course a matter of politics, rather than arguments of principle. Practice has therefore changed over the years. During the early years, where astrophysics was defined as part of physics, no prize was awarded in this area. Bethe's belated prize of 1967, awarded for his 1938 theory of stellar energy production, was the first prize

TABLE 28.1
Nobel Prizes Awarded to Physicists from or Associated with Private Industrial Laboratories

Year	*Scientists*	*Institutions*
1937	C. J. Davisson	AT&T / Bell Laboratories
1956	J. Bardeen, W. Brattain, W. Shockley	AT&T / Bell Laboratories
1977	P. W. Anderson	AT&T / Bell Laboratories
1978	A. Penzias, R. Wilson	AT&T / Bell Laboratories
1986	E. Ruska	Siemens & Halske
	H. Rohrer, G. Binnig	IBM Research Laboratory
1987	K. A. Müller, J. G. Bednorz	IBM Research Laboratory

ever in astrophysics. It coincided with the rapid development of nuclear and relativistic astrophysics, which brought astronomy and physics closely together and may have caused a change in attitude of what constituted Nobel physics. At any rate, between 1974 and 1993, no fewer than seven prizes were awarded for work that must be classified as closer to astronomy than physics. Also worth noticing is that at least one physicist, Max Delbrück, received the Nobel prize in biology. With regard to the border line between physics and chemistry, matters were rather different, among other reasons because there is a separate prize in chemistry but none in astronomy. With one possible exception, Jean Perrin's 1926 prize, no Nobel physics prize has been awarded for a work in chemistry, including physical chemistry and chemical physics. This may seem natural enough—chemists are awarded prizes in chemistry, physicists in physics—but the relationship is not symmetrical. Without entering a discussion of the changed relationships between physics and chemistry over time, it is noticeable that a large number of Nobel chemistry prizes have been awarded to scientists who were either physicists or whose work would be normally counted as belonging to physics. Most of the prize works listed in table 28.2 could equally well, or perhaps better, be categorized as physics. Radioactivity, isotopes, molecular structure, and the discovery of chemical elements were traditionally seen as belonging to chemistry, which explains some of the prizes. The best known case is perhaps the prize awarded to Rutherford, who considered his work in radioactivity to be strictly physical and held chemistry in low esteem. The interdisciplinary nature of many discoveries in the area of radioactivity and nuclear chemistry has, on occasion, led to controversial decisions. One of them concerns the discovery of nuclear fission, which was investigated by the chemistry committee as early as 1941. In 1945, the chemistry prize for 1944 was awarded to Hahn alone, whereas Strassmann, Meitner, and Frisch were passed over. Meitner and Frisch, who had first explained the Berlin experiments correctly, were nominated for a prize in physics but the commit-

TABLE 28.2
Chemistry Prizes Awarded to Physicists or Areas of Research of a Physics Nature

Year	*Laureate*	*Nobel work*	*Laureate's background*
1908	Ernest Rutherford	radioactive decay	physics
1911	Marie Curie	discovery of Ra, Po	physics
1920	Walther Nernst	chemical thermodynamics	physical chemistry
1921	Frederick Soddy	radioactivity, isotopes	physical chemistry
1922	Francis Aston	mass spectroscopy	physics
1932	Irving Langmuir	surface chemistry	physical chemistry
1934	Harold C. Urey	deuterium	chemical physics
1935	Frédéric Joliot	artificial radioactivity	physics
	Irène Joliot-Curie		physics
1936	Peter Debye	molecular structure	physics
1944	Otto Hahn	discovery of fission	physical chemistry
1949	William Giauque	low-temperature methods	physical chemistry
1951	Edwin McMillan	transuranium elements	physics
	Glenn Seaborg		nuclear chemistry
1960	Willard Libby	C-14 dating method	nuclear chemistry
1968	Lars Onsager	irreversible thermodynamics	physical chemistry
1971	Gerhard Herzberg	molecular structure	physics
1977	Ilya Prirogine	dissipative structures	physical chemistry

tee report was negative. It was judged that Meitner and Frisch's explanation was too much of a lucky guess and that, if the theoretical explanation was to be awarded, it was rather the non-nominated Bohr who deserved the honor.

Individual members of the science committees were sometimes very influential in Nobel science policy and successful in, for example, promoting their favorites or particular branches of physics research. In the 1920s, the theorist Carl Oseen was largely responsible for the gradual turn toward theoretical physics that marked a break with the experimentalist past. The physical chemist Svante Arrhenius exerted a strong influence on early Nobel decisions in both chemistry and physics. He used his position to highlight the atomic theory, and was instrumental in the prizes awarded to Rutherford and Planck. Arrhenius also succeeded in blocking, for a long period, Nernst's chemistry prize, in this case not so much for scientific as for personal reasons.

A work worthy of a Nobel prize had to be a "discovery or invention"; this was originally taken to mean that purely theoretical work did not qualify for a prize. For the first two decades or so, the physics committee was dominated by experimenters and had a decided preference for high-precision experiments. This reflected the general experimentalist tendency in Swedish physics at the time. Thus, between 1890 and 1920, seventy-seven Swedes graduated with doctorates in physics from the country's universities in

Stockholm, Lund, and Uppsala. Sixty-seven of the dissertations were mainly experimental, and only ten theoretical. The physics committee was unwilling to consider theoretical physics seriously if the theory had not been directly verified by experiment. Theoretical and mathematical physics was often looked upon as "speculative." The ideal of science was expressed by a Swedish physicist in a 1918 comment on Einstein's theory of relativity. This theory, he wrote, cultivated theory for its own sake and attracted young people. But "[o]lder persons, who have lived through many theories . . . are more inclined to be satisfied only with factual knowledge, i.e., experimentally demonstrated theories, and they are sceptical when it comes to theories which cannot be verified. . . . [It] is a simple and uncultivated taste that gets carried away by such [abstract] theories" (Elzinga 1995, 84). Most Nobel committee members, whatever their age, were "older persons." It is well known that Einstein did not receive his Nobel prize for the theory of relativity, but for his 1905 explanation of the photoelectric effect; or rather, the prize was given for the prediction of the correct photoelectric law, not the photon theory on which the prediction was based: to be precise, "for his contributions to theoretical physics, especially for his discovery of the law of the photoelectric effect"—note the magical word "discovery." Until 1922, Einstein had been nominated no fewer than 62 times, and only one of the nominations mentioned the photoelectric effect specifically. The Swedish physicist who wrote the report on Einstein's theory of relativity concluded that acceptance was "a matter of faith"; and another committee member asserted that "it is highly improbable that Nobel considered speculations such as these to be an object for his prizes" (Friedman 1981, 795). From that time, the committee began to judge theoretical work in a more favorable light, especially with the election of Oseen in 1923. But the general attitude was still conservative and distrustful of theoretical physics without a strong connection to experimental work. When Heisenberg and Schrödinger were proposed for the 1929 prize, the committee resolved that quantum mechanics "[has] not as yet given rise to any new discovery of a more fundamental nature." And in his 1933 evaluation of Dirac, Oseen concluded that the young British theorist had not, so far, done any "really great innovative work" (Kragh 1990, 116). All the same, the three quantum pioneers were rewarded.

Table 28.3 lists the Nobel prizes in physics from 1901 through 1998. Altogether, 160 scientists have received the physics prize, undivided or divided. Of the 160 recipients, only two (1.3 percent) were women, namely, Marie Curie and Maria Goeppert-Mayer. Since they both received a quarter of a prize, the grand total of women's Nobel prizes is one-half. Also, there have only been two recipients from the third world; three, if Chandrasekhar is included. No African or Latin American physicist has ever been rewarded. Most of the Nobel prize winners have been Europeans (54.4 percent) or

TABLE 28.3
Nobel Prizes in Physics

Year	Name(s)	Country	Subject	Year of discovery	
1901	Wilhelm K. Röntgen	Germany	x-rays	1895	E
1902	Pieter Zeeman	Netherlands	Zeeman effect	1896	E
	Hendrik A. Lorentz	Netherlands	—	1896	T
1903	Antoine H. Becquerel	France	radioactivity	1896	E
	Pierre Curie	France	—	~ 1898	E
	Marie Curie	France	—	~ 1898	E
1904	Lord Rayleigh	England	argon	1895	E
1905	Philipp Lenard	Germany	cathode rays	~ 1902	E
1906	Joseph J. Thomson	England	electron	1897	E
1907	Albert A. Michelson	USA	precision interferometry	~ 1890	E
1908	Gabriel Lippmann	France	color photography	~ 1895	E
1909	Guglielmo Marconi	Italy	wireless telegraphy	~ 1895	E
	C. Ferdinand Braun	Germany	—	~ 1897	E
1910	Johannes van der Waals	Netherlands	gases and liquids	1873	T
1911	Wilhelm Wien	Germany	heat radiation	1896	T
1912	Nils Gustav Dalén	Sweden	automatic regulators	1907	E
1913	Heike Kammerlingh Onnes	Netherlands	low temperatures	~ 1908	E
1914	Max von Laue	Germany	x-ray diffraction	1912	E/T
1915	William H. Bragg	England	crystal structure	1912	E
	William L. Bragg	England	—		
1917	Charles Barkla	England	secondary x-rays	1905	E
1918	Max Planck	Germany	quantum theory	1900	T
1919	Johannes Stark	Germany	Stark effect	1913	E
1920	Charles-Edouard Guillaume	France	nickel-steel alloys	~ 1895	E
1921	Albert Einstein	Switzerland	photoelectric effect	1905	T
1922	Niels Bohr	Denmark	atomic theory	1913	T
1923	Robert Millikan	USA	electron charge and photoelectric effect	1911–15	E
1924	Karl M. Siegbahn	Sweden	x-ray spectroscopy	~ 1922	E
1925	James Franck	Germany	electron-atom collisions	1915	E
	Gustav Hertz	Germany	—		

TABLE 28.3 (*Continued*)

Year	*Name(s)*	*Country*	*Subject*	*Year of discovery*	
1926	Jean Perrin	France	equilibrium of suspensions	1908	E
1927	Arthur H. Compton	USA	Compton effect	1923	E
	Charles T. R. Wilson	England	cloud chamber	1911	E
1928	Owen Richardson	England	thermionic phenomena	~ 1910	T
1929	Louis de Broglie	France	wave nature of electrons	1923	T
1930	Chandrasekhara Raman	India	Raman effect	1928	E
1932	Werner Heisenberg	Germany	quantum mechanics	1925	T
1933	Paul Dirac	England	quantum mechanics	1925–28	T
	Erwin Schrödinger	Austria	wave mechanics	1926	T
1935	James Chadwick	England	neutron	1932	E
1936	Victor Hess	Austria	cosmic radiation	~ 1911	E
	Carl D. Anderson	USA	positron	1932	E
1937	Clinton J. Davisson	USA	wave nature of electrons	1927	E
	George P. Thomson	England	—	1927	E
1938	Enrico Fermi	Italy	neutron nuclear reactions	1934	E
1939	Ernest Lawrence	USA	cyclotron	1932	E
1943	Otto Stern	Germany	molecular beam method	~ 1920	E
1944	Isidor Rabi	USA	magnetic resonance method	1930s	E
1945	Wolfgang Pauli	Austria	exclusion principle	1925	T
1946	Percy Bridgman	USA	high-pressure physics	~ 1930	E
1947	Edward Appleton	England	Appleton layer	1926	E
1948	Patrick M. S. Blackett	England	cloud chamber physics	~ 1933	E
1949	Hideki Yukawa	Japan	meson theory	1935	T
1950	Cecil F. Powell	England	meson discoveries	1947	E
1951	John D. Cockcroft	England	high-voltage accelerator	1932	E
	Ernest Walton	England	—		
1952	Felix Bloch	USA	nuclear magnetism	~ 1945	E/T
	Edward Purcell	USA	—		E

TABLE 28.3 (*Continued*)

Year	Name(s)	Country	Subject	Year of discovery	
1953	Frits Zernike	Netherlands	phase-contrast method	~ 1932	E
1954	Max Born	Germany	quantum mechanics	1926	T
	Walther Bothe	Germany	coincidence method	1924	E
1955	Willis Lamb	USA	Lamb effect in hydrogen	1947	E
	Polykarp Kusch	USA	magnetic moment of electron	1947	E
1956	John Bardeen	USA	transistor effect	1948	E
	Walter Brattain	USA	—		
	William Shockley	USA	junction transistor	1949	E
1957	Chen Ning Yang	USA	parity nonconservation	1957	T
	Tsung Dao Lee	USA	—		
1958	Pavel Cerenkov	USSR	Cerenkov effect	1935	E
	Ilya M. Frank	USSR	—	1937	T
	Igor Tamm	USSR	—		T
1959	Owen Chamberlain	USA	antiproton	1955	E
	Emilio Segré	USA	—		
1960	Donald Glaser	USA	bubble chamber	1952	E
1961	Robert Hofstadter	USA	electron scattering by nuclei	~ 1955	E
	Rudolf Mössbauer	Germany	Mössbauer effect	1958	E
1962	Lev Landau	USSR	liquid helium theory	~ 1941	T
1963	Eugene Wigner	USA	symmetry principles	~ 1930	T
	Maria Goeppert-Mayer	USA	shell structure of nucleus	1948	T
	J. Hans D. Jensen	Germany	—	1948	T
1964	Charles Townes	USA	quantum electronics (laser)	1954	E
	Nikolai G. Basov	USSR	—	1955	E
	Alexander Prokhorov	USSR	—		E
1965	Richard Feynman	USA	quantum electrodynamics	1949	T
	Julian Schwinger	USA	—	1948	T
	Sin-Itiro Tomonaga	Japan	—	1943	T
1966	Alfred Kastler	France	double resonance	1950	E
1967	Hans Bethe	USA	stellar energy production	1938	T

TABLE 28.3 (*Continued*)

Year	*Name(s)*	*Country*	*Subject*	*Year of discovery*	
1968	Luis Alvarez	USA	elementary particles	~ 1962	E
1969	Murray Gell-Mann	USA	theory of elementary particles	1962–64	T
1970	Hannes Alfvén	Sweden	magnetohydrodynamics	~ 1948	T
	Louis E. F. Néel	France	models for magnetism	1932–48	T
1971	Dennis Gabor	England	holography	1948	E
1972	Leon Cooper	USA	superconductivity theory	1957	T
	John Schrieffer	USA	—		
	John Bardeen	USA	—		
1973	Leo Esaki	Japan	semiconductor tunneling	1957	E
	Ivar Giaver	USA	—		E
	Brian Josephson	England	Josephson effect	1962	T
1974	Antony Hewish	England	pulsars	1967	E
	Martin Ryle	England	radio astronomy techniques	1954	E
1975	Aage Bohr	Denmark	collective nuclear model	1953	T
	Ben Mottelson	USA	—		
	L. James Rainwater	USA	—	1950	T
1976	Burton Richter	USA	J/ϕ particle	1974	E
	Samuel Ting	USA	—		E
1977	John van Vleck	USA	theories of magnetism	1930s	T
	Nevill Mott	England	metal theory	~ 1936	T
	Philip Anderson	USA	solid state theory	~ 1960	T
1978	Peter Kapitza	USSR	low-temperature physics	~ 1940	E
	Arno Penzias	USA	3K background radiation	1965	E
	Robert Wilson	USA	—		
1979	Steven Weinberg	USA	electroweak theory	1967	T
	Abdus Salam	Pakistan	—	1967	T
	Sheldon Glashow	USA	—	1961	T
1980	James Cronin	USA	CP nonconservation	1964	E
	Val Fitch	USA	—		

TABLE 28.3 (*Continued*)

Year	*Name(s)*	*Country*	*Subject*	*Year of discovery*	
1981	Kai M. Siegbahn	Sweden	electron spectroscopy	1950s	E
	Nicolas Bloembergen	Netherlands	laser spectroscopy	1956	E
	Arthur Schawlow	USA	—	~ 1956	E
1982	Kenneth Wilson	USA	critical phenomena	1970	T
1983	William Fowler	USA	nuclear astrophysics	~ 1960	T
	Subrahmanyan Chandrasekhar	India	structure of stars	1934–39	T
1984	Carlo Rubbia	Italy	discovery of W and Z	1983	E
	Simon van der Meer	Netherlands	—		
1985	Klaus von Klitzing	Germany	quantized Hall effect	1980	E
1986	Ernst Ruska	Germany	electron microscopy	1930s	E
	Heinrich Rohrer	Switzerland	scanning tunneling microscopy	1978	E
	Gerd Binnig	Germany	—		
1987	Karl Alex Müller	Switzerland	high-T superconductivity	1986	E
	J. Georg Bednorz	Germany	—		
1988	Leon Lederman	USA	muon neutrino	1962	E
	Melvin Schwartz	USA	—		
	Jack Steinberger	USA	—		
1989	Norman Ramsey	USA	atomic clocks	1950s	E
	Hans Dehmelt	Germany	ion trap technique	1973	E
	Wolfgang Paul	Germany		1950s	E
1990	Richard Taylor	USA	deep inelastic scattering	1960s	E
	Henry Kendall	USA	—		
	Jerome Friedman	USA	—		
1991	Pierre-Gilles de Gennes	France	liquid crystals	1974	E
1992	Georges Charpak	France	detector devices	1970s	E
1993	Russell Hulse	USA	discovery of binary pulsar	1974	E
	Joseph Taylor	USA	—		
1994	Bertram Brockhouse	Canada	neutron spectroscopy	~ 1960	E
	Clifford Shull	USA	—		

TABLE 28.3 (*Continued*)

Year	*Name(s)*	*Country*	*Subject*	*Year of discovery*	
1995	Martin Perl	USA	discovery of tau lepton	1974	E
	Frederick Reines	USA	discovery of neutrino	1956	E
1996	Robert Lee	USA	superfluidity of He-3	1972	E
	Robert Richardson	USA	—		
	Douglas Osherfoff	USA	—		
1997	Steven Chu	USA	laser cooling of atoms	1980s	E
	William Phillips	USA	—		
	Claude Cohen-Tannoudji	France	—		
1998	Robert Laughlin	USA	electron quasi-particles	1982	T
	Horst Störmer	Germany	—		E
	Daniel Tsui	USA	—		E

Note: The year of award refers to the year for which the prize was awarded, not the year in which the laureate received the prize. The letters "E" and "T" indicate whether the work was primarily experimental or theoretical. Nationalities refer to the period of the prize work or, if this covered an extended period, to the earliest part of the period.

Americans (41.9 percent) at the time they did their work. Whereas by 1940 physicists from Germany and England were the most rewarded (with each nation receiving 22 percent of the prizes), by 1970 American physicists had collected more Nobel prizes than German and British physicists together.

Chapter 29

A CENTURY OF PHYSICS IN RETROSPECT

Growth and Progress

THE NINETEENTH CENTURY has often be labeled the age of science, and it was indeed during this century that modern science emerged as more than an intellectual activity for a small group of natural philosophers; only during this century was science organized nationally, and to a lesser extent internationally, and only then did it prove its worth for society at large. Yet the following century can, with even more justification, be called the century of science, for it was during this period that science came to dominate not only the intellectual sphere, but also a large part of the social, economic, and military spheres of life. Science took off in the nineteenth century, was fully established and ready for action by 1900, and then embarked on a course of rapid development that has been unprecedented in the entire history of humankind. It is quantitative growth, first of all, that has characterized the development of science during the last century of the second millennium. Of course, physics is not the only science that has experienced a remarkable growth. In general, all the classical branches of science have exploded during the century, although with rather different growth patterns. For example, whereas physics leveled off during the last two or three decades, this has been a period of sustained growth within the biological and environmental sciences.

Around 1900, as we saw in chapter 2, physics was a thriving but small business, with a total number of academic physicists less than 1,500. The old German Physical Society included about 350 members, and membership in the new American Physical Society was less than one hundred. The number of physics publications, as reported in the major abstract journals, was around 2,400. In the 1990s, the world of physics had grown very substantially in size. Membership in the two physical societies had increased to about 25,000 (Germany) and 40,000 (United States), and the number of research papers abstracted in *Physics Abstracts* approached 200,000. The world population of physicists was more than 150,000. Rounded off, manpower and paper output has risen by a factor of one hundred in less than a century. Expenditures for physics research and teaching have undoubtedly risen by an even greater factor.

With the growth in physics that took off especially after the end of World War I followed, slowly at first, a different way of organizing and doing physics. The change was already visible in the late 1920s, but the real watershed was clearly World War II and the great changes that the war and its aftermath caused in a science that had proved so eminently useful for military purposes. One result of the postwar generational shift and the general turn toward instrumentalist and pragmatic modes of thought was that philosophy lost its place in physics. Many members of the earlier generation of physicists had a deep interest in philosophical questions and were sometimes inspired by philosophers in their innovative work, or they discussed competently the philosophical implications of the new physics. For Planck, Bohr, Schrödinger, Weyl, Heisenberg, Einstein, Eddington, and many of their colleagues, philosophy was an important aspect of physics. Few of the leaders of postwar physics cared about philosophy or had more than a superficial knowledge of the field. In the 1960s, with the deaths of Bohr and Schrödinger, the once proud and vital tradition of physicist-philosophers came to an end.

After 1945, physics began its full-speed march toward the big science era, although it should be kept in mind that the big science of the glamorous kind found in experimental high-energy physics is not and never was typical for physics as a whole. Yet even small-science physics was influenced by the trend, one important result of which was the shift from the individual physicist as the unit producer of knowledge to collectives, teams, or cooperations. On the level of bibliometrics, the trend was manifest in a shift from individually authored to multiauthored research papers. The inevitable trend toward collectivization—inevitable because of the growing size and complexity of instruments and also because of the increasing transdisciplinarity of many fields of physics—does not mean that the days of the great individual have gone. It is still with the genius of individual scientists that true creativity lies, a claim that is supported by the history of postwar physics to no lesser extent than in the more heroic prewar physics, with its apparent wealth of genius from Planck and Einstein to Bohr and Yukawa.

Throughout the century, physics has continued to be a truly international science. Or perhaps cosmopolitan is a better word, for it is a characteristic feature that a great many physicists have moved effortlessly from country to country, from city to city, from laboratory to laboratory, with the aim of working under the best possible conditions and without paying much attention to the national locations. We may like to think of physics in the late twentieth century as much more cosmopolitan than it was previously, but in fact, young physicists have always moved (more or less) freely from country to country and have thought of their science as transnational. On the other hand, to characterize physics as international in the social sense would be an exaggeration. In spite of the great political changes that have occurred in the relationship between the rich northern part of the world and the so-called third world, physics is still completely dominated by Europe and North

America. The fact that the former colonies have become independent nations and some of the third-world countries (especially in east Asia and South America) have moved from developing to developed countries is hardly reflected at all in the international composition of physics. Of the eighty papers presented during the Paris world conference in physics in 1900, two were delivered by physicists from what later would be called third-world countries. If a corresponding conference had been held in 1995 (unthinkable as it would be), the share would not have been higher.

What has been said about the third world issue can be said about the gender issue as well. By 1900, physics was almost totally dominated by men, as were all branches of science and, indeed, all of public and professional life. Ninety years later, the number of female physicists has increased very considerably and relatively faster than the number of male physicists. But the increase has not made much difference. Worldwide, the proportion of women is still less than 10 percent, a figure that compares unfavorably with corresponding figures in the biological and chemical sciences. Moreover, women have rarely made the move to the élite of physics, and in frontier research they are even less visible than would be expected from their proportion of teaching posts. Marie Curie remains the female physicist of the century—indeed, of all time.

By and large, physics in the earliest part of the century was dominated by Germany, England, and France (for some details, see chapter 2). Germany, contrary to France, continued to play a leading role well into the 1930s, and British physicists never ceased to contribute significantly to world physics. By the 1930s, the United States had already become the new world power in physics, as it had in several other sciences and especially in technological development. In spite of the rise of a strong Soviet physics community, for decades after the war American physics had an almost hegemonic position. Indeed, one of the most notable changes in twentieth-century physics, especially if compared with the previous century, has been the rise to dominance of American physics. During the last quarter of the century, the dominance has become less marked and Europe has, to some extent, come back on the scene, but it is still the United States that is unquestionably the world's leading nation in physics. On the other hand, the increased transnational or cosmopolitan nature of physics, as caused in part by technological advances in transportation and telecommunications, have made it somewhat less relevant to speak of physics in national terms. Technological advances have had a major influence on physics. Recall that there were only two American representatives at the 1900 Paris conference—at that time, transatlantic voyages were slow and expensive; and recall how American physicists had troubles joining the quantum revolution after 1925 because of the delay in communications from Europe—it took time before the German physics journals arrived in the United States. The conditions of doing physics and following the research frontier have changed in an age in which jet traffic, worldwide telephony, electronic mail, and Internet service are the order of the day.

It has often been claimed that science and democracy go hand in hand and that science can thrive only under democratic political conditions. This claim does not receive much support from the history of twentieth-century physics, and even less from the history of earlier centuries. Whatever relation there may be between democracy and science development, if any, it is not a simple one. To be sure, physics declined in Germany after 1933, but was it hardly because the Nazi dictatorship strangled the free spirit of democracy that was supposed to be so necessary for the healthy development of science. Given the large number of élite physicists who fled Germany the decline was remarkably modest. And, as we noted in chapter 16, Italian physics experienced its golden years during the fascist dictatorship. Again, in spite of the purges in the 1930s and the terrible loss of people and resources during the war, physics in the Soviet Union continued to be at a high level, both qualitatively and quantitatively, during the Stalin era and later.

Physics in the twentieth century has not only increased in terms of manpower, organization, apparatus, research output, and economic support, but it has also progressed scientifically, that is, produced much new knowledge about nature. There is no automatic relationship between quantitative growth and qualitative progress in knowledge and it is, in fact, difficult to find any specific correlation between the two in twentieth-century physics. It is noteworthy that what most people would probably single out as the two most profound and wide-ranging conceptual innovations of the century, relativity and quantum mechanics, emerged without relying on large-scale economic support or expensive experiments. This is not to say that there is no correlation at all, for it seems reasonable to assume that interesting discoveries will more likely be made in a large and well-funded community of physicists than in a small and poorly funded community.

Progress in knowledge, as distinct from sheer growth, comes in different kinds. One kind, which may be called the extensive mode of progress, consists in extending the knowledge base in areas of nature already opened up for research. This will happen typically by means of more precise measurements or by developing instruments that make it possible to investigate broader areas of a certain field. Spectroscopy, in its wide sense, is a prime example of the extensive mode of progress, and so is solid state physics. Originally confined to metal physics, solid state physics covered larger and larger areas and eventually became the modern and very broad field of condensed-matter physics. Extensive progress is often low-rated because it is believed merely to produce new data and instrumental knowledge; yet, in reality it is a powerful agent for qualitative conceptual change as well. This is illustrated by many cases mentioned in this book, such as the careful measurements of blackbody radiation in the late 1890s and the extension of telescopic observation in the 1950s to include radio waves.

Another kind of progress consists of opening new windows through which nature's secrets can be studied, that is, the discovery of qualitatively new phenomena. Modern physics has also been extremely successful in this respect, perhaps especially before 1940. What is generally known as modern physics owes its very existence to such discoveries, of which x-rays and radioactivity were the earliest and most important. There is, last, a third kind of progress, which is essentially theoretical, namely the introduction of new frameworks and principles that reorganize and give new meaning to already obtained knowledge. Such principles do more than merely bring order to what initially seemed to be disjunct bodies of knowledge. They are also heuristically powerful, and throughout the century they have been rated highly by theoretical physicists.

Progress, whatever the meaning of the term, has been as conspicuous a feature of twentieth-century physics as has growth. Although the progress in knowledge has not been uniform and has varied from field to field, it has occurred continually, largely cumulatively, over a broad front, and often dramatically. In the 1990s, physicists knew vastly more about how nature works than they did in the 1890s. To single out just one area, the entire knowledge about atomic and subatomic structures and the forces that keep the basic particles together was harvested during the twentieth century. Before the 1890s, practically nothing was known about the submicroscopic constitution of matter, and most of what little was known, or believed to be known, was due to chemists rather than physicists. When the issue of progress is worth mentioning at all, it is not because physicists have any doubt about the matter, but because it has become fashionable in some quarters of sociology and philosophy to question the reality of scientific progress. There are scholars who seriously deny that one can speak meaningfully about progress in science. The simple answer to such folly would be to take a look at the development of some area of modern physics, say, electrical conductivity in metallic bodies or stellar energy processes, and compare the state of knowledge in the 1990s with that of a century earlier. Some sociologists of scientific knowledge claim to have shown that "scientists at the research front cannot settle their disagreements through better experimentation, more knowledge, more advanced theories, or clearer thinking" (Collins and Pinch 1992, 144). Perhaps it is needless to point out that the strange claim lacks support from the history of twentieth-century physics where disagreements have been routinely settled in exactly this way.

Physics and the Other Sciences

During the first half of the twentieth century, physics emerged as the number-one glamour science, a position it probably still has in spite of the troubles that it has faced more recently. The public came to see the great physicists as wizards with some kind of a direct connection to either God or nature. The most famous of the great thinkers who so fascinated the public was, of course, Einstein, but he was followed by other celebrated physicists, such as Bohr, Feynman, Gell-Mann and Hawking. The great and visionary theorists represented one side of the fascination of physics, the other and darker side being the power of physics as most dramatically symbolized by the mushroom clouds caused by nuclear bombs. Physics seemed to cover the entire spectrum, from deep quantum philosophy to technological devices such as radar and the laser. With respect to public fascination, no other science could compete with physics. A recent book undertakes to rank the world's one hundred most influential scientists in history, including psychologists and social scientists (Simmons 1997). Although the ranking should not be taken too seriously—how can one meaningfully compare Archimedes with Oppenheimer?—it is interesting to see how highly the author has ranked the physicists. The first three on the list (Newton, Einstein, and Bohr) are all physicists and among the twenty-five "most influential" scientists, twelve are physicists, eight of them belonging to the twentieth century.

Physics' dominant position among the sciences in the first three quarters of the century can be illustrated by the impact that physics exerted on other of the classical sciences such as astronomy, chemistry, geology, and biology. The impact occurred mainly through three channels, the most direct of which was the migration of physicists to other scientific disciplines. In many cases, young physicists successfully migrated to, or did important work in, one of the other sciences. It is noteworthy that there have been very few cases of the reverse traffic. Another channel of impact has been the adoption of physical modes of thought in sciences that were traditionally foreign or even hostile to such attitudes to scientific work. No less important was the influence on the nonphysics sciences by the instruments and techniques provided by experimental physics. In the case of chemistry, in many ways a sister discipline to physics, this was not a new feature. It was well known at the time of Lavoisier and was an important part of the physical chemistry that emerged in the late 1880s. But in the twentieth century, chemistry came to rely even more closely on new experimental methods originating in physics, such as x-ray and electron diffraction, NMR spectroscopy, and mass spectrometry. On a more fundamental level, chemistry was even threatened with becoming a branch of physics, namely in the sense that some atomic and quantum physicists (including Born and Dirac) claimed that chemistry was merely applied quantum theory. Five years before the advent of quantum mechanics, Born wrote that "we have not penetrated far into the vast territory of chemistry," yet "we have travelled far enough to see before us in the distance the passes which must be traversed before physics can impose her laws upon her sister science" (Nye 1993, 229). With the emergence in the late 1920s of quantum chemistry—a theory first developed by physicists, rather than chemists—Born's imperialistic hope (and the nightmare of many chemists) seemed to become a reality. However, it sooned turned out that not even simple molecules could be reduced to quantum physics without empirical input from the chemists. All the same, theoretical chemistry was deeply affected by quantum mechanics (and other branches of physics) and the field can, to some extent, be regarded as "applied physics." We have previously dealt with the development of astrophysics and the fundamental changes in the astronomical sciences that followed the stormy development of physics (chapters 12 and 23).

In the case of geology, the impact was less direct, but nonetheless led to a drastic reorientation of this science, from its traditional status as natural history to a new "earth science" that was modeled on the standards of physics and made use of instruments and reasoning characteristic of physics. Part of the geological sciences, such as geophysics and seismology, were already "physicalized" in the early part of the century, especially under the influence of the German physicist Emil Wiechert. With the plate-tectonic revolution of the 1960s, the physics-inspired transformation was complete.

A somewhat similar story can be told about the impact of physics on biology, where the advent of molecular biology in the 1930s marked a further intrusion of physical and reductionist thought in the life sciences. This is hardly surprising, for several of the early leaders of molecular biology were trained as physicists, including Max Delbrück and Walter Elsasser, who had both made valuable contributions to physics before they left the field for biology (and, in Elsasser's case, the earth sciences). Francis Crick, of double-helix fame, graduated in physics in 1938 and turned to biology only after the end of the war, in part under the inspiration of Schrödinger's 1944 book *What is Life*? The elucidation of the structure of DNA in 1953, widely seen as the most important discovery of modern biology, was to a large extent the result of analysis of x-ray diffraction patterns made by Maurice Wilkins, another physicist-turned-biologist. The general trend of biology in this century, and molecular biology in particular, has been greatly inspired by physics and a reductionist thinking taken over from this science. In 1966, Crick wrote: "The ultimate aim of the modern movement in biology is in fact to explain *all* biology in terms of physics and chemistry. There is a very good reason for this. Since the revolution in physics in the mid-twenties, we have had a sound theoretical basis for chemistry and the relevant parts of physics. . . . And it is the realization that our knowledge on the atomic level is secure which has led to the great influx of physicists and chemists into biology" (Olby 1994, 425). All in all, it would not be an exaggeration to claim that the overall pace and direction of the sciences in the twentieth century have been heavily influenced by the development of physics.

One of the most important results of this century's science is what appears to be the unlimited validity of the basic laws of physics. In the 1890s, it was still a matter of debate whether the second law of thermodynamics applied to living cells and, more generally, whether the laws of physics applied everywhere in nature and at any time. It is far from obvious that the laws have this wide range of validity, but many years of research seem to confirm that this is, in fact, the case. Not only do the laws apply to living organisms, but they also apply to the most distant parts of the universe, to the centers of stars, and to the supercompact state of the very early universe some ten billion years ago. All attempts to provide separate laws for separate strata of the world have failed. Physicists at the end of the twentieth century can claim with some confidence (not to be confused with certainty) that they know the fundamental laws and that these apply to of all of nature. This does not mean that all of nature has been explained by physics, nor that the other sciences have been reduced to physics. But it does mean that there are no phenomena in nature whose explanation requires principles or laws that stand in contradiction to those accepted by the physicists. (Such phenomena may turn up, but so far they have not and we have no reason to assume that they will.) So, without suggesting any sort of simplistic reductionism, there is a sense in which physics can be said to be the most fundamental and general of all the sciences. This "imperialist" point of view is far from new, but it is only in this century that it has been substantiated and has become more than an article of faith and self-congratulation.

Conservative Revolutions

As mentioned, the role played by physics in areas outside physics has changed completely during the twentieth century and turned the science into an integrated part of postindustrial society. This, and the effects it has had on the organization and performance of physics, is perhaps the biggest change that has occurred. When we look toward other aspects, it is fairly clear that the general picture has been one of both continuity and discontinuity, both permanence and revolutionary changes. On the ontological level, the changes have been deep indeed, largely a result of the quantum revolution—according to Philip Anderson, "a dislocation which is yet to be mentally healed even for many physicists" (Anderson 1995, 2018). Quantum mechanics has provided us with fundamental structures that have no resemblance at all to what can be perceived or measured directly. Our present beliefs in what the world ultimately consists of are a far cry from the beliefs of the 1890s, when it still made sense to think of matter as a collection of miniature blocks. The vacuum has turned out to be anything but "nothingness" and to be full of life, activity, and properties. This is a very important result of the new physics, but in itself it would not have shocked a physicist in 1900, who was accustomed to thinking of the vacuum as being filled with ether.

In some other respects, physics and physicists have not changed very much during the century. Thus, the basic rules of the game—the methodology of research—are much the same in the 1990s as they were in the 1890s. How to evaluate a claim, what counts as a good experiment, testing procedures, the function of mathematics in physical reasoning, and the use of thought experiments—these and other methodological topics have largely remained the same, although since the 1970s computer experiments have been added to the methods of modern physics. Had young Rutherford or Sommerfeld been catapulted into our world, they would have had great, but not insuperable, trouble in understanding many things about the theories and experiments of physics; they would easily have appreciated the methods of modern physics, so very close to those used in their own time. The same kind of continuity holds for the dreams and ultimate aspirations of physicists. Ideas of unification, mathematical beauty, and general principles as the basis of physics are not products of late twentieth-century physics. Although Planck or Mie would not have understood either the mathematics or the physics of GUT theories, they would have fully appreciated the general idea and aim of this class of modern theories.

I do not want to claim that there have been no changes in the methods of physics, only that methods and ideas widely different from those known in the nineteenth century have been relatively unimportant. As we noted in chapter 27, certain fields of high-energy physics (such as superstring theory and inflation cosmology) are so remote from experiment that they cannot be tested empirically. Mathematical consistency and aesthetic arguments therefore tend to become the means of demonstrating the "truth" of these theories. This is certainly an aberration from the commonly accepted methodology of science, and a potentially dangerous one at that. However, the situation should not be overdramatized. For one thing, this is only a tendency in a small corner of theoretical physics, and it does not affect the 99 percent of physics in which theory and experiment are in healthy contact. Moreover, it is not really a new problem. The ether vortex theory of the nineteenth century, the unified field theories of the early twentieth century, Eddington's fundamental theory of the 1930s, and most postwar theories of quantum gravity made use of standards that did not rely on experiment. Many years before the superstring theorists, there were physicists who argued for pure rationalism. For example, in a famous statement of 1933, Einstein suggested that "Nature is the realization of the simplest conceivable mathematical ideas . . . [and] we can discover, by means of purely mathematical constructions, those concepts and those lawful connections between them which furnish the key to the understanding of natural phenomena" (Holton 1988, 252).

As the methods of doing physics have essentially remained the same, so have the ideals of what physics should be and how physicists should behave. Science has its uncodified cultural norms—what the sociologist Robert Merton in 1942 called the scientific ethos or set of institutional imperatives. For example, scientists generally adhere to the idea of "universalism" (that the evaluation of scientific claims should be impersonal and objective), believe that secrecy should be avoided (a part of "communalism"), and accept "organized skepticism" as the proper attitude toward claims of new knowledge. These and other rules are occasionally violated, but they are nonetheless accepted as rules. The norms that contributed to end the N-ray affair in 1903 were largely the same norms that entered the scene when cold fusion was announced in 1989.

The great changes that have occurred in twentieth-century physics have built on existing knowledge and a healthy respect for traditions. There have been several attempts to base physics on an entirely new worldview (such as those proposed by Eddington and Milne in the 1930s), but they have all failed. It may seem strange that respect for traditions can produce revolutionary changes, but this is just what Thomas Kuhn described in 1962 under the label "normal science." On the other hand, the changes that sometimes follow paradigm-ruled or "normal" science are not revolutions in the strong sense that Kuhn suggested in 1962, namely, new paradigms incompatible with and totally different from the old ones. No such revolution has occurred in twentieth-century physics. After all, a theoretical physicist of the 1990s will have no trouble in understanding the spirit and details of Planck's work of 1900 in which the quantum discontinuity was introduced, nor will a modern experimentalist fail to appreciate J. J. Thomson's classical paper of 1897 in which the electron was announced. There is no insurmountable gap of communication, no deep incommensurability, between the physics of the 1990s and that of a century earlier.

The lesson to be extracted from the latest century of physics is that physical knowledge has greatly expanded and resulted in new and much-improved theories, but that these have been produced largely cumulatively and without a complete break with the past. It has always been important to be able to reproduce the successes of the old theories, and this sensible requirement guarantees a certain continuity in theoretical progress. The great discoveries and theories of our century have not, of course, left earlier knowledge intact, but neither have they turned it wholesale into non-knowledge. Most experimental facts continue to be facts even in the light of the new theories. The observation that Mercury's perihelion excess is 0.43″ per year was explained, not overthrown, by Einstein's theory of relativity, and any future theory of gravitation will have to accommodate the observational fact.

A large part of physics seems to be firmly stabilized. It becomes increasingly difficult to imagine that these parts, so thoroughly tested and so closely bound in a larger network of theories and experiments, will change drastically in the future. For several decades, it has been considered heretical, even ridiculous, to suggest that science develops "teleologically," that is, toward a certain state of knowledge that reflects the true structure of nature. It is correct, as the philosopher Nicholas Rescher pointed out, that "[s]ignificant scientific progress is generally a matter not of adding further facts—on the order of filling in of a crossword puzzle—but of changing the framework itself" (Rescher 1978, 48). Relativity, quantum mechanics, and the electroweak theory are examples of such changed frameworks that did not at first either rely on or suggest new experimental facts. Yet, not only does the view underrate the value of "adding further facts," but it also leaves open the question of whether or not there is a best possible framework that will leave a theory in a stable or "finished" state. We can smile at the naïveté of the fin-de-siècle physicists who believed that physics had essentially reached its final state, but their failure does not imply that no such final state exists. Because most physical theories proposed during history have turned out to be wrong, it does not follow that those accepted today are wrong as well and will be replaced by entirely new theories.

One might speculate that history might repeat itself and that tomorrow's physicists might find quite new phenomena in nature that would demand a major reframing of theoretical physics—a kind of analogy to the surprising discoveries of 1895–97. Is such a scenario plausible? It seems that although one can never preclude the possibility, it becomes still more unlikely that physicists have missed some big and important aspect of nature. The modern army of physicists and their arsenal of sophisticated high-precision instruments makes it much more difficult for such phenomena to remain hidden than in the case of radioactivity a century ago. It is many decades since a new discovery squarely contradicted fundamental theory. In 1986 the discovery of a "fifth force" was announced, a force of intermediate range that could not be accounted for within established theory. Had the discovery claim been accepted, it might have led to a major conceptual change in theoretical physics. But this was not what happened. After a few years of experiments and intense debate, it turned out that the fifth force did not pose a threat to the standard physics operating with four forces of nature. The fifth force does not exist.

Physics will undoubtedly continue to develop and make many interesting discoveries in the new century. But it is possible that the pattern of progress in physics will change, and that many of the most fundamental aspects will remain as they are now known. There will always be exciting work to do and discoveries to make, but it is far from certain that the development of physics in the twenty-first century will be as explosive as it has been in the twentieth century. Feynman, as we quoted him in chapter 26, believed that "[t]he age in which we live is the age in which we are discovering the fundamental laws of nature, and that day will never come again." Whether Feynman's prophecy was correct or not can be shown only by developments in the next century. Perhaps a historian writing the history of physics in the twenty-first century will quote Feynman in order to demonstrate his wisdom; or perhaps he will quote him in order to show how utterly wrong he was.

APPENDIX

FURTHER READING

CHAPTER 1

The chapter relies in part on Kragh 1996a. Surveys of nineteenth-century physics include Harman 1982 and Purrington 1997. See also Brush 1978, especially for the broader contexts. Jungnickel and McCormmach 1986 is recommended for the entire period up to 1925, and is especially informative on German physics. Useful overviews of fin-de-siècle physics can be found in Hiebert 1979 and Heilbron 1982. The question of completeness of science is discussed in Badash 1972, and the problems of the mechanical world picture in Klein 1973. On speculations about four-dimensional and non-Euclidean spaces, see Bork 1964 and Beichler 1988. Scientific and quasi-scientific aspects of the ether concept are discussed in Kragh 1989a and Cantor and Hodge 1981. On Ostwald's energetics, see Hiebert 1971 and Hakfoort 1992. LeBon's ideas and the spiritual climate at the turn of the century are analyzed in Nye 1974.

CHAPTER 2

Statistical data on physics around 1900 are presented and discussed in Forman, Heilbron, and Weart 1975. See also Hirosige and Nisio 1986 on subdisciplines in physics, and Pyenson and Skopp 1977 for a detailed analysis of physics education in Germany. Kevles 1976 compares the physics, chemistry, and mathematics communities in the United States. The account of Nagaoka's description of European physics in 1910 is based on Badash 1967 and Carazza and Kragh 1991. Italian physics between 1900 and 1904 is analyzed quantitatively in Galdabini and Giuliani 1988. Jungnickel and McCormmach 1986 is rich on information on German theoretical physics. On physics in America, see Kevles 1987 and Reingold and Reingold 1981.

CHAPTER 3

The history of x-rays, radioactivity, the electron, and other aspects of atomic, nuclear, and particle physics is described in Pais 1986. A very readable but less detailed account can be found in Segré 1980. On Becquerel's route to the discovery of radioactivity, see Martins 1997, and for early attempts to explain the origin of radioactivity, Kragh 1997a. The discovery of the electron is treated in Falconer 1987, Feffer 1989, Dahl 1997, and Davis and Falconer 1997, of which the last work includes reprints of many of Thomson's papers. Arabatzis 1996 summarizes the complex discovery histories. The pseudodiscoveries of black light and N-rays are examined in Nye 1974 and 1980. On magnetic rays, see Carazza and Kragh 1990, and on the discovery of the cosmic radiation, Xu and Brown 1987 and De Maria, Ianniello, and Russo 1991.

CHAPTER 4

Conn and Turner 1965 includes extracts or full reproductions of many of the important papers in atomic theory between 1895 and 1914. Aspects of the Thomson atomic model are treated in Sinclair 1987 and Kragh 1997a and 1997b. Nicholson's model is analyzed in McCormmach 1966, and the birth of Rutherford's nuclear atom in Heilbron 1968. On Bohr's atomic theory, see Heilbron and Kuhn 1969, Heilbron 1981, and also French and Kennedy 1985. Bohr's 1913 papers and the Manchester memorandum are reproduced in Bohr 1963, with an introduction by Léon Rosenfeld.

CHAPTER 5

Very much has been written about the early development of quantum theory. The most comprehensive work is probably volume 1 of Mehra and Rechenberg 1982. Kangro 1976 contains full information about the early experiments, and Kuhn 1978 is a detailed analysis of theoretical development from Boltzmann to Planck. It should, however, be mentioned that Kuhn's interpretation is controversial and not generally accepted. For the complex question of the relationship between the methods of Boltzmann and Planck, see also Darrigol 1988a. Other recommendable books on the history of quantum theory include Hermann 1971, Hund 1974, Jammer 1966, and Darrigol 1992. A concise history is presented in Klein 1970. The Solvay congresses are surveyed in Mehra 1975, and the 1911 congress is analyzed in Barkan 1993.

CHAPTER 6

Mendelssohn 1977 is a semipopular history of the development of low-temperature physics. The subject is treated more scholarly in Dahl 1992, which offers a complete history of superconductivity, from the late nineteenth century to the early 1990s. See also Dahl 1984 for the discovery of superconductivity. Other aspects of the development of cryogenics are described in Scurlock 1992. Theories of electrical conduction from 1898 to the 1920s are the subject of Kaiser 1987. The history of superconductivity and superfluidity is also analyzed in detail in Gavroglu and Goudaroulis 1989, where emphasis is on more general and methodological aspects. A Leiden physicist's account of the discovery of superconductivity and Kammerlingh Onnes's laboratory can be found in Casimir 1983.

CHAPTER 7

There is a rich literature on the special theory of relativity and its precursors, including Holton 1988, Goldberg 1984, and Darrigol 1996. On pre-Einsteinian theories, see Hirosige 1976, Nersessian 1986, and Darrigol 1994. Einstein's 1905 theory is analyzed in great detail in Miller 1981. The best of the many Einstein biographies is Pais 1982. Some of the early developments are analyzed in Goldberg 1976 and Galison 1979. On the early history of general relativity, see Mehra 1974, and the detailed analysis in Norton 1985 and Earman and Glymour 1978. The three classical tests of general relativity are analyzed in, for example, Roseveare 1982 (perihelion advance), Earman and Glymour 1980a (light bending), and Earman and Glymour 1980b (gravitational redshift). Hentschel 1992 is a useful analysis of Einstein's attitude toward experiments, and Hentschel 1990 provides a detailed account of scientific and non-scientific reactions to the theory of relativity. On the reception of relativity in different countries, see also Glick 1987.

CHAPTER 8

The electromagnetic worldview is treated in McCormmach 1970, and Jammer 1961 includes a condensed account of the electromagnetic concept of mass. Miller 1986, a collection of essays, deals with electrodynamics without relativity, including a detailed analysis of Poincaré's 1906 theory of the electron. A discussion of Abraham's rigid electron can be found in Goldberg 1970. For the mass-variation experiments, see Miller 1981, Cushing 1981, and Batimelli 1981. The role of electromagnetic mass in early atomic and nuclear physics is discussed in Siegel 1978, and Kragh 1985 contains information about mass variation experiments in the 1920s. The best source of Mie's theory and other unified theories in the first third of the century is Vizgin 1994.

CHAPTER 9

Atherton 1984 is a general account of the history of electrical technology. On coil loading and long-distance telephony, see Wasserman 1985, and, for the continuous loading method, Kragh 1994. Early AT&T research is covered in Hoddeson 1981a and Fagen 1975, and Russo 1981 tells the interesting story about Davisson's and Germer's discovery of electron diffraction. For details about the development of vacuum tubes, see Tyne 1977. On Langmuir and General Electric research, see Reich 1983 and 1985. Aspects of physics in the World War I are dealt with in Hartcup 1988, Schröder-Gudehus 1978, Kevles 1987, and Cardwell 1975.

CHAPTER 10

On science policy and funding in the Weimar republic, see Schröder-Gudehus 1978 and Forman 1974. On international science policy in the 1920s, see also Cock 1983. Heilbron 1986 and Cassidy 1992 give interesting accounts of German science in the period as seen through the lives of Planck and Heisenberg, respectively. Details about physics and science policy in the young Soviet Union can be found in Josephson 1991. The ideology of German physicists is analyzed in Forman 1973, and Forman 1971 argues that the intellectual climate of the Weimar republic decisively influenced physicists' thinking about quantum phenomena. Forman's controversial thesis is criticized in Hendry 1980 and Kraft and Kroes 1984.

CHAPTER 11

Among the many works that deal with the history of quantum theory in the 1910s and 1920s are the books by Jammer 1966, Hendry 1984a, MacKinnon 1982, Darrigol 1992, and Mehra and Rechenberg 1982 and 1987. For aspects of Schrödinger's theory, see Kragh 1982b and, further, the contributions in Bitbol and Darrigol 1992. Also worth consulting are the biographies by Pais 1991 (on Bohr), Cassidy 1992 (on

Heisenberg), Dresden 1987 (on Kramers), Moore 1989 (on Schrödinger), and Kragh 1990 (on Dirac). De Broglie's theory is analyzed in Darrigol 1993 and the experimental history of wave-particle dualism is detailed in Wheaton 1983. The origin of the Dirac equation is dealt with in Kragh 1981a and 1990 and in Moyer 1981. On the reception and transmission of quantum mechanics, see volume 4 of Mehra and Rechenberg 1982, Sopka 1988, Heilbron 1985, Cartwright 1987, and Kojevnikov and Novik 1989. Of these sources, Heilbron 1985 and Cartwright 1987 deal with the Americans' lack of interest in philosophical aspects. Sources of classical papers in English translations include ter Haar 1967, Van der Waerden 1967, and Ludwig 1968.

CHAPTER 12

Sources on nuclear physics are reproduced in Beyer 1949 and Brink 1965. On early nuclear models, see especially Stuewer 1983 and 1986a. A concise review is presented in Badash 1983. Stuewer 1979 is a volume of conference proceedings on the history of nuclear physics in the 1930s. For a detailed account of the Cambridge-Vienna controversy, see Stuewer 1985a. For the proton-electron model and its problems, as well as other aspects of nuclear physics, see, e.g., Pais 1986, Brown and Rechenberg 1996, Bromberg 1971, and Weiner 1972. The early neutrino story is covered in Brown 1972, and Heisenberg's lattice world idea in Carazza and Kragh 1995. On the discovery of the deuteron, see Stuewer 1986c and Brickwedde 1982. My account of early nuclear astrophysics is taken from Kragh 1996b. A more detailed analysis of the earliest phase can be found in Hufbauer 1981. The early cyclotrons are described in great detail in Heilbron and Seidel 1989, and information about the Cavendish high-voltage machines can be found in Hendry 1984b. An interesting perspective on Bohr's involvement in nuclear physics is provided in Aaserud 1990.

CHAPTER 13

The section on Dirac and the positron is based on Kragh 1990. On this subject, see also Hanson 1963, Moyer 1981, Kragh 1989, De Maria and Russo 1985, and Roqué 1997a. Among the works that deal comprehensively with early particle physics, Brown and Hoddeson 1983 is very informative; especially regarding theory, so are Pais 1986 and Brown and Rechenberg 1996. Rueger 1992 gives a good survey of the problems in quantum electrodynamics in the 1930s. Attempts to solve the problems by introducing a smallest length are analyzed in Kragh 1995. Cosmic ray physics in the period, the discovery of the muon, and the relationship between theory and experiment are analyzed in De Maria and Russo 1989, Galison 1983, 1987, and 1997, and Cassidy 1981. For the discovery of the mesons and early meson theory, see Brown 1981, Darrigol 1988a, Brown and Hoddeson 1983, and Brown and Rechenberg 1996. Reprints of several of the important papers of the period can be found in Cahn and Goldhaber 1989 and, on cosmic-ray research, in Hillas 1972.

CHAPTER 14

The literature on quantum philosophy is vast, but most of it is nonhistorical. Among the historical works dealing with the situation in the 1920s and 1930s are Hendry 1984a, MacKinnon 1982, and Jammer 1966. Whitaker 1996 is a good semihistorical and largely nontechnical account. Jammer 1974 is still the best historical-philosophical survey. On different aspects of the history of complementarity, see Beller 1992, Heilbron 1985, and Holton 1988. Von Neumann's impossibility proof is critically reviewed in Pinch 1977 and Caruana 1995. Cushing 1995 is a detailed historical-philosophical examination of interpretations of quantum mechanics, with particular emphasis on the position of Bohm. Wheeler and Zurek 1983 includes reproductions of many of the important sources from the period between 1926 and 1980.

CHAPTER 15

The cosmophysical trend in the 1930s is analyzed in Kragh 1982a, on which part of the chapter is based. On Eddington's theories, see Kilmister 1994 and, for his philosophy of science, Yolton 1960. Schrödinger's inspiration from Eddington is documented in Rueger 1988. Milne's system is treated in Harder 1974 and Urani and Gale 1993, and Dirac's cosmological views in Kragh 1990. Numerical relationships between natural constants are surveyed in Barrow 1981 and Barrow and Tipler 1986.

CHAPTER 16

Beyerchen 1977 was the first, and is still one of the best, detailed accounts of physics in the Third Reich. See also Walker 1995, which focuses on the German nuclear power project. Among the anthologies, Renneberg and Walker 1994 and Mehrtens and Richter 1980 are valuable collections of analyses of German science and technology during 1933–45. An excellent collection of annotated primary sources, translated into English, can be found in Hentschel 1996. Physicists' conditions in Nazi Germany are also covered in many biographies, such as Heilbron 1986 (Planck), Cassidy 1992 (Heisenberg), and Sime 1996 (Meitner). The impact of the Nazi regime on physics research in Germany is examined quantitatively in Fischer 1988. Graham 1972 includes chapters on physics and cosmology in the Soviet Union under the Stalin era. Josephson 1991 gives a detailed analysis of Soviet physics in the period, covering both ideological and institutional contexts. On this topic, see also Gorelik 1995, Vucinich 1980, and Kojevnikov 1991. On Italian physics in the Mussolini era, see Holton 1978.

CHAPTER 17

On American physics in the 1930s, see Weiner 1970 and Kevles 1987. Weiner 1969 deals with the physics refugees in America, and Weart 1979a is a detailed quantitative account of American physics mainly in the interwar period. Among the best studies of the migration of physicists in the 1930s are Holton 1983, Hoch 1983, and Rider 1984. See also Fermi 1971, written by the widow of one of the most famous physics emigrants. Stuewer 1984 examines the case of nuclear physics in an emigration perspective.

CHAPTER 18

The discovery of fission is treated in many places, e.g., Graetzer and Anderson 1971, Weart 1983, and Krafft 1983. A detailed study of the reaction in 1939 to the news of the discovery of fission is given in Badash, Hodes, and Tiddens 1986. Graetzer and

Anderson 1971 includes translated reprints of several of the key papers from 1934–45, as does Wohlfarth 1979 (in English and German). Personal recollections by two of the centrally involved physicists are given in Frisch and Wheeler 1967. On the transmission of and early work on fission, see Stuewer 1985b and Weart 1976. Stuewer 1994 provides details about the nuclear-physical background in the 1930s. The French work in nuclear energy is fully described in Weart 1979b, and the British work in Gowing 1964. A complete and very readable history of the path leading to the atomic explosions of 1945 is given in Rhodes 1986. Details of the German and Russian nuclear projects, mentioned only briefly in the text, can be found in Powers 1993, Walker 1995 (both on Germany) and Holloway 1994 and Rhodes 1995 (on Russia). For more on the Japanese attempt to develop a nuclear bomb, see Wilcox 1985.

CHAPTER 19

There is no good history of postwar nuclear physics, but a few peaks of its development can be followed through the Nobel lectures and the survey in Brink 1995. See also Lee and Wiringa 1990 on the nuclear shell model. Mladjenovič 1998 is a comprehensive and useful survey. For the early discoveries of transuranic elements, one may consult Weeks 1968, which is still the standard work on the discovery of chemical elements. On the history of nuclear energy in a global perspective, see Goldschmidt 1982. Nuclear weapons and nuclear policies in the postwar period are the subjects of many books and articles. Badash 1995 is a good introduction. A more detailed account can be found in Hewlett 1989, for example, and a good collection of sources on nuclear policy from 1939 to 1990 is included in Cantelon, Hewlett, and Williams 1991. For the development of the hydrogen bomb, see Rhodes 1995. The history of controlled fusion is given in Hendry 1987 (prehistory), Hendry and Lawson 1993 (British project), Post 1995 (mostly scientific aspects), and Bromberg 1982 (American project).

CHAPTER 20

There is a vast body of literature on political and military aspects of postwar physics (or science in general), mostly concentrating on the American scene in the first two decades after 1945. Forman 1987 is a fine example. More general analyses can be found in Dickson 1984 and Greenberg 1967. On big science physics, see Galison and Hevly 1992 and Weinberg 1967. Schweber 1989 is an interesting essay on the political and ideological contexts of particle physics in the 1950s, and Hoddeson et al. 1997 includes aspects of high-energy big science in the 1960s and 1970s. On accelerators and big science in Japan, only briefly mentioned in the text, see Hoddeson 1983. The history of CERN is detailed in Hermann et al. 1987–90 and Krige 1996. An interesting bibliometric comparison between American and European accelerators around 1960–85 is provided in Irvine et al. 1986 and, in greater detail, Irvine and Martin 1985.

CHAPTER 21

The secondary literature on particle physics is comprehensive. Pais 1986 is useful, especially on theory. Doncel et al. 1987 focuses on symmetry principles, and Cahn and Goldhaber 1989 provides a condensed description supported by reprints of ninety-eight notable experimental papers. Brown, Dresden, and Hoddeson 1989, Hoddeson et al. 1997, and Foster and Fowler 1988 are very informative collections. Newman and Ypsilantis 1996 is a massive collection of historical papers on modern high-energy physics and is addressed mostly to physicists. Different aspects of weak interaction physics are treated in Cline and Riedasch 1984 and Franklin 1986. Comprehensive bibliometric analyses of weak interaction physics between 1950 and 1975 are included in White, Sullivan, and Barboni 1979 and Koester, Sullivan, and White 1982. On experimental aspects of quark history, see, for example, Pickering 1981 and the more popular Riordan 1987. Details on detectors and many other aspects of "the material culture" of high-energy physics are given in Galison 1997. For other and more specialized literature, see Hovis and Kragh 1991.

CHAPTER 22

A complete and detailed historical treatment of QED up to about 1952 can be found in Schweber 1994a. Other aspects of renormalization physics are analyzed in Brown 1993. See also the historical accounts of QED in Weinberg 1977 and Aramaki 1987 and 1989. A selection of the most important QED papers between 1927 and 1953 is reproduced in Schwinger 1958. The development of the *S*-matrix program and related topics from about 1943 to 1985 are analyzed expertly in Cushing 1990 and, less demandingly, in Freundlich 1980. Cao 1997 is a synthetic conceptual history, tracing the development of field theories up to about 1990. The discovery of neutral currents are analyzed in Galison 1987 and, from a different point of view, Pickering 1984b. Pickering 1984a and Pais 1986 are valuable surveys of particle physics in the 1960s and 1970s, and Cahn and Goldhaber 1989 includes reprints of many of the pioneering experimental papers. Hoddeson et al. 1997 is a rich source with respect to particle physics between 1965 and 1980. For the 1983 discovery of the intermediate vector bosons, see, for example, the popular account in Watkins 1986. Further literature is listed in Hovis and Kragh 1991.

CHAPTER 23

Parts of the chapter rely on Kragh 1996b, which gives a full description of cosmology between 1930 and 1967. For aspects of the development of modern cosmology, see also Bertotti et al. 1990 and Lightman and Brawer 1990. Hetherington 1993 is a useful collection of articles on the history and philosophy of cosmology, and Bernstein and Feinberg 1986 includes a selection of classical papers. Harwit 1981 deals with astrophysical discoveries in an original and fascinating way. The fate of general relativity between 1925 and 1955 is discussed in Eisenstaedt 1989. On postwar developments in general relativity, see Thorne 1994, Will 1993, and contributions in Hawking and Israel 1987. The controversy over Weber's claim of having detected gravitational waves is examined in Collins 1992.

CHAPTER 24

The best and most comprehensive history of solid state physics is Hoddeson et al. 1992, a collaborative project that covers the development of the discipline from its nineteenth-century roots until about 1960. I have used, in particular, the contributions of Hoddeson, Baym and Eckert 1992, Braun 1992, and Weart 1992. Also valuable is Eckert and Schubert 1990, which emphasizes the applied-science aspects, and Braun and MacDonald 1978, which deals mostly with semiconductor electronics. For a detailed account of the invention of the transistor, see Hoddeson 1981b and Riordan and Hoddeson 1997. On post-1930 superconductivity research, see Dahl 1996, Gavroglu and Goudaroulis 1989, and Bardeen 1973. The discovery of and earliest developments in high-T_c superconductivity are covered in Hazen 1988.

CHAPTER 25

The evolution in microelectronics is covered in, for example, Eckert and Schubert 1990, Braun and MacDonald 1978, and Riordan and Hoddeson 1997. On the early development of microwave technologies, see Forman 1992 and 1995. The history of the laser is described in Bromberg 1991 and Bertolotti 1983. For optical fibers, see Faltas 1988 and Bray 1995. The history of the Bell Telephone Laboratories, Smits 1985, is a source of general interest for many aspects of communications technology. For Bell Labs research, see also Bernstein 1984.

CHAPTER 26

Data on U.S. physics circa 1965–85 are discussed in Brinkman 1986. For the crises in American physics, see Kevles 1987 and Schweber 1994b and 1995. The more general wave of counterculture, including versions of antiscience and alternative sciences, is analyzed in Easlea 1973, Nowotny and Rose 1979, and Burnham 1987. The modern discussion of the end of science is the subject of Horgan 1996 and contributions in Elvee 1992. For a philosophical perspective, see Rescher 1978.

CHAPTER 27

There is no good historical work on postwar unified field and particle theories. For the early phase, see Vizgin 1994. The history of theories of quantum gravity is reviewed briefly in Ashtekar 1991. Grand unification theories until the early 1980s are described in Pickering 1984a and, on a popular level, in Trefil 1983. A collection of primary sources of modern unified theories can be found in Zee 1982 and, focusing on the astrophysical and cosmological aspects, in Schramm 1996. On superstring theories, see Davies and Brown 1988, Galison 1995, and Schwarz 1996.

CHAPTER 28

The Nobel motivations and lectures, together with biographical essays, are published in a series of, so far, seven volumes; see [Nobel] 1967–97. The origin and early development of the Nobel institution is described in Crawford 1984, and a complete list of Nobel nominators and nominees from 1901 to 1937 is included in Crawford, Heilbron, and Ullrich 1987. For analyzes of the Nobel physics prize processes, see Friedmann 1981 and 1989. Crawford, Sime, and Walker 1997 is one of the few works on postwar Nobel physics (in this case, a nonprize rather than a prize).

BIBLIOGRAPHY

ABBREVIATIONS USED FOR PERIODICALS

AHES	*Archive for the History of Exact Sciences*
AJP	*American Journal of Physics*
AS	*Annals of Science*
BW	*Berichte zur Wissenschaftsgeschichte*
HS	*Historia Scientiarum*
HSPS	*Historical Studies in the Physical Sciences* (since 1988: *Historical Studies in the Physical and Biological Sciences)*
NS	*New Scientist*
PS	*Perspectives on Science*
PRSA	*Proceedings of the Royal Society of London*, series A
PT	*Physics Today*
RHS	*Revue d'Histoires des Sciences*
RMP	*Reviews of Modern Physics*
RSS	*Rivista di Storia della Scienze*
SHPMP	*Studies in the History and Philosophy of Modern Physics*
SHPS	*Studies in the History and Philosophy of Science*
SIC	*Science in Context*
SSS	*Social Studies of Science*
T&C	*Technology and Culture*

Aaserud, Finn. 1990. *Redirecting Science: Niels Bohr, Philanthropy, and the Rise of Nuclear Physics*. Cambridge: Cambridge University Press.

Anderson, Philip. 1971. "Are the big machines necessary?" *NS* 82: 510–14.

———. 1972. "More is different," *Science* 177: 393–96.

———. 1995. "Historical overview of the twentieth century in physics," in Brown, Pais, and Pippard, eds., pp. 2017–32.

Anthony, L. J., H. East, and M. J. Slater. 1969. "The growth of the literature in physics," *Reports on Progress in Physics* 32: 709–67.

Arabatzis, Theodore. 1992. "The discovery of the Zeeman effect: A case study of the interplay between theory and experiment," *SHPS* 23: 365–88.

———. 1996. "Rethinking the 'discovery' of the electron," *SHPMP* 27: 405–35.

Aramaki, Seiya. 1987. "Formation of the renormalization theory in quantum electrodynamics," *HS* 32: 1–42.

———. 1989. "Development of the renormalization theory in quantum electrodynamics," *HS* 37: 91–113.

Aronovitch, Lawrence. 1989. "The spirit of investigation: Physics at Harvard University, 1870–1910," in Frank A. J. L. James, ed. *The Development of the Laboratory*. London: Macmillan, pp. 83–100.

Ashtekar, A. 1991. "The winding road to quantum gravity," in A. Ashtekar and John Stachel, eds., *Einstein Studies. Vol. 3: Conceptual Problems of Quantum Gravity*. Boston: Birkhäuser, pp. 1–9.

Atherton, W. A. 1984. *From Compass to Computer: A History of Electrical and Electronics Engineering*. San Francisco: San Francisco Press.

Badash, Lawrence. 1967. "Nagaoka to Rutherford, 22 February 1911," *PT* 20 (April): 55–60. Reprinted in Weart and Phillips, eds., 1985, pp. 103–107.

———. 1972. "The completeness of nineteenth-century science," *Isis* 63: 48–58.

———. 1983. "Nuclear physics in Rutherford's laboratory before the discovery of the neutron," *AJP* 51: 884–88.

———. 1995. *Scientists and the Development of Nuclear Weapons*. Atlantic Highlands, N.J.: Humanities Press.

Badash, Lawrence, Elisabeth Hodes, and Adolph Tiddens. 1986. "Nuclear fission: Reactions to the discovery in 1939," *Proceedings of the American Philosophical Society* 130: 196–231.

Bardeen, John. 1973. "History of superconductivity research," in B. Kursunoglu and A. Perlmutter, eds., *Impact of Basic Research on Technology*. New York: Plenum Press, pp. 15–57.

Barkan, Diana K. 1993. "The witches' Sabbath: The first international Solvay congress in physics," *SIC* 6: 59–82.

Barrow, John D. 1981. "The lore of large numbers: Some historical background to the anthropic principle," *Quarterly Journal of the Royal Astronomical Society* 22: 388–420.

Barrow, John D., and Frank J. Tipler. 1986. *The Anthropic Cosmological Principle*. Oxford: Oxford University Press.

Battimelli, G. 1981. "The electromagnetic mass of the electron: A case study of a non-crucial experiment," *Fundamenta Scientiae* 2: 137–50.

Beichler, James E. 1988. "Ether/or: Hyperspace models of the ether in America," in Stanley Goldberg and Roger H. Stuewer, eds., *The Michelson Era in American Science 1870–1930*. New York: AIP Conference Proceedings no. 179, pp. 206–23.

Beller, Mara. 1992. "The birth of Bohr's complementarity: The context and the dialogues," *SHPS* 23: 147–80.

Bernal, John D. 1939. *The Social Function of Science*. London: Routledge & Sons.

Bernstein, Jeremy. 1984. *Three Degrees Above Zero: Bell Labs in the Information Age*. New York: Scribner's.

Bernstein, Jeremy, and Gerald Feinberg, eds. 1986. *Cosmological Constants: Papers in Modern Cosmology*. New York: Columbia University Press.

Bertolotti, Mario. 1983. *Masers and Lasers: An Historical Approach*. Bristol, England: Adam Hilger.

Bertotti, Bruno, et al., eds. 1990. *Modern Cosmology in Retrospect*. Cambridge: Cambridge University Press.

Beyer, Robert T., ed. 1949. *Foundations of Nuclear Physics*. New York: Dover Publications.

Beyerchen, Alan D. 1977. *Scientists under Hitler: Politics and the Physics Community in the Third Reich*. New Haven: Yale University Press.

Bishop, Amasa S. 1958. *Project Sherwood—The U.S. Program in Controlled Fusion*. Reading, Mass.: Addison-Wesley.

Bitbol, Michel, and Oliver Darrigol, eds. 1992. *Erwin Schrödinger: Philosophy and the Birth of Quantum Mechanics*. Paris: Editions Frontières.

Boag, J. W., P. E. Rubinin, and D. Shoenberg, eds. 1990. *Kapitza in Cambridge and Moscow: Life and Letters of a Russian Physicist*. Amsterdam: North-Holland.

Bohr, Niels. 1963. *On the Constitution of Atoms and Molecules*. With an introduction by L. Rosenfeld. Copenhagen: Munksgaard.

Boltzmann, Ludwig. 1992. "A German professor's trip to El Dorado," *PT* 45 (January): 44–51.

Bork, Alfred M. 1964. "The fourth dimension in nineteenth-century physics," *Isis* 55: 326–38.

Born, Max. 1978. *My Life: Recollections of a Nobel Laureate*. New York: Charles Scribner's Sons.

Braun, Ernest. 1992. "Selected topics from the history of semiconductor physics and its applications," in Hoddeson et al., eds., pp. 443–88.

Braun, Ernest, and Stuart MacDonald. 1978. *Revolution in Miniature: The History and Impact of Semiconductor Electronics*. Cambridge: Cambridge University Press.

Bray, John. 1995. *The Communications Miracle*. New York: Plenum Press.

Brickwedde, Ferdinand G. 1982. "Harold Urey and the discovery of deuterium," *PT* 35 (September): 34–39.

Bridgman, Percy. 1950. *Reflections of a Physicist*. New York: Philosophical Library.

Brink, David M., ed. 1965. *Nuclear Forces*. New York: Pergamon.

———. 1995. "Nuclear dynamics," in Brown, Pais and Pippard, eds., pp. 1183–1232.

Brinkman, William F., et al., eds. 1986. *Physics Through the 1990s: An Overview*. Washington, D.C.: Washington Academy Press.

Bromberg, Joan. 1971. "The impact of the neutron: Bohr and Heisenberg," *HSPS* 3: 307–41.

———. 1976. "The concept of particle creation before and after quantum mechanics," *HSPS* 7: 161–91.

———. 1982. *Fusion: Science, Politics, and the Invention of a New Energy Source*. Cambridge, Mass.: MIT Press.

———. 1991. *The Laser in America, 1950–1970*. Cambridge, Mass.: MIT Press.

Brown, Laurie. 1972. "The idea of the neutrino," *PT* 31 (September): 23–28.

———. 1981. "Yukawa's prediction of the meson," *Centaurus* 25: 71–132.

———. ed. 1993. *Renormalization: From Lorentz to Landau (and Beyond)*. New York: Springer-Verlag.

Brown, Laurie, Max Dresden, and Lillian Hoddeson, eds. 1989. *Pions to Quarks: Particle Physics in the 1950s*. New York: Cambridge University Press.

Brown, Laurie, and Lillian Hoddeson, eds. 1983. *The Birth of Particle Physics*. Cambridge: Cambridge University Press.

Brown, Laurie, Abraham Pais, and Brian Pippard, eds. 1995. *Twentieth Century Physics*. 3 vols. New York: AIP Press.

Brown, Laurie, and Helmut Rechenberg. 1996. *The Origin of the Concept of Nuclear Forces*. Bristol, England: Institute of Physics Publishing.

Brush, Stephen G. 1978. *The Temperature of History: Phases of Science and Culture in the Nineteenth Century*. New York: Burt Franklin & Co..

———. 1986. *The Kind of Motion We Call Heat: A History of the Kinetic Theory of Gases in the 19th Century*. Amsterdam: North-Holland.

Buchwald, Jed Z. 1994. *The Creation of Scientific Effects: Heinrich Hertz and Electric Waves*. Chicago: University of Chicago Press.

Bullard, Edward. 1975. "The effect of World War II on the development of knowledge in the physical sciences," *PRSA* 342: 519–36.

Bunge, Mario, and William R. Shea, eds. 1979. *Rutherford and Physics at the Turn of the Century*. New York: Science History Publications.

Burnham, J. C. 1987. *How Superstition Won and Science Lost*. New Brunswick, N.J.: Rutgers University Press.

Cahan, David. 1985. "The institutional revolution in German physics, 1865–1914," *HSPS* 15: 1–65.

———. 1989. *An Institute for an Empire: The Physikalisch-Technische Reichsanstalt 1871–1918*. Cambridge: Cambridge University Press.

Cahn, Robert N., and Gerson Goldhaber. 1989. *The Experimental Foundation of Particle Physics*. New York: Cambridge University Press.

Cantelon, Philip L., Richard G. Hewlett, and Robert C. Williams, eds. 1991. *The American Atom: A Documentary History of Nuclear Policies from the Discovery of Fission to the Present*. Philadelphia: University of Pennsylvania Press.

Cantor, G. M., and M. J. S. Hodge, eds. 1981. *Conceptions of Ether: Studies in the History of Ether Theories 1740–1900*. Cambridge: Cambridge University Press.

Cao, Tian Yu. 1997. *Conceptual Developments of 20th Century Field Theories*. Cambridge: Cambridge University Press.

Carazza, Bruno, and Helge Kragh 1990. "Augusto Righi's magnetic rays: A failed research program in early 20th-century physics," *HSPS* 21: 1–28.

———. 1991. "An Oriental in Europe: Nagaoka, Righi, and the state of physics 1910," *HS* 41: 37–44.

———. 1995. "Heisenberg's lattice world: The 1930 theory sketch," *AJP* 63: 595–605.

Cardwell, Donald S. L. 1975. "Science and World War I," *PRSA* 342: 447–56.

Cartwright, Nancy. 1987. "Philosophical problems of quantum theory: The response of American physicists," in Lorenz Krüger, Gerd Grigerenzer, and Mary Morgan, eds., *The Probabilistic Revolution. Vol. 2. Ideas in the Sciences*. Cambridge, Mass.: MIT Press, pp. 417–35.

Caruana, Louis. 1995. "John von Neumann's 'impossibility proof' in a historical perspective," *Physis* 32: 109–24.

Casimir, Hendrik. 1983. *Haphazard Reality*. New York: Harper and Row.

Cassidy, David. 1981. "Cosmic ray showers, high energy physics, and quantum field theories: Programmatic interactions in the 1930s," *HSPS* 12: 1–39.

———. 1992. *Uncertainty: The Life and Science of Werner Heisenberg*. New York: W. H. Freeman and Co.

Chew, Geoffrey F. 1970. "Hadron bootstrap: Triumph or frustration?" *PT* 23 (November): 23–28.

Cline, David, and Gail Riedasch, eds. 1984. *50 Years of Weak Interactions*. Madison: University of Wisconsin.

Cock, A. G. 1983. "Chauvinism and internationalism in science: The International Research Council, 1919–1926," *Notes & Records of the Royal Society* 37: 249–88.

Collins, Harry M. 1992. *Changing Order: Replication and Induction in Scientific Practice*. Chicago: University of Chicago Press.

Collins, Harry, and Trevor Pinch. 1992. *The Golem: What Everyone Should Know about Science*. Cambridge: Cambridge University Press.

Conn, G. K. T. and H. D. Turner. 1965. *The Evolution of the Nuclear Atom*. London: Iliffe Books.

Crawford, Elisabeth. 1984. *The Beginnings of the Nobel Institution: The Science Prizes, 1901–1915*. Cambridge: Cambridge University Press.

Crawford, Elisabeth, John L. Heilbron, and Rebecca Ullrich. 1987. *The Nobel Population 1901–1937: A Census of the Nominators and Nominees for the Prizes in Physics and Chemistry*. Berkeley, Calif.: Office for History of Science and Technology.

Crawford, Elisabeth, Ruth L. Sime, and Mark Walker. 1997. "A Nobel tale of postwar injustice," *PT* 50 (September): 26–32.

Crease, Robert P., and Charles C. Mann. 1986. *The Second Creation: Makers of the Revolution in 20th-Century Physics*. New York: Collier Books.

Cushing, James T. 1981. "Electromagnetic mass, relativity, and the Kaufmann experiments," *AJP* 49: 1133–49.

———. 1990. *Theory Construction and Selection in Modern Physics: The S Matrix*. Cambridge: Cambridge University Press.

———. 1995. *Quantum Mechanics: Historical Contingency and the Copenhagen Hegemony*. Chicago: University of Chicago Press.

Dahl, Per F. 1984. "Kammerlingh Onnes and the discovery of superconductivity," *HSPS* 15: 1–38.

———. 1992. *Superconductivity: Its Historical Roots and Development from Mercury to the Ceramic Oxides*. New York: AIP Press.

———. 1997. *Flash of the Cathode Rays: A History of J. J. Thomson's Electron*. Bristol, England: Institute of Physics Publishing.

Darrigol, Olivier. 1988a. "The quantum electrodynamical analogy in early nuclear theory or the roots of Yukawa's theory," *RHS* 41: 225–97.

———. 1988b. "Statistics and combinatorics in early quantum theory," *HSPS* 19: 17–80.

———. 1992. *From c-Numbers to q-Numbers: The Classical Analogy in the History of Quantum Theory*. Berkeley: University of California Press.

———. 1993. "Strangeness and soundness in Louis de Broglie's early works," *Physis* 30: 303–72.

———. 1994. "The electron theories of Larmor and Lorentz: A comparative study," *HSPS* 24: 265–336.

———. 1996. "The electrodynamic origins of relativity theory," *HSPS* 26: 241–312.

Davies, P. C. W. and Julian Brown. 1988. *Superstrings: A Theory of Everything?* Cambridge: Cambridge University Press.

Davis, E. A., and Isobel Falconer. 1997. *J. J. Thomson and the Discovery of the Electron*. London: Taylor & Francis.

De Maria, M., M. G. Ianniello, and A. Russo. 1991. "The discovery of cosmic rays: Rivalries and controversies between Europe and the United States," *HSPS* 22: 165–92.

De Maria, M., and A. Russo. 1985. "The discovery of the positron," *RSS* 2: 237–86.

———. 1989. "Cosmic ray romancing: The discovery of the latitude effect and the Compton-Millikan controversy," *HSPS* 19: 211–66.

Dickson, David. 1984. *The New Politics of Science*. New York: Pantheon Books.

Doncel, Manuel, et al., eds. 1987. *Symmetries in Physics (1600–1980)*. Barcelona: Seminari d'Historia de les Ciències.

Dresden, Max. 1987. *H. A. Kramers: Between Tradition and Revolution.* New York: Springer.

Earman, John, and Clark Glymour. 1978. "Lost in tensors: Einstein's struggles with covariance principles, 1912–1916," *SHPS* 9: 251–78.

———. 1980a. "Relativity and eclipses: The British eclipse expeditions of 1919 and their predecessors," *HSPS* 11: 49–85.

———. 1980b. "The gravitational redshift as a test of general relativity," *SHPS* 11: 175–214.

Easlea, Brian. 1973. *Liberation and the Aims of Science.* London: Chatto & Windus.

Eckert, Michael. and Helmut Schubert. 1990. *Crystals, Electrons, Transistors: From Scholar's Study to Industrial Research.* New York: American Institute of Physics.

Einstein, Albert. 1982. "How I created the theory of relativity," *PT* 35 (August): 45–47.

Eisenstaedt, Jean. 1989. "The low water mark of general relativity," in Donald Howard and John Stachel, eds. *Einstein and the History of General Relativity.* Boston: Birkhäuser, pp. 277–92.

Eisenstaedt, Jean, and Ana J. Kox, eds. 1992. *Einstein Studies. Vol. 3: Studies in the History of General Relativity.* Boston: Birkhäuser.

Elvee, Richard Q., ed. 1992. *The End of Science? Attack and Defense.* Lanham, Md.: University Press of America.

Elzinga, Aant. 1995. "Einstein in the land of Nobel: An episode in the interplay of science, politics, epistemology and popular culture," in Kostas Gavroglu, John Stachel, and Marx Wartofsky, eds. *Physics, Philosophy and the Scientific Community.* New York: Kluwer, pp. 73–103.

Exhela, V. V., et al. 1996. *Particle Physics: One Hundred Years of Discoveries. An Annotated Chronological Bibliography.* Woodbury, N.Y.: AIP Press.

Fadner, W. L. 1988. "Did Einstein really discover '$E = mc^2$'?" *AJP* 56: 114–22.

Fagen, M. D., ed. 1975. *A History of Engineering and Science in the Bell System: The Early Years, 1875–1925.* Whippany, N.J.: Bell Telephone Laboratories.

Falconer, Isobel. 1987. "Corpuscles, electrons and cathode rays: J. J. Thomson and the 'discovery of the electron'," *British Journal for the History of Science* 20: 241–76.

Feffer, Stuart. 1989. "Arthur Schuster, J. J. Thomson, and the discovery of the electron," *HSPS* 20: 33–61.

Fermi, Laura. 1971. *Illustrious Immigrants: The Intellectual Migration from Europe 1930–41.* Chicago: University of Chicago Press.

Feyerabend, Paul. 1978. *Science in a Free Society.* London: New Left Books.

Feynman, Richard P. 1992. *The Character of Physical Law.* London: Penguin Books.

Fischer, Klaus. 1988. "Der quantitative Beitrag der nach 1933 emigrierten Naturwissenschaftler zur deutschsprachigen physikalischen Forschung," *BW* 11: 83–104.

———. 1993. *Changing Landscapes of Nuclear Physics: A Scientometric Study on the Social and Cognitive Position of German-Speaking Emigrants Within the Nuclear Physics Community, 1921–1947.* Berlin: Springer-Verlag.

Forman, Paul. 1968. "The doublet riddle and atomic physics *circa* 1924," *Isis* 59: 156–74.

———. 1971. "Weimar culture, causality, and quantum theory, 1918–1927: Adaption by German physicists and mathematicians to a hostile intellectual environment," *HSPS* 3: 1–115.

———. 1973. "Scientific internationalism and the Weimar physicists: The ideology and its manipulation in Germany after World War I," *Isis* 64: 151–78.

———. 1974. "The financial support and political alignment of physicists in Weimar Germany," *Minerva* 12: 39–66.

———. 1987. "Behind quantum electronics: National security as basis for physical research in the United States, 1940–1960," *HSPS* 18: 149–229.

———. 1992. "Inventing the maser in postwar America," *Osiris* 7: 105–34.

———. 1995. "'Swords into ploughshares': Breaking new ground with radar hardware and technique in physical research after World War II," *RMP* 67: 397–455.

Forman, Paul, John L. Heilbron, and Spencer Weart. 1975. "Physics circa 1900: Personnel, funding, and productivity of the academic establishments," *HSPS* 5: 1–185.

Foster, B., and P. H. Fowler, eds. 1988. *40 Years of Particle Physics*. Bristol, England: Adam Hilger.

Franklin, Allan. 1986. *The Neglect of Experiment*. New York: Cambridge University Press.

French, Anthony P., and P. J. Kennedy, eds. 1985. *Niels Bohr: A Centenary Volume*. Cambridge, Mass.: MIT Press.

Frenkel, Victor Y. 1997. "Yakov Ilich Frenkel: Sketches toward a civic portrait," *HSPS* 27: 197–236.

Freundlich, Yehudah. 1980. "Theory evaluation and the bootstrap hypothesis," *SHPS* 11: 267–77.

Friedman, Robert M. 1981. "Nobel physics prize in perspective," *Nature* 292: 793–98.

———. 1989. "Text, context, and quicksand: Method and understanding in studying the Nobel science prizes," *HSPS* 20: 63–78.

Frisch, Otto R., and John A. Wheeler. 1967. "The discovery of fission," *PT* 20 (November): 43–52. Reprinted in Weart and Phillips, eds., 1985, pp. 272–81.

Galdabini, Silvana, and Giuseppe Giuliani. 1988. "Physics in Italy between 1900 and 1940: The universities, physicists, funds, and research," *HSPS* 19: 115–36.

Galison, Peter. 1979. "Minkowski's space-time: From visual thinking to the absolute world," *HSPS* 10: 85–121.

———. 1983. "The discovery of the muon and the failed revolution against quantum electrodynamics," *Centaurus* 26: 262–316.

———. 1987. *How Experiments End*. Chicago: University of Chicago Press.

———. 1995. "Theory bound and unbound: Superstrings and experiments," in Friedel Weinert, ed., *Laws of Nature. Essays on the Philosophical, Scientific and Historical Dimensions*. Berlin: Walter de Gruyter, pp. 369–408.

———. 1997. *Image and Logic: A Material Culture of Microphysics*. Chicago: University of Chicago Press.

Galison, Peter, and Bruce Hevly, eds. 1992. *Big Science: The Growth of Large-Scale Research*. Stanford, Calif.: Stanford University Press.

Gaston, Jerry. 1973. *Originality and Competition in Science: A Study of the British High Energy Physics Community*. Chicago: University of Chicago Press.

Gavroglu, Kostas. 1995. *Fritz London: A Scientific Biography*. Cambridge: Cambridge University Press.

Gavroglu, Kostas, and Yorgos Goudaroulis. 1989. *Methodological Aspects of the Development of Low Temperature Physics 1881–1956*. Dordrecht: Kluwer Academic.

Glick, Thomas F., ed. 1987. *The Comparative Reception of Relativity*. Boston: Reidel.

Goenner, Hubert. 1992. "The reception of the theory of relativity in Germany as reflected by books published between 1908 and 1945," in Eisenstaedt and Kox, eds., pp. 15–38.

Goldberg, Stanley. 1970. "The Abraham theory of the electron: The symbiosis of experiment and theory," *AHES* 7: 7–25.

———. 1976. "Max Planck's philosophy of nature and his elaboration of the special theory of relativity," *HSPS* 7: 125–60.

———. 1984. *Understanding Relativity: Origin and Impact of a Scientific Revolution*. Boston: Birkhäuser.

Goldschmidt, Bertrand. 1982. *The Atomic Complex: A Worldwide Political History of Nuclear Energy*. La Grange Park, Ill.: American Nuclear Society.

Gorelik, Gennady. 1995. *"Meine Antisowjetische Tätigkeit. . .": Russische Physiker unter Stalin*. Braunschweig, Germany: Vieweg.

Gorelik, Gennady, and Victor Y. Frenkel. 1994. *Matvei Petrovich Bronstein and Soviet Theoretical Physics in the Thirties*. Basel: Birkhäuser Verlag.

Gowing, Margaret M. 1964. *Britain and Atomic Energy 1939–1945*. London: Macmillan.

Graetzer, Hans G., and David L. Anderson. 1971. *The Discovery of Nuclear Fission*. New York: Van Nostrand Reinhold.

Graham, Loren R. 1972. *Science and Philosophy in the Soviet Union*. New York: Alfred A. Knopf.

Greenberg, Daniel. 1967. *The Politics of Pure Science*. New York: New American Library.

Hahn, Otto. 1970. *My Life: The Autobiography of a Scientist*. New York: Herder and Herder.

Hakfoort, C. 1992. "Science deified: Wilhelm Ostwald's energeticist world-view and the history of scientism," *AS* 49: 525–44.

Hanson, Norwood Russell. 1963. *The Concept of the Positron*. Cambridge: Cambridge University Press.

Harder, A. J. 1974. "E. A. Milne, scientific revolutions and the growth of knowledge," *AS* 31: 351–63.

Harman, Peter M. 1982. *Energy, Force, and Matter: The Conceptual Development of Nineteenth-Century Physics*. Cambridge: Cambridge University Press.

Hartcup, G. 1988. *The War of Invention: Scientific Developments, 1914–1918*. London: Brassey's Defense Publishers.

Harwit, Martin. 1981. *Cosmic Discovery: The Search, Scope and Heritage of Astronomy*. New York: Basic Books.

Hawking, Stephen W. 1988. *A Brief History of Time*. Toronto: Bantam Books.

Hawking, Stephen W., and Werner Israel, eds. 1987. *Three Hundred Years of Gravitation*. Cambridge: Cambridge University Press.

Hazen, Robert. 1988. *Breakthrough: The Race for the Superconductor*. New York: Summit Books.

Hecht, Jeff. 1992. *Laser Pioneers*. Boston: Academic Press.

Heilbron, John L. 1968. "The scattering of α and β particles and Rutherford's atom," *AHES* 4: 247–307.

———. 1977. "Lectures on the history of atomic physics 1900–1922," in Charles

Weiner, ed. *History of Twentieth Century Physics*. New York: Academic Press, pp. 40–108.

———. 1981. "Rutherford-Bohr atom," *AJP* 49: 223–31.

———. 1982. "Fin-de-siècle physics," in C. G. Bernhard, et al., eds., *Science, Technology, and Society in the Time of Alfred Nobel*. New York: Pergamon Press, pp. 51–73.

———. 1985. "The earliest missionaries of the Copenhagen spirit," *RHS* 38: 194–230.

———. 1986. *The Dilemmas of an Upright Man: Max Planck as Spokesman for German Science*. Berkeley: University of California Press.

———. 1992. "Creativity and big science," *PT* 45 (November): 42–47.

Heilbron, John L., and Thomas S. Kuhn. 1969. "The genesis of the Bohr atom," *HSPS* 1: 211–90.

Heilbron, John L., and Robert W. Seidel. 1989. *Lawrence and His Laboratory: A History of the Lawrence Berkeley Laboratory*. Vol. 1. Berkeley: University of California Press.

Hendry, John. 1980. "Weimar culture and quantum causality," *History of Science* 18: 155–80.

———. 1984a. *The Creation of Quantum Mechanics and the Bohr-Pauli Dialogue*. Dordrecht, the Netherlands: Reidel.

———. ed. 1984b. *Cambridge Physics in the Thirties*. Bristol, England: Adam Hilger.

———. 1987. "The scientific origins of controlled fusion energy," *AS* 44: 143–68.

Hendry, John, and J. D. Lawson. 1993. *Fusion Research in the UK 1945–1960*. AEA Technology.

Hentschel, Klaus. 1990. *Interpretationen und Fehlinterpretationen der Speziellen und der Allgemeinen Relativitätstheorie durch Zeitgenossen Albert Einsteins*. Basel: Birkhäuser.

———. 1992. "Einstein's attitude towards experiments: Testing relativity theory 1907–1927," *SHPS* 23: 593–624.

———. ed. 1996. *Physics and National Socialism: An Anthology of Primary Sources*. Basel: Birkhäuser.

Hermann, Armin. 1971. *The Genesis of the Quantum Theory (1899–1913)*. Cambridge, Mass.: MIT Press.

Hermann, Armin, John Krige, Dominique Pestre, and Ulrike Mersits, eds. 1987–1990. *History of CERN*. Vol. 1: *Launching the European Organization for Nuclear Research*. Vol. 2: *Building and Running the Laboratory*. New York: North-Holland.

Hetherington, Norriss S., ed. 1993. *Encyclopedia of Cosmology: Historical, Philosophical, and Scientific Foundations of Modern Cosmology*. New York: Garland Publishing.

Hewlett, Richard G. 1989. *Atoms for Peace and War 1953–1961*. Berkeley: University of California Press.

Hiebert, Erwin N. 1971. "The energetics controversy and the new thermodynamics," in D. H. D. Roller, ed., *Perspectives in the History of Science and Technology*. Norman: University of Oklahoma Press, pp. 67–86.

———. 1979. "The state of physics at the turn of the century," in Bunge and Shea, eds., pp. 3–22.

Hillas, Anthony M., ed. 1972. *Cosmic Rays*. New York: Pergamon Press.

Hirosige, Tetu. 1976. "The ether problem, the mechanistic worldview, and the origins of the theory of relativity," *HSPS* 7: 3–82.

Hirosige, Tetu, and Sigeko Nisio. 1986. "Rise and fall of various fields of physics at the turn of the century," *Japanese Studies in the History of Science* 7: 93–113.

Hoch, Paul K. 1983. "The reception of central European refugee physicists of the 1930s: USSR, UK, USA," *AS* 40: 217–46.

Hoddeson, Lillian. 1981a. "The emergence of basic research in the Bell Telephone System, 1875–1915," *T&C* 22: 512–44.

———. 1981b. "The discovery of the point-contact transistor," *HSPS* 12: 41–76.

———. 1983. "Establishing KEK in Japan and Fermilab in the U.S.: Internationalism, nationalism, and high energy accelerators," *SSS* 13: 1–48.

Hoddeson, Lillian, Gordon Baym, and Michael Eckert. 1992. "The development of the quantum mechanical electron theory of metals, 1926–1933," in Hoddeson et al., eds., pp. 88–181.

Hoddeson, Lillian, et al., eds. 1992. *Out of the Crystal Maze: Chapters from the History of Solid-State Physics.* New York: Oxford University Press.

Hoddeson, Lillian et al., eds. 1997. *The Rise of the Standard Model: Particle Physics in the 1960s and 1970s*. New York: Cambridge University Press.

Hoffmann, Banesh. 1972. *Albert Einstein: Creator and Rebel.* New York: Viking Press.

Holloway, David. 1994. *Stalin and the Bomb: The Soviet Union and Atomic Energy 1939–1956*. New Haven: Yale University Press.

Holton, Gerald. 1978. "Fermi's group and the recapture of Italy's place in physics," in G. Holton, *The Scientific Imagination: Case Studies*. Cambridge: Cambridge University Press, pp. 155–98.

———. 1983. "The migration of physicists to the United States," in Jarrell C. Jackman and Carla M. Borden, eds., *The Muses Flee Hitler: Cultural Transfer and Adaption 1930–1945*. Washington, D.C.: Smithsonian Institution Press, pp. 169–88.

———. 1988. *Thematic Origins of Scientific Thought, Kepler to Einstein*. Cambridge, Mass.: Harvard University Press.

———. 1994. *Science and Anti-Science*. Cambridge, Mass.: Harvard University Press.

Horgan, John. 1996. *The End of Science*. Reading, Mass.: Addison-Wesley.

Hovis, R. Corby, and Helge Kragh. 1991. "Resource letter HEPP-1: History of elementary-particle physics," *AJP* 59: 779–807.

Hufbauer, Karl. 1981. "Astronomers take up the stellar-energy problem, 1917–1920," *HSPS* 11: 277–303.

Hund, Friedrich. 1974. *The History of Quantum Theory*. London: Harrap.

Illy, József. 1981. "Revolutions in a revolution," *SHPS* 12: 173–210.

Irvine, John, et al. 1986. "The shifting balance of power in experimental particle physics," *PT* 39 (November): 27–34.

Irvine, John, and Ben R. Martin. 1985. "Basic research in the East and West: A comparison of the scientific performance of high energy physics accelerators," *SSS* 15: 293–341.

Jammer, Max. 1961. *Concepts of Mass in Classical and Modern Physics*. Cambridge, Mass.: Harvard University Press.

———. 1966. *The Conceptual Development of Quantum Mechanics*. New York: McGraw-Hill.

———. 1974. *The Philosophy of Quantum Mechanics: The Interpretations of Quantum Mechanics in Historical Perspective*. New York: Wiley-Interscience.

Josephson, Paul R. 1991. *Physics and Politics in Revolutionary Russia*. Berkeley: University of California Press.

Jungnickel, Christa, and Russell McCormmach. 1986. *Intellectual Mastery of Nature: Theoretical Physics from Ohm to Einstein*. Vol. 2. Chicago: University of Chicago Press.

Kaiser, Walter. 1987. "Early theories of the electron gas," *HSPS* 17: 271–97.

Kangro, Hans. 1976. *Early History of Planck's Radiation Law*. London: Taylor & Francis.

Kargon, Robert. 1982. *The Rise of Robert Millikan: Portrait of a Life in American Science*. Ithaca, N.Y.: Cornell University Press.

Kayser, Heinrich. 1938. "Statistik der Spektroskopie," *Physikalische Zeitschrift* 39: 466–68.

Kevles, Daniel J. 1971. "'Into hostile political camps': The reorganisation of international science in World War I," *Isis* 62: 47–60.

———. 1976. "The physics, mathematics and chemistry communities: A comparative analysis," in A. Oleson and J. Voss, eds., *The Organization of Knowledge in Modern America, 1860–1920*. Baltimore: Johns Hopkins University Press, pp. 139–79.

———. 1987. *The Physicists: The History of a Scientific Community in Modern America*. Cambridge, Mass.: Harvard University Press.

———. 1997. "Big Science and big politics in the United States: Reflections on the death of the SSC and the life of the Human Genome Project," *HSPS* 27: 269–98.

Kilmister, Clive W. 1994. *Eddington's Search for a Fundamental Theory: A Key to the Universe*. Cambridge: Cambridge University Press.

Kirsten, Christa, and Hans-Jürgen Treder, eds. 1979. *Albert Einstein in Berlin 1913–1933*. 2 vols. Berlin: Akademie-Verlag.

Klein, Martin J. 1966. "Thermodynamics and quanta in Planck's work," *PT* 19 (November): 23–32. Reprinted in Weart and Phillips, eds., 1985, pp. 294–302.

———. 1970. *Paul Ehrenfest: The Making of a Theoretical Physicist*. Amsterdam: North-Holland.

———. 1973. "Mechanical explanation at the end of the nineteenth century," *Centaurus* 17: 58–82.

Kleinert, Andreas. 1978. "Von der *science allemande* zur deutschen Physik," *Francia* 8: 509–25.

Koester, David, Daniel Sullivan, and D. Hywel White. 1982. "Theory selection in particle physics: A quantitative case study of the evolution of weak-electromagnetic unification theory," *SSS* 12: 73–100.

Koizumi, K. 1975. "The emergence of Japan's first physicists," *HSPS* 6: 3–108.

Kojevnikov, Alexei B. 1991. "Piotr Kapitza and Stalin's government: A study in moral choice," *HSPS* 22: 131–64.

Kojevnikov, Alexei B., and O. I. Novik. 1989. "Analysis of informational ties dynamics in early quantum mechanics (1925–1927)," *Acta Historiae Rerum Naturalium Necnon Technicarum*, No. 20: 115–60.

Kolb, Adrienne, and Lillian Hoddeson. 1993. "The mirage of the 'world accelerator for world peace' and the origins of the SSC, 1953–1983," *HSPS* 24: 101–24.

Konuma, Michiji. 1989. "Social aspects of Japanese particle physics in the 1950s," in Brown, Dresden, and Hoddeson, eds., pp. 536–50.

Krafft, Fritz. 1983. "Internal and external conditions for the discovery of fission by the Berlin team," in Shea, ed., pp. 135–66.

Kraft, P., and P. Kroes. 1984. "Adaption of scientific knowledge to an intellectual environment: Paul Forman's 'Weimar culture, causality, and quantum theory, 1918–1927': Analysis and criticism," *Centaurus* 27: 76–99.

Kragh, Helge. 1979. "Niels Bohr's second atomic theory," *HSPS* 10: 123–86.

———. 1980a. "Anatomy of a priority conflict: The case of element 72," *Centaurus* 23: 275–301.

———. 1980b. *On Science and Underdevelopment*. Roskilde, Denmark: RUC Forlag.

———. 1981a. "The genesis of Dirac's relativistic theory of electrons," *AHES* 24: 31–67.

———. 1981b. "The concept of the monopole: A historical and analytical case-study," *SHPS* 12: 141–72.

———. 1982a. "Cosmo-physics in the thirties: Towards a history of Dirac cosmology," *HSPS* 13: 70–108.

———. 1982b. "Erwin Schrödinger and the wave equation: The crucial phase," *Centaurus* 26: 154–97.

———. 1985. "The fine structure of hydrogen and the gross structure of the physics community, 1916–26," *HSPS* 16: 67–125.

———. 1989a. "The aether in late nineteenth century chemistry," *Ambix* 36: 49–65.

———. 1989b. "Concept and controversy: Jean Becquerel and the positive electron," *Centaurus* 32: 203–40.

———. 1990. *Dirac: A Scientific Biography*. Cambridge: Cambridge University Press.

———. 1992. "Relativistic collisions: The work of Christian Møller in the early 1930s," *AHES* 43: 299–328.

———. 1994. "The Krarup cable: Invention and early development," *T&C* 35: 129–57.

———. 1995. "Arthur March, Werner Heisenberg, and the search for a smallest length," *RHS* 48: 401–34.

———. 1996a. "The new rays and the failed anti-materialistic revolution," in Dieter Hoffmann, Fabio Bevilacqua, and Roger Stuewer, eds., *The Emergence of Modern Physics*. Pavia, Italy: La Goliardica Pavese, pp. 61–78.

———. 1996b. *Cosmology and Controversy: The Historical Development of Two Theories of the Universe*. Princeton, N.J.: Princeton University Press.

———. 1997a. "The origin of radioactivity: From solvable problem to unsolved non-problem," *AHES* 50: 331–58.

———. 1997b. "J. J. Thomson, the electron, and atomic architecture," *The Physics Teacher* 35: 328–32.

———. 1997c. "The electrical universe: Grand cosmological theory versus mundane experiments," *PS* 5: 199–231.

Kragh, Helge, and Bruno Carazza. 1994. "From time atoms to space-time quantization: The idea of discrete time, ca. 1925–1936," *SHPS* 25: 437–62.

Krige, John, ed. 1996. *History of CERN*. Vol. 3. Amsterdam: Elsevier.

Kuhn, Thomas. 1978. *Black-Body Theory and the Quantum Discontinuity, 1894–1912*. Oxford: Clarendon Press.

Larmor, Joseph. 1900. *Aether and Matter*. Cambridge: Cambridge University Press.

———. 1927. *Mathematical and Physical Papers*, Vol. 1. Cambridge: Cambridge University Press.

LeBon, Gustave. 1905. *The Evolution of Matter*. New York: Charles Scribner's Sons.

Lee, T.-S. H., and R. B. Wiringa, eds. 1990. *Proceedings of a Symposium on the Occasion of the 40th Anniversary of the Nuclear Shell Model*. (Special issue of *Nuclear Physics A* 507, 1–321).

Lévy-Leblond, Jean-Marc. 1976. "Ideology of/in contemporary physics," in Hilary Rose and Steven Rose, eds., *The Radicalisation of Science*. London: Macmillan, pp. 136–75.

Lightman, Alan, and Roberta Brawer. 1990. *Origins: The Lives and Worlds of Modern Cosmologists*. Cambridge, Mass.: Harvard University Press.

Lindqvist, Svante, ed. 1993. *Center on the Periphery: Historical Aspects of 20th-Century Swedish Physics*. Canton, Mass.: Science History Publications.

Lorentz, Hendrik A. 1952. *The Theory of Electrons*. New York: Dover Publications.

Ludwig, Gunther, ed. 1968. *Wave Mechanics*. New York: Pergamon Press.

MacKinnon, Edward M. 1982. *Scientific Explanation and Atomic Physics*. Chicago: University of Chicago Press.

MacLeod, Roy. 1982. "The 'bankruptcy of science' debate: The creed of science and its critics, 1885–1900," *Science, Technology, & Human Values* 7: 2–15.

Mann, Alfred K., and David B. Cline, eds. 1994. *Discovery of Weak Neutral Currents: The Weak Interaction Before and After*. New York: AIP Press.

Martins, Roberto de Andrade. 1997. "Becquerel and the choice of uranium compounds," *AHES* 51: 67–81.

Mayer-Kuckuk, Theo, ed. 1995. *150 Jahre Deutsche Physikalische Gesellschaft*. Weinheim, Germany: VCH Verlagsgesellschaft.

McCormmach, Russell 1966. "The atomic theory of John William Nicholson," *AHES* 3: 160–84.

———. 1970. "H. A. Lorentz and the electromagnetic world view of nature," *Isis* 61: 459–97.

Mehra, Jagdish. 1974. *Einstein, Hilbert, and the Theory of Gravitation*. Dordrecht, the Netherlands: Reidel.

———. 1975. *The Solvay Conferences on Physics: Aspects of the Development of Physics Since 1911*. Dordrecht, the Netherlands: Reidel.

———. 1994. *The Beat of a Different Drum: The Life and Science of Richard Feynman*. New York: Oxford University Press.

Mehra, Jagdish, and Helmut Rechenberg. 1982. *The Historical Development of Quantum Theory*. Vols. 1–4. Berlin: Springer-Verlag.

———. 1987. *The Historical Development of Quantum Theory*. Vol. 5. Berlin: Springer-Verlag

Mehrtens, Herbert, and Steffens Richter, eds. 1980. *Naturwissenschaft, Technik und NS-Ideologie*. Frankfurt am Main: Suhrkamp.

Menard, Henry W. 1971. *Science: Growth and Change*. Cambridge, Mass.: Harvard University Press.

Mendelssohn, Kurt. 1977. *The Quest for Absolute Zero*. London: Taylor & Francis.

Merton, Robert K. 1973. *The Sociology of Science: Theoretical and Empirical Investigations*. Chicago: University of Chicago Press.

Michelson, Albert. 1902. *Light Waves and Their Uses*. Chicago: University of Chicago Press.

Miller, Arthur I. 1981. *Albert Einstein's Special Theory of Relativity: Emergence (1905) and Early Interpretation (1905–1911)*. London: Addison-Wesley.

———. 1986. *Frontiers of Physics: 1900–1911. Selected Essays*. Boston: Birkhäuser.

Millikan, Robert A. 1951. *The Autobiography of Robert A. Millikan*. London: MacDonald.

Mladjenovič, Milovad. 1998. *The Defining Years in Nuclear Physics, 1932–1960s*. Bristol, England: Institute of Physics Publishing.

Moore, Walter. 1989. *Schrödinger: Life and Thought*. Cambridge: Cambridge University Press.

Moravcsik, Michael J. 1977. "The crisis in particle physics," *Research Policy* 6: 78–107.

Morrison, Philip. 1946. "The laboratory demobilizes," *Bulletin of the Atomic Scientists* 2 (November) : 5–6.

Moyer, D. F. 1981. "Origins of Dirac's electron," *AJP* 49: 944–49.

Nakayama, Shigeru, David L. Swain, and Eri Yagi, eds. 1974. *Science and Society in Modern Japan*. Cambridge, Mass.: MIT Press.

Nersessian, Nancy J. 1986. "Why wasn't Lorentz Einstein: An examination of the scientific method of H. A. Lorentz," *Centaurus* 29: 37–73.

Newman, Harvey B., and Thomas Ypsilantis, eds. 1996. *History of Original Ideas and Basic Discoveries in Particle Physics*. New York: Plenum Press.

Nicolai, Georg. 1918. *The Biology of War*. New York: Century.

[Nobel] 1967–1997. *Nobel Lectures. Physics*. Vols. 1–4. Amsterdam: Elsevier. Vols. 5–7. Singapore: World Scientific.

Norton, John. 1985. "How Einstein found his field equations: 1912–1915," *HSPS* 14: 253–316.

Nowotny, Helga, and Hilary Rose, eds. 1979. *Counter-Movements in the Sciences*. Dordrecht, the Netherlands: Reidel.

Nye, Mary Jo. 1974. "Gustave LeBon's black light: A study of physics and philosophy in France at the turn of the century," *HSPS* 4: 163–95.

———. 1975. "The nineteenth-century atomic debates and the dilemma of an 'indifferent hypothesis'," *SHPS* 7: 245–68.

———. 1980. "N-rays: An episode in the history and psychology of science," *HSPS* 11: 125–56.

———. ed. 1984. *The Question of the Atom, from the Karlsruhe Congress to the First Solvay Conference, 1860–1911*. Los Angeles: Tomash.

———. 1993. *From Chemical Philosophy to Theoretical Chemistry: Dynamics of Matter and Dynamics of Disciplines 1800–1950*. Berkeley: University of California Press.

Olby, Robert. 1994. *The Path to the Double Helix: The Discovery of DNA*. New York: Dover Publications.

Pais, Abraham. 1982. *'Subtle is the Lord. . .': The Science and the Life of Albert Einstein*. Oxford: Oxford University Press.

———. 1986. *Inward Bound: Of Matter and Forces in the Physical World*. Oxford: Clarendon Press.

———. 1991. *Niels Bohr's Times, in Physics, Philosophy, and Polity*. Oxford: Clarendon Press.

Panofsky, Wolfgang. 1994. *Particles and Policy*. Woodbury, N.Y.: AIP Press.

Pickering, Andrew. 1981. "The hunting of the quark," *Isis* 72: 216–36.

———. 1984a. *Constructing Quarks: A Sociological History of Particle Physics*. Edinburgh: Edinburgh University Press.

———. 1984b. "Against putting the phenomena first: The discovery of the weak neutral current," *SHPS* 15: 85–117.

Pinch, Trevor J. 1977. "What does a proof do if it does not prove?" in E. Mendelsohn, P. Weingart, and R. Whitley, eds., *The Social Production of Scientific Knowledge*. Dordrecht, the Netherlands: Reidel, pp. 171–215.

Post, Richard F. 1995. "Plasma physics in the twentieth century," in Brown, Pais, and Pippard, eds., pp. 1617–90.

Powers, Thomas. 1993. *Heisenberg's War: The Secret History of the German Bomb*. New York: Knopf.

Purrington, Robert D. 1997. *Physics in the Nineteenth Century*. New Brunswick, N.J.: Rutgers University Press.

Pyenson, Lewis. 1979. "Mathematics, education, and the Göttingen approach to physical reality, 1890–1914," *Europa* 2: 91–128.

Pyenson, Lewis, and Douglas Skopp. 1977. "Educating physicists in Germany *circa* 1900," *SSS* 7: 329–66.

Redhead, Paul A., ed. 1994. *Vacuum Science and Technology: Pioneers of the 20th Century*. New York: AIP Press.

Reich, Leonard. 1983. "Irving Langmuir and the pursuit of science and technology in the corporate environment," *T&C* 24: 199–221.

———. 1985. *The Making of American Industrial Research: Science and Business at GE and Bell, 1876–1926*. New York: Cambridge University Press.

Reingold, Nathan, and Ida Reingold, eds. 1981. *Science in America: A Documentary History 1900–1939*. Chicago: University of Chicago Press.

Renneberg, Monika, and Mark Walker, eds. 1994. *Science, Technology and National Socialism*. Cambridge: Cambridge University Press.

Rescher, Nicholas. 1978. *Scientific Progress*. Oxford: Blackwell.

Rhodes, Richard. 1986. *The Making of the Atomic Bomb*. New York: Simon and Schuster.

———. 1995. *Dark Sun: The Making of the Hydrogen Bomb*. New York: Simon and Schuster.

Rider, Robin E. 1984. "Alarm and opportunity: Emigration of mathematicians and physicists to Britain and the United States, 1933–1945," *HSPS* 15: 107–76.

Riordan, Michael. 1987. *The Hunting of the Quark: A True Story of Modern Physics*. New York: Simon & Schuster.

Riordan, Michael, and Lillian Hoddeson. 1997. *Crystal Fire: The Birth of the Information Age*. New York: Norton.

Robertson, Peter. 1979. *The Early Years: The Niels Bohr Institute 1921–1930*. Copenhagen: Akademisk Forlag.

Roqué, Xavier 1997a. "The manufacture of the positron," *SHPMP 28*: 73–129.

———. 1997b. "Marie Curie and the radium industry: A preliminary sketch," *History and Technology* 13: 267–92.

Rosenthal-Schneider, Ilse. 1980. *Reality and Scientific Truth: Discussions With Einstein, von Laue and Planck*. Detroit: Wayne State University Press.

Roseveare, N. T. 1982. *Mercury's Perihelion from Le Verrier to Einstein*. Oxford: Clarendon Press.

Rothman, Tony, and George Ellis. 1987. "Has cosmology become metaphysical?" *Astronomy* 15 (February): 6–22.

Rowland, Henry. 1902. *The Physical Papers of Henry Augustus Rowland*. Baltimore: Johns Hopkins University Press.

Rueger, Alexander. 1988. "Atomism from cosmology. Erwin Schrödinger's work on wave mechanics and space-time structure," *HSPS* 18: 377–401.

———. 1992. "Attitudes towards infinities: Responses to anomalies in quantum electrodynamics, 1927–1947," *HSPS* 22: 309–38.

Russo, Arturo. 1981. "Fundamental research at Bell Laboratories: The discovery of electron diffraction," *HSPS* 12: 117–60.

Ryan, M. P., and L. C. Shepley. 1976. "Resource letter RC-1: Cosmology," *AJP* 44: 223–30.

Sànchez-Ron, José. 1987. "The reception of special relativity in Great Britain," in Glick, ed., pp. 27–58.

———. 1992. "The reception of general relativity among British physicists and mathematicians (1915–1930)," in Eisenstaedt and Kox, eds., pp. 57—88.

Schilpp, Paul A., ed. 1949. *Albert Einstein: Philosopher-Scientist*. Evanston, Ill.: Library of Living Philosophers.

Schramm, David N. 1996. *The Big Bang and Other Explosions in Nuclear and Particle Astrophysics*. Singapore: World Scientific.

Schramm, David N., and Gary Steigman. 1988. "Particle accelerators test cosmological theory," *Scientific American* 262 (June): 66–72.

Schröder-Gudehus, Brigitte. 1966. *Deutsche Wissenschaft und Internationale Zusammenarbeit 1914–1928*. Geneva: Dumaret & Golay.

———. 1972. "The argument for the self-government and public support of science in Weimar Germany," *Minerva* 10: 537–70.

———. 1978. *Les Scientifiques et la Paix: La Communauté Scientifique Internationale au Cours des Années 20*. Montréal: Presses de l'Université de Montréal.

Schwarz, John H. 1996. "Superstring—a brief history," in Newman and Ypsilantis, eds., pp. 695–706.

Schweber, Sylvan S. 1988. "The mutual embrace of science and the military: ONR and the growth of physics in the United States after World War II," in E. Mendelsohn, M. R. Smith, and P. Weingart, eds., *Science, Technology and the Military*. Dordrecht, the Netherlands: Kluwer, pp. 1–45.

———. 1989. "Some reflections on the history of particle physics in the 1950s," in Brown, Dresden, and Hoddeson, eds., pp. 668–93.

———. 1990. "The young John Clarke Slater and the development of quantum chemistry," *HSPS* 20: 339–406.

———. 1994a. *QED and the Men Who Made It: Dyson, Feynman, Schwinger, and Tomonaga*. Princeton, N.J.: Princeton University Press.

———. 1994b. "Some reflections on big science and high energy physics in the United States," *RSS* 2: 127–89.

———. 1995. "Physics, community and the crisis in physical theory," in Kostas Gavroglu, John Stachel and Marx W. Wartofsky, eds., *Physics, Philosophy and the Scientific Community*. Dordrecht, the Netherlands: Kluwer Academic, pp. 125–52.

———. 1997. "A historical perspective on the rise of the standard model," in Hoddeson et al., eds., pp. 645–84.

Schwinger, Julian, ed. 1958. *Selected Papers on Quantum Electrodynamics*. New York: Dover Publications.

Scurlock, Ralph G., ed. 1992. *History and Origins of Cryogenics*. Oxford: Clarendon Press.

Segré, Emilio. 1970. *Enrico Fermi: Physicist*. Chicago: University of Chicago Press.

———. 1980. *From X-rays to Quarks: Modern Physicists and Their Discoveries*. San Francisco: W. H. Freman and Co.

Serber, Robert. 1992. *The Los Alamos Primer. The First Lectures on How to Build an Atomic Bomb*. Berkeley: University of California Press.

Shea, William R., ed. 1983. *Otto Hahn and the Rise of Nuclear Physics*. Dordrecht, the Netherlands: Reidel.

Siegel, Daniel M. 1978. "Classical electromagnetic and relativistic approaches to the problem of nonintegral atomic masses," *HSPS* 9: 323–60.

Sime, Ruth L. 1996. *Lise Meitner: A Life in Physics*. Berkeley: University of California Press.

Simmons, John. 1997. *The 100 Most Influential Scientists*. London: Robinson.

Sinclair, S. B. 1987. "J. J. Thomson and the chemical atom: From ether vortex to atomic decay," *Ambix* 34: 89–116.

Smith, Alice K., and Charles Weiner, eds. 1980. *Robert Oppenheimer: Letters and Recollections*. Cambridge, Mass.: Harvard University Press.

Smits, F. M., ed. 1985. *A History of Engineering and Science in the Bell System: Electronics Technology (1925–1975)*. Murray Hill, N.J.: Bell Telephone Laboratories.

Smyth, Henry. 1945. *Atomic Energy for Military Purposes*. Princeton, N.J.: Princeton University Press.

Sopka, Katherine R. 1988. *Quantum Physics in America, 1920–1935*. New York: Arno Press.

Sopka, Katherine R., and Albert E. Moyer, eds. 1986. *Physics for a New Century: Papers Presented at the 1904 St. Louis Congress*. New York: American Institute of Physics.

Stehle, Philip. 1994. *Order, Chaos, Order: The Transition from Classical to Quantum Physics*. New York: Oxford University Press.

Stern, Alexander W. 1964. "The third revolution in 20th century physics," *PT* 17 (April): 42–45.

Stuewer, Roger H. 1975. *The Compton Effect: Turning Point in Physics*. New York: Science History.

———. ed. 1979. *Nuclear Physics in Retrospect: Proceedings of a Symposium on the 1930s*. Minneapolis: University of Minnesota Press.

———. 1983. "The nuclear electron hypothesis," in Shea, ed., pp. 19–68.

———. 1984. "Nuclear physicists in a new world: The emigrés of the 1930s in America," *BW* 7: 23–40.

———. 1985a. "Artificial disintegration and the Cambridge-Vienna controversy," in Peter Achinstein and Owen Hannaway, eds., *Observation, Experiment, and Hypothesis in Modern Physical Science*. Cambridge, Mass.: MIT Press, pp. 239–307.

———. 1985b. "Bringing the news of fission to America," *PT* 38 (November): 48–56.

———. 1986a. "Rutherford's satellite model of the nucleus," *HSPS* 16: 321–52.

———. 1986b. "Gamow's theory of alpha-decay," in E. Ullmann-Margalit, ed., *The Kaleidoscope of Science*. Dordrecht, the Netherlands: Reidel, pp. 147–86.

———. 1986c. "The naming of the deuteron," *AJP* 54: 206–18.

———. 1994. "The origin of the liquid-drop model and the interpretation of nuclear fission," *PS* 2: 76–129.

ter Haar, Dirk, ed. 1967. *The Old Quantum Theory*. London: Pergamon Press.

Thomson, J. J. 1896. "The Röntgen rays," *Nature* 54: 302–6.

Thorne, Kip S. 1994. *Black Holes and Time Warps: Einstein's Outrageous Legacy*. New York: Norton.

Trefil, James S. 1983. *The Moment of Creation*. New York: Macmillan.

Trigg, George L. 1995. *Landmark Experiments in Twentieth Century Physics*. New York: Dover Publications.

Tyne, Gerald. 1977. *Saga of the Vacuum Tube*. Indianapolis: H. W. Sams & Co.

Urani, J., and George Gale. 1993. "E. A. Milne and the origins of modern cosmology: An essential presence," in John Earman, Michel Janssen, and John D. Norton, eds., *The Attraction of Gravitation: New Studies in the History of General Relativity*. Boston: Birkhäuser, pp. 390–419.

Van der Waerden, Bartel L., ed. 1967. *Sources of Quantum Mechanics*. New York: Dover Publications.

Vizgin, Vladimir P. 1994. *Unified Field Theories in the First Third of the 20th Century*. Basel: Birkhäuser.

Vlachy, Jan. 1982. "World publication output in particle physics," *Czechoslovakian Journal of Physics* B32: 1065–72.

Vucinich, Alexander. 1980. "Soviet physicists and philosophers in the 1930s: Dynamics of a conflict," *Isis* 71: 236–50.

Walker, Mark. 1995. *Nazi Science: Myth, Truth and the German Atomic Bomb*. New York: Plenum.

Wang, Zuoyue. 1995. "The politics of big science in the Cold War: PSAC and the funding of SLAC," *HSPS* 25: 329–56.

Wasserman, Neil. 1985. *From Invention to Innovation: Long-Distance Telephone Transmision at the Turn of the Century*. Baltimore: Johns Hopkins University Press.

Watkins, Peter. 1986. *Story of the W and Z*. New York: Cambridge University Press.

Weart, Spencer R. 1976a. "The rise of 'prostituted' physics," *Nature* 262: 13–17.

———. 1976b. "Scientists with a secret," *PT* 29 (February): 23–30. Reprinted in Weart and Phillips, eds., pp. 123–29.

———. 1979a. "The physics business in America 1919–1940: A statistical reconaissance," in Nathan Reingold, ed., *The Sciences in the American Context: New Perspectives*. Washington, D.C.: Smithsonian Institution Press, pp. 295–328.

———. 1979b. *Scientists in Power*. Cambridge, Mass.: Harvard University Press.

———. 1983. "The discovery of fission and a nuclear physics paradigm," in Shea, ed., pp. 91–134.

———. 1992. "The solid community," in Hoddeson et al., eds., pp. 617–68.

Weart, Spencer R. and Melba Phillips, eds. 1985. *History of Physics: Readings from Physics Today*. New York: American Institute of Physics.

Weeks, Mary E. 1968. *Discovery of the Elements*. Easton, Pa.: Journal of Chemical Education.

Weinberg, Alvin. 1967. *Reflections on Big Science*. Cambridge, Mass.: MIT Press.

Weinberg, Steven. 1977. "The search for unity: Notes for a history of quantum field theory," *Dædalus* 106 (April): 17–35.

———. 1993. *Dreams of a Final Theory: The Scientist's Search for the Ultimate Laws of Nature*. New York: Vintage Books.

Weiner, Charles. 1969. "A new site for the seminar: The refugees and American physics in the thirties," in Donald Fleming and Bernard Bailyn, eds., *The Intellectual Migration: Europe and America, 1930–1960*. Cambridge, Mass.: Harvard University Press, pp. 190–234.

———. 1970. "Physics in the great depression," *PT* 23 (January): 31–38. Reprinted in Weart and Phillips, eds., pp. 115–121.

———. 1972. "1932—Moving into the new physics," *PT* 25 (May): 40–49.

Wheaton, Bruce R. 1983. *The Tiger and the Shark: Empirical Roots of Wave-Particle Dualism*. Cambridge: Cambridge University Press.

Wheeler, John A., and W. H. Zurek, eds. 1983. *Quantum Theory and Measurement*. Princeton, N.J.: Princeton University Press.

Whitaker, Andrew. 1996. *Einstein, Bohr and the Quantum Dilemma*. Cambridge: Cambridge University Press.

White, D. Hywel, Daniel Sullivan, and Edward J. Barboni. 1979. "The interdependence of theory and experiment in revolutionary science: The case of parity violation," *SSS* 9: 303–27.

Whitehead, Alfred N. 1925. *Science and the Modern World*. New York: Macmillan.

Widmalm, Sven. 1995. "Science and neutrality: The Nobel Prizes of 1919 and scientific internationalism in Sweden," *Minerva* 33: 339–60.

Wilcox, Robert K. 1985. *Japan's Secret War: Japan's Race Against Time to Build its Own Atomic Bomb*. New York: William Morris and Company.

Will, Clifford M. 1993. *Was Einstein Right? Putting General Relativity to the Test*. Oxford: Oxford University Press.

Williams, L. Pearce, ed. 1968. *Relativity Theory: Its Impact on Modern Thought*. New York: John Wiley & Sons.

Wilson, David. 1983. *Rutherford: Simple Genius*. Cambridge, Mass.: MIT Press.

Wohlfarth, Horst. 1979. *40 Jahre Kernspaltung: Eine Einführung in die Originallitteratur*. Darmstadt, Germany: Wissenschaftliche Buchgesellschaft.

Wynne, Brian. 1979. "Physics and psychics: Science, symbolic action, and social control in late Victorian England," in Barry Barnes and Steven Shapin, eds., *Natural Order: Historical Studies of Scientific Culture*. Beverly Hills, Calif.: Sage Publications, pp. 167–90.

Xu, Qiaozhen, and Laurie Brown. 1987. "The early history of cosmic ray research," *AJP* 55: 23–33.

Yaes, Robert. 1974. "Physics fads and finance," *NS* 85: 462–63.
Yang, Chen Ning. 1989. "Particle physics in the early 1950s," in Brown, Dresden, and Hoddeson, eds., pp. 40–47.
Yolton, John W. 1960. *The Philosophy of Science of A. S. Eddington*. The Hague: Martinus Nijhoff.
Zee, A., ed. 1982. *Unity of Forces in the Universe*. 2 vols. Singapore: World Scientific.

INDEX